Lecture Notes in Mathematics

Edited by A. Dold and B. Eckmann

533

Frederick R. Cohen
Thomas J. Lada
J. Peter May

The Homology of Iterated Loop Spaces

Springer-Verlag
Berlin · Heidelberg · New York 1976

Authors
Frederick R. Cohen
Department of Mathematics
Northern Illinois University
DeKalb, Illinois 60115/USA

Thomas J. Lada
Department of Mathematics
North Carolina State University
Raleigh, North Carolina 27607/USA

J. Peter May
Department of Mathematics
University of Chicago
Chicago, Illinois 60637/USA

Library of Congress Cataloging in Publication Data

Cohen, Frederick Ronald, 1946-
The homology of iterated loop spaces.

(Lecture notes in mathematics ; 533)
Bibliography: p.
Includes index.
1. Loop spaces. 2. Classifying spaces.
3. Homology theory. I. Lada, Thomas Joseph,
1946- joint author. II. May, J. Peter, joint
author. II. May, J. Peter, joint author.
III. Title. IV. Series: Lecture notes in mathematics (Berlin) ; 533.
QA3.L28 vol. 533 [QA612.76] 510'.8s [514'.2]

AMS Subject Classifications (1970): 18F25, 18H10, 55D35, 55D40, 55F40, 55G99

ISBN 3-540-07984-X Springer-Verlag Berlin · Heidelberg · New York
ISBN 0-387-07984-X Springer-Verlag New York · Heidelberg · Berlin

Printed in Germany.
Printing and binding: Beltz Offsetdruck, Hemsbach/Bergstr.

Preface

This volume is a collection of five papers (to be referred to as I-V). The first four together give a thorough treatment of homology operations and of their application to the calculation of, and analysis of internal structure in, the homologies of various spaces of interest. The last studies an up to homotopy notion of an algebra over a monad and the role of this notion in the theory of iterated loop spaces. I have established the algebraic preliminaries necessary to the first four papers and the geometric preliminaries necessary for all of the papers in the following references, which shall be referred to by the specified letters throughout the volume.

[A]. A general algebraic approach to Steenrod operations. Springer Lecture Notes in Mathematics Vol. 168, 1970, 153-231.

[G]. The Geometry of Iterated Loop Spaces. Springer Lecture Notes in Mathematics Vol. 271, 1972.

[G']. E_∞ spaces, group completions, and permutative categories. London Math. Soc. Lecture Note Series Vol. 11, 1974, 61-93.

In addition, the paper II here is a companion piece to my book (contributed to by F. Quinn, N. Ray, and J. Tornehave)

[R]. E_∞ Ring Spaces and E_∞ Ring Spectra.

With these papers, this volume completes the development of a comprehensive theory of the geometry and homology of iterated loop spaces. There are no known results in or applications of this area of topology which do not fit naturally into the framework thus established. However, there are several papers by other authors which seem to me to add significantly to the theory developed in [G]. The relevant references will be incorporated in the list of errata and addenda to [A], [G], and [G'] which concludes this volume.

The geometric theory of [G] was incomplete in two essential respects. First, it worked well only for connected spaces (see [G, p. 156-158]). It was the primary purpose of [G'] to generalize the theory to non-connected spaces. In particular, this allowed it to be applied to the classifying spaces of permutative categories and thus to algebraic K-theory. More profoundly, the ring theory of [R] and II was thereby made possible.

Second, the theory of [G] circumvented analysis of homotopy invariance (see [G, p. 158-160]). It is the purpose of Lada's paper V to generalize the theory of [G] to one based on homotopy invariant structures on topological spaces in the sense of Boardman and Vogt [Springer Lecture Notes in Mathematics, Vol. 347] [1]. In Boardman and

[1] Incidentally, the claim there (p. VII) that [G] failed to apply to non Σ-free operads is based on a misreading; see [G, p. 22].

Vogt's work, an action up to homotopy by an operad (or PROP) on a space was essentially an action by a larger, but equivalent, operad on the same space. In Lada's work, an action up to homotopy is essentially an action by the given operad on a larger, but equivalent, space. In both cases, the expansion makes room for higher homotopies. While these need not be made explicit in the first approach, it seems to me that the second approach is nevertheless technically and conceptually simpler (although still quite complicated in detail) since the expansion construction is much less intricate and since the problem of composing higher homotopies largely evaporates.

We have attempted to make the homological results of this volume accessible to the reader unfamiliar with the geometric theory in the papers cited above. In I, I set up the theory of homology operations on infinite loop spaces. This is based on actions by E_∞ operads $\mathcal{C}$ on spaces and is used to compute $H_*(CX; Z_p)$ and $H_*(QX; Z_p)$ as Hopf algebras over the Dyer-Lashof and Steenrod algebras, where CX and QX are the free $\mathcal{C}$-space and free infinite loop space generated by a space X. The structure of the Dyer-Lashof algebra is also analyzed. In II, I set up the theory of homology operations on E_∞ ring spaces, which are spaces with two suitably interrelated E_∞ space structures. In particular, the mixed Cartan formula and mixed Adem relations are proven and are shown to determine the multiplicative homology operations of the free E_∞ ring space $C(X^+)$ and the free E_∞ ring infinite loop space $Q(X^+)$ generated by an E_∞ space X. In the second half of II, homology operations on E_∞ ring spaces associated to matrix groups are analyzed and an exhaustive study is made of the homology of BSF and of such related classifying spaces as BTop (at $p > 2$) and BCoker J. Perhaps the most interesting feature of these calculations is the precise homological analysis of the infinite loop splitting BSF = BCoker J $\times$ BJ at odd primes and of the infinite loop fibration BCoker J $\to$ BSF $\to$ $BJ_\otimes$ at $p = 2$.

In III, Cohen sets up the theory of homology operations on n-fold loop spaces for $n < \infty$. This is based on actions by the little cubes operad $\mathcal{C}_n$ and is used to compute $H_*(C_nX; Z_p)$ and $H_*(\Omega^n\Sigma^nX; Z_p)$ as Hopf algebras over the Steenrod algebra with three types of homology operations. While the first four sections of III are precisely parallel to sections 1, 2, 4, and 5 of I, the construction of the unstable operations (for odd p) and the proofs of all requisite commutation formulas between them (which occupies the rest of III) is several orders of magnitude more difficult than the analogous work of I (most of which is already contained in [A]). The basic ingredient is a homological analysis of configuration spaces, which should be of independent interest. In IV, Cohen computes

$H_*(SF(n); Z_p)$ as an algebra for p odd and n even, the remaining cases being determined by the stable calculations of II. Again, the calculation is considerably more difficult than in the stable case, the key fact being that $H_*(SF(n); Z_p)$ is commutative even though SF(n) is not homotopy commutative. Due to the lack of internal structure on BSF(n), the calculation of $H_*(BSF(n); Z_p)$ is not yet complete.

In addition to their original material, I and III properly contain all work related to homology operations which antedates 1970, while II contains either complete information on or at least an introduction to most subsequent work in this area, the one major exception being that nothing will be said about BTop and BPL at the prime 2. Up to minor variants, all work since 1970 has been expressed in the language and notations established in I §1-§2 and II §1.

Our thanks to Maija May for preparing the index.

J.P. May
August 20, 1975

TABLE OF CONTENTS

The Homology of E_∞ Spaces

J.P. May

Homology operations on iterated loop spaces were first introduced, mod 2, by Araki and Kudo [1] in 1956; their work was clarified and extended by Browder [2] in 1960. Homology operations mod p, $p > 2$, were first introduced by Dyer and Lashof [6] in 1962. The work of Araki and Kudo proceeded in analogy with Steenrod's construction of the Sq^n in terms of $\cup_i$-products, whereas that of Browder and of Dyer and Lashof proceeded in analogy with Steenrod's later construction of the P^n in terms of the cohomology of the symmetric group Σ_p. The analogy was closest in the case of infinite loop spaces and, in [A], I reformulated the algebra behind Steenrod's work in a sufficiently general context that it could be applied equally well to the homology of infinite loop spaces and to the cohomology of spaces. Later, in [G], I introduced the notions of E_∞ operad and E_∞ space. Their use greatly simplifies the geometry required for the construction and analysis of the homology operations and, in the non-connected case, yields operations on a wider class of spaces than infinite loop spaces. These operations, and further operations on the homology of infinite loop spaces given by the elements of $H_*\tilde{F}$, will be analyzed in section 1.

Historically, the obvious next step after introduction of the homology operations should have been the introduction of the Hopf algebra of all homology operations and the analysis of geometrically allowable modules (and more complicated structrues) over this algebra, in analogy with the definitions in cohomology given by Steenrod [22] in 1961. However, this step seems not to have been taken until lectures of mine in 1968-69. The requisite definitions will be given in section 2. Since the idea that homology operations should satisfy Adem relations first appears in [6] (although these relations were not formulated or proven there), we call the resulting algebra of operations the Dyer-Lashof algebra; we denote it by R. The main point of section 2 is the explicit construction of free allowable structures over R.

During my 1968-69 lectures, Madsen raised and solved at the prime 2 the problem of carrying out for R the analog of Milnor's calculation of the dual of the Steenrod algebra A. His solution appears in [8]. Shortly after, I solved the problem at odd primes, where the structure of R^* turned out to be surprisingly complicated. The details of this computation ($p = 2$ included) will be given in section 3.

In section 4, we reformulate (and extend to general non-connected spaces X) the calculation of H_*QX, $QX = \varinjlim \Omega^n\Sigma^n X$, given by Dyer and Lashof [6]. Indeed, the definitions in section 2 allow us to describe H_*QX as the free allowable Hopf algebra with conjugation over R and A. With the passage of time, it has become possible to give considerably simpler details of proof than were available in 1962. We also compute the Bockstein spectral sequence of QX (for each prime) in terms of that of X.

Just as QX is the free infinite loop space generated by a space X, so CX, as constructed in [G, §2], is the free $\mathcal{C}$-space generated by X (where $\mathcal{C}$ is an E_∞ operad). In section 5, we prove that H_*CX is the free allowable Hopf algebra (without conjugation) over R and A. The proof is quite simple, especially since the geometry of the situation makes half of the calculation an immediate consequence of the calculation of H_*QX. Although the result here seems to be new, in this generality, special cases have long been known. When X is connected, CX is weakly equivalent to QX by [G, 6.3]. When $X = S^0$, $CX = \coprod_j K(\Sigma_j, 1)$ and the result thus contains Nakaoka's calculations [16, 17, 18] of the homology of symmetric groups. We end section 5 with a generalization (from S^0 to arbitrary spaces X) of Priddy's homology equivalence $B\Sigma_\infty \to Q_0S^0$ [20].

In section 6, we describe how the iterated homology operations of an infinite loop space appear successively in the stages of its Postnikov decomposition.

In section 7, we construct and analyze homology operations analogous to the Pontryagin p^{th} powers in the cohomology of spaces. When $p = 2$, these operations were first introduced by Madsen [9].

Most of the material of sections 1-4 dates from my 1968-69 lectures at Chicago and was summarized in [12]. The material of section 5 dates from my 1971-72 lectures at Cambridge. The long delay in publication, for which I must apologize, was caused by problems with the sequel II (to be explained in its introduction).

CONTENTS

§1. Homology operations

We first define and develop the properties of homology operations on E_∞ spaces. We then specialize to obtain further properties of the resulting operations on infinite loop spaces. In fact, the requisite geometry has been developed in [G, §1,4,5, and 8] and the requisite algebra has been developed in [A, §1-4 and 9]. The proofs in this section merely describe the transition from the geometry to the algebra.

All spaces are to be compactly generated and weakly Hausdorff; $\mathcal{T}$ denotes the category of spaces with non-degenerate base-point [G, p.1]. All homology is to be taken with coefficients in Z_p for an arbitrary prime p; the modifications of statements required in the case p=2 are indicated inside square brackets.

We require some recollections from [G] in order to make sense of the following theorem. Recall that an E_∞ space (X,θ) is a $\mathcal{C}$-space over any E_∞ operad $\mathcal{C}$ [G, Definitions 1.1, 1.2, and 3.5]; θ determines an H-space structure on X with the base-point $* \in X$ as identity element and with $\theta_2(c)\colon X\times X \to X$ as product for any $c \in \mathcal{C}(2)$ [G, p.4]. Recall too that the category $\mathcal{C}[\mathcal{T}]$ of $\mathcal{C}$-spaces is closed under formation of loop and path spaces [G, Lemma 1.5] and has products and fibred products [G, Lemma 1.7].

__Theorem 1.1.__ Let $\mathcal{C}$ be an E_∞ operad and let (X,θ) be a $\mathcal{C}$-space. Then there exist homomorphisms $Q^s\colon H_*X \to H_*X$, $s \geq 0$, which satisfy the following properties:

(1) The Q^s are natural with respect to maps of $\mathcal{C}$-spaces.

(2) Q^s raises degrees by $2s(p-1)$ [by s].

(3) $Q^s x = 0$ if $2s <$ degree (x) [if $s <$ degree (x)], $x \in H_*X$

(4) $Q^s x = x^p$ if $2s =$ degree (x) [if $s =$ degree (x)], $x \in H_*X$

(5) $Q^s \phi = 0$ if $s > 0$, where $\phi \in H_0(X)$ is the identity element.

(6) The external, internal, and diagonal Cartan formulas hold:

$$Q^s(x\otimes y) = \sum_{i+j=s} Q^i x \otimes Q^j y \text{ if } x\otimes y \in H_*(X\times Y),\ (Y,\theta')\in \mathcal{C}[\mathcal{T}];$$

$$Q^s(xy) = \sum_{i+j=s} (Q^i x)(Q^j y) \text{ if } x,\ y\in H_*X; \text{ and}$$

$$\psi(Q^s x) = \sum_{i+j=s} \Sigma Q^i x' \otimes Q^j x'' \text{ if } \psi x = \Sigma x'\otimes x'',\ x\in H_*X.$$

(7) The Q^s are stable and the Kudo transgression theorem holds: $Q^s\sigma_* = \sigma_* Q^s$, where $\sigma_*\colon \tilde{H}_*\Omega X \to H_*X$ is the homology suspension; if X is simply connected and if $x\in H_q X$ transgresses to $y\in H_{q-1}\Omega X$ in the Serre spectral sequence of the path space fibration $\pi\colon PX \to X$, then $Q^s x$ and $\beta Q^s x$ transgress to $Q^s y$ and $-\beta Q^s y$ and, if $p > 2$ and $q = 2s$, $x^{p-1}\otimes y$ "transgresses" to $-\beta Q^s y$, $d^{q(p-1)}(x^{p-1}\otimes y) = -\beta Q^s y$.

(8) The Adem relations hold: If $p \geq 2$ and $r > ps$, then

$$Q^r Q^s = \sum_i (-1)^{r+i}(pi-r,\ r-(p-1)s-i-1)Q^{r+s-i}Q^i;$$

if $p > 2$, $r \geq ps$, and β denotes the mod p Bockstein, then

$$Q^r\beta Q^s = \sum_i (-1)^{r+i}(pi-r,\ r-(p-1)s-i)\beta Q^{r+s-i}Q^i$$

$$-\sum_i (-1)^{r+i}(pi-r-1,\ r-(p-1)s-i)Q^{r+s-i}\beta Q^i$$

(9) The Nishida relations hold: Let $P^r_*\colon H_*X \to H_*X$ be dual to P^r where $P^r = Sq^r$ if $p = 2$ (thus $P^r = (P^r_*)^*$ on $H^*X = (H_*X)^*$). Then

$$P^r_* Q^s = \sum_i (-1)^{r+i}(r-pi,\ s(p-1)-pr+pi)Q^{s-r+i}P^i_*;$$

in particular, $\beta Q^s = (s-1)Q^{s-1}$ if $p=2$; if $p > 2$,

$$P^r_*\beta Q^s = \sum_i (-1)^{r+i}(r-pi,\ s(p-1)-pr+pi-1)\beta Q^{s-r+i}P^i_*$$

$$+\sum_i (-1)^{r+i}(r-pi-1,\ s(p-1)-pr+pi)Q^{s-r+i}P^i_*\beta.$$

(In (8) and (9), $(i,j) = (i+j)!/i!j!$ if $i > 0$ and $j > 0$, $(i,0)=1=(0,i)$ if $i \geq 0$, and $(i,j) = 0$ if $i < 0$ or $j < 0$; the sums run over the integers.)

Proof: The symmetric group Σ_p acts freely on the contractible space $\mathcal{C}(p)$, hence the normalized singular chain complex $C_*\mathcal{C}(p)$ is a Σ_p-free resolution of Z_p [7, IV 11]. Let W be the standard π-free resolution of Z_p [A, Definition 1.2], where π is cyclic of order p, and let $j\colon W \to C_*\mathcal{C}(p)$ be a morphism of π-complexes over Z_p. Let $(C_*X)^p$ denote the p-fold tensor product. We are given a Σ_p-equivariant map $\theta_p\colon \mathcal{C}(p)\times X^p \to X$, and we define $\theta_*\colon W\otimes(C_*X)^p \to C_*X$ to be the following composite morphism of π-complexes:

$$W\otimes(C_*X)^p \xrightarrow{j\otimes\eta} C_*\mathcal{C}(p)\otimes C_*(X^p) \xrightarrow{\eta} C_*(\mathcal{C}(p)\times X^p) \xrightarrow{C_*\theta_p} C_*X.$$

Here η is the shuffle map; for diagram chases, it should be recalled that $\eta\colon C_*X\otimes C_*Y \to C_*(X\times Y)$ is a commutative and associative natural transformation which is chain homotopy inverse to the Alexander-Whitney map ξ. In view of [G, Lemmas 1.6 and 1.9 (i)], (C_*X,θ_*) is a unital and mod p reduced object of the category $\mathcal{C}(p,\infty)$ defined in [A, Definitions 2.1]. Moreover, $(X,\theta) \to (C_*X,\theta_*)$ is clearly the object map of a functor from $\mathcal{C}[\mathcal{T}]$ to the subcategory $\mathcal{P}(p,\infty)$ of $\mathcal{C}(p,\infty)$ defined in [A, Definitions 2.1]. Let $x\in H_qX$. As in [A, Definitions 2.2], we define

(i) $Q_i(x) = \theta_*(e_i\otimes x^p)$, $\theta_*\colon H(W\otimes_\pi(H_*X)^p)\cong H(W\otimes_\pi(C_*X)^p) \to H_*X$,

and we define the desired operations Q^s by the formulas

(ii) $p = 2$: $Q^s x = 0$ if $s < q$ and $Q^s x = Q_{s-q}(x)$ if $s \geq q$; and

(iii) $p > 2$: $Q^s x = 0$ if $2s < q$ and $Q^s x=(-1)^s v(q) Q_{(2s-q)(p-1)}(x)$ if $2s > q$, where $v(q) = (-1)^{q(q-1)m/2}(m!)^q$, with $m = \frac{1}{2}(p-1)$.

The Q^s are homomorphisms which satisfy (1) through (5) by [A, Proposition 2.3 and Corollary 2.4]. Note that [A, Proposition 2.3] also implies that if $p > 2$, then $\beta Q^s x = (-1)^s v(q)\, Q_{(2s-q)(p-1)-1}(x)$ and the Q^s and βQ^s account for all non-trivial operations Q_i. For (6), recall that the product of $\mathcal{C}$-spaces (X,θ) and (Y,θ') is $(X\times Y,\tilde{\theta})$, where $\tilde{\theta}_p$ is the composite

$$\mathcal{C}(p)\times(X\times Y)^p \xrightarrow{\Delta\times u} \mathcal{C}(p)\times\mathcal{C}(p)\times X^p \xrightarrow{1\times t\times 1} \mathcal{C}(p)\times X^p\times\mathcal{C}(p)\times Y^p \xrightarrow{\theta_p\times\theta'_p} X\times Y$$

(Here Δ, u, and t are the diagonal and the evident shuffle and interchange maps.) Similarly, the tensor product of objects (K,θ_*) and (L,θ'_*) in $\mathcal{C}(p,\infty)$ is $(K\times L,\tilde{\theta}_*)$, where $\tilde{\theta}_*$ is the composite

$$W\otimes(K\otimes L)^p \xrightarrow{\psi\otimes U} W\otimes W\otimes K^p\otimes L^p \xrightarrow{1\otimes T\otimes 1} W\otimes K^p\otimes W\otimes L^p \xrightarrow{\theta_*\otimes\theta'_*} K\otimes L.$$

(Here ψ, U, and T are the coproduct on W and the evident shuffle and interchange homomorphisms.) Since $(j\otimes j)\psi$ is π-homotopic to $(\xi\circ C_*\Delta)j$, an easy diagram chase demonstrates that $\eta: C_*X\otimes C_*Y \to C_*(X\times Y)$ is a morphism in the category $\mathcal{C}(p,\infty)$. The external Cartan formula now follows from [A, Corollary 2.7]. By [G, Lemmas 1.7 and 1.9 (ii)], $\Delta: X\to X\times X$ is a map of $\mathcal{C}$-spaces and (C_*X,θ_*) is a Cartan object of $\mathcal{C}(p,\infty)$; the diagonal and internal Cartan formulas follow by naturality. Part (7) is an immediate consequence of [G, Lemma 1.5] and [A, Theorems 3.3 and 3.4]; the simple connectivity of X serves to ensure that $E^2 = H_*X\otimes H_*\Omega X$ in the Serre spectral sequence of $\pi: PX \to X$. For (8), note that the following diagram is commutative by [G, Lemma 1.4]:

$$\begin{array}{ccccc}
\mathcal{C}(p)\times\mathcal{C}(p)^p\times X^{p^2} & \xrightarrow{\gamma\times 1} & \mathcal{C}(p^2)\times X^{p^2} & & \\
 & & & \searrow^{\theta_{p^2}} & \\
\big\downarrow{\scriptstyle 1\times u} & & & & X \\
 & & & \nearrow_{\theta_p} & \\
\mathcal{C}(p)\times[\mathcal{C}(p)\times X^p]^p & \xrightarrow{1\times(\theta_p)^p} & \mathcal{C}(p)\times X^p & &
\end{array}$$

An easy diagram chase demonstrates that (C_*X,θ_*) is an Adem object, in the sense of [A, Definition 4.1], and (8) follows by [A, Theorem 4.7]. Part (9) follows by the naturality of the Steenrod operations from [A, Theorem 9.4], which computes the Steenrod operations in $H_*(\mathcal{C}(p)\times_\pi X^p)$. As explained in [A, p.209], our formula differs by a sign from that obtained by Nishida [19].

Let $\mathcal{L}_\infty$ be the category of infinite loop sequences. Recall that an object $Y = \{Y_i\}$ in $\mathcal{L}_\infty$ is a sequence of spaces with $Y_i = \Omega Y_{i+1}$ and a morphism $g = \{g_i\}$ in $\mathcal{L}_\infty$ is a sequence of maps with $g_i = \Omega g_{i+1}$. Y_0 is said to be an infinite loop space, g_0 an infinite loop map. By the results of [10], these notions are equivalent for the purposes of homotopy theory to the more usual ones in which equalities are replaced by homotopies. By [G, Theorem 5.1], there is a functor $W_\infty\colon \mathcal{L}_\infty \to \mathcal{C}_\infty[\mathcal{T}]$, with $W_\infty Y = (Y_0,\theta_\infty)$ and $W_\infty g = g_0$, where $\mathcal{C}_\infty$ is the infinite little cubes operad of [G, Definition 4.1]. The previous theorem therefore applies to H_*Y_0; the resulting operations Q^s will be referred to as the loop operations. The relevant Pontryagin product is that induced by the loop product on $Y_0 = \Omega Y_1$. Note that there are two different actions of $\mathcal{C}_\infty$ on ΩY_0, one coming from [G, Lemma 1.5] and the other from the fact that ΩY_0 is again an infinite loop space; by [G, Lemma 5.6], these two actions are equivariantly homotopic, hence part (7) of the theorem does apply to the loop operations. Similarly, part (6) applies to the loop operations since, by [G, Lemma 5.7], the two evident actions of $\mathcal{C}_\infty$ on the product of two infinite loop spaces are in fact the same.

The recognition theorem [G, Theorem 14.4; G'] gives a weak homotopy equivalence between any given grouplike E_∞ space X and an infinite loop space B_0X; moreover, as explained in [G, p.153-155], the homology operations on H_*X coming via Theorem 1.1 from the given E_∞ space structure agree under the equivalence with the loop operations on H_*B_0X. Thus, in principle, it is only for non grouplike E_∞ spaces that the operations of Theorem 1.1 are more general than loop operations. In practice, the theorem gives considerable geometric freedom in the construction of the operations, and this freedom is often essential to the calculations.

The following additional property of the loop operations, which is implied by [G, Remarks 5.8], will be important in the study of non-connected infinite loop spaces. Recall that the conjugation χ on a Hopf algebra, if present, is related to the unit η, augmentation ε, product ϕ, and coproduct ψ by the formula $\eta\varepsilon = \phi(1\times\chi)\psi$.

Lemma 1.2. For $Y\in\mathcal{L}_\infty$, $Q^s\chi = \chi Q^s$ on H_*Y_0, where the conjugation is induced from the inverse map on $Y_0 = \Omega Y_1$.

In the next two sections, we will define and study the global algebraic structures which are naturally suggested by the results above. We make a preliminary definition here.

Definition 1.3. Let A be a Hopf algebra. Let A act on Z_p through its augmentation, $a\cdot 1 = \varepsilon(a)$, and let A act on the tensor product $M\otimes N$ of two left A-modules through its coproduct,

$$a(m\otimes n) = \Sigma(-1)^{\deg a''\deg m}a'm\otimes a''n \text{ if } \psi a = \Sigma a'\otimes a''.$$

A left or right structure (algebra, coalgebra, Hopf algebra, Hopf algebra with conjugation, etc.) over A is a left or right

A-module and a structure of the specified type such that all of the structure maps are morphisms of A-modules.

We shall define a Hopf algebra R of homology operations in the next section and, if $Y \in \mathcal{L}_\infty$, H_*Y_0 will be a left Hopf algebra with conjugation over R. For any space X, H_*X is a left coalgebra over the opposite algebra A^0 of the Steenrod algebra; here the opposite algebra enters because dualization is contravariant. Henceforward, although we shall continue to write the Steenrod operations P^r_* on the left, we shall speak of right A-modules rather than of left A^0-modules. Thus H_*Y_0 is a right Hopf algebra with conjugation over A, and the Nishida relations give commutation formulas between the A and R operations on H_*Y_0.

There is yet another Hopf algebra which acts naturally and stably on H_*Y_0 namely $H_*\tilde{F}$ where $\tilde{F}$ is the monoid (under composition) of based maps of spheres. The precise definition of $\tilde{F}$ is given in [G, p.74], and it is shown there that composition of maps defines a natural action $c_\infty: Y_0 \times \tilde{F} \to Y_0$ of $\tilde{F}$ on infinite loop spaces. The following theorem gives the basic properties of the induced action of $H_*\tilde{F}$ on H_*Y_0.

Theorem 1.4. For $Y \in \mathcal{L}_\infty$, $c_{\infty *}: H_*Y_0 \otimes H_*\tilde{F} \to H_*Y_0$ gives H_*Y_0 a structure of right Hopf algebra with conjugation over $H_*\tilde{F}$. Moreover, $c_{\infty *}$ satisfies the following properties, where $c_{\infty *}(x \otimes f) = xf$:

(1) $c_{\infty *}$ is natural with respect to maps in $\mathcal{L}_\infty$.

(2) $\sigma_*(xf) = (\sigma_* x)f$, where $\sigma_*: \tilde{H}_*\Omega Y_0 \to H_*Y_0$ is the suspension.

(3) $P^r_*(xf) = \sum_{i+j=r} (P^i_* x)(P^j_* f)$ and $\beta(xf) = (\beta x)f + (-1)^{\deg x} x(\beta f)$.

(4) $(Q^s x)f = \sum_i Q^{s+i}(xP^i_* f)$ and, if $p > 2$,

$$(\beta Q^s x)f = \sum_i \beta Q^{s+i}(xP^i_* f) - \sum_i (-1)^{\deg x} Q^{s+i}(x\, P^i_* \beta f).$$

<u>Proof</u>: Part (1) is trivial. The maps $* \to Y_0$ and $Y_0 \to *$ are infinite loop maps, hence the unit and augmentation of H_*Y_0 are morphisms of $H_*\tilde{F}$-modules. The loop product is a morphism of $H_*\tilde{F}$-modules by a simple diagram chase from [G, Lemma 8.8], and a similar lemma for the inverse map implies that the conjugation is a morphism of $H_*\tilde{F}$-modules. The coproduct on H_*Y_0 is a morphism of $H_*\tilde{F}$-modules and formula (3) holds because $c_{\infty *}$ is induced by a map. Formula (2) is an immediate consequence of [G, Lemmas 8.4 and 8.5]. For (4), consider the following diagram, in which we have abbreviated $\mathcal{C}$ for $\mathcal{C}_\infty(p)$ and X, $X^{(p)}$, and X^p for C_*X, $(C_*X)^p$, and $C_*(X^p)$:

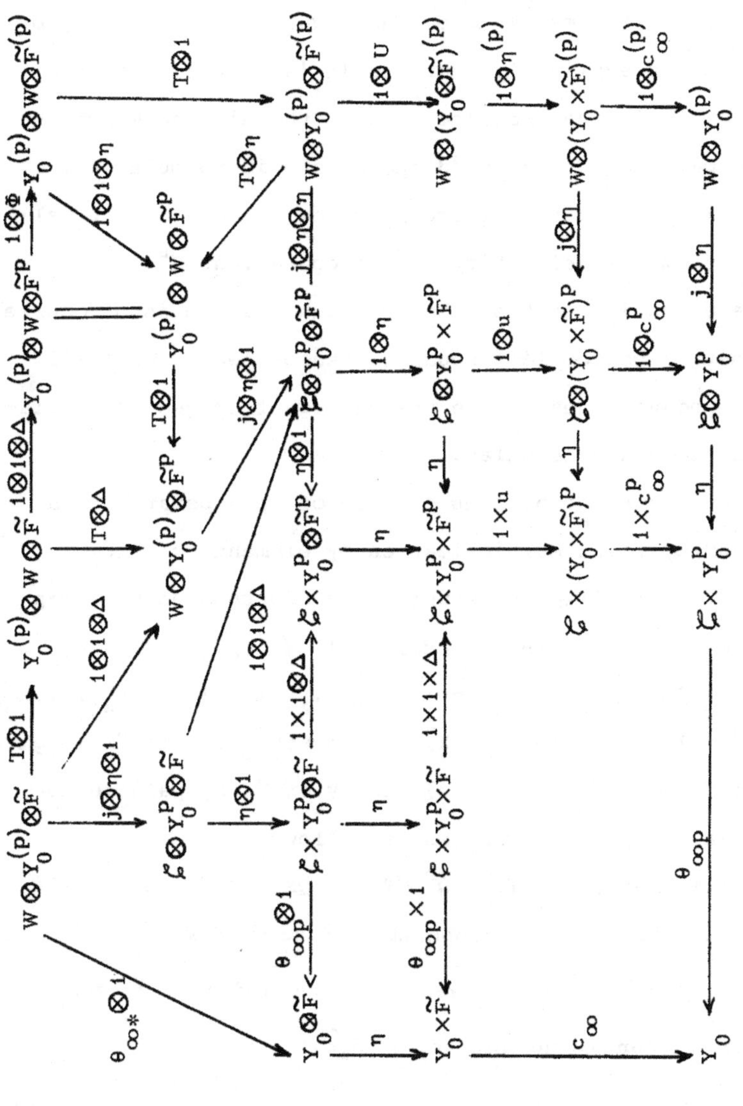

Here Φ: $W \otimes C_*(X^p) \to W \otimes (C_*X)^p$ is given by [A, Lemma 7.1], and $(1 \otimes \eta)\,\Phi$ is π-homotopic to the identity by [A, Remarks 7.2]. The bottom left square is Σ_p-homotopy commutative by [G, Proposition 8.9], and the remaining parts of the diagram commute trivially. The shuffle and interchange homomorphisms U and T merely involve signs, the composite $c_\infty \eta$: $C_*Y_0 \otimes C_*\tilde{F} \to C_*Y_0$ induces $c_{\infty *}$, and the map induced on π-equivariant homology by $d = \Phi(1 \otimes \Delta)$ is explicitly computed in [A, Proposition 9.1]. Of course, it is the presence of d in the diagram which leads to the appearance of Steenrod operations in formula (4). The verification of this formula is now an easy direct calculation from the definition of the operations Q^s.

The essential part of the previous diagram is of course the geometric bottom left square. Henceforward, we shall omit the pedantic details in the passage from geometric diagrams to algebraic formulas.

We evaluate one obvious example of the operations on H_*Y_0 given by right multiplication by elements of $H_*\tilde{F}$.

Lemma 1.5. Let $[i] \in H_0\tilde{F}$ be the class represented by a map of degree i. Then $x\,[-1] = \chi x$ for all $x \in H_*Y_0$, $Y \in \mathcal{L}_\infty$.

Proof: Define f: $S^n \to S^n$ by $f(s_1, \ldots, s_n) = (1-s_1, s_2, \ldots, s_n)$, where $S^n = I^n/\partial I^n$. For any X, the inverse map on $\Omega^n X$ is given by $g \to g \circ f$, g: $S^n \to X$, and similarly on Y_0 by passage to limits via [G, p.74]. The result follows.

Recall that $QX = \varinjlim \Omega^n \Sigma^n X$ and $QX = \Omega Q \Sigma X$ [G, p.42]. As we shall see in II § 5, application of the results above to $Y_0 = QS^0$, where c_∞ reduces to the product on $\tilde{F}$, completely determines the composition product on $H_*\tilde{F}$.

Remarks 1.6. A functorial definition of a smash product between objects of $\mathcal{L}_\infty$ is given in [13], in which a new construction of the stable homotopy category is given. (In the language of [13], $\mathcal{L}_\infty$ is a category of coordinatized spectra; the smash product is constructed by passing to the category $\mathcal{S}$ of coordinate-free spectra, applying the smash product there, and then returning to $\mathcal{L}_\infty$.) For objects $Y, Z \in \mathcal{L}_\infty$ and elements

$x, y \in H_*Y_0$ and $z \in H_*Z_0$, [G, Lemma 8.1] and a similar lemma for the inverse map imply the formulas

$$(x * y) \wedge z = \sum (-1)^{\deg y \deg z'} (x \wedge z') * (y \wedge z'') \quad \text{if } \psi z = \sum z' \otimes z''$$

and

$$(\chi y) \wedge z = \chi(y \wedge z),$$

where $*$ and $\wedge$ denote the loop and smash products respectively. Via a diagram chase precisely analogous to that in the proof of Theorem 1.4, [G, Proposition 8.2] implies the formulas

$$(Q^s y) \wedge z = \sum_i Q^{s+i}(y \wedge P_*^i z)$$

and, if $p > 2$,

$$(\beta Q^s y) \wedge z = \sum_i \beta Q^{s+i}(y \wedge P_*^i z) - \sum_i (-1)^{\deg y} Q^{s+i}(y \wedge P_*^i \beta z).$$

In particular, these results apply to $\wedge : QX \times QX' \to Q(X \wedge X')$ for any spaces X and X'. By [G, Lemma 8.7], the smash and composition products coincide and are commutative on $H_* QS^0 = H_* \tilde{F}$.

§2. Allowable structures over the Dyer-Lashof algebra

We here describe the Hopf algebra of homology operations on E_∞ spaces generated by the Q^s and βQ^s and develop analogs of the notions of unstable modules and algebras over the Steenrod algebra. The following definition determines the appropriate "admissible monomials".

Definition 2.1. (i) $p = 2$: Consider sequences $I = (s_1,\ldots,s_k)$ such that $s_j \geq 0$. Define the degree, length, and excess of I by

$$d(I) = \sum_{j=1}^{k} s_j;\quad \ell(I) = k;\ \text{and}$$

$$e(I) = s_k - \sum_{j=2}^{k} (2s_j - s_{j-1}) = s_1 - \sum_{j=2}^{k} s_j.$$

The sequence I determines the homology operation $Q^I = Q^{s_1}\ldots Q^{s_k}$. I is said to be admissible if $2s_j \geq s_{j-1}$ for $2 \leq j \leq k$.

(ii) $p > 2$: Consider sequences $I = (\varepsilon_1, s_1,\ldots,\varepsilon_k, s_k)$ such that $\varepsilon_j = 0$ or 1 and $s_j \geq \varepsilon_j$. Define the degree, length, and excess of I by

$$d(I) = \sum_{j=1}^{k} [2s_j(p-1)-\varepsilon_j];\quad \ell(I) = k;\ \text{and}$$

$$e(I) = 2s_k-\varepsilon_1 - \sum_{j=2}^{k} [2ps_j-\varepsilon_j-2s_{j-1}] = 2s_1-\varepsilon_1- \sum_{j=2}^{k} [2s_j(p-1)-\varepsilon_j].$$

The sequence I determines the homology operation $Q^I = \beta^{\varepsilon_1}Q^{s_1}\ldots\beta^{\varepsilon_k}Q^{s_k}$. I is said to be admissible if $ps_j - \varepsilon_j \geq s_{j-1}$ for $2 \leq j \leq k$.

(iii) Conventions: $b(I) = \varepsilon_1$ if $p > 2$ and $b(I) = 0$ if $p = 2$. The empty sequence I is admissible and satisfies $d(I) = 0$, $\ell(I) = 0$, $e(I) = \infty$, and $b(I) = 0$; it determines the identity homology operation $Q^I = 1$.

Definition 2.2. Let F denote the free associative algebra

generated by $\{Q^s | s \geq 0\}$ if $p = 2$ or by $\{Q^s | s \geq 0\} \cup \{\beta Q^s | s > 0\}$ if $p > 2$ (not β itself). For $q \geq 0$, define $J(q)$ to be the two-sided ideal of F generated by the Adem relations (and, if $p > 2$, the relations obtained by applying β to the Adem relations, with $\beta^2 = 0$) and by the relations $Q^I = 0$ if $e(I) < q$. Define $R(q)$ to be the quotient algebra $F/J(q)$, and observe that there are successive quotient maps $R(q) \to R(q+1)$. Let $R = R(0)$; R will be called the Dyer-Lashof algebra.

To avoid circularity, we have defined the $R(q)$ purely algebraically. The following theorem implies that this definition agrees with that naturally suggested by the geometry.

Theorem 2.3. (i) Let $i_q \in H_q S^q \subset H_q QS^q$ be the fundamental class if $q > 0$ and the class of the point other than the base-point if $q = 0$; then

$$\{Q^I i_q | I \text{ is admissible and } e(I) \geq q\}$$

is a linearly independent subset of $H_* QS^q$.

(ii) $J(q)$ coincides with the set $K(q)$ of all elements of F which annihilate every homology class of degree $\geq q$ of every E_∞ space (equivalently, of every infinite loop space).

(iii) $\{Q^I | I \text{ is admissible and } e(I) \geq q\}$ is a Z_p-basis for $R(q)$.

(iv) $R(q)$ admits a unique structure of right A-module such that the Nishida relations are satisfied.

(v) $R = R(0)$ admits a structure of Hopf algebra and of unstable right coalgebra over A with coproduct defined on generators by

$$\psi Q^s = \sum_{i+j=s} Q^i \otimes Q^j \quad \text{and} \quad \psi \beta Q^{s+1} = \sum_{i+j=s} (\beta Q^{i+1} \otimes Q^j + Q^i \otimes \beta Q^{j+1})$$

and with augmentation defined on $R_0 = P\{Q^0\}$ by $\varepsilon(Q^{0^k}) = 1$, $k \geq 0$.

Proof: We shall prove (i) in §4. It is obvious from the Adem relations that $R(q)$ is generated as a Z_p-space by the set specified in (iii), and $J(q)$ is contained in $K(q)$ by (3) and (8) of Theorem 1.1. Therefore (i) implies (ii) and (iii). For (iv), the A operations on $1 \in R_0(q)$ are determined by $P_*^0 = 1$ and $R_i(q)=0$ for $i < 0$ and the A operations on all elements $Q^I = Q^I \cdot 1$ with $\ell(I) > 0$ are determined from the Nishida relations by induction on $\ell(I)$. This action does give an A-module structure since if $f(q)\colon\ R(q) \to H_*(QS^q)$ is defined by $f(q)\ Q^I = Q^I i_q$, then $f(q)$ is a monomorphism which commutes with the Steenrod operations. Let $f(0) = f$; since $\psi(1) = 1 \otimes 1$ and $\psi(i_0) = i_0 \otimes i_0$ and since $\varepsilon(i_0^k) = 1$, f commutes with the coproduct and augmentation. Here ψ is well-defined on R and R is a Hopf algebra since $J = J(0)$ is a Hopf ideal, $\psi(J) \subset F \otimes J + J \otimes F$, by commutativity of the following diagram (where π is the quotient map):

$$\begin{array}{ccccc} F & \xrightarrow{\ \pi\ } & R & \xrightarrow{\ f\ } & H_*QS^0 \\ \downarrow{\scriptstyle\psi} & & & & \downarrow{\scriptstyle\psi} \\ F \otimes F & \xrightarrow{\ \pi\otimes\pi\ } & R \otimes R & \xrightarrow{\ f\otimes f\ } & H_*QS^0 \otimes H_*QS^0 \end{array}$$

Observe that this argument fails for $q > 0$ since $\psi i_q = i_q \otimes 1 + 1 \otimes i_q$. Since H_*QS^0 is an unstable right coalgebra over A, so is R.

Of course, we understand unstable right structures over A in the sense of homology: the dual object (if of finite type) is an unstable A-structure of the dual type, as defined by Steenrod [22,23]. We shall study the structure of R itself in the next section. The remainder of this section will be devoted to the study of structures over R. In order to deal with non-connected structures, we need some preliminaries.

Definition 2.4. A component coalgebra is a unital (and augmented) coalgebra C such that C is a direct sum of connected coalgebras. Given such a C, define

$$\pi C = \{g \mid g \in C,\ \psi g = g \otimes g \text{ and } g \neq 0\}.$$

Clearly πC is a basis for C_0. For $g \in \pi C$, define C_g to be the connected sub-coalgebra of C such that $g \in C_g$ and the set of positive degree elements of C_g is

$$\overline{C}_g = \{x \mid \psi x = x \otimes g + \Sigma x' \otimes x'' + g \otimes x,\ \deg x' > 0 \text{ and } \deg x'' > 0\}.$$

Then C is the direct sum of its components C_g for $g \in \pi C$. Note that $\varepsilon g = 1$ for $g \in \pi C$. πC contains the distinguished element $\phi = \eta(1)$, and $JC = \operatorname{Coker} \eta$ may be identified with $\overline{C}_\phi \oplus (\bigoplus_{g \neq \phi} C_g) \subset C$.

If X is a based space, then H_*X is a component coalgebra; the base-point determines η and the components determine the direct sum decomposition. Indeed, there is an obvious identification of $\pi_0 X$ with $\pi H_* X$. As another example, we have the following observations on the structure of R.

<u>Lemma 2.5</u>. R is a component coalgebra. πR is the free monoid generated by Q^0 and the component $R[k]$ of $(Q^0)^k$, $k \geq 0$, is the sub unstable A-coalgebra of R spanned by

$$\{Q^I \mid I \text{ is admissible, } e(I) \geq 0, \text{ and } \ell(I) = k\}.$$

The product on R sends $R[k] \otimes R[\ell]$ to $R[k+\ell]$ for all k and ℓ, and the elements Q^s and βQ^s are all indecomposable.

<u>Definition 2.6</u>. A component Hopf algebra B is said to be monoidal (resp., grouplike) if πB is a monoid (resp., group) under the product of B. Equivalently, B is monoidal if all pairwise products of elements of πB are non-zero.

The proof of the following lemma requires only the defining formula $\eta\varepsilon = \phi(1 \otimes \chi)\psi$ for a conjugation.

<u>Lemma 2.7</u>. A component Hopf algebra B admits a conjugation if and only if B is grouplike, and then $\chi g = g^{-1}$ if $g \in \pi B$ and

$$\chi x = -g^{-1}xg^{-1} - \sum g^{-1}x'(\chi x'')$$

if $\deg x > 0$ and $\psi x = x \otimes g + \Sigma x' \otimes x'' + g \otimes x$ with $\deg x' > 0$ and $\deg x'' > 0$.

We can now define allowable structures over R, by which we simply mean those kinds of R-structures which satisfy the algebraic constraints dictated by the geometry.

Definition 2.8. A left R-module D is allowable if $J(q)\ D_q = 0$ for all $q \geq 0$. The category of allowable R-modules is the full subcategory of that of R-modules whose objects are allowable; it is an Abelian subcategory which is closed under the tensor product. An allowable R-algebra is an allowable R-module and a commutative algebra over R such that $Q^s x = x^p$ if $2s = \deg x$ [$Q^s x = x^2$ if $s = \deg x$]. An allowable R-coalgebra is an allowable R-module and a cocommutative component coalgebra over R. An allowable R-Hopf algebra (with conjugation) is a monoidal Hopf algebra (with conjugation) over R which is allowable both as an R-algebra and as an R-coalgebra. For any of these structures, an allowable AR-structure is an allowable R-structure and an unstable right A-structure of the same type such that the A and R operations satisfy the Nishida relations.

Theorem 1.1 implies that the homology of an E_∞ space is an allowable AR-Hopf algebra. Lemma 1.2 implies that the homology of an infinite loop space is an allowable AR-Hopf algebra with conjugation. Observe that a connected allowable AR-Hopf algebra is automatically an allowable AR-Hopf algebra with conjugation.

In order to take advantage of these definitions, we require five basic free functors, D, E, V, W, and G, of which E and W are essentially elaborations of D and V in the presence of coproducts. In addition, each of these functors has a more elaborate counter-part, to be defined parenthetically, in the presence of Steenrod operations. The composite functors WE and GWE will describe $H_*C_\infty X$ and H_*QX, with all structure in sight, as functors of H_*X.

We shall describe our functors on objects and shall show that the given internal structures uniquely determine the required internal structures. The verifications (not all of which are trivial) that these structures are in fact well-defined and satisfy all of the requisite algebraic indentities will be left to the reader, since these consistency statements obviously hold for those structures which can be realized geometrically. It is trivial to verify that our functors are indeed free, in the sense that they are adjoint to the forgetful functors going the other way. The functor V, which is a special case of the universal enveloping algebra functor on Abelian restricted Lie algebras, and the functor W occur in many other contexts in algebraic topology; they are discussed in detail in [11].

D: Z_p-modules (resp., unstable A-modules) to allowable R-modules (resp., AR-modules): Given M, define

$$DM = \bigoplus_{q\geq 0} R(q)\otimes M_q.$$

R acts on DM via the quotient maps $R \to R(q)$; thus DM is just the obvious quotient of the free R-module $R\otimes M$. The inclusion of M in DM is given by $m \to 1\otimes m$. If A acts on M, then this action and the Nishida relations determine the action of A on DM by induction on the length of admissible monomials.

E: Cocommutative component coalgebras (resp., unstable A-coalgebras) to allowable R-coalgebras (resp., AR-coalgebras): Given C, define EC as an R-module, and as an A-module if A acts on C, by

$EC = DC/IR \cdot Im\eta = Z_p \oplus DJC$, $IR = Ker\ \varepsilon$ and $JC = Coker\ \eta$.
The inclusion of C in EC is induced by that of C in DC. The coproduct on C and the diagonal Cartan formula determine the coproduct on EC. The unit of C and the augmentations of R and C determine the unit and augmentation of EC. Equivalently, EC is the obvious quotient component coalgebra of $R \otimes C$; thus

$$\pi EC = \{(Q^0)^k \otimes g \mid k \geq 0 \text{ and } g \in \pi C,\ k=0 \text{ if } g=\phi=\eta(1)\},$$

and the component of $(Q^0)^k \otimes g$ is the image of $R[k] \otimes C_g$ in EC if $g \neq \phi$ while the component of $1 \otimes \phi$ is the image of $R \otimes C_\phi$.

V: Allowable R-modules (resp., AR-modules) to allowable R-algebras (resp., AR-algebras): Given D, define

$$VD = AD/K$$

where AD is the free commutative algebra generated by D and K is the ideal of AD generated by

$$\{x^p - Q^s x \mid 2s = \deg x \text{ if } p>2 \text{ or } s = \deg x \text{ if } p=2\}.$$

The R-action, and the A-action if A acts on D, are determined from the actions on $D \subset VD$ by the internal Cartan formulas (for R and A) and the properties required of the unit.

W: Allowable R-coalgebras (resp., AR-coalgebras) to allowable R-Hopf algebras (resp., AR-Hopf algebras): Given E, define WE as an R-algebra, and as an A-algebra if A acts on E, by

$$WE = VJE,\ JE = Coker\ \eta.$$

The inclusion of E in WE is given by $E = Z_p \oplus JE$ and $JE \subset VJE$. The coproduct and augmentation of WE are determined by those of E and the requirement that WE be a Hopf algebra (it is a well-defined Hopf algebra by [11, Proposition 12]). The components of WE are easily read off from the definition of $V_0 JE$.

G: Allowable R-Hopf algebras (resp., AR-Hopf algebras) to

<u>allowable R-Hopf algebras (resp., AR-Hopf algebras) with conjugation</u>:

Given W, define GW as follows. πW is a commutative monoid under the product in W and W_0 is its monoid ring. Let πGW be the commutative group generated by πW and let G_0W be its group ring. Let $\phi = \eta(1)$, let $\overline{W}$ be the set of positive degree elements of W, and let $\overline{W}^+$ be the connected subalgebra $Z_p\phi \oplus \overline{W}$ of W. Define

$$GW = \overline{W}^+ \otimes G_0W \cong W \otimes_{\pi W} G_0 W$$

as an augmented algebra. Embed W in GW as the subalgebra

$$(\overline{W} \otimes Z_p\ \phi) \oplus (Z_p\phi \otimes W_0)$$

The coproduct on GW is determined by the requirements that W and G_0W be subcoalgebras and that GW be a Hopf algebra. The conjugation is given by Lemma 2.7. The R-action, and the A-action if A acts on W, are determined from the actions on $W \subset GW$ by commutation with χ and the Cartan formulas. If the product in WG is denoted by $*$, then the positive degree elements of the component of $f \in \pi GW$ are given by

$$(\overline{GW})_f = \bigoplus_{g \in \pi W} (\overline{W}_g \otimes g^{-1} * f) = \bigoplus_{g \in \pi W} \overline{W}_g * g^{-1} * f.$$

Observe that GW = W if W is connected and that, as a ring, GW is just the localization of the ring W at the monoid πW.

§3. The dual of the Dyer-Lashof algebra

Since EH_*S^0 is the allowable AR-coalgebra $Z_p \oplus R$ (which should be thought of as $Z_p \cdot [0] \oplus R \cdot [1]$), a firm grasp on the structure of R is important to the understanding of $H_*C_\infty S^0$ and of H_*QS^0. The coproduct and A-action on R are determined by the diagonal Cartan formula and the Nishida relations, but these merely give recursion formulas with respect to length, the explicit evaluation of which requires use of the Adem relations. To obtain precise information, we proceed by analogy with Milnor's computation of the dual of the Steenrod algebra [14]. In the case p = 2, the analogy is quite close; in the case p > 2, the Bocksteins introduce amusing complications. The structure of R*, in the case p = 2, was first determined by Madsen [8]; his proofs are closer to the spirit of Milnor's work, but do not generalize readily to the case of odd primes.

By Lemma 2.5, $R = \bigoplus_{k \geq 0} R[k]$ as an A-coalgebra. Of course, $R[0] = Z_p$. We must first determine the primitive elements $PR[k]$ of the connected coalgebras $R[k]$, $k \geq 1$. To this end, define $P[k] = \{I \mid I \text{ is admissible, } e(I) \geq 0,\ \ell(I) = k, \text{ and } I \text{ ends with } 1\}$. We shall see that $\{Q^I \mid I \in P[k]\}$ is a basis for $PR[k]$. Define (inductively and explicitly) certain elements of $P[k]$ as follows:

(I) I_{jk}, $1 \leq j \leq k$, $p=2$: $I_{11} = (1)$, $I_{j,k+1} = (2^k - 2^{k-j}, I_{jk})$ if $j \leq k$, and $I_{k+1,k+1} = (2^k, I_{kk})$; then $d(I_{jk}) = 2^k - 2^{k-j}$, $e(I_{jk}) = 0$ if $j < k$, $e(I_{kk}) = 1$; $I_{jk} = (2^{k-1} - 2^{k-1-j}, 2^{k-2} - 2^{k-2-j}, \ldots, 2^j - 1, 2^{j-1}, \ldots, 1)$.

(II) I_{jk}, $1 \leq j \leq k$, $p>2$: $I_{11} = (0,1)$, $I_{j,k+1} = (0, p^k - p^{k-j}, I_{jk})$ if $j < k$, and $I_{k+1,k+1} = (0, p^k, I_{kk})$; then $d(I_{jk}) = 2(p^k - p^{k-j})$, $e(I_{jk}) = 0$ if $j < k$, $e(I_{kk}) = 2$; $I_{jk} = (0, p^{k-1} - p^{k-1-j}, \ldots, 0, p^j - 1, 0, p^{j-1}, 0, p^{j-2}, \ldots, 0, 1)$.

(III) J_{jk}, $1\le j\le k$, $p>2$: $J_{11}=(1,1)$, $J_{j,k+1}=(0,p^k-p^{k-j}, J_{jk})$ if $j<k$, and $J_{k+1,k+1}=(1,p^k,I_{kk})$; then $d(J_{jk}) = 2(p^k-p^{k-j})-1$, $e(J_{jk})=1$; $J_{jk} = (0,p^{k-1}-p^{k-1-j},\ldots,0,p^j-1,1, p^{j-1},0,p^{j-2},\ldots,0,1)$.

(IV) K_{ijk}, $1\le i<j\le k$, $p>2$: $K_{i,j,k+1}=(0,p^k-p^{k-i}-p^{k-j},K_{ijk})$ if $j\le k$, and $K_{i,k+1,k+1}=(1,p^k-p^{k-i},J_{ik})$; then $d(K_{ijk}) = 2(p^k-p^{k-i}-p^{k-j})$, $e(K_{ijk})=0$; K_{ijk}

$$=(0,p^{k-1}-p^{k-1-i}-p^{k-1-j},\ldots,0,p^j-p^{j-i}-1,1,p^{j-1}-p^{j-1-i},J_{i,j-1}).$$

If we look back at the definition of the Q^s in terms of the Q_i in the proof of Theorem 1.1, we see that, when acting on a zero-dimensional class, our four classes of sequences correspond to sequences of operations of the respective forms

(I) $Q_0\cdots Q_0 Q_1\cdots Q_1$

(II) $Q_0\cdots Q_0 Q_{2(p-1)}\cdots Q_{2(p-1)}$

(III) $Q_{p-1}\cdots Q_{p-1}\beta Q_{2(p-1)}\cdots Q_{2(p-1)}$

(IV) $Q_0\cdots Q_0\beta Q_{p-1}\cdots Q_{p-1}\beta Q_{2(p-1)}\cdots Q_{2(p-1)}$.

Many arguments in this section and the next can be illuminated by translation to lower indices.

Lemma 3.1. $P[k] = \{I_{jk} \mid 1\le j\le k\}$ if $p=2$; $P[k]=\{I_{jk},J_{jk},K_{ijk} \mid 1\le j\le k, 1\le i<j\}$ if $p > 2$. If $I\in P[k]$, then Q^I is primitive, $\psi Q^I=Q^I\otimes (Q^0)^k+(Q^0)^k\otimes Q^I$.

Proof: Proceed by induction on k, the case k = 1 being trivial. Consider $I = (\varepsilon,s,J)\in P[k]$, $k\ge 2$. Then, since I is admissible and $e(I) \ge 0$, $J\in P[k-1]$, $pr-\delta\ge s$ if $J = (\delta,r,K)$, and $2s-\varepsilon \ge d(J)$. The first part follows inductively from these inequalities by a trivial examination of cases. The second part is an easy calculation based on the facts that $\beta^{\gamma}Q^{i}Q^{J}=0$ if $2i-\gamma<d(J)$ and that $\beta^{\gamma}Q^{1}Q^{0}=0$ by the first Adem relation.

The computation of $R[k]^*$ as an algebra is based on a correspondence between addition of admissible sequences and multiplication of duals of admissible monomials. We first set up the required calculus of admissible sequences.

Definitions 3.2. The sum I+J and difference I-J of two sequences (as in Definition 2.1) of length k is defined termwise, under the conventions that I+J is undefined if $p>2$ and the ith "Bockstein entry" ε_i is one in both I and J and that I-J is undefined if any entry is less than zero. Observe that $e(I\pm J) = e(I)\pm e(J)$ and $d(I\pm J) = d(I)\pm d(J)$. If I and J are admissible, then I+J is admissible but I-J need not be admissible. In order to enumerate the admissible monomials when $p>2$, consider all sequences $e = \{e_1,\ldots,e_j\}$ with $1 \leq e_1<\ldots<e_j\leq k$ and define

$$I_e[k]=\{I \mid I \text{ is admissible, } e(I)\geq 0,\ \ell(I)=k, \text{ and } \varepsilon_{k+1-a}=1 \Leftrightarrow a\in e\}.$$

Write $I_e[k] = I[k]$ when e is empty. When $j\geq 1$, define $L_{e,k}\in I_e[k]$ by

$$L_{e,k} = \begin{cases} K_{e_1e_2k}+\ldots+K_{e_{j-1}e_jk} & \text{if } j \text{ is even} \\ K_{e_1e_2k}+\ldots+K_{e_{j-2}e_{j-1}k}+J_{e_jk} & \text{if } j \text{ is odd} \end{cases}$$

If $p=2$, write $I[k] = \{I \mid I$ is admissible, $e(I)\geq 0$, and $\ell(I)=k\}$.

With these notations, we have the following two counting lemmas.

Lemma 3.3. Let N denote the set of non-negative integers. For $p\geq 2$ and $k\geq 1$, define $f\colon N^k \to I[k]$ by $f(n_1,\ldots,n_k) = \sum_{j=1}^{k} n_jI_{jk}$. Then f is an isomorphism of sets.

Proof: For p>2, omit the irrelevant zeroes corresponding to absence of Bocksteins. Then f is given explicitly by $f(n_1,\ldots,n_k)=(s_1,\ldots,s_k)$, where $s_{k+1-j}=\sum_{q=1}^{j-1} n_q(p^{j-1}-p^{j-1-q})+\sum_{q=j}^{k} n_q p^{j-1}$.

The required inverse to f is given by

$$f^{-1}(s_1,\ldots,s_k)=(n_1,\ldots,n_k) \text{ where } n_j=\begin{cases} ps_{k+1-j}-s_{k-j} & \text{if } 1\leq j<k \\ s_k-\sum_{q=2}^{k}(p\,s_q-s_{q-1}) & \text{if } j=k \end{cases}$$

Lemma 3.4. For p>2, $k\geq 1$, and each non-empty e, define $f_e\colon I[k]\to I_e[k]$ by $f_e(I)=I+L_{ek}$. Then f_e is an isomorphism of sets.

Proof: Obviously f_e^{-1} must be given by $f_e^{-1}(J)=J-L_{ek}$, $J\in I_e[k]$, and it suffices to show that $J-L_{ek}$ is defined, admissible, and has non-negative excess. Write $L_{ek}=(\delta_1,r_1,\ldots,\delta_k,r_k)$ and $J=(\varepsilon_1,s_1,\ldots,\varepsilon_k,s_k)$. Observe that

$$e(J)=2s_k-\sum_{q=1}^{k}\varepsilon_q-2\sum_{q=2}^{k}(p\,s_q-\varepsilon_q-s_{q-1})\geq 0 \text{ and } ps_q-\varepsilon_q\geq s_{q-1}.$$

L_{ek} is the unique element of $I_e[k]$ such that, if e has $j=2i-\varepsilon$ elements, then $e(L_{ek})=\varepsilon$ and L_{ek} ends with i. Explicitly, $\delta_q=\varepsilon_q$ is determined by e, $r_k=i$, and $r_{q-1}=pr_q-\delta_q$ for $q\leq k$. The result follows from the inequalities satisfied by the entries of J.

As a final preliminary, we require an ordering of sequences.

Definition 3.5. For a sequence $I=(\varepsilon_1,s_1,\ldots,\varepsilon_k,s_k)$, define $I_j=(\varepsilon_j,s_j,\ldots,\varepsilon_k,s_k)$, $1\leq j\leq k$, and similarly when p=2. Note that $e(I_j)=e(J_j)$ for all j implies I=J, and define a total ordering of the sequences of length k by I<J if $e(I_j)<e(J_j)$ for the smallest j such that $e(I_j)\neq e(J_j)$. Observe that $I\leq I'$ and $J<J'$ implies $I+J<I'+J'$.

An easy inspection demonstrates the following lemma.

Lemma 3.6. If I is inadmissible and $Q^I = \Sigma\lambda_J Q^J$ where the J are admissible, then $\lambda_J \neq 0$ implies J<I. If $P^r_* Q^I = \Sigma\lambda_J Q^J$, where r>0 and the J are admissible, then $\lambda_J \neq 0$ implies J<I.

$R^* = \prod_{k\geq 0} R[k]^*$ as an A-algebra. In the dual basis to that of admissible monomials, define elements of $R[k]^*$ by

$$\xi_{0k} = (Q^{0k})^* \quad \text{if } 0 \leq k$$

$$\xi_{jk} = (Q^{I_{jk}})^* \quad \text{if } 1 \leq j \leq k$$

$$\tau_{jk} = (Q^{J_{jk}})^* \quad \text{if } 1 \leq j \leq k$$

$$\sigma_{ijk} = (Q^{K_{ijk}})^* \quad \text{if } 1 \leq i < j \leq k.$$

To simplify statements of formulas, define $\xi_{jk}=0$ if $j<0$ or $j>k$, $\tau_{jk}=0$ if $j<1$ or $j>k$, and $\sigma_{ijk}=0$ if $i<1$, $j\leq i$, or $j>k$.

ξ_{0k} is the identity element of $R[k]^*$ and $\prod_{k\geq 0} \xi_{0k}$ is the identity element of R^*. The augmentation of R^* is given by $\varepsilon(\prod_{k\geq 0} \lambda_k \xi_{0k}) = \lambda_0$. Of course, R^* is not a coalgebra since R_0 is not finite dimensional (although R_q is finite dimensional for $q>0$). However, R^* does have a well-defined coproduct on positive degree elements and on finite linear combinations of the ξ_{0k}; the latter is evidently given by

$$\psi\xi_{0k} = \Sigma\xi_{0,k-i} \otimes \xi_{0,i} .$$

It is perhaps worth observing that although $\prod_{k=0}^{n} R[k]^*$ is a quotient augmented A-algebra of R^* and a coalgebra (dual to the quotient algebra $R/\sum_{m>n} R[m]$ of R) such that the product is a morphism of coalgebras, $\prod_{k=0}^{n} R[k]^*$ is nevertheless not a Hopf algebra because its unit fails to be a morphism of coalgebras (dually, $(Q^0)^{n+1} = 0$ but $\varepsilon Q^0 = 1$).

We shall successively compute R[k]* as an algebra, compute the Steenrod operations on generators, and compute the coproduct on generators.

Theorem 3.7. If p=2, $R[k]^* = P\{\xi_{1k},\ldots,\xi_{kk}\}$ as an algebra. If p>2, let M[k] be the subspace of R[k]* spanned by the set consisting of ξ_{0k} together with the monomials

$$\sigma_{e_1e_2k}\cdots\sigma_{e_{j-1}e_jk},\quad 1\le e_1<\ldots<e_j\le k \text{ and } j \text{ even},$$

and

$$\sigma_{e_1e_2k}\cdots\sigma_{e_{j-2}e_{j-1}k}\tau_{e_jk},\quad 1\le e_1<\ldots<e_j\le k \text{ and } j \text{ odd}.$$

This set is linearly independent, and the product defines an isomorphism of Z_p-spaces

$$P\{\xi_{1k},\ldots,\xi_{kk}\}\otimes M[k] \to R[k]^*.$$

R[k]* is determined as an algebra by commutativity and the following relations:

(i) $\tau_{ik}\,\tau_{jk} = \xi_{kk}\sigma_{ijk}$ if $i<j$ (and $\tau_{ik}\tau_{ik} = 0$);

(ii) $\sigma_{ijk}\,\tau_{nk} = (\tau_{ik}\tau_{jk}\tau_{nk})/\xi_{kk}$; and

(iii) $\sigma_{ijk}\sigma_{mnk} = (\tau_{ik}\tau_{jk}\tau_{mk}\tau_{nk})/\xi_{kk}^2$.

(In (ii) and (iii), the right sides are to be evaluated in terms of the basis monomials by use of (i); the numerators, if non-zero, are divisible by the non zero-divisor ξ_{kk} or ξ_{kk}^2.)

Proof: By the counting lemmas, an admissible monomial I with $\ell(I)=k$ and $e(I)\ge 0$ can be uniquely expressed in the form

$$I = n_1I_{1k}+\ldots+n_kI_{kk}+L_{ek},\quad n_q\ge 0 \text{ and } e = \{e_1,\ldots,e_j\},$$

where L_{ek} is the sequence of all zeroes if j=0 (or if p=2).

Let $j=2i-\varepsilon$, $\varepsilon=0$ or 1, and define $n(I)=i+\sum_q n_q$. Let λ_e denote the monomial in M[k] corresponding to the sequence e. Let <, > be the Kronecker product (that is, the evaluation pairing $R[k]^* \otimes R[k] \to Z_p$). We claim that

(1) $\langle \xi_{1k}^{n_1} \dots \xi_{kk}^{n_k} \lambda_e, Q^I \rangle = 1$, and

(2) $\langle \xi_{1k}^{n_1} \dots \xi_{kk}^{n_k} \lambda_e, Q^J \rangle \neq 0$ and $J \neq I$ imply $J>I$.

Let $\psi: R[k] \to R[k]^{n(I)}$ be the iterated coproduct. For any J,

(3) $\psi Q^J = \Sigma \pm Q^{J_1} \otimes \dots \otimes Q^{J_{n(I)}}$, $\Sigma J_i = J$.

Now (1) is immediate from the definition of the L_{ek}. Given J as in (2), we can obtain a summand

$$\lambda Q^{I_1} \otimes \dots \otimes Q^{I_{n(I)}}, \ \Sigma I_i = I \text{ and } \lambda \neq 0.$$

on the right side of (3) by applying Adem relations to put the Q^{J_i} in admissible form, and J>I follows. If we express the monomials $\xi_I = \xi_{1k}^{n_1} \dots \xi_{kk}^{n_k} \lambda_e$ in the ordered basis dual to that of admissible monomials,

$$\xi_I = \sum_J a_{IJ} (Q^J)^*,$$

then (1) and (2) state that (a_{IJ}) is an upper triangular matrix with ones along the main diagonal. Therefore $\{\xi_I\}$ is a basis for $R[k]^*$. It remains only to prove (i), (ii), and (iii). By inspection of the definitions, we have

$$I_{kk} + K_{ijk} = J_{ik} + J_{jk} \text{ if } i<j.$$

An easy dimensional argument shows that $\xi_{kk}\sigma_{ijk}$ is the only possible summand of $\tau_{ik}\tau_{jk}$, and this proves (i). Since

$$\xi_{kk}\sigma_{ijk}\tau_{nk} = \tau_{ik}\tau_{jk}\tau_{nk} \text{ and } \xi_{kk}^2\sigma_{ijk}\sigma_{mnk} = \tau_{ik}\tau_{jk}\tau_{mk}\tau_{nk},$$

formulas (ii) and (iii) follow immediately from (i).

In order to determine the Steenrod operations on the generators of R[k]*, we need to know all operations in R[k] which hit any of the Q^I, $I \in P[k]$, from above; of course, we may restrict attention to the generators P^{p^r} and β of A. For dimensional arguments, it should be observed that R can be given a second grading by the number of Bocksteins which occur in monomials and that all structure (except, of course, action by β) preserves this grading.

<u>Lemma 3.8</u>. The following formulas are valid in R[k], $k \geq 1$, and these formulas specify all operations βQ^J and $P_*^{p^r} Q^J$, $r \geq 0$, on basis elements Q^J, which have a summand of the form λQ^I with $0 \neq \lambda \in Z_p$ and $I \in P[k]$:

(i) $p>2$: $\beta Q^{I_{kk}} = Q^{J_{kk}}$ and $\beta Q^{J_{ik}} = Q^{K_{ikk}}$ if $1 \leq i<k$

(ii) $p \geq 2$: $P_*^{p^{k-1-j}} Q^{I_{j+1,k}} = -Q^{I_{jk}}$ if $1 \leq j<k$

(iii) $p>2$: $P_*^{p^{k-1-j}} Q^{J_{j+1,k}} = -Q^{J_{jk}}$ if $1 \leq j<k$

(iv) $p>2$: $P_*^{p^{k-1-i}} Q^{K_{i+1,j,k}} = -Q^{K_{ijk}}$ if $1 \leq i<j-1<k$

(v) $p>2$: $P_*^{p^{k-1-j}} Q^{K_{i,j+1,k}} = -Q^{K_{ijk}}$ if $1 \leq i<j<k$

(vi) $p \geq 2$: $P_*^{p^{k-1}} Q^{I_{1k}+I_{jk}} = Q^{I_{jk}}$ if $1 \leq j \leq k$

(vii) $p>2$: $P_*^{p^{k-1}} Q^{I_{1k}+J_{jk}} = Q^{J_{jk}}$ if $2 \leq j \leq k$

(viii) $p>2$: $P_*^{p^{k-1}} Q^{I_{jk}+J_{1k}} = 2Q^{J_{jk}}$ if $1 \leq j \leq k$

(ix) $p>2$: $P_*^{p^{k-1}} Q^{I_{1k}+K_{ijk}} = Q^{K_{ijk}}$ if $2 \leq i<j \leq k$

(x) $p>2$: $P_*^{p^{k-1}} Q^{I_{ik}+K_{1jk}} = 2Q^{K_{ijk}}$ if $1 \leq i<j \leq k$

Proof: The statements about β are obvious. For the rest, we first reduce the problem to manageable proportions by a search of dimensions. Observe that

(a) $I \in P[k]$ implies $2p^{k-2}(p^2-p-1) \leq d(I) \leq 2(p^k-1)$ $[2^{k-1} \leq d(I) \leq 2^k-1]$.

Since $R[k]^*$ is an unstable A-module, (a) implies that

$P^{p^r}(Q^{I*}) = 0$ if $r \geq k$ and $I \in P[k]$. For $r<k$ and $I \in P[k]$, we have

(b) $d(I) + 2p^r(p-1) \leq 2(2p^k - p^{k-1} - 1) < 6p^{k-2}(p^2-p-1)[d(I) + 2^r < 3 \cdot 2^{k-1}]$.

Clearly (a) and (b) imply that if $P_*^{p^r} Q^J$ has a summand λQ^I with $\lambda \neq 0$ and $I \in P[k]$, then either $J \in P[k]$ or $J = J' + J''$ with both J' and J'' in $P[k]$. Observe further that

$$d(I) + 2p^r(p-1) < 4p^{k-2}(p^2-p-1), \quad I \in P[k],$$

if either $p>3$ and $r \leq k-2$ or $p=3$ and $r \leq k-3$; thus the possibility $J = J'+J''$ is also ruled out in these cases. Simple dimensional arguments in the few remaining cases demonstrate that our list will be exhaustive provided that the following formulas also hold:

(xi) $p>2$: $P_*^{p^{k-1}} Q^{I_{jk}+K_{1ik}} = 0$ if $1<i<j \leq k$,

(xii) $p=2$: $P_*^{2^{k-j}} Q^{I_{1k}+K_{1k}} = 0$ if $2 \leq j \leq k$.

To prove (ii) through (xii), observe that the Nishida relations can be described as follows on admissible monomials Q^J.

(4) Let $J = (\varepsilon, s, K)$; if $r \geq 0$ and $s < p^r + \varepsilon$, then $P_*^{p^r} Q^J = 0$; if $r \geq 1$ and $s \geq p^r + \varepsilon$, then $P_*^{p^r} Q^J \equiv \beta^{\varepsilon} Q^{s-p^r+p^{r-1}} P_*^{p^{r-1}} Q^K$ modulo linear combinations of admissible monomials Q^L such that $e(L) \leq e(J) - 2(p-\varepsilon)$ $[e(L) \leq e(J) - 2]$.

Further, we have the particular Nishida relations

(5) $P^1_*Q^s = (s-1)Q^{s-1}$ and $P^1_*\beta Q^s = s\beta Q^{s-1} - Q^{s-1}\beta$ if $s \geq 1$.

Formulas (4) and (5) clearly imply

(6) If $e(J_j) < 2(p-\varepsilon)$, $1 \leq j < k$, then $P_*^{p^{k-1}} Q^J = (s_k - 1 + \varepsilon_k) Q^{J - I_{1k}}$

[If $e(J_j) < 2$, $1 \leq j < k$, then $P_*^{p^{k-1}} Q^J = (s_k - 1) Q^{J - I_{1k}}$],

where $J = (\varepsilon_1, s_1, \ldots, \varepsilon_k, s_k)$ and $J_j = (\varepsilon_j, s_j, \ldots, \varepsilon_k, s_k)$.

In all cases (vi) through (xi), the hypothesis of (6) is satisfied, and this proves (vi), (vii), and (ix). In (viii), (x), and (xi), the sequence $J-I_{1k}$ obtained on the right side of (6) is not admissible. However, the only Adem relation required to reduce $J-I_{1k}$ to admissible form is

(7) $Q^{ps}\beta Q^s = \beta Q^{ps} Q^s$ if $s \geq 1$.

The proofs of (ii) through (v) and (xii) are similar applications of (4) and (5); they are simplified by use of induction on k. The following Adem relation is needed in the proof of (xii).

(8) $Q^{ps+1}Q^s = 0$ if $s \geq 0$.

Because of the change of basis involved in our description of $R[k]^*$, our formulas simplify slightly upon dualization.

<u>Theorem 3.9</u>. The following list of relations specifies all non-trivial actions of the generators P^{p^r}, $r \geq 0$, and β of the Steenrod algebra on the generators $\{\xi_{jk}, \tau_{jk}, \sigma_{ijk}\}$ of $R[k]^*$.

(i) $p>2$: $\beta\tau_{kk} = \xi_{kk}$ and $\beta\sigma_{ikk} = -\tau_{ik}$ if $1 \leq i \leq k$

(ii) $p \geq 2$: $P^{p^{k-1-j}} \xi_{jk} = -\xi_{j+1,k}$ if $1 \leq j < k$

(iii) $p > 2$: $P^{p^{k-1-j}} \tau_{jk} = -\tau_{j+1,k}$ if $1 \leq j < k$

(iv) $p > 2$: $P^{p^{k-1-i}} \sigma_{ijk} = -\sigma_{i+1,j,k}$ if $1 \leq i < j-1 < k$

(v) $p > 2$: $P^{p^{k-1-j}} \sigma_{ijk} = -\sigma_{i,j+1,k}$ if $1 \leq i < j < k$

(vi) $p \geq 2$: $P^{p^{k-1}} \xi_{jk} = \xi_{1k}\xi_{jk}$ if $1 \leq j \leq k$

(vii) $p > 2$: $P^{p^{k-1}} \tau_{jk} = \xi_{1k}\tau_{jk} + \xi_{jk}\tau_{1k}$ if $1 \leq j \leq k$

(viii) $p > 2$: $P^{p^{k-1}} \sigma_{ijk} = \xi_{1k}\sigma_{ijk} + \xi_{ik}\sigma_{1jk} - \xi_{jk}\sigma_{1ik}$ if $1 \leq i < j \leq k$, $\sigma_{11k} = 0$.

<u>Proof.</u> (i) is trivial since, as explained in [A, p. 207], the cohomology and homology Bocksteins are related by

$$<\beta\alpha, a> \;=\; (-1)^{\deg \alpha+1} <\alpha, \beta a>.$$

Relations (ii) through (vi) are immediate from the corresponding numbered relations of the lemma, since $R[k]^*$ is one-dimensional in the degrees in which these relations occur. For (vii), we can certainly write

$$P^{p^{k-1}} \tau_{jk} = a\xi_{1k}\tau_{jk} + b\xi_{jk}\tau_{1k} \qquad \text{(with } a = 0 \text{ if } j = 1\text{)}.$$

For $j \geq 2$, $I_{1k} + J_{jk} < J_{1k} + J_{jk}$ and $<\xi_{1k}\tau_{jk}, Q^{J_{1k}+I_{jk}}> \;=\; 1$, as can be seen by examination of $\psi Q^{J_{1k}+I_{jk}}$. By (vii) and (viii) of the lemma, and by formulas (1) and (2), we find

$$1 = <P^{p^{k-1}} \tau_{jk}, Q^{I_{1k}+J_{jk}}> \;=\; a \quad \text{if } 2 \leq j \leq k$$

and

$$2 = <P^{p^{k-1}} \tau_{jk}, Q^{J_{1k}+I_{jk}}> \;=\; a+b \quad \text{if } 1 \leq j \leq k.$$

This proves (vii). Similarly, for (viii), we can certainly write

$$P^{p^{k-1}} \sigma_{ijk} = a\xi_{1k}\sigma_{ijk} + b\xi_{ik}\sigma_{1jk} + c\xi_{jk}\sigma_{1ik} \qquad \text{(with } a = 0 \text{ and } c = 0 \text{ if } i = 1\text{)}.$$

$I_{1k} + K_{ijk} < I_{ik} + K_{1jk} < I_{jk} + K_{1ik}$, and $<\xi_{ik}\sigma_{1jk}, Q^{I_{jk}+K_{1ik}}> \;=\; 0$,

$$<\xi_{1k}\sigma_{ijk}, Q^{I_{ik}+K_{1jk}}> = 1 \quad\text{and}\quad <\xi_{ik}\sigma_{1jk}, Q^{I_{jk}+K_{1ik}}> = 1$$

by examinations of coproducts. Now (ix), (x), and (xi) of the lemma, together with (1) and (2), imply (viii) by evaluation of $P^{p^{k-1}}\sigma_{ijk}$ on $Q^{I_{1k}+K_{ijk}}$, $Q^{I_{ik}+K_{1jk}}$, and $Q^{J_{jk}+K_{1ik}}$. Note that (viii) can be predicted from (vi) and (vii) by application of $P^{2p^{k-1}}$ to the relation $\tau_{ik}\tau_{jk} = \xi_{kk}\sigma_{ijk}$.

We have the following immediate corollary.

Corollary 3.10. If $p = 2$, $R[k]^*$ is generated as an A-algebra by ξ_{1k}. If $p > 2$, $R[1]^*$ is generated as an A-algebra by τ_{11} and $R[k]^*$, $k \geq 2$, is generated as an A-algebra by ξ_{1k} and σ_{12k}.

In other words, $R[k]^*$ is a quotient A-algebra of $H^*K(Z_p, n)$ or of $H^*K(Z_p, m) \otimes H^*K(Z_p, n)$ for appropriate integers m and n.

Remarks 3.11. In order to obtain an upper bound on the spherical classes of H_*QS^0 by determination of its A-annihilated primitive elements, it would be desirable to have complete information on the A-module (rather than the A-algebra) generators of $R[k]^*$: we can add classes not in R to elements of R to obtain primitive classes of H_*QS^0; we cannot so obtain A-annihilated classes of H_*QS^0 unless the given class in R was A-annihilated. I have not carried out the necessary calculations. Madsen [8] has obtained considerable information in the case $p = 2$ and has used this information to retrieve Browder's results [4] on the Arf invariant.

It remains to compute the coproduct on generators of $R[k]^*$, and we need information about the products in R which hit any of the Q^I, $I \in P[k]$. Fortunately, we do not need complete information when $I = K_{ijk}$.

Lemma 3.12. Let I_{0k} denote the sequence of length k, $k \geq 0$, with all entries zero. Suppose that J and K are admissible sequences such that $Q^J Q^K$ has a summand λQ^I with $\lambda \neq 0$ and $I \in P[k]$. Then either $K \in P[i]$ or $K = I_{0i}$ for some $i < k$ and, in the latter case, $I = I_{jk}$ for some j. All possible choices for J and K when $I = I_{jk}$ or $I = J_{jk}$ are specified in the following relations; in (i) and (ii), if $h < i$ and $h < j$, then the asserted relations merely hold modulo the subspace of $R[k]$ spanned by the admissible monomials which do not end with Q^1.

(i) $Q^{(p^i - p^{i-h})I_{k-i,k-i} + p^{i-h}I_{j-h,k-i}} Q^{I_{hi}} \equiv Q^{I_{jk}}$, $0 \leq h \leq i \leq k$, $0 \leq j-h \leq k-i$.

(ii) $Q^{(p^i - p^{i-h})I_{k-i,k-i} + p^{i-h}I_{j-h,k-i}} Q^{J_{hi}} \equiv Q^{J_{jk}}$, $1 \leq h \leq i \leq k$, $0 \leq j-h \leq k-i$.

(iii) $Q^{(p^i - 1)I_{k-i,k-i} + J_{j-i,k-i}} Q^{I_{ii}} = Q^{J_{jk}}$, $0 \leq i < j \leq k$.

Proof. If (J, K) is admissible and in P[k], then $K \in P[i]$ for some i; such decompositions of I_{jk} and J_{jk} account for the relations with $h = i$ in (i) and with $h = j$ in (i) and (ii) and for all relations in (iii). If (J, K) is inadmissible, then $K \in P[i]$ or $K = I_{0i}$ for some i since the only Adem relations which have Q^1 appearing on the right side are

$$Q^p Q^0 = Q^{p-1} Q^1 \quad \text{and} \quad Q^p \beta Q^1 = \beta Q^p Q^1 .$$

We claim that if (t, I_{hi}) is inadmissible, $0 \leq h \leq i$, then $Q^t Q^{I_{hi}}$ has no non-zero summand ending with Q^1 unless $t = p^i$ and $h < i$, when $Q^{I_{h+1,i+1}}$ is the only such summand. Indeed, $Q^t Q^{I_{ii}}$ can have no such summand because, in the Adem relation $Q^r Q^s = \Sigma \lambda_j Q^{r+s-j} Q^j$ for $r > ps$, $\lambda_j = 0$ unless $j > s$. The claim now follows by upwards induction on i and, for fixed i, downwards induction on h, via explicit calculation from the Adem relations and the inductive definition of the I_{hi}. The essential fact is that $Q^{p^i}Q^{p^{i-1}-p^{i-1-h}}$ has the summand $Q^{p^i-p^{i-1-h}}Q^{p^{i-1}}$, $0 \leq h < i$. Note that, since $\beta Q^{I_{hi}} = 0$ for $h < i$, it follows that if (J, I_{hi}) is inadmissible, $0 \leq h \leq i$, and if any Bockstein entry ϵ_j in J is non-zero, then $Q^J Q^{I_{hi}}$ has no non-zero summand ending with Q^1. We claim also that if (t, J_{hi}) is inadmissible, $1 \leq h \leq i$, then $Q^t Q^{J_{hi}}$ has no non-zero summand ending with Q^1 unless $t = p^i$, when $Q^{J_{h+1,i+1}}$ is the only such summand. The proof is again an easy double induction; the Adem relation (7) is used to prove the claim when $h = i$. A straightforward bookkeeping argument from our claims shows that the relations of (i) and (ii) with $h < i$ and $h < j$ give all possibilities for $Q^J Q^K$ to have a non-zero summand $\lambda Q^{I_{jk}}$ or $\lambda Q^{J_{jk}}$ when (J, K) is inadmissible.

In our formulas for the coproduct in R^*, the sums are to range over the integers; this makes sense in view of our convention that ξ_{jk}, τ_{jk}, and σ_{ijk} are zero except where explicitly specified otherwise. The formula for $\psi\sigma_{ijk}$ announced in [12] is incorrect; the correct formula given here is in fact somewhat simpler.

<u>Theorem 3.13.</u> The following formulas specify the coproduct on the generators of R^*.

(i) $$\psi\xi_{jk} = \sum_{(h,i)} \xi_{k-i,k-i}^{p^i-p^{i-h}} \xi_{j-h,k-i}^{p^{i-h}} \otimes \xi_{hi}$$

(ii) $$\psi\tau_{jk} = \sum_{(h,i)} \xi_{k-i,k-i}^{p^i-p^{i-h}} \xi_{j-h,k-i}^{p^{i-h}} \otimes \tau_{hi} + \sum_{i} \xi_{k-i,k-i}^{p^i-1} \tau_{j-i,k-i} \otimes \xi_{ii}$$

(iii) $$\psi\sigma_{ijk} = \sum_{(f,g,h)} \xi_{k-h,k-h}^{p^h-p^{h-f}-p^{h-g}} (\xi_{j-g,k-h}^{p^{h-g}} \xi_{i-f,k-h}^{p^{h-f}} - \xi_{j-f,k-h}^{p^{h-f}} \xi_{i-g,k-h}^{p^{h-g}}) \otimes \sigma_{fgh}$$

$$+ \sum_{(g,h)} \xi_{k-h,k-h}^{p^h-p^{h-g}-1} (\xi_{j-g,k-h}^{p^{h-g}} \tau_{i-h,k-h} - \xi_{i-g,k-h}^{p^{h-g}} \tau_{j-h,k-h}) \otimes \tau_{gh}$$

$$+ \sum_h \xi_{k-h,k-h}^{p^h-1} \sigma_{i-h,j-h,k-h} \otimes \xi_{hh}$$

<u>Proof.</u> Observe first that if $J = \Sigma n_i I_{ik} + L_{ek}$, then $e(J) = n_k + \epsilon$, where $\epsilon = e(L_{ek})$ is zero or one. In view of the lemma, (i) and (ii) will hold provided that the monomials to the left of the tensor signs are precisely dual to the corresponding admissible monomials Q^J. By (1) and (2) in the proof of Theorem 3.7, this will certainly hold if the J are maximal among all admissible sequences of the requisite degrees. A dimensional argument shows that, due to the multiple $p^i - p^{i-h}$ of $I_{k-i,k-i}$ which appears, the J actually have maximal excess among the admissible sequences of the requisite degrees. We prove (iii) by a trick. By the lemma, we can certainly write

$$\psi\sigma_{ijk} = \sum_{(f,g,h)} \alpha_{fgh} \otimes \sigma_{fgh} + \sum_{(g,h)} \beta_{gh} \otimes \tau_{gh} + \sum_h \gamma_h \otimes \xi_{hh}$$

(ξ_{gh} for $g < h$ cannot appear on the right because, as noted in the proof of the lemma, $Q^J Q^{I_{gh}}$ cannot have a non-zero summand $\lambda Q^{K_{ijk}}$ unless (J, I_{gh}) is admissible, when $g = h$.) We have $\tau_{ik}\tau_{jk} = \xi_{kk}\sigma_{ijk}$ and therefore $(\psi\tau_{ik})(\psi\tau_{jk}) = (\psi\xi_{kk})(\psi\sigma_{ijk})$. After expanding both sides by use of Theorem 3.7 and the fact that $R[k]^* \cdot R[\ell]^* = 0$ for $k \neq \ell$, we find that there is a unique solution for the unknowns α_{fgh}, β_{gh}, and γ_h, namely that specified in (iii).

§4. The homology of QX

In this section and the next, we shall compute H_*QX and H_*CX for any space X, where C is the monad associated to an E_∞ operad $\mathcal{C}$ [see G, Construction 2.4]. We shall also compute the mod p Bockstein spectral sequences of QX and CX, hence our results will determine the integral homology groups of these spaces.

QX is the free infinite loop space generated by X in the sense that if $Y \in \mathcal{L}_\infty$ and $f: X \to Y_0$ is any map in $\mathcal{J}$, then there is a unique map $g: \{Q\Sigma^i X\} \to Y$ in $\mathcal{L}_\infty$ such that $g_0 \circ \eta = f$, where $\eta: X \to QX$ is the natural inclusion [see G, p. 43]. Since, for all finite n, the composite

$$\Sigma^n X \xrightarrow{\Sigma^n \eta} \Sigma^n \Omega^n \Sigma^n X \xrightarrow{\lambda} \Sigma^n X$$

is the identity, where λ is the evaluation map, $\eta_*: H_*X \to H_*QX$ is a monomorphism. It is therefore reasonable to expect H_*QX to be an appropriate free object generated by H_*X.

Similarly, for any operad $\mathcal{C}$, (CX, μ) is the free $\mathcal{C}$-space generated by X in the sense that if (Y, θ) is a $\mathcal{C}$-space and $f: X \to Y$ is a map in $\mathcal{J}$, then there is a unique map $g: CX \to Y$ of $\mathcal{C}$-spaces such that $g\eta = f$, $\eta: X \to CX$ [see G, p. 13, 16, 17]. Again, it is reasonable to expect H_*CX to be an appropriate free object generated by H_*X, at least for nice operads $\mathcal{C}$.

We have constructed certain free functors WE and GWE in section 2 and, by freeness, there are unique morphisms $\bar{\eta}_*$ of allowable AR-Hopf algebras and $\tilde{\eta}_*$ of allowable AR-Hopf algebras with conjugation such that the following diagrams are commutative:

$$\begin{array}{ccc} & & WEH_*X \\ & \nearrow & \downarrow \bar{\eta}_* \\ H_*X & \xrightarrow{\eta_*} & H_*CX \end{array} \qquad \text{and} \qquad \begin{array}{ccc} & & GWEH_*X \\ & \nearrow & \downarrow \tilde{\eta}_* \\ H_*X & \xrightarrow{\eta_*} & H_*QX \end{array}$$

We have the following two theorems.

<u>Theorem 4.1.</u> For every space $X \in \mathcal{T}$ and every E_∞ operad $\mathcal{C}$, $\bar{\eta}_*: WEH_*X \to H_*CX$ is an isomorphism of AR-Hopf algebras.

<u>Theorem 4.2.</u> For every space $X \in \mathcal{T}$, $\tilde{\eta}_*: GWEH_*X \to H_*QX$ is an isomorphism of AR-Hopf algebras with conjugation.

The second theorem is a reformulation (and generalization) of the calculations of Dyer and Lashof [6].

By [G, Lemma 8.11], $CS^0 = \coprod_{j \geq 0} \mathcal{C}(j)/\Sigma_j$ for any operad $\mathcal{C}$ (where $\coprod$ denotes disjoint union). If $\mathcal{C}$ is an E_∞ operad, the orbit space $\mathcal{C}(j)/\Sigma_j$ is just a $K(\Sigma_j, 1)$. Thus, as a very special case, Theorem 4.1 contains a concise reformulation of Nakaoka's results [16,17,18] on the homology of symmetric groups. An E_∞ operad $\mathcal{C}$ should be thought of as a suitably coherent construction of universal bundles for symmetric groups; the simple statement that CS^0 is a $\mathcal{C}$-space contains a great deal of information that is usually obtained by more cumbersome algebraic techniques.

The elements of $H_*X \subset H_*CX$ and of $H_*X \subset H_*QX$ play a role in the homology of E_∞ spaces and of infinite loop spaces which is analogous to that played by the fundamental classes of $K(\pi, n)$'s in the cohomology of spaces. In particular, the following corollaries are analogs of the statement that the cohomology of any space can be represented, via the morphism induced by a map, as a quotient of a free unstable A-algebra.

<u>Corollary 4.3.</u> If (X, θ) is a $\mathcal{C}$-space, where $\mathcal{C}$ is an E_∞ operad, then $\theta_*: H_*CX \to H_*X$ represents H_*X as a quotient AR-Hopf algebra of the free allowable AR-Hopf algebra WEH_*X.

<u>Proof.</u> $\theta: CX \to X$ is the unique map of $\mathcal{C}$-spaces such that $\theta\eta = 1$.

<u>Corollary 4.4.</u> If Y is an infinite loop sequence, then $\xi_{\infty*}: H_*QY_0 \to H_*Y_0$ represents H_*Y_0 as a quotient AR-Hopf algebra with conjugation of the free allowable AR-Hopf algebra with conjugation $GWEH_*X$.

<u>Proof.</u> $\xi_\infty: QY_0 \to Y_0$ is the unique infinite loop map such that $\xi_\infty\eta = 1$; ξ_∞ is defined explicitly in [G, p. 43].

Of course, Theorems 4.1 and 4.2 are not unrelated. By [G, Theorem 4.2], there is a morphism of monads $\alpha_\infty: C_\infty \to Q$. Thus $\alpha_\infty\eta = \eta$, $\alpha_\infty: C_\infty X \to QX$ is a map of $\mathcal{C}_\infty$-spaces for all X, and we have the following commutative diagram:

$$\begin{array}{ccccc}
WEH_*X & & \xrightarrow{\quad\bar{\eta}_*\quad} & & H_*C_\infty X \\
 & \nwarrow & & \nearrow{\scriptstyle \eta_*} & \\
\Big\downarrow{\scriptstyle \iota} & & H_*X & & \Big\downarrow{\scriptstyle \alpha_{\infty*}} \\
 & \swarrow & & \searrow{\scriptstyle \eta_*} & \\
GWEH_*X & & \xrightarrow{\quad\tilde{\eta}_*\quad} & & H_*QX
\end{array}$$

Here ι is the natural inclusion. Since ι is the identity if X is connected, Theorems 4.1 and 4.2, coupled with the Whitehead theorem for connected H-spaces, imply the following result.

<u>Corollary 4.5.</u> $\alpha_\infty: C_\infty X \to QX$ is a weak homotopy equivalence for all connected spaces X.

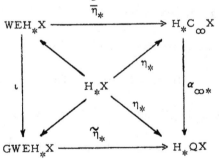

The corollary was proven geometrically in [G, Theorem 6.1] by use of the much deeper fact that $\alpha_n : C_nX \to \Omega^n\Sigma^nX$ is a weak homotopy equivalence for all n and all connected X. We shall prove Theorem 4.1 and shall generalize the corollary by obtaining a homology approximation to QX, for arbitrary X, in the next section. We prove Theorem 4.2 and compute the Bockstein spectral sequence of CX and QX here.

For counting arguments, it will be useful to have explicit bases for WEH_*X and $GWEH_*X$. Let tX be a basis for JH_*X which contains the set of components of X, other than the component ϕ of the base-point, regarded as homology classes of degree zero. Thus $tX \cup \{\phi\}$ is a basis for H_*X. Let $N\pi_0X$ and $\tilde{N}\pi_0X$ denote the free commutative monoid and the free commutative group generated by π_0X, each subject to the single relation $\phi = 1$; let $Z_pN\pi_0X$ and $Z_p\tilde{N}\pi_0X$ denote their monoid and group rings. Let ATX be the free commutative algebra generated by the set

(1) $TX = \{Q^Ix \mid x \in tX,\ I \text{ is admissible},\ e(I) + b(I) > \deg x,\ \deg Q^Ix > 0\}$

(Recall the conventions, Definition 2.1(iii).) Then, as algebras,

(2) $WEH_*X = ATX \otimes Z_pN\pi_0X$ and $GWEH_*X = ATX \otimes Z_p\tilde{N}\pi_0X$.

Note that the Q^Ix with $e(I) = \deg x$, $b(I) = 0$, and $\deg Q^Ix > 0$ precisely account for all p-th powers of positive degree elements. Note also that Theorems 4.1 and 4.2 are correct in degree zero by comparison of (2) with [G, Proposition 8.14].

We need some preliminaries in order to prove Theorem 4.2 for non-connected spaces. The following well-known lemma clearly implies that Theorem 4.2 will hold provided that it correctly describes the homology of the component $Q_\emptyset X$ of the base-point of QX.

Lemma 4.6. Let X be a homotopy associative H-space such that $\pi_0 X$ is a group under the induced product. Choose a point $a \in [a]$ for each component $[a]$ of X, write a^{-1} for the chosen point in $[a]^{-1}$, and let $X_\emptyset$ denote the component of the identity element. Define $f: X \to X_\emptyset \times \pi_0 X$ by $f(x) = (x \cdot a^{-1}, [a])$ if $x \in [a]$. Then f is a homotopy equivalence with homotopy inverse g given by $g(y, [a]) = ya$. If left translation by any given element of X is homotopic to right translation by the same element, then f and g are H-maps.

To study $Q_\emptyset X$, which is the component $\Omega_\emptyset Q\Sigma X$ of the trivial loop in $Q\Sigma X$, observe that we may assume, without loss of generality, that all connected spaces Y in sight are sufficiently well-behaved locally to have universal covers $\pi: UY \to Y$. Of course, $\Omega\pi: \Omega UY \to \Omega_\emptyset Y$ is then a weak homotopy equivalence. We require two simple lemmas on universal covers.

Lemma 4.7. Let X be a homotopy associative H-space such that X is connected and $\pi_1 X$ is a free Abelian group. Then there exists a map $\rho: K(\pi_1 X, 1) \to X$ such that ρ induces an isomorphism on π_1. The composite of the product on X and $\pi \times \rho$ is therefore a weak homotopy equivalence $UX \times K(\pi_1 X, 1) \to X$.

Proof. $K(\pi_1 X, 1)$ is the restricted Cartesian product of one copy of S^1 for each generator of $\pi_1 X$, restricted in the sense that all but finitely many coordinates of each point are at a chosen base-point in S^1. Now ρ can be constructed by (transfinite) induction and use of the product on X from any chosen representatives $S^1 \to X$ for the generators of $\pi_1 X$. Of course, if X is a monoid, we can use the product directly rather than inductively.

Lemma 4.8. Let (X, θ) be a connected $\mathcal{C}$-space, where $\mathcal{C}$ is any operad. Then UX admits a structure of $\mathcal{C}$-space such that $\pi: UX \to X$ is a map of $\mathcal{C}$-spaces.

Proof. $UX = PX/(\sim)$, where two paths in X which start at $*$ are equivalent if they end at the same point and are homotopic with end-points fixed, and π is induced by the end-point projection. It is trivial to verify that the pointwise $\mathcal{C}$-space structure on PX of [G, Lemma 1.5] passes to the quotient space UX.

As a final preliminary, we have the following observation concerning the homology suspension.

Lemma 4.9. Let X be a space. Let $x \in H_0 \Omega X$ and $y \in H_* \Omega X$. Then, if $\varepsilon y = 0$, the loop product $x * y$ suspends to $(\varepsilon x)(\sigma_* y)$.

Proof. Let a and b be representative cycles in $C_* \Omega X$ for x and y. Let $i: \Omega X \to PX$ and $\pi: PX \to X$ be the inclusion and end-point projection. $\sigma_* y$ is the homology class of $\pi_* c$, where $c \in C_* PX$ is a chain such that $i_* b = dc$. ΩX acts on the left of PX by composition of paths, and $\pi(f * g) = \pi g$ for a loop f and path g. Now $d(i_* a * c) = i_*(a * b)$ and $\pi_*(i_* a * c) = (\varepsilon a)(\pi_* c)$. The result follows.

Proof of Theorem 4.2. If X is $(q-1)$-connected, $q > 1$, then $\eta_*: H_* X \to H_* QX$ is an isomorphism in degrees less than $2q$, as can easily be verified by inductive calculation of $H_* \Omega^i \Sigma^n X$ for $i \leq n$ in low degrees (by use of the Serre spectral sequence). Indeed, this is just the standard proof that $\eta_*: \pi_* X \to \pi_* QX = \pi_*^s X$ is an isomorphism in degrees less than $2q-1$ and an epimorphism in degree $2q-1$. Thus the theorem is trivially true in degrees less than $2q$ if X is $(q-1)$-connected. We claim that if the theorem is true for ΣX in degrees less than n, then the theorem is true for X in degrees less than $n-1$. This will complete the proof since it will follow that the theorem for $\Sigma^q X$ in degrees less than $2q$ implies the theorem for X in degrees less than q, for all integers $q > 1$. We shall prove our claim by constructing a model spectral sequence $\{'E^r\}$, mapping it into the Serre spectral sequence $\{E^r\}$ of the path space fibration over $UQ\Sigma X$, and invoking the comparison theorem [7, XI 11.1]. By [G, Proposition 8.14] and Lemma 4.7, we may write

$$H_*Q\Sigma X = H_*UQ\Sigma X \otimes H_*K(\tilde{N}\pi_0 X, 1).$$

Let $x' = x - (\varepsilon x)\phi$ for $x \in H_*X$. We take $t\Sigma X = \{\Sigma_* x' \mid x \in tX\}$ as our basis for $JH_*\Sigma X$, $\Sigma_*: \tilde{H}_*X \cong \tilde{H}_*\Sigma X$. We may then write

$$H_*K(\tilde{N}\pi_0 X, 1) = E\{\Sigma_* x' \mid x \in t_0 X\} \subset H_*Q\Sigma X \ .$$

Of course, if $p = 2$, this is not a sub-algebra and the squares $(\Sigma_* x')^2 = Q^1\Sigma_* x'$ lie in $H_*UQ\Sigma X$. Define $\widetilde{W}EH_*\Sigma X$ to be the sub-algebra of $WEH_*\Sigma X$ generated by the elements of $T\Sigma X$ of degree greater than one and, if $p = 2$, the squares $Q^1\Sigma_* x'$, $x \in t_0 X$. Define

$$'E^2 = \widetilde{W}EH_*\Sigma X \otimes (GWEH_*X)_\phi \ .$$

(If X is connected, $'E^2$ reduces to $WEH_*\Sigma X \otimes WEH_*X$.) The differentials of $\{'E^r\}$ are specified by requiring $\{'E^r\}$ to be a spectral sequence of differential algebras such that if $Q^I x \in TX$, then

$$\tau Q^I\Sigma_* x' = (-1)^{d(I)} Q^I x * [-p^{\ell(I)} ax]$$

and, if $p > 2$ and $\deg Q^* x = 2s-1$,

$$\tau((Q^I\Sigma_* x')^{p-1} \otimes Q^I x * [-p^{\ell(I)} ax]) = (-1)^{d(I)+1}\beta Q^s Q^I x * [-p^{\ell(I)+1} ax].$$

Here $ax \in \pi_0 X$ denotes the component in which the homology class x lies ($\psi x = x \otimes ax + ax \otimes x$ plus other terms if $\deg x > 0$) and, for $a \in \pi_0 X$ and $n \in Z$, $[na]$ denotes the n-th power of a in the group $\tilde{N}\pi_0 X \subset GWEH_* X$. An easy counting argument demonstrates that $\{'E^r\}$ is isomorphic to a tensor product of elementary spectral sequences of the forms $E\{y\} \otimes P\{\tau y\}$ and, if $p > 2$, $P\{z\}/(z^p) \otimes [E\{\tau z\} \otimes P\{\tau(z^{p-1} \otimes \tau z)\}]$, where E and P denote exterior and polynomial algebras. Here y runs through

$$\{Q^I\Sigma_* x' \mid I \text{ admissible, } e(I) > \deg x,\ \deg Q^I\Sigma_* x' > 1 \text{ and odd if } p > 2\}$$

and, if $p > 2$, z runs through

$$\{Q^I\Sigma_* x' \mid I \text{ admissible, } e(I) > \deg x,\ \deg Q^I\Sigma_* x' \text{ even}\}$$

(Note that, if $p > 2$, $e(I) \equiv d(I) \bmod 2$, hence $e(I) = \deg x + 1$ implies that $\deg Q^I\Sigma_* x'$ is even.) Of course, to the eyes of $\{'E^r\}$, the base $\widetilde{W}EH_*\Sigma X$ looks like a tensor product of exterior and truncated polynomial algebras rather than like a free commutative algebra. Clearly $'E^\infty = Z_p$. By construction, there is a unique morphism of algebras $f\colon 'E^2 \to E^2$ such that the following diagram is commutative:

$$\begin{array}{ccccc}
'E^2 = \widetilde{W}EH_*\Sigma X \otimes (GWEH_* X)_\emptyset & \xrightarrow{f} & H_*UQ\Sigma X \otimes H_*\Omega UQ\Sigma X = E^2, \\
\cap\big\downarrow & & \big\downarrow \pi_* \otimes (\Omega\pi)_* \\
WEH_*\Sigma X \otimes GWEH_* X & \xrightarrow{\tilde{\eta}_* \otimes \tilde{\eta}_*} & H_*Q\Sigma X \otimes H_*QX
\end{array}$$

Since $Q^I x = Q^I x'$ if $d(I) > 0$, by Theorem 1.1 (5), Lemma 4.9 implies that, for $\sigma_*\colon H_*QX \to H_*Q\Sigma X$,

$$\sigma_*(Q^I x * [p^{\ell(I)} ax]) = (-1)^{d(I)} Q^I \Sigma_* x'$$

(the sign comes from $\sigma\beta = -\beta\sigma$). By the naturality of σ_*, the same formula holds for $\sigma_*: H_*\Omega UQ\Sigma X \to H_*UQ\Sigma X$, although here the elements $Q^s\Sigma_*x'$, $x \in t_0X$, are of course not operations because the elements Σ_*x' are not present in $H_*UQ\Sigma X$. By Theorem 1.1 (7) and the definition of $\{{}'E^r\}$, f induces a morphism of spectral sequences. Since $f = f(\text{base}) \otimes f(\text{fibre})$, our claim and the theorem now follow directly from the comparison theorem.

The following observation on the structure of H_*QX is sometimes useful. Note that $H_*Q_\emptyset X$ is the free commutative algebra generated by $\{y*(ay)^{-1} \mid y \in TX\}$, where ay is the component in which y lies. This description uses operations which occur in various components of QX. We can instead use just those operations which actually occur in the component $Q_\emptyset X$.

Lemma 4.10. $H_*Q_\emptyset X$ is the free commutative algebra generated by the union of the following three sets:

$$\{Q^I x \mid Q^I x \in TX \text{ and } ax = \emptyset\}$$

$$\{x*[-ax] \mid x \in tX, \deg x > 0, \text{ and } ax \neq \emptyset\}$$

$$\{Q^J(\beta^\varepsilon Q^s x * [-p\cdot ax]) \mid Q^J\beta^\varepsilon Q^s x \in TX \text{ and } ax \neq \emptyset\} .$$

Proof. $[-p\cdot ax] = [-ax] * \ldots * [-ax]$, and we therefore have

$$Q^J(\beta^\varepsilon Q^s x * [-p\cdot ax]) \equiv (Q^J\beta^\varepsilon Q^s x) * [-p^{\ell(J)+1}\cdot ax] ,$$

modulo decomposable elements of $H_*Q_\emptyset X$, by the Cartan formula.

When $X = S^0$, the first two sets above are clearly empty.

We complete this section by computing the Bockstein spectral sequences of H_*CX and of H_*QX. Let $\{E^rX\}$ denote the mod p Bockstein spectral sequence of a space X. A slight variant (when $p > 2$ and $r = 2$) of the proof of [A, Proposition 6.8] yields the following lemma.

Lemma 4.11. If (X, θ) is a $\mathcal{C}$-space, where $\mathcal{C}$ is an E_∞ operad, then $\{E^rX\}$ is a spectral sequence of differential algebras such that if $y \in E^{r-1}_{2q}X$, then $\beta_r y^p = y^{p-1}\beta_{r-1}y$ if $p > 2$ or if $p = 2$ and $r > 2$, and $\beta_2 y^2 = y\beta y + Q^{2q}\beta y$ if $p = 2$ and $r = 2$.

Let $Y = CX$ or $Y = QX$, and let $\{E^rATX\}$ denote the restriction of $\{E^rY\}$ to ATX; in both cases, we clearly have $E^rY = E^rATX \otimes H_0Y$ for all $r \geq 1$. To describe E^rATX explicitly, we require some notations.

Definition 4.12. Let C_r, $r \geq 1$, be a basis for the positive degree elements of E^rX, and assume the C_r to be so chosen that

$$C_r = D_r \cup \beta_r D_r \cup C_{r+1},$$

where D_r, $\beta_r D_r$, and C_{r+1} are disjoint linearly independent subsets of E^rX such that $\beta_r D_r = \{\beta_r y \mid y \in D_r\}$ and C_{r+1} is a set of cycles under β_r which projects to the chosen basis for $E^{r+1}X$. Define A^rX, $r \geq 2$, to be the free strictly commutative algebra generated by the following set (strictness requires the squares of odd degree elements to be zero):

$$\bigcup_{1 \leq j < r} (S_{jr} \cup \beta_r S_{jr}) \cup C_r,$$

where $S_{jr} = \{y^{p^{r-j}} \mid y \in D_j, \text{ deg } y \text{ even}\}$ and

$$\beta_r S_{jr} = \begin{cases} \{y^{p^{r-j}-1}\beta_j y \mid y \in D_j, \text{ deg } y \text{ even}\} & \text{if } p > 2 \text{ or } j \geq 2 \\ \{y^{2^{r-j}-2}(y\beta y + Q^{2q}\beta y) \mid y \in D_1, \text{ deg } y = 2q\} & \text{if } p = 2 \text{ and } j = 1. \end{cases}$$

The proof of the following theorem is precisely analogous to the computation of the cohomology Bockstein spectral sequence of $K(Z_{p^t}, n)$ given in [A, Theorem 10.4] and will therefore be omitted. It depends only on Lemma 4.11, the fact that $\beta Q^{2s} = Q^{2s-1}$ if $p = 2$, and counting arguments.

Theorem 4.13. Define a subset SX of TX as follows:

(a) $p = 2 : SX = \{Q^I x \mid I = (2s, J),\ \deg Q^I x \text{ is even},\ \ell(I) > 0\}$

(b) $p > 2 : SX = \{Q^I x \mid b(I) = 0,\ \deg Q^I x \text{ is even},\ \ell(I) > 0\}$

Then $E^{r+1}ATX = P\{y^{p^r} \mid y \in SX\} \otimes E\{\beta_{r+1} y^{p^r} \mid y \in SX\} \otimes A^{r+1}X$ for all $r \geq 1$, where

$$\beta_{r+1} y^{p^r} = \begin{cases} y^{p^r - 1} \beta y & \text{if } p > 2 \\ y^{2^r - 2}(y\beta y + Q^{2q}\beta y) & \text{if } p = 2 \text{ and } \deg y = 2q \end{cases}$$

Therefore $E^{\infty}ATX = A^{\infty}X$ is the free strictly commutative algebra generated by the positive degree elements of $E^{\infty}X$.

§5. The homology of CX and the spaces $\overline{C}X$

We first prove Theorem 4.1 and then construct a homology approximation $\overline{\alpha}_\infty : \overline{C}_\infty X \to Q_\emptyset X$ for arbitrary spaces X. The space $\overline{C}_\infty S^0$ will be a $K(\Sigma_\infty, 1)$, and this special case of our approximation theorem was first obtained by Priddy [20].

Observe that the maps $\overline{\eta}_*$ of Theorem 4.1 are natural in $\mathcal{C}$ as well as in X. In particular, the following result holds.

Lemma 5.1. If $\mathcal{C}$ and $\mathcal{C}'$ are E_∞ operads, then the following is a commutative diagram of morphisms of AR-Hopf algebras:

$$\begin{array}{ccccc} & & H_*(C \times C')X & & \\ & \swarrow^{\pi_{1*}} & \uparrow \overline{\eta}_* & \searrow^{\pi_{2*}} & \\ H_*CX & & & & H_*C'X \\ & \nwarrow_{\overline{\eta}_*} & & \nearrow_{\overline{\eta}_*} & \\ & & WEH_*X & & \end{array}$$

Moreover, π_{1*} and π_{2*} are isomorphisms for all spaces X.

Proof. $\mathcal{C} \times \mathcal{C}'$ is an E_∞ operad by [G, Definitions 3.5 and 3.8], and π_{1*} and π_{2*} are isomorphisms by [G, Proposition 3.10] and the proof of [G, Proposition 3.4].

By Theorem 4.2 and Figure I, we already know that $\overline{\eta}_* : WEH_*X \to H_*C_\infty X$ is a monomorphism (since $\alpha_{\infty *}\overline{\eta}_*\iota$ is a monomorphism); by the lemma, we know this for every E_∞ operad $\mathcal{C}$. In order to prove that $\overline{\eta}_*$ is an epimorphism, we need the following standard consequence of the properties of the transfer in the (mod p) homology of finite groups; a proof may be found in [5, p. 255].

Lemma 5.2. If π is a subgroup of the finite group Π and if the index of π in Π is prime to p, then the restriction

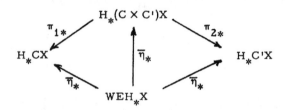

$$i_*: H_*(\pi; M) \rightarrow H_*(\Pi; M)$$

is an epimorphism for every $Z_p\Pi$-module M.

We shall also need the definition of wreath products.

Definition 5.3. Let π be a subgroup of Σ_n and let G be any monoid. Then the wreath product $\pi \int G$ is the semi-direct product of π and G^n determined by the permutation action of π on G^n; explicitly if $\sigma \in \pi$ and $\tau_i \in G$, then, in $\pi \int G$,

$$(\tau_1, \ldots, \tau_n)\sigma = \sigma(\tau_{\sigma(1)}, \ldots, \tau_{\sigma(n)}).$$

Embed G^n in G^{n+1} as $G^n \times \{e\}$ and embed Σ_n in Σ_{n+1} as the subgroup fixing the last letter; this fixes an embedding of $\Sigma_n \int G$ in $\Sigma_{n+1} \int G$, and $\Sigma_\infty \int G$ is defined to be the union of the $\Sigma_n \int G$ for finite n.

Proof of Theorem 4.1. Consider the monad (C, μ, η) associated to an E_∞ operad $\mathcal{C}$. As in [G, p. 17], we write μ both for the $\mathcal{C}$-action on CX and for the monad product $\mu: CCX \rightarrow CX$. Recall that, by [G, p. 13 and 14], CX is a filtered space such that the product $*$ and $\mathcal{C}$-action μ restrict to give

$$*: F_jCX \times F_kCX \rightarrow F_{j+k}CX$$

and

$$\mu_k: \mathcal{C}(k) \times F_{j_1}CX \times \ldots \times F_{j_k}CX \rightarrow F_jCX, \quad j = j_1 + \ldots + j_k.$$

Indeed, $*$ is $\mu_2(c)$ for any fixed $c \in \mathcal{C}(2)$ and, if γ denotes the structural map of the operad [G, Definition 1.1], then

$$\mu_k(d; [e_1, y_1], \ldots, [e_k, y_k]) = [\gamma(d; e_1, \ldots, e_k); y_1, \ldots, y_k]$$

for $d \in \mathcal{C}(k)$, $e_i \in \mathcal{C}(j_i)$, and $y_i \in X^{j_i}$. We define a corresponding algebraic filtration of WEH_*X by giving all elements of the image of

$R[k]\otimes JH_*X$ in WEH_*X filtration precisely p^k and by requiring WEH_*X to be a filtered algebra. Then F_0WEH_*X is spanned by $\emptyset$, $F_1WEH_*X = H_*X$, and each F_kWEH_*X is a sub A-coalgebra of WEH_*X. Visibly, the restriction of $\bar{\eta}_*: WEH_*X \to H_*CX$ to F_kWEH_*X factors through H_*F_kCX. $H_*F_1CX = H_*X$ since $F_1CX = \mathcal{C}(1)\times X$ and $\mathcal{C}(1)$ is contractible. Assume inductively that $\bar{\eta}_*: F_jWEH_*X \to H_*F_jCX$ is an isomorphism for all $j < k$. Define

$$E_k^0CX = F_kCX/F_{k-1}CX \quad \text{and} \quad E_k^0WEH_*X = F_kWEH_*X/F_{k-1}WEH_*X.$$

Consider the following commutative diagram with exact rows and columns:

$$\begin{array}{ccccccc}
 & 0 & & 0 & & 0 & \\
 & \downarrow & & \downarrow & & \vdots & \\
0 \to & F_{k-1}WEH_*X & \longrightarrow & F_kWEH_*X & \longrightarrow & E_k^0WEH_*X & \to 0\\
 & \downarrow \bar{\eta}_* & & \downarrow \bar{\eta}_* & & \downarrow \lambda & \\
\cdots \longrightarrow & H_*F_{k-1}CX & \xrightarrow{\iota_*} & H_*F_kCX & \xrightarrow{\pi_*} & H_*E_k^0CX & \xrightarrow{\partial} \cdots\\
 & \downarrow & & & & & \\
 & 0 & & & & &
\end{array}$$

Here $\iota: F_{k-1}CX \to F_kCX$ is the inclusion, which is a cofibration by [G, Proposition 2.6], and π is the quotient map. The maps $\bar{\eta}_*$ are known to be monomorphisms and the left map $\bar{\eta}_*$ is assumed to be an epimorphism. It follows that ι_* is a monomorphism, hence that $\partial = 0$ and π_* is an epimorphism. Define λ by commutativity of the right-hand square; then λ is a monomorphism by the five lemma. If we can prove that λ is an epimorphism, it will follow that the middle arrow $\bar{\eta}_*$ is an isomorphism, as required. By [G, p. 14], E_k^0CX is the equivariant half-smash product

$$E_k^0CX = \mathcal{C}(k) \times_{\Sigma_k} X^{[k]}/\mathcal{C}(k)\times_{\Sigma_k} * ,$$

where $X^{[k]}$ denotes the k-fold smash product of X with itself. By

[A, Lemma 1.1(iii) and Remarks 7.2], there is a composite chain homotopy equivalence

$$C_*\mathcal{C}(k)\otimes_{\Sigma_k}(H_*X)^k \to C_*(\mathcal{C}(k)\times_{\Sigma_k} X^k),$$

hence we may identify $H_*(\mathcal{C}(k)\times_{\Sigma_k} X^k)$ with $H_*(\Sigma_k;(H_*X)^k)$. Let $\pi': \mathcal{C}(k)\times_{\Sigma_k} X^k \to E_k^0 CX$ be the evident quotient map and let $\nu: \mathcal{C}(k)\times_{\Sigma_k} X^k \to F_kCX$ be the sub-quotient map given by the definition of CX. Then, since $\eta(x) = [1,x]$, where $1 \in \mathcal{C}(1)$ is the identity element and since $\mu \circ C\eta = 1$ on CX, the following diagram is commutative:

$$\begin{array}{ccc}
\mathcal{C}(k)\times_{\Sigma_k} X^k & \xrightarrow{\;1\times\eta^k\;} & \mathcal{C}(k)\times_{\Sigma_k}(F_1CX)^k \\
\downarrow{\scriptstyle \pi'} & \searrow{\scriptstyle \nu} & \downarrow{\scriptstyle \mu_k} \\
E_k^0CX & \xleftarrow{\;\pi\;} & F_kCX
\end{array}$$

Of course, π' induces an epimorphism on homology. If $k < p$, then $H_*E_k^0CX = H_*(*\times_{\Sigma_k} X^{[k]})$ and λ is an epimorphism since, by the diagram, $H_*E_k^0CX$ is spanned by images under π_* of k-fold products. Let $k = p$. Since $i_*: H_*(\pi;(H_*X)^p) \to H_*(\Sigma_p;(H_*X)^p)$ is an epimorphism, where π is cyclic of order p, $H_*(\Sigma_p;(H_*X)^p)$ is spanned by images under i_* of elements of the forms $e_0\otimes x_1\otimes\ldots\otimes x_p$ and $e_i\otimes x^p$, by [A, Lemma 1.3]. By the diagram, $H_*E_p^0CX$ is therefore spanned by images under π_* of p-fold products $x_1 * \ldots * x_p$ and operations $\beta^\epsilon Q^s x$, hence λ is an epimorphism. We now have that $\overline{\eta}_*: F_pWEH_*X \to H_*F_pCX$ is an isomorphism of A-coalgebras. Let

$$\xi: WEH_*F_pCX \to WEH_*X$$

be the unique morphism of allowable AR-Hopf algebras such that ξ restricts to $\overline{\eta}_*^{-1}$ on H_*F_pCX; observe that the restriction of ξ to $F_jWEH_*F_pCX$ has image in $F_{pj}WEH_*X$. Suppose that $k = pj$, $j > 1$.

The index of $\Sigma_j \int \Sigma_p$ in Σ_k is prime to p since $k! = \prod_{i=1}^{j} \prod_{n=0}^{p-1} (pi - n) = p^j (j!) q$, where q is prime to p. Consider the following commutative diagram:

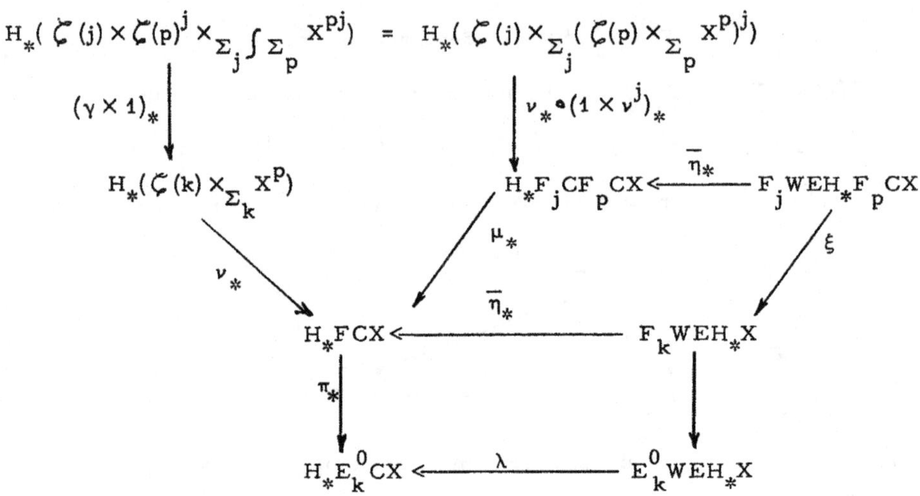

Here $\bar{\eta}_* \xi = \mu_* \bar{\eta}_*$ since both maps restrict to the inclusion induced by $\iota : F_p CX \to F_k CX$ on $H_* F_p CX$. The map γ is $\Sigma_j \int \Sigma_p$-equivariant by the very definition of an operad, hence $(\gamma \times 1)_*$ may be identified with the restriction i_* and is therefore an epimorphism. $\bar{\eta}_*$ on $F_j W E H_* F_p CX$ is an epimorphism since our induction hypothesis can be applied to any space, and in particular to $F_p CX$. Since $\pi_* \nu_* = \pi'_*$ is also an epimorphism, it follows from the diagram that λ is an epimorphism. Finally, suppose that k is prime to p. Let $\rho(1): \mathcal{C}(k-1) \to \mathcal{C}(k)$ be the Σ_{k-1}-equivariant map defined by $\rho(1)(d) = d * 1 = \gamma(c; d, 1)$ and consider the following commutative diagram:

$$
\begin{array}{ccccc}
H_*(\mathcal{C}(k-1)\times_{\Sigma_{k-1}} X^k) & = & H_*(\mathcal{C}(k-1)\times_{\Sigma_{k-1}} X^{k-1})\otimes H_*X & & \\
\downarrow{\scriptstyle (\rho(1)\times 1)_*} & & \downarrow{\scriptstyle \nu_*\otimes\eta_*} & & \\
H_*(\mathcal{C}(k)\times_{\Sigma_k} X^k) & & H_*F_{k-1}CX\otimes H_*F_1CX & \xleftarrow{\ \overline{\eta}_*\otimes\overline{\eta}_*\ } & F_{k-1}WEH_*X\otimes F_1WEH_*X \\
{\scriptstyle \nu_*}\searrow & & \swarrow{\scriptstyle *} & & \swarrow{\scriptstyle *} \\
 & H_*F_kCX & & \xleftarrow{\ \overline{\eta}_*\ } & F_kWEH_*X \\
 & \downarrow{\scriptstyle \pi_*} & & & \downarrow \\
 & H_*E_k^0CX & & \xleftarrow{\ \lambda\ } & E_k^0WEH_*X
\end{array}
$$

$(\rho(1)\times 1)_*$ may be identified with the restriction i_* and is therefore an epimorphism, $\overline{\eta}_*\otimes\overline{\eta}_*$ is an epimorphism by the induction hypothesis, $\pi_*\nu_*$ is an epimorphism, and therefore λ is an epimorphism. The proof is complete.

Our homology approximation to $Q_\emptyset X$ realizes geometrically the obvious algebraic isomorphism from $H_*CX\otimes_{H_0CX} Z_p$ to $H_*QX\otimes_{H_0QX} Z_p$; indeed, non-invariantly, each of these is just the connected free commutative algebra ATX. Of course, via $Q^Ix \to Q^Ix*[-p^{\ell(I)}\cdot ax]$ on generators, ATX is isomorphic as an algebra to $H_*Q_\emptyset X$.

Henceforward in this section, we restrict attention to the full subcategory $\mathcal{V}$ of $\mathcal{J}$ which consists of spaces of the based homotopy type of CW-complexes. By [G', Corollary A.3], $CX\in\mathcal{V}$ if $X\in\mathcal{V}$ and $\mathcal{C}$ is a suitably nice operad (as we tacitly assume below).

Construction 5.4. Let $\mathcal{C}$ be an operad and let $X\in\mathcal{V}$. Construct a space $\overline{C}X$ as follows. Choose a point a in each component $[a]$ of X.

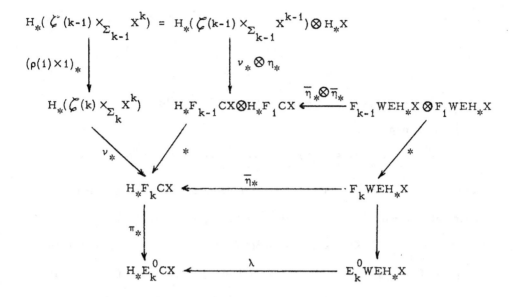

Choose a point $c_i \in \mathcal{C}(i)$ for $i \geq 1$, with $c_1 = 1$, and let $ia = [c_i; a^i] \in CX$. (Thus, by abuse, a is identified with $\eta(a) = [1; a]$.) Let $(CX)_{ia}$ denote the component of CX in which ia lies. Define $\rho(a): CX \to CX$ to be right translation by a, $\rho(a)(x) = x * a$. Define $(\overline{C}X)_a$ to be the telescope of the sequence of maps

$$* \xrightarrow{\rho(a)} (CX)_a \xrightarrow{\rho(a)} (CX)_{2a} \longrightarrow \cdots \longrightarrow (CX)_{ia} \xrightarrow{\rho(a)} \cdots$$

Define $\overline{C}X$ to be the restricted Cartesian product (all but finitely many coordinates of each point are $*$) of the spaces $(\overline{C}X)_a$ for $[a] \in \pi_0 X$. The homotopy type of $(\overline{C}X)_a$ is independent of the choice of $a \in [a]$, and $\overline{C}$ is the object function of a functor from the homotopy category of $\mathcal{V}$ to itself.

Remarks 5.5. (i) $\overline{C}X$ is a functor of $\mathcal{C}$ as well as of X.

(ii) If X is connected, then $\overline{C}X$ is homotopy equivalent to CX.

(iii) If $[a] \neq \emptyset$, then $(CX)_{ia} = \mathcal{C}(i) \times_{\Sigma_i} X_a^i$, where $X_a = [a]$, and

$\rho(a): (CX)_{ia} \to (CX)_{(i+1)a}$ is given by the formula

$$\rho(a)[c; x_1, \ldots, x_i] = [\gamma(c_2; c, 1); x_1, \ldots, x_i, a] .$$

Lemma 5.6. Let $\mathcal{C}$ be an E_∞ operad. Then $H_*\overline{C}X$ is naturally isomorphic to the connected algebra $H_*CX \otimes_{H_0CX} Z_p$.

Proof. Since $(x*a)*(y*a) = (x*y)*a*a$ for $x, y \in H_*CX$, each $H_*(\overline{C}X)_a$, hence also $H_*\overline{C}X$, is a well-defined algebra. The result is obvious from Theorem 4.1 and the construction.

Lemma 5.7. Let $\mathcal{C}$ be an E_∞ operad. Let G be any discrete group and let $X = K(G,1)^+$, the union of a $K(G,1)$ and a disjoint base-point. Then $\overline{C}X$ is a $K(\Sigma_\infty \int G, 1)$. In particular, with $G = \{e\}$, $\overline{C}S^0$ is a $K(\Sigma_\infty, 1)$.

Proof. $\mathscr{C}(i) \times_{\Sigma_i} K(G,1)^i$ is clearly a $K(\Sigma_i \int G, 1)$, and $\overline{C}X$ is the limit of the $\mathscr{C}(i) \times_{\Sigma_i} K(G,1)^i$ under appropriate maps by Remarks 5.5 (iii).

We now consider the functors $\overline{C}_n$ derived from the little cubes operads.

Construction 5.8. Fix n, $1 \leq n$ or $n = \infty$ (when $\Omega^\infty\Sigma^\infty = Q$). With the same notations as in Construction 5.4, let ia also denote the image of ia under $\alpha_n : C_n X \to \Omega^n\Sigma^n X$ and let $\Omega^n_{ia}\Sigma^n X$ denote the component of $\Omega^n\Sigma^n X$ in which ia lies. Define $\rho(a): \Omega^n\Sigma^n X \to \Omega^n\Sigma^n X$ by $\rho(a)(x) = x * a$ and let $\overline{\Omega}^n_a\Sigma^n X$ denote the telescope of the sequence of inclusions

$$\Omega^n_\emptyset\Sigma^n X \xrightarrow{\rho(a)} \Omega^n_a\Sigma^n X \xrightarrow{\rho(a)} \Omega^n_{2a}\Sigma^n X \longrightarrow \cdots \longrightarrow \Omega^n_{ia}\Sigma^n X \xrightarrow{\rho(a)} \cdots$$

The inclusion of $\Omega^n_\emptyset\Sigma^n X$ in $\overline{\Omega}^n_a\Sigma^n X$ is a homotopy equivalence (since each $\rho(a)$ is); choose an inverse homotopy equivalence $\emptyset_a : \overline{\Omega}^n_a\Sigma^n X \to \Omega^n_\emptyset\Sigma^n X$. Observe that $\alpha_n\ \rho(a) = \rho(a)\ \alpha_n$ and let $\overline{\alpha}_{n,a} : (\overline{C}_n X)_a \to \overline{\Omega}^n_a\Sigma^n X$ be the map obtained from α_n by passage to limits. Either by (possibly transfinite) induction and use of the ordinary loop product or by direct construction in terms of the monoid structure on the Moore loop space of $\Omega^{n-1}\Sigma^n X$, the maps $\emptyset_a \circ \overline{\alpha}_{n,a} : (\overline{C}_n X)_a \to \Omega^n_\emptyset\Sigma^n X$ determine a map $\overline{\alpha}_n : \overline{C}_n X \to \Omega^n_\emptyset\Sigma^n X$. Up to homotopy, $\overline{\alpha}_n$ is natural in X.

Lemma 5.9. Let $\iota : (C_n X)_{ia} \to \overline{C}_n X$ be the inclusion. Then, for $x \in H_*(C_n X)_{ia}$, $\overline{\alpha}_{n*}\iota_*(x) = \alpha_{n*}(x) * [-ia]$, where $[-ia]$ is the component inverse to $[ia]$ in the group $\pi_0\Omega^n\Sigma^n X$.

Proof. The restriction of $\emptyset_a$ to $\Omega^n_{ia}\Sigma^n X$ is homotopic to right translation by any chosen point of $[-ia]$.

Our approximation theorem is now an immediate consequence of Theorems 4.1 and 4.2 and Lemmas 5.6 and 5.9.

<u>Theorem 5.10.</u> For $X \in \mathcal{V}$, $\bar{\alpha}_{\infty *} : H_* \bar{C}_\infty X \to H_* Q_\emptyset X$ is an isomorphism of algebras.

Since the result holds for all primes p, $\bar{\alpha}_\infty$ also induces an isomorphism an integral homology. Via $\bar{\alpha}_{\infty *}^{-1}$, $H_* \bar{C}X$ is an allowable AR-Hopf algebra. However, if X is not connected, then $\bar{C}_\infty X$ is not an H-space, let alone an E_∞ space, hence much of this structure is purely algebraic. As illustrated in Lemma 5.7, $\bar{C}_\infty X$ generally has a non-Abelian fundamental group. Clearly $\pi_1 \alpha_\infty : \pi_1 \bar{C}_\infty X \to \pi_1 Q_\emptyset X$ is the Abelianization homomorphism.

As explained in [G', §2], Theorems 4.1, 4.2, and 4.13 imply that $\alpha_\infty : C_\infty X \to QX$ is a group completion in the sense of [G', Definition 1.3]. Theorem 5.10 is a reflection of this fact (compare [G', Proposition 3.9]). This fact also suggests that any natural group completion of CX for any E_∞ operad $\mathcal{C}$ should yield a homotopy approximation to QX. $\Omega C \Sigma X$ is one example [G, Corollary 4.6], and ΩBDX is another (but the second result labeled Theorem 3.7 in [G'], about the monoid structure on DX, is incorrect; see [R, VII. 2.7]). Yet another example is $B_0 CX$, the infinite loop space obtained by application of the recognition principle to the E_∞ space CX [G', Theorem 2.3 (vii)]. As explained in [R, VII §4], this last construction often admits a multiplicative elaboration and yields the most structured version of the Barratt-Quillen theorem to the effect that QS^0 is equivalent to the group completion of $\coprod_{j \geq 0} K(\Sigma_j, 1)$.

In III, Cohen will prove that $\alpha_n : C_n X \to \Omega^n \Sigma^n X$ is a group completion for all spaces X by proving the analogs of Theorems 4.1, 4.2, and 4.13. Thus his calculations will imply the following unstable analog of Theorem 5.10.

Theorem 5.11. For $X \in \mathcal{V}$, $\bar{\alpha}_{n*} : H_*\bar{C}_n X \to H_*\Omega^n_\phi \Sigma^n X$ is an isomorphism of algebras.

§6. A remark on Postnikov systems

Infinite loop spaces can be approximated by stable Postnikov towers, and it is natural to ask whether there is a relationship between the homology operations and the Postnikov decomposition of such a space. We present such a result here, and we begin with the following easy (and well-known) lemma.

Lemma 6.1. Let X be a $K(\pi, n)$ for some Abelian group π and integer $n \geq 1$. Then the Nishida relations and diagonal Cartan formula imply that the operations Q^s (no matter how constructed geometrically) are all trivial on $\widetilde{H}_*X$.

Proof. Assume the contrary and let i be minimal such that $Q^r x \neq 0$ for some $x \in \widetilde{H}_i X$ and some r and let r be minimal such that $Q^r x \neq 0$. Then $Q^r x$ is primitive and is annihilated by all Steenrod operations P_*^r. It follows that $Q^r x = 0$, which is a contradiction.

It should be emphasized that this result fails for products of $K(\pi, n)$'s. The loop operations on such a product are certainly trivial, but such a product can also be the zeroth space of a spectrum the higher terms of which have non-trivial k-invariants. If we wish to analyze infinite loop spaces in terms of Postnikov systems, then we must use the Postnikov systems of all of the de-loopings (or pass to spectra).

The lemma admits the following generalization, which may also be regarded as a generalization of [15, Theorem 6.2].

Proposition 6.2. Let X be a stable k-stage Postnikov system, as an infinite loop space. Then $Q^I x = 0$ if $x \in \widetilde{H}_*X$ and $\ell(I) \geq k$.

Proof. A 1-stage stable Postnikov system is a product of $K(\pi, n)$'s

π Abelian and $n \geq 1$. Inductively, a k-stage stable Postnikov system X is the pullback from the path space fibration over a simply connected 1-stage Postnikov system Z of an infinite loop map $f: Y \to Z$, where Y is a (k-1)-stage stable Postnikov system. The natural map $\pi: X \to Y$ and the inclusion $i: \Omega Z \to X$ of the fibre of π are infinite loop maps. Direct calculation by Hopf algebra techniques (see [15, Theorem 6.1]) demonstrates that $E_p = E_\infty$ in the Eilenberg-Moore spectral sequence of the fibre square and that the following is an exact sequence of Hopf algebras:

$$Z_p \to i_* H_* \Omega Z \xrightarrow{\subset} H_* X \xrightarrow{\pi_*} H_* Y \backslash\backslash f_* \to Z_p \ .$$

Here $H_* Y \backslash\backslash f_*$ is the kernel of the composite

$$H_* Y \xrightarrow{\psi} H_* Y \otimes H_* Y \xrightarrow{f_* \otimes 1} \tilde{H}_* Z \otimes H_* Y$$

($\tilde{H}_* Z = H_* Z / H_0 Z$) and is isomorphic to $H_* Z // i_*$ by the displayed exact sequence. If $x \in \tilde{H}_* X$, then $\pi_* Q^J x = 0$ for $\ell(J) \geq k-1$ by induction. Thus consider $Q^s x$ where $\pi_* x = 0$, say $x = \sum x_i z_i$ with $z_i \in i_* \tilde{H}_* \Omega Z$. $Q^s x = \sum Q^s(x_i z_i) = 0$ by the Cartan formula, since all $Q^r z_i = 0$ by the lemma, and the conclusion follows.

§7. The analogs of the Pontryagin p^{th} powers

As was first exploited by Madsen [9] (at the prime 2), the homology of E_∞ spaces carries analogs of the Pontryagin p^{th} powers defined and analyzed by Thomas [25] on the cohomology of spaces. These operations are often useful in the study of torsion. Indeed, in favorable cases, they serve to replace the fuzzy description of torsion classes in $Z_{(p)}$-homology derived by use of the Bockstein spectral sequence by precise information in terms of primary homology operations, with no indeterminacy.

We revert to the general context of §1, except that the coefficient groups in homology will vary, and we first list the various Bockstein operations that will appear in this section in the following diagram:

$$
\begin{array}{lcccccccccc}
\tilde\beta & : & 0 & \longrightarrow & Z_{(p)} & \xrightarrow{\subset} & Q & \xrightarrow{\rho} & Z_{p^\infty} & \longrightarrow & 0 \\
 & & & & \| & & \uparrow{\scriptstyle p^{-r}} & & \uparrow{\scriptstyle i_r} & & \\
\tilde\beta_r = \tilde\beta i_{r*} & : & 0 & \longrightarrow & Z_{(p)} & \xrightarrow{p^r} & Z_{(p)} & \xrightarrow{\pi} & Z_{p^r} & \longrightarrow & 0 \\
 & & & & \downarrow{\scriptstyle \pi} & & \downarrow{\scriptstyle \pi} & & \| & & \\
\bar\beta_r = \pi_*\tilde\beta_r & : & 0 & \longrightarrow & Z_{p^r} & \xrightarrow{p^r} & Z_{p^{2r}} & \xrightarrow{\pi} & Z_{p^r} & \longrightarrow & 0 \\
 & & & & \downarrow{\scriptstyle \pi} & & \downarrow{\scriptstyle \pi} & & \| & & \\
\beta_r = \pi_*\bar\beta_r & : & 0 & \longrightarrow & Z_p & \xrightarrow{p^r} & Z_{p^{r+1}} & \xrightarrow{\pi} & Z_{p^r} & \longrightarrow & 0 \\
 & & & & \downarrow{\scriptstyle p^{r-1}} & & \| & & \downarrow{\scriptstyle \pi} & & \\
\partial_r = \pi_*\tilde\beta_1 & : & 0 & \longrightarrow & Z_{p^r} & \xrightarrow{p} & Z_{p^{r+1}} & \xrightarrow{\pi} & Z_p & \longrightarrow & 0 \\
 & & & & \uparrow{\scriptstyle \pi} & & \uparrow{\scriptstyle \pi} & & \| & & \\
\tilde\beta_1 & : & 0 & \longrightarrow & Z_{(p)} & \xrightarrow{p} & Z_{(p)} & \xrightarrow{\pi} & Z_p & \longrightarrow & 0
\end{array}
$$

In each row, the notation at the left specifies the homology Bockstein derived from the short exact sequence at the right. All of the homomorphisms labelled π are natural quotient maps. $Z_{p^\infty} = \varinjlim Z_{p^r}$ and $i_r : Z_{p^r} \to Z_{p^\infty}$ is the natural inclusion; $\rho: Q \to Z_{p^\infty}$ is specified by $\rho(\frac{a}{p^r} + \frac{b}{q}) = i_r(a)$ for $a, b \in Z$, $r \geq 1$, and q prime to p. Since $\tilde{\beta}$ determines all of the remaining Bocksteins listed, it should be thought of as the universal Bockstein operation. Clearly $\beta_1 = \beta = \partial_1$ and $(p^{r-1})_* \beta_r = \partial_r \pi_*$, $\pi: Z_{p^r} \to Z_p$, if $r > 1$. At least if $H_*(X; Z_{(p)})$ is of finite type over $Z_{(p)}$, so that the natural homomorphism $j^r: H_*(X; Z_{p^r}) \to E^r X$ is an epimorphism, β_r determines the rth differential d^r (previously denoted β_r) of the mod p Bockstein spectral sequence $\{E^r \mathbf{X}\}$. Explicitly, for $x \in H_*(X; Z_{p^r})$, $\beta_r x \in H_*(X; Z_p) = E^1 X$ survives to $d^r j^r(x) \in E^r X$; alternatively, $d^r j^r(x) = j^r \overline{\beta}_r(x)$.

<u>Theorem 7.1.</u> Let $\mathcal{C}$ be an E_∞ operad and let (X, θ) be a $\mathcal{C}$-space. Then there exist functions

$$\mathcal{Q}: H_q(X; Z_{p^r}) \to H_{pq}(X; Z_{p^{r+1}})$$

for all $q \geq 0$ and $r \geq 1$ which satisfy the following properties:

(1) The $\mathcal{Q}$ are natural with respect to maps of $\mathcal{C}$-spaces.

(2) If $x \in H_{2q+1}(X; Z_{p^r})$, then $\mathcal{Q}x = 0$ if $p > 2$ and, if $p = 2$,

$$\mathcal{Q}x = \begin{cases} \partial_2 Q^{2q+2} x + 2_* Q^{2q+2} \beta x & \text{if } r = 1 \\ \partial_{r+1} Q^{2q+2} \pi_* x & \text{if } r > 1 . \end{cases}$$

(3) The following diagram is commutative:

$$\begin{array}{ccc} H_q(X;Z_{p^r}) & \xrightarrow{p_*} & H_q(X;Z_{p^{r+1}}) \\ \downarrow \mathcal{Q} & & \downarrow \mathcal{Q} \\ H_{pq}(X;Z_{p^{r+1}}) & \xrightarrow{(p^p)_*} & H_{pq}(X;Z_{p^{r+2}}) \end{array}$$

(4) The following composites both coincide with p^{th} power operations:

$$H_q(X;Z_{p^{r+1}}) \xrightarrow{\pi_*} H_q(X;Z_{p^r}) \xrightarrow{\mathcal{Q}} H_{pq}(X;Z_{p^{r+1}})$$

and

$$H_q(X;Z_{p^r}) \xrightarrow{\mathcal{Q}} H_{pq}(X;Z_{p^{r+1}}) \xrightarrow{\pi_*} H_{pq}(X;Z_{p^r}) .$$

(5) If $x \in H_{2q}(X;Z_{p^r})$, then

$$\beta_{r+1}\mathcal{Q}x = \begin{cases} x\cdot(\beta x) + Q^{2q}(\beta x) & \text{if } p = 2 \text{ and } r = 1 \\ x^{p-1}\cdot(\beta_r x) & \text{if } p > 2 \text{ or } r > 1 \end{cases}$$

where the product $H_*(X;Z_{p^r}) \otimes H_*(X;Z_p) \to H_*(X;Z_p)$ is understood;

if $x \in H_{2q+1}(X;Z_{2^r})$, then

$$\beta_{r+1}x = \begin{cases} Q^{2q+1}(\beta x) & \text{if } r = 1 \\ 0 & \text{if } r > 1 \end{cases} .$$

(6) $\mathcal{Q}(x+y) = \mathcal{Q}x + \mathcal{Q}y + \sum_{i=1}^{p-1} (i, p-i)_* x^i y^{p-i}$ for $x, y \in H_*(X;Z_{p^r})$,

where $(i, p-i)_*$ is induced from $(i, p-i)\colon Z_{p^r} \to Z_{p^{r+1}}$.

(7) If $x \in H_s(X;Z_{p^r})$ and $y \in H_t(X;Z_{p^r})$, then

$$\mathcal{Q}(xy) = (\mathcal{Q}x)(\mathcal{Q}y) \quad \text{if } p > 2$$

and

$$\mathcal{Q}(xy) = (\mathcal{Q}x)(\mathcal{Q}y) + (2^r)_*[x\cdot\beta_r x\cdot Q^{t+1}\pi_* y + y\cdot\beta_r y\cdot Q^{s+1}\pi_* x] \quad \text{if } p = 2.$$

Here s and t are not assumed to be even; when $p = 2$, $r > 1$, and s and t are even, the error term vanishes (since $Q^{2q-1} = \beta Q^{2q}$ and $2^r_*\beta = 0$).

(8) $\sigma_* \mathcal{Q} x = 0$ if $p > 2$ and $\sigma_* \mathcal{Q} x = (2^r)_*[(\overline{\beta}\sigma_* x)(\sigma_* x)]$ if $p = 2$,

where $\sigma_*: \widetilde{H}_*(\Omega X; ?) \to H_*(X; ?)$ is the homology suspension.

Proof. Precisely as in the proof of Theorem 1.1, except that W is here taken as the standard π-free resolution of Z and C_* is taken to mean chains with integer coefficients, θ induces $\theta_*: W \otimes (C_*X)^p \to C_*X$. In the language of [A, Definition 2.1], (C_*X, θ_*) is a unital Cartan object of the category $\mathcal{C}(\pi, \infty, Z)$. The rest is elementary chain level algebra, the details of which are the same for the present homology operations as for the cohomology Pontryagin p^{th} powers. Given $x \in H_q(X; Z_{p^r})$, let x be represented by a chain $a \in C_*X$ such that $da = p^r b$. Let α generate the cyclic group π, let $M = \sum_{i=1}^{p-1} i\alpha^i$, and define

$$\mathcal{Q} x = \{\theta_*(e_0 \otimes a^p + p^r M e_1 \otimes a^{p-1} b)\} \in H_{pq}(X; Z_{p^{r+1}}) .$$

Here $e_0 \otimes a^p + p^r M e_1 \otimes a^{p-1} b$ is a cycle modulo p^{r+1} by explicit computation (see [25, p. 32]). The salient facts are that $TM = p - N = MT$, where $T = \alpha - 1$ and $N = \sum_{i=0}^{p-1} \alpha^i$, that $da^p = p^r N a^{p-1} b$ if $p > 2$, and that $de_1 = Te_0$. The same calculation carried out mod p^{r+2} rather than mod p^{r+1} yields (5). $\mathcal{Q} x$ is well-defined and $\mathcal{Q}$ is natural by the method of proof of [21, 3.1] or [A, Lemma 1.1]. When $p > 2$, the fact that $\mathcal{Q} x = 0$ if q is odd depends on the factorization of θ_* through $\mathcal{C}_* \mathcal{C}(p) \otimes (C_*X)^p$; see [24, §9-10] for details. When $p = 2$, see [25, p. 42] for the verification of (2). Parts (3) and (4) are trivial to verify, and part (8) is straightforward. See [25, §9] for the verification of (6) and [25, §8] (or [9] when $p = 2$) for the verification of (7).

We have chosen the notation $\mathcal{Q}$ since it goes well with Q^s and since the Pontryagin p^{th} powers are often denoted by $\mathcal{P}$.

Remarks 7.2. Let $\mathcal{Q}^1 = \mathcal{Q}$ and $\mathcal{Q}^r = \mathcal{Q} \circ \mathcal{Q}^{r-1}$. For an E_∞ space X, define

$$\mathcal{Q}X = i_{1*}H_*(X; Z_p) + \sum_{r \geq 1} (i_{r+1})_* \mathcal{Q}^r H_{2*}(X; Z_p) \subset H_*(X; Z_{p^\infty})$$

and consider

$$\tilde{\beta}\mathcal{Q}X = \tilde{\beta}_1 H_*(X; Z_p) + \sum_{r \geq 1} \tilde{\beta}_{r+1} \mathcal{Q}^r H_{2*}(X; Z_p) \subset H_*(X; Z_{(p)}) .$$

Madsen [9] suggested the term "Henselian at p" for E_∞ spaces X such that the torsion subgroup of the ring $H_*(X; Z_{(p)})$ coincides with the ideal generated by $\tilde{\beta}\mathcal{Q}X$. In view of (4) and (5) of the theorem, this will be the case if $H_*(X; Z_{(p)})$ is of finite type over $Z_{(p)}$ and all non-trivial differentials d^r, $r \geq 2$, in the mod p Bockstein spectral sequence of X are determined by the general formulas for differentials on p^{th} powers specified in Lemma 4.11. In particular, by Theorem 4.13, QY is Henselian at p for any space Y such that $H_*(Y; Z_{(p)})$ is of finite type over $Z_{(p)}$ and has no p^2-torsion.

Remarks 7.3. The operations $\mathcal{Q}$ can already be defined on the homology of $\mathcal{C}_2$-spaces, where $\mathcal{C}_2$ is the little 2-cubes operad, and thus on the homology of second loop spaces. All of the properties listed in Theorem 7.1 are valid for $\mathcal{C}_3$-spaces, and most of the properties are valid for $\mathcal{C}_2$-spaces. The exceptions, (2) and (6), are those properties the proof of which requires use of the element $e_2 \in W$, and they have more complicated versions with error terms which involve the two variable operation λ_1 discussed in Theorem 1.2 of Cohen's paper III.

Bibliography

1. S. Araki and T. Kudo. Topology of H_n-spaces and H-squaring operations. Mem. Fac. Sci. Kyusyu Univ. Ser. A 10(1956), 85-120.

2. W. Browder. Homology operations and loop spaces. Illinois J. Math. 4 (1960), 347-357.

3. W. Browder. Torsion in H-spaces. Annals of Math. 74 (1961), 24-51.

4. W. Browder. The Kervaire invariant for framed manifolds and its generalizations. Annals of Math. 90 (1969), 157-186.

5. H. Cartan and S. Eilenberg. Homological Algebra. Princeton University Press. 1956.

6. E. Dyer and R. K. Lashof. Homology of iterated loop spaces. Amer. J. Math. 84 (1962), 35-88.

7. S. Mac Lane. Homology. Springer-Verlag. 1963.

8. I. Madsen. On the action of the Dyer-Lashof algebra in H_*G. Pacific J. Math. To appear.

9. I. Madsen. Higher torsion in SG and BSG. Math. Zeitschrift. 143 (1975), 55-80.

10. J. P. May. Categories of spectra and infinite loop spaces. Springer Lecture Notes in Mathematics Vol. 99, 1969, 448-479.

11. J. P. May. Some remarks on the structure of Hopf algebras. Proc. Amer. Math. Soc. 23 (1969), 708-713.

12. J. P. May. Homology operations on infinite loop spaces. Proc. Symp. Pure Math. Vol. 22, 171-185. Amer. Math. Soc. 1971.

13. J.P. May. The stable homotopy category. To appear.

14. J. Milnor. The Steenrod algebra and its dual. Annals of Math. 67 (1958), 150-171.

15. J.C. Moore and L. Smith. Hopf algebras and multiplicative fibrations I. Amer. J. Math. 90 (1968), 752-780.

16. M. Nakaoka. Decomposition theorem for homology groups of symmetric groups. Annals of Math. 71 (1960), 16-42.

17. M. Nakaoka. Homology of the infinite symmetric group. Annals of Math. 73 (1961), 229-257.

18. M. Nakaoka. Note on cohomology algebras of symmetric groups. J. Math. Osaka City Univ. 13 (1962), 45-55.

19. G. Nishida. Cohomology operations in iterated loop spaces. Proc. Japan Acad. 44 (1968), 104-109.

20. S. Priddy. On $\Omega^{\infty}S^{\infty}$ and the infinite symmetric group. Proc. Symp. Pure Math. Vol. 22, 217-220. Amer. Math. Soc. 1971.

21. N.E. Steenrod. Cohomology operations derived from the symmetric group. Comment. Math. Helv. 31 (1957), 195-218.

22. N.E. Steenrod. The cohomology algebra of a space. L'Enseignement Mathématique 7 (1961), 153-178.

23. N.E. Steenrod and D.B.A. Epstein. Cohomology Operations. Princeton University Press. 1962.

24. N.E. Steenrod and E. Thomas. Cohomology operations derived from cyclic groups. Comment. Math. Helv. 32 (1957), 129-152.

25. E. Thomas. The generalized Pontryagin cohomology operations and rings with divided powers. Memoirs Amer. Math. Soc. No. 27 (1957).

The Homology of E_∞ Ring Spaces

J. P. May

The spaces $Q(X^+)$ for an E_∞ space X, the zeroth spaces of the various Thom spectra MG, the classifying spaces of bipermutative categories and the zeroth spaces of their associated spectra are all examples of E_∞ ring spaces. The last example includes models for $BO \times Z$ and $BU \times Z$ as E_∞ ring spaces. A complete geometric theory of such spaces and of their relationship to E_∞ ring spectra is given in [R], along with the examples above (among others) and various applications. On the level of (mod p) homology, the important fact is that all of the formulas developed by Milgram, Madsen, Tsuchiya, and myself for the study of $H_*\tilde{F}$ (where $\tilde{F}$ denotes QS^0 regarded as an E_∞ space under the smash product) are valid in H_*X for an arbitrary E_∞ ring space X. Moreover, the general setting leads to very much simpler proofs than those originally obtained from the geometry of $\tilde{F}$.

The first three sections are devoted to these formulas. Thus section 1 establishes notations and gives formulas for the evaluation of the "multiplicative" Pontryagin product $\#$ on elements decomposable in terms of the "additive" Pontryagin product $*$ or of its homology operations Q^S. The formula for $(x * y) \# z$ is due to Milgram [22] and that for $(Q^s x) \# y$ is due to me [20]; both date to 1968. Section 2 gives the mixed Cartan formula for the evaluation of the multiplicative homology operations $\tilde{Q}^s$ on elements $x * y$. A partial result and the basic geometric idea are due to Tsuchiya [36], but the complete formula is due to Madsen [15] when

$p = 2$ and to myself [20] when $p > 2$; it dates to 1970. Section 3 gives the mixed Adem relations for the evaluation of $\tilde{Q}^r Q^s x$. Again, the basic geometric idea is due to Tsuchiya. These relations are incredibly complicated when $p > 2$, and Tsuchiya and I arrived at the correct formulas, for $x = [1]$, by a sequence of successive approximations. I obtained the complete formula, for arbitrary x, in 1973 but it is published here for the first time. As we point out formally in section 4, the formulas we obtain are exhaustive in the sense that # and the $\tilde{Q}^s$ are completely determined in $H_*C(X^+)$ and $H_*Q(X^+)$ from # and the $\tilde{Q}^s$ on H_*X, where X is a (multiplicative) E_∞ space. Indeed, $H_*C(X^+)$ is the free AR-Hopf bialgebra generated by H_*X and $H_*Q(X^+)$ is the free AR-Hopf bialgebra with conjugation generated by H_*X.

Section 5, which is independent of sections 3 and 4 and makes minimal use of section 2, is devoted to analysis of the sequence of Hopf algebras

$$H_*SO \to H_*SF \to H_*F/O \to H_*BSO \to H_*BSF .$$

At the prime 2, this material is due to Milgram [22] and Madsen [15] and has also appeared in [5]. At $p > 2$, this material is due to Tsuchiya [36, 38] and myself [20], but the present proofs are much simpler (and the results more precise) than those previously published.

I made a certain basic conjecture about the R-algebra structure of H_*SF in 1968 (stated in [20]). It was the primary purpose of Madsen's paper [15] to give a proof of this conjecture when $p = 2$. Similarly, it was the primary purpose of Tsuchiya's paper [38] to give a proof of this conjecture when $p > 2$. Unfortunately, Tsuchiya's published proof, like several of my unpublished ones, contains a gap and the conjecture is at present still open when $p > 2$. It was my belief that this paper would be incomplete without a proof of the conjecture that has so long delayed its publication. Since the proof has been reduced to pure algebra, which I am unlikely to carry out, and since more recent geometric results make the conjecture inessential to our later calculations, further delay now seems pointless. This reduction will be given in section 6. It consists of a sequence of lemmas which analyze the decomposable elements of H_*SF. These results are generalizations to the case of odd primes of the lemmas used by Madsen to prove the conjecture when $p = 2$, and we shall see why these lemmas complete the proof when $p = 2$ but are only the beginning of a proof when $p > 2$. (To reverse a dictum of John Thompson, the virtue of 2 is not that it is so even but that it is so small.)

The material described so far, while clarified and simplified by the theory of E_∞ ring spaces, entirely antedates the development of that theory. In the last seven sections, which form a single unit wholly independent of sections 3, 4, and 6 and with minimal dependence on section 2, we exploit the constructions of [R] to obtain a conceptual series of calculations. Various familiar categories of matrix groups are bipermutative, hence give rise via [R] to E_∞ ring spectra whose zeroth spaces are E_∞ ring spaces. In section 7, we give a general discussion of procedures for the computation of the two kinds of homology operations on these spaces. For the additive operations Q^s, the basic idea and the mod 2 calculations are due to Priddy [27] while the mod p calculations are due to Moore [24]. Examples for which the procedures discussed in principle give complete information are the categories $\mathcal{O}$ and $\mathcal{U}$ of classical orthogonal and unitary groups (section 8), the category $\mathcal{GL}k_r$ of general linear groups of the field with r elements (section 9), and the category $\mathcal{O}k_q$ of orthogonal groups of the field with q elements, q odd (section 11). In the classical group case, we include a comparison with the earlier results of Kochman [13] (which are by no means rendered superfluous by the present procedures). The cases $\mathcal{GL}k_r$ and $\mathcal{O}k_q$ are entirely based on the calculations of Quillen [29] and Fiedorowicz and Priddy [6], respectively.

In sections 10, 12, and 13, we put everything together to analyze the homologies of B Coker J and the classifying space B(SF; kO) for kO-oriented stable spherical fibrations. At odd primes, $B(SF;kO) \simeq BTop$ as an infinite loop space and our results therefore include determination of the p-primary characteristic classes for stable topological (or PL) bundles. The latter calculation was first obtained by Tsuchiya [37] and myself [20], independently, but the present proofs are drastically simpler and yield much more precise information. In particular, we obtain a precise hold on the image of $H_*BCoker J$ in H_*BSF. This information rather trivially implies Peterson's conjecture [25] that the kernel of the natural map $A \rightarrow H^*MTop$ is the left ideal generated by Q_0 and Q_1, a result first proven by Tsuchiya [37] by analysis of the p-adic construction on certain 5-cell Thom complexes.

The essential geometry behind our odd primary calculations is the splitting $BSF = BJ \times B\,Coker\,J$ of infinite loop spaces at p. The 2 primary analysis of sections 12 and 13 is more subtle because, at 2, we only have a non-splittable fibration of infinite loop spaces $B\,Coker\,J \rightarrow BSF \xrightarrow{Be} BJ_{\otimes}$. Indeed, we shall see that, with the model for J relevant to this fibration, it is a triviality that SF cannot split as $J \times Coker\,J$ as an H-space. Section 12 is primarily devoted to analysis of

$e_*: H_*SF \to H_*J_\otimes$ and the internal structure of $H_*J_\otimes$. The main difficulty is that, until the calculation of e_* is completed, we will not even know explicit generators for the algebra $H_*J_\otimes$. An incidental consequence of our computations will be the determination of explicit polynomial generators for $H_*BSO_\otimes$. In section 13, we give a thorough analysis of the homological behavior of the fibration cited above. On the level of mod 2 homology, complete information falls out of the calculation of e_*, and the bulk of the section is devoted to analysis of higher torsion via the calculation of the Bockstein spectral sequences of all spaces in sight. The key ingredients, beyond our mod 2 calculations, are a new calculation of the torsion in BBSO, which was first computed by Stasheff [31], and Madsen's determination [16] of the crucial differentials in the Bockstein spectral sequence of BSF. The results we obtain are surprisingly intricate, one interesting new phenomenon uncovered being an exact sequence of the form $0 \to Z_4 \to Z_2 \oplus Z_8 \to Z_4 \to 0$ contained in the integral homology sequence $H_{4i}BCoker\ J \to H_{4i}BSF \to H_{4i}BJ_\otimes$ for $i \neq 2^j$, the Z_8 in $H_{4i}BSF$ being in the image of $H_{4i}BSO$.

Because of the long delay in publication of the first few sections, various results and proofs originally due to myself have long since appeared elsewhere. Conversely, in order to make this paper a useful summary of the field, I have included proofs of various results originally due to Milgram, Madsen, Tsuchiya, Kochman, Herrero, Stasheff, Peterson, Priddy, Moore, and Fiedorowicz, to all of whom I am also greatly indebted for very helpful discussions of this material.

CONTENTS

§1. E_∞ ring spaces and the # product

Before proceeding to the analysis of their homological structure, we must recall the definition of E_∞ ring spaces. This notion is based on the prior notion of an action λ of an E_∞ operad $\mathcal{H}$ on an E_∞ operad $\mathcal{C}$. Such an action consists of maps

$$\lambda: \mathcal{H}(k) \times \mathcal{C}(j_1) \times \ldots \times \mathcal{C}(j_k) \rightarrow \mathcal{C}(j_1 \cdots j_k), \quad k \geq 1 \text{ and } j_r \geq 1,$$

subject to certain axioms which state how the λ relate to the internal structure of $\mathcal{C}$ and $\mathcal{H}$. Only the equivariance formulas are relevant to the homological calculations, and we shall not give the additional axioms required for theoretical purposes here. We require some notations in order to state the equivariance formulas.

Definition 1.1. For $j_r \geq 1$, let $S(j_1, \ldots, j_k)$ denote the set of all sequences $I = \{i_1, \ldots, i_k\}$ such that $1 \leq i_r \leq j_r$ and order $S(j_1, \ldots, j_k)$ lexicographically. This fixes an action of Σ_j on $S(j_1, \ldots, j_k)$, where $j = j_1 \cdots j_k$. For $\sigma \in \Sigma_k$, define

$$\sigma\langle j_1, \ldots, j_k\rangle : S(j_1, \ldots, j_k) \rightarrow S(j_{\sigma^{-1}(1)}, \ldots, j_{\sigma^{-1}(k)})$$

by

$$\sigma\langle j_1, \ldots, j_k\rangle\{i_1, \ldots, i_k\} = \{i_{\sigma^{-1}(1)}, \ldots, i_{\sigma^{-1}(k)}\}.$$

Via the given isomorphisms of $S(j_1, \ldots, j_k)$ and $S(j_{\sigma^{-1}(1)}, \ldots, j_{\sigma^{-1}(k)})$ with $\{1, 2, \ldots, j\}$, $\sigma\langle j_1, \ldots, j_k\rangle$ may be regarded as an element of Σ_j. For $\tau_r \in \Sigma_{j_r}$, define $\tau_1 \otimes \ldots \otimes \tau_k \in \Sigma_j$ by

$$(\tau_1 \otimes \ldots \otimes \tau_k)\{i_1, \ldots, i_k\} = \{\tau_1 i_1, \ldots, \tau_k i_k\}.$$

Observe that these are "multiplicative" analogs of the permutations

$\sigma(j_1,\ldots,j_k)$ and $\tau_1 \oplus \ldots \oplus \tau_k$ in $\Sigma_{j_1+\ldots+j_k}$ which were used in the definition, [G, 1.1], of an operad.

The equivariance formulas required of the maps λ are

$$\lambda(g\sigma; c_1,\ldots,c_k) = \lambda(g; c_{\sigma^{-1}(1)},\ldots,c_{\sigma^{-1}(k)})\sigma\langle j_1,\ldots,j_k\rangle$$

and

$$\lambda(g; c_1\tau_1,\ldots,c_k\tau_k) = \lambda(g; c_1,\ldots,c_k)(\tau_1 \otimes \ldots \otimes \tau_k)$$

for $g \in \mathcal{G}(k)$, $c_r \in \mathcal{C}(j_r)$, $\sigma \in \Sigma_k$, and $\tau_r \in \Sigma_{j_r}$.

We require two other preliminary definitions.

Definition 1.2. Let $X \in \mathcal{J}$. For $j_r \geq 1$, define $\delta: X^{j_1} \times \ldots \times X^{j_k} \to (X^k)^{j_1 \cdots j_k}$ by

$$\delta(y_1,\ldots,y_k) = \mathop{\times}_{I \in S(j_1,\ldots,j_k)} y_I ,$$

where if $y_r = (x_{r1},\ldots,x_{rj_r})$ and $I = \{i_1,\ldots,i_k\}$, then $y_I = (x_{1i_1},\ldots,x_{ki_k})$.

Definition 1.3. A $\mathcal{G}_0$-space (X,ξ) is a $\mathcal{G}$-space with basepoint 1 together with a second basepoint 0 such that

$$\xi_k(g, x_1,\ldots,x_k) = 0$$

for all $g \in \mathcal{G}(k)$ if any $x_r = 0$. Let $\mathcal{G}_0[\mathcal{J}]$ denote the category of $\mathcal{G}_0$-spaces.

Definition 1.4. Let $\mathcal{G}$ act on $\mathcal{C}$. A $(\mathcal{C},\mathcal{G})$-space (X,ξ,θ) is a $\mathcal{G}_0$-space (X,ξ) and a $\mathcal{C}$-space (X,θ) with basepoint 0 such that the following distributivity diagram is commutative for all $k \geq 1$ and $j_r \geq 1$, where $j = j_1 \cdots j_k$:

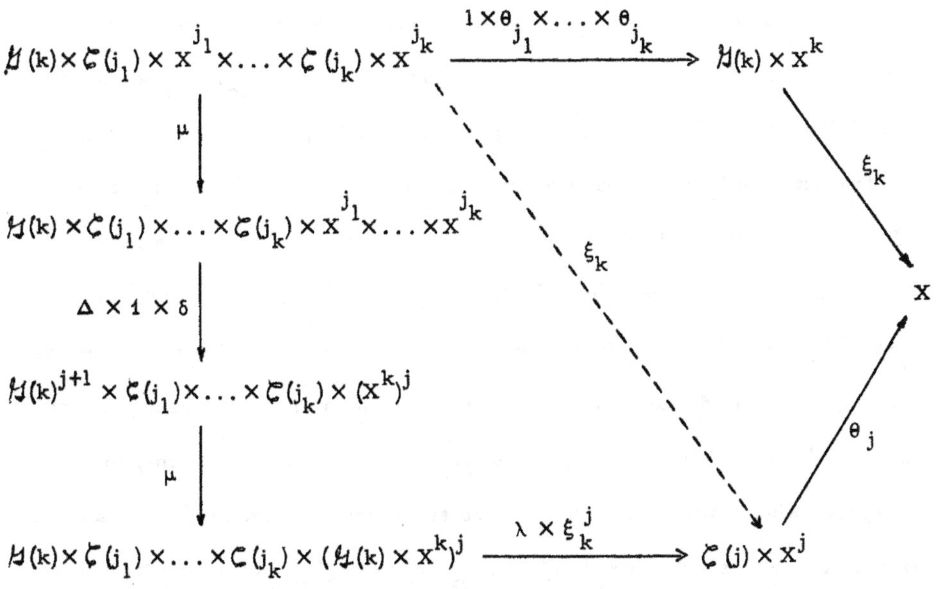

Here the maps μ are shuffle homeomorphisms and Δ is the iterated diagonal.

Let $\xi_k: \mathcal{G}(k)\times\mathcal{C}(j_1)\times X^{j_1}\times\ldots\times\mathcal{C}(j_k)\times X^{j_k}\to\mathcal{C}(j)\times X^j$ be defined by commutativity of the left-hand side of the distributivity diagram. This definition makes sense for any $\mathcal{G}_0$-space (X,ξ), and the omitted parts of the definition of an action of $\mathcal{G}$ on $\mathcal{C}$ serve to ensure that the ξ_k induce an action of $\mathcal{G}$ on CX such that $\eta: X\to CX$ and $\mu: CCX\to CX$ are morphisms of $\mathcal{G}_0$-spaces. In other words, if $\mathcal{G}$ acts on $\mathcal{C}$, then C defines a monad in $\mathcal{G}_0[\mathcal{J}]$. The distributivity diagram states that $\theta: CX\to X$ gives (X,ξ) a structure of an algebra over this monad. It is this more conceptual formulation of the previous definition which is central to the geometric theory of [R]. We refer the reader to [R, VI §4 and VII § 2 and 4] for examples of suitable pairs $(\mathcal{C},\mathcal{G})$ and to [R, IV §1-2, VI § 4-5, and VII §2] for examples of $(\mathcal{C},\mathcal{G})$-spaces.

We should perhaps mention one technical problem which arose in [R], if only to indicate its irrelevance to the calculations here. In practice, one is given a pair $(\mathcal{C}', \mathcal{G}')$ of locally contractible (e.g., E_∞) operads such that $\mathcal{G}'$ acts on $\mathcal{C}'$. In order to pass from $(\mathcal{C}', \mathcal{G}')$-spaces to ring spectra with similar internal structure on their zeroth spaces, one replaces $(\mathcal{C}', \mathcal{G}')$ by $(\mathcal{C}, \mathcal{G}) = (\mathcal{C}' \times \mathcal{K}_\infty, \mathcal{G} \times \mathcal{L})$, where $\mathcal{K}_\infty$ is the infinite little convex bodies operad and $\mathcal{L}$ is the linear isometries operad. However, $\mathcal{K}_\infty$ and $\mathcal{C}$ are in fact only partial operads, in that their structural maps γ are only defined on appropriate subspaces of the relevant product spaces. (See [R, VII §1 and §2] for details.) The results of I apply to $\mathcal{C}$-spaces since, when only the additive structure is at issue, $\mathcal{C}$ may as well be replaced by its sub operad $\mathcal{C}' \times \mathcal{C}_\infty$, where $\mathcal{C}_\infty$ is the infinite little cubes operad used in I. On the other hand, since the maps γ do not appear in the distributivity diagram, they play little role in the study of multiplicative operations and their interrelationship with additive operations which is our concern here. (In the few places the γ do appear, in the proofs of 2.2 and 3.3, we can again replace $\mathcal{C}$ by $\mathcal{C}' \times \mathcal{C}_\infty$.)

Henceforward, throughout this and the following two sections, we tacitly assume given a fixed pair of E_∞ operads $\mathcal{C}$ and $\mathcal{G}$ such that $\mathcal{G}$ acts on $\mathcal{C}$ and a fixed $(\mathcal{C}, \mathcal{G})$-space (X, ξ, θ). (This general hypothesis remains in force even when results are motivated by a discussion of their consequences for $H_*\widetilde{F}$.) We refer to such spaces X as E_∞ ring spaces. It should be noted that E_∞ semi-ring would be a more accurate term: we have built in all of the axioms for a ring, up to all possible higher coherence homotopies, except for the existence of additive inverses.

We intend to analyze the interrelationships between the two R-algebra structures on H_*X, and we must first fix notations. For $i \in \pi_0 X$, we write

X_i for i considered as a subspace of X and write [i] for i regarded as an element of H_0X. Of course, π_0X is a semi-ring, with addition and multiplication derived from θ and ξ. Fix $c_r \in \mathcal{C}(r)$ and define the r-fold " additive " product on X to be $\theta_r(c_r): X^r \to X$. Write * for this product both on the level of spaces and on homology. Note that * takes $X_i \times X_j$ to X_{i+j} and that $[i]*[j] = [i+j]$. Write Q^s for the homology operations determined by θ; Q^s takes H_*X_i to H_*X_{pi} and $Q^0[i] = [pi]$. Fix $g_r \in \mathcal{G}(r)$ and define the r-fold "multiplicative" product on X to be $\xi_r(g_r): X^r \to X$. Write # for this product both on the level of spaces and on homology; however, to abbreviate, we write # on elements by juxtaposition, $x \# y = xy$. Note that # takes $X_i \times X_j$ to X_{ij} and that $[i][j] = [ij]$. Write $\tilde{Q}^s$ for the homology operations determined by ξ; $\tilde{Q}^s$ takes H_*X_i to $H_*X_{i^p}$ and $\tilde{Q}^0[i] = [i^p]$. Let $\varepsilon: H_*X \to Z_p$ be the augmentation and note that $\varepsilon[i] = 1$.

Let $\psi: H_*X \to H_*X \otimes H_*X$ be the coproduct and note that $\psi[i] = [i] \otimes [i]$. For $x \in H_*X$, we shall write $\psi x = \sum x' \otimes x''$, as usual; the iterated coproduct $\psi: H_*X \to (H_*X)^r$ will sometimes be written in the form $\psi x = \sum x^{(1)} \otimes \dots \otimes x^{(r)}$, with the index of summation understood.

If π_0X is a ring (additive inverses exist), then H_*X admits the conjugation χ (for *) defined in Lemma I.2.7; of course, $\eta\varepsilon = *(1 \otimes \chi)\psi$, where η is the unit for *, $\eta(1) = [0]$. Moreover, $Q^s\chi = \chi Q^s$ by inductive calculation or by Lemma I.1.2 and the fact that the operations Q^s agree with the loop operations on the weakly homotopy equivalent infinite loop space B_0X. In contrast, multiplicative inverses do not exist in π_0X in the interesting examples, hence the product # does not admit a conjugation.

We complete this section by obtaining formulas for the evaluation of # in terms of * and the Q^s. Formulas for the evaluation of the $\tilde{Q}^s$ in terms of $*$ and the Q^s will be obtained in the following two sections. All of our formulas will be derived by analysis of special cases of the distributivity diagram. The following result was first proven, for $X = \tilde{F}$, by Milgram [22].

<u>Proposition 1.5.</u> Let $x, y, z \in H_*X$ and $i, j \in \pi_0 X$. Then

(i) $[0]x = (\varepsilon x)[0]$ (and $[1]x = x$);

(ii) if $\pi_0 X$ is a ring, $[-1]x = \chi x$;

(iii) $(x * y)z = \sum (-1)^{\deg y \deg z'} xz' * yz''$;

(iv) $(x * [i])(y * [j]) = \sum (-1)^{\deg y' \deg x''} x'y' * x''[j] * y''[i] * [ij]$.

<u>Proof.</u> Since (X, ξ) is given to be a $\mathcal{H}_0$-space, we have that $0 \# x = 0$ for $x \in X$, and (i) follows. Formula (iii) holds since the following diagram is homotopy commutative:

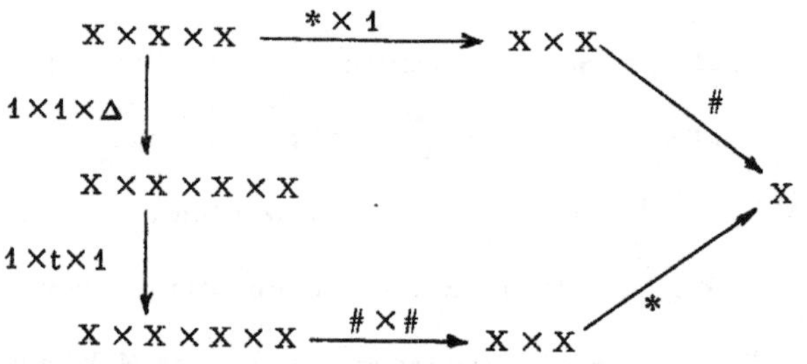

Indeed, by the distributivity diagram applied to elements $(g_2, c_2, x, y, 1, z)$, where $1 \in \mathcal{C}(1)$ is the identity, the diagram would actually commute if $*$ on the bottom right were replaced by the product $\theta_2(c'_2)$, $c'_2 = \lambda(g_2; c_2, 1)$, and any path from c_2 to c'_2 in $\mathcal{C}(2)$ determines a homotopy from $* = \theta_2(c_2)$ to $\theta_2(c'_2)$. Formula (iv) follows formally from (iii) and the commutativity of the product $\#$ on homology. Formula (ii) follows by induction on the degree of x from (i), (iii), the fact that $[0] = [1] * [-1]$, and the equation $\eta\varepsilon = *(1 \otimes \chi)\psi$, since these formulas give

$$\sum x' * [-1]x'' = ([1]*[-1])x = [0]x = (\eta\varepsilon)(x) = \sum x' * \chi x''.$$

<u>Proposition 1.6.</u> Let $x, y \in H_* X$. Then

$$(Q^s x)y = \sum_i Q^{s+i}(xP^i_* y) \quad \text{and , if } p > 2,$$

$$(\beta Q^s x)y = \sum_i \beta Q^{s+i}(xP^i_* y) - \sum_i (-1)^{\deg x} Q^{s+i}(xP^i_* \beta y) .$$

<u>Proof.</u> The following diagram is Σ_p-homotopy commutative:

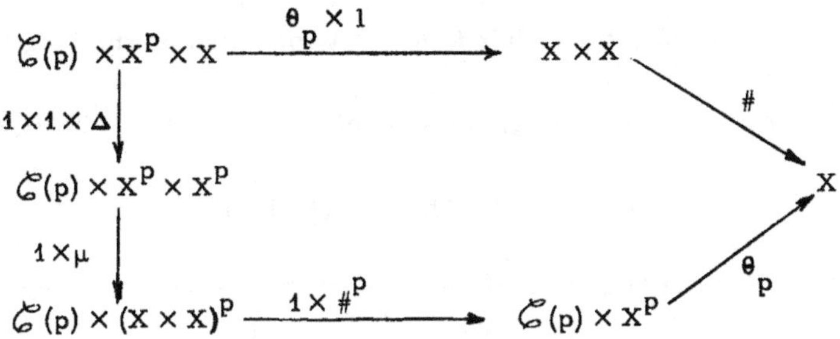

Indeed, by the distributivity diagram applied to elements $(g_2, c, y, 1, x)$ for $(c, y) \in \mathcal{C}(p) \times X^p$ and $x \in X$, the diagram would actually commute if the identity of $\mathcal{C}(p)$ on the bottom arrow were replaced by the map $c \to \lambda(g_2; c, 1)$, and these two maps $\mathcal{C}(p) \to \mathcal{C}(p)$ are Σ_p-homotopic since $\mathcal{C}(p)$ is Σ_p-free and contractible. The rest of the argument is identical to the proof of Theorem I.1.4.

Of course, in the case of $QS^0 = \tilde{F}$, the previous two results are contained in Theorem I.1.4 and Lemma I.1.5. Since, by Theorem I. 42, $H_* QS^0$ is generated by $[\pm 1]$ as an R- algebra, the previous results completely determine the smash product on $H_* \tilde{F}$; all products $Q^I[1]x$ can be computed by induction on $\ell(I)$. The following observation implies that $H_* \tilde{F}$ is generated under the loop and smash products by $[-1]$ and the elements $\beta^\varepsilon Q^s[1]$ (or $Q^s[1]$ if $p = 2$).

Proposition 1.7. Every element $Q^I[1]$ of H_*X is decomposable as a linear combination of products $\beta^{\varepsilon_1}Q^{s_1}[1]\cdots\beta^{\varepsilon_k}Q^{s_k}[1]$ (or $Q^{s_1}[1]\cdots Q^{s_k}[1]$ if $p = 2$), where $k = \ell(I)$.

Proof. If $I = (s, J)$, then $Q^I[1] = Q^s[1]Q^J[1] - \sum_{i>0} Q^{s+i}(P^i_*Q^J[1])$. On the right, $Q^{s+i}(P^i_*Q^J[1]) = Q^{s+i}[1]P^i_*Q^J[1] - \sum_{j>0} Q^{s+i+j}(P^j_*P^i_*Q^J[1])$. Iterating, we reach terms where the error summation is zero after finitely many steps. Since $P^i_*Q^J[1]$ is a linear combination of monomials $Q^K[1]$ such that $\ell(K) = \ell(J)$, the result follows by induction on the length of I.

When π_0X is a ring, we can define a product $\underline{*}$ in H_*X_1 by

$$x \mathbin{\underline{*}} y = (x*[-1]) * (y*[-1]) * [1].$$

Thus $\underline{*}$ is just the translate of the product $*$ from the zero component to the one component. Since

$$(\beta^{\varepsilon_1}S^{s_1}[1]\cdots\beta^{\varepsilon_k}Q^{s_k}[1]) * [1-p^k] \equiv (\beta^{\varepsilon_1}Q^{s_1}[1] * [1-p])\cdots(\beta^{\varepsilon_k}Q^{s_k}[1] * [1-p])$$

modulo elements of H_*X_1 which are decomposable under the product $\underline{*}$, we have the following corollary when $X = \tilde{F}$.

Corollary 1.8. The elements $\beta^{\varepsilon}Q^s[1] * [1-p]$ (or $Q^s[1] * [-1]$ if $p = 2$) generate H_*SF under the products $\#$ and $\underline{*}$.

The following explicit calculation is due to Milgram [22]; his proof depended on use of the known structure of the cohomology algebras of the groups $\Sigma_2\int\Sigma_2$ and Σ_4.

Lemma 1.9. Let $p = 2$. Then $Q^s[1]Q^s[1] = Q^sQ^s[1]$ and, if $s > 0$ and π_0X is a ring, $(Q^s[1] * [-1])(Q^s[1] * [-1]) = 0$.

Proof. By Proposition 1.6 and the Nishida relations, we have

$Q^s[1]Q^s[1] = Q^sQ^s[1] + a_s$, where $a_s = \sum_{i>0} (i, s(p-1) - pi) Q^{s+i}Q^{s-i}[1]$.

Visibly $a_1 = 0$. Assume inductively that $a_k = 0$ for $0 < k < s$, $s \geq 2$. Rather than use the Adem relations to compute a_s directly, we observe that, by an easy calculation, the induction hypothesis implies that a_s is primitive. Since a_s is a linear combination of length two elements of R acting on [1], Theorem I.3.7 implies that $a_s = 0$, $a_s = Q^1Q^1[1]$, or $a_s = Q^2Q^1[1]$. Since $\deg a_s \geq 4$, the first alternative must hold. For the second formula,

$$(Q^s[1]*[-1])^2 = \sum_i \sum_j Q^{s-i}[1]Q^{s-j}[1]*Q^i[1]\cdot[-1]*Q^j[1]\cdot[-1]*[1]$$

$$= \sum_i (Q^{s-i}[1]*Q^{s-i}[1])*\chi(Q^i[1]*Q^i[1])*[1]$$

$$= *(1\otimes\chi)\ \psi(Q^s[1]*Q^s[1])*[1] = 0.$$

Here the first equation follows from Proposition 1.5 (iv), the second from our first formula and symmetry (the terms with $i \neq j$ cancel in pairs), and the last from the definition of χ.

§ 2. The mixed Cartan formula

We shall compute $\tilde{Q}^s(x * y)$ in this section and $\tilde{Q}^r\beta^\varepsilon Q^s x$ in the next, $x, y \in H_*X$. To do so, we shall have to decompose special cases of the distributivity diagram. We need the following notations.

Definition 2.1. Let S be a subset of $S(j_1, \ldots, j_k)$ and give S the ordering induced by that of $S(j_1, \ldots, j_k)$. Let $n(S)$ denote the number of elements in S. Define

$$\sigma(S): \mathcal{C}(j_1 \ldots j_k) \to \mathcal{C}(n(S))$$

to be the iterated degeneracy (as in [G, Notations 2.3]) given by

$$\sigma(S)(c) = \gamma(c; s), \quad c \in \mathcal{C}(j_1 \ldots j_k),$$

where s is that element of $[\mathcal{C}(0) \coprod \mathcal{C}(1)]^{j_1 \ldots j_k}$ whose I^{th} coordinate is $* \in \mathcal{C}(0)$ if $I \notin S$ and is $1 \in \mathcal{C}(1)$ if $I \in S$. For example, if $\mathcal{C} = \mathcal{K}_\infty$, then $\sigma(S)$ deletes those little convex bodies of c which are indexed on $I \notin S$. Define

$$\lambda(S) = \sigma(S) \cdot \lambda : \mathcal{G}(k) \times \mathcal{C}(j_1) \times \ldots \times \mathcal{C}(j_k) \to \mathcal{C}(n(S))$$

and define

$$\delta(S): X^{j_1} \times \ldots \times X^{j_k} \to (X^k)^{n(S)}$$

by letting $\delta(S)(y_1, \ldots, y_k)$ have I^{th} coordinate $(x_{1,i_1}, \ldots, x_{k,i_k})$ if $y_r = (x_{r1}, \ldots, x_{rj_r})$ and $I = \{i_1, \ldots, i_k\} \in S$. Then define a map $\xi(S)$ by commutativity of the following diagram:

$$
\begin{array}{ccc}
\mathcal{G}(k)\times\mathcal{C}(j_1)\times X^{j_1}\times\dots\times\mathcal{C}(j_k)\times X^{j_k} & \xrightarrow{\quad\xi(S)\quad} & \mathcal{C}(n(S))\times X^{n(S)} \\
\Big\downarrow{\scriptstyle\mu} & & \\
\mathcal{G}(k)\times\mathcal{C}(j_1)\times\dots\times\mathcal{C}(j_k)\times X^{j_1}\times\dots\times X^{j_k} & & \\
\Big\downarrow{\scriptstyle\Delta\times 1\times\dots\times 1\times\delta(S)} & & \\
\mathcal{G}(k)^{n(S)+1}\times\mathcal{C}(j_1)\times\dots\times\mathcal{C}(j_k)\times(X^k)^{n(S)} & & \\
\Big\downarrow{\scriptstyle\mu} & & \\
\mathcal{G}(k)\times\mathcal{C}(j_1)\times\dots\times\mathcal{C}(j_k)\times(\mathcal{G}(k)\times X^k)^{n(S)} & \xrightarrow{\lambda(S)\times\xi_k^{n(S)}} & \mathcal{C}(n(S))\times X^{n(S)}
\end{array}
$$

Define $\tilde{\xi}(S)=\theta_{n(S)}\circ\xi(S)\colon \mathcal{G}(k)\times\mathcal{C}(j_1)\times X^{j_1}\times\dots\times\mathcal{C}(j_k)\times X^{j_k}\to X$.

Abbreviate $S(j;k)=S(j_1,\dots,j_k)$ when $j_1=\dots=j_k=j$. To compute $\tilde{Q}^s(x*y)$ and $\tilde{Q}^r\beta^\varepsilon Q^s x$, we must analyze $\tilde{\xi}S(2;p)$ and $\tilde{\xi}S(p;p)$. The definition suggests the procedure to be followed: we break the relevant set $S(j;k)$ into an appropriate union of disjoint subsets in order to decompose $\tilde{\xi}S(j;k)$ into pieces we can analyze. The following result gives the general pattern. Observe that the evident action of the wreath product $\Sigma_k\int\Sigma_j$ (defined in Definition I.5.3) on the set $S(j;k)$ fixes an inclusion of $\Sigma_k\int\Sigma_j$ in Σ_{jk}. The distributivity diagram is clearly $\Sigma_k\int\Sigma_j$-equivariant when $j_1=\dots=j_k=j$.

<u>Proposition 2.2.</u> Let G be a subgroup of $\Sigma_k\int\Sigma_j$ and let $S\subset S(j;k)$ be the disjoint union of subsets $S_1,\dots,S_q$ such that each S_i is closed under the action of G. Then each $\xi(S_i)$ is G-equivariant, and $\tilde{\xi}(S)$ is G-equivariantly homotopic to the composite

$$\mathcal{G}(k)\times(\mathcal{C}(j)\times X^j)^k\xrightarrow{\Delta}(\mathcal{G}(k)\times(\mathcal{C}(j)\times X^j)^k)^q\xrightarrow{\overset{q}{\underset{i=1}{\times}}\tilde{\xi}(S_i)}X^q\xrightarrow{*}X.$$

<u>Proof.</u> Consider the following diagram:

$$\begin{array}{ccccc}
\mathcal{G}(k) \times (\mathcal{C}(j) \times X^j)^k & \xrightarrow{\xi(S)} & \mathcal{C}(n(S)) \times X^{n(S)} & \xrightarrow{\theta_{n(S)}} & X \\
 & & \uparrow{\scriptstyle \hat{\nu}} & \nearrow{\scriptstyle \theta_{n(S)}} & \\
 & & \mathcal{C}(n(S)) \times X^{n(S)} & & \\
\downarrow{\scriptstyle \Delta} & & \uparrow{\scriptstyle * \times 1} & & \uparrow{\scriptstyle *} \\
 & & \mathop{\times}\limits_{i=1}^{q} \mathcal{C}(n(S_i)) \times X^{n(S)} & & \\
 & & \uparrow{\scriptstyle \mu} & & \\
(\mathcal{G}(k) \times (\mathcal{C}(j) \times X^j)^k)^q & \xrightarrow{\mathop{\times}\limits_{i=1}^{q} \xi(S_i)} & \mathop{\times}\limits_{i=1}^{q} (\mathcal{C}(n(S_i)) \times X^{n(S_i)}) & \xrightarrow{\mathop{\times}\limits_{i=1}^{q} \theta_{n(S_i)}} & X^q
\end{array}$$

Here μ is a shuffle homeomorphism and $\hat{\nu}$ denotes right action by that permutation ν of $n(S)$ letters which corresponds to changing the ordering of the set S from that obtained by regarding S as the ordered union $S_1 \cup \dots \cup S_q$ (where each S_i is ordered as a subset of $S(j;k)$) to that obtained by restricting the ordering of $S(j;k)$ to S. The map $*$ in the operad $\mathcal{C}$ is defined by

$$(d_1, \dots, d_q) \to \gamma(c_q; d_1, \dots, d_q), \quad d_i \in \mathcal{C}(n(S_i)),$$

for our fixed $c_q \in \mathcal{C}(q)$. By [G, Lemma 1.4], the right-hand triangle and trapezoid commute. In the left-hand rectangle, the coordinates in $X^{n(S)}$, in order, given by $\xi(S)$ and by the specified composite are the same. To study the coordinate in $\mathcal{C}(n(S))$, consider the diagram

$$\begin{array}{ccccc}
\mathcal{G}(k) \times \mathcal{C}(j)^k & \xrightarrow{\lambda} & \mathcal{C}(j^k) & \xrightarrow{\sigma(S)} & \mathcal{C}(n(S)) \\
 & & & & \uparrow{\scriptstyle \hat{\nu}} \\
\downarrow{\scriptstyle \Delta} & & \downarrow{\scriptstyle \Delta} & & \mathcal{C}(n(S)) \\
 & & & & \uparrow{\scriptstyle *} \\
(\mathcal{G}(k) \times \mathcal{C}(j)^k)^q & \xrightarrow{\lambda^q} & \mathcal{C}(j^k)^q & \xrightarrow{\mathop{\times}\limits_{i=1}^{q} \sigma(S_i)} & \mathop{\times}\limits_{i=1}^{q} \mathcal{C}(n(S_i))
\end{array}$$

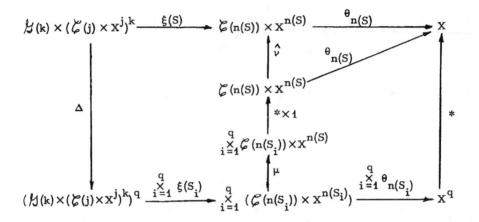

Obviously the left-hand square commutes. Let $\iota(S): G \to \Sigma_{n(S)}$ be the homomorphism (not necessarily an inclusion) determined by the action of G on S. Clearly $\sigma(S)$ is G-equivariant, where G acts on $\mathcal{C}(j^k)$ via the given inclusion of G in $\Sigma_k \int \Sigma_j \subset \Sigma_{j^k}$ and G acts on $\mathcal{C}(n(S))$ via $\iota(S)$. If $\tilde{\nu}$ is defined by $\tilde{\nu}(\tau) = \nu^{-1}\tau\nu$, then $\iota(S)$ coincides with the composite

$$G \xrightarrow{\Delta} G^q \xrightarrow{\times_{i=1}^{q} \iota(S_i)} \times_{i=1}^{q} \Sigma_{n(S_i)} \xrightarrow{\oplus} \Sigma_{n(S)} \xrightarrow{\tilde{\nu}} \Sigma_{n(S)}$$

since $\tilde{\nu}\circ\oplus$ is the inclusion of $\times_{i=1}^{q} \Sigma_{n(S_i)}$ in $\Sigma_{n(S)}$ determined by the inclusions of the S_i in S as ordered subsets. Therefore the composite $\hat{\nu}\circ * \circ \times_{i=1}^{q} \sigma(S_i) \circ \Delta$ is also G-equivariant. Since $\mathcal{C}(j^k)$ is G-free and $\mathcal{C}(n(S))$ is contractible, the right-hand square is G-equivariantly homotopy commutative and the proof is complete.

We give two lemmas which will aid in the homological evaluation of the composite appearing in the proposition. It will sometimes be the case that all $\xi(S_i)$ induce the same map on G-equivariant homology. The following lemma will then be used to simplify formulas.

<u>Lemma 2.3.</u> Let $\phi_i: Y \to X$, $1 \leq i \leq q$, be maps such that $\phi_{i*} = \phi_{j*}$. Then the map on homology induced by the composite

$$Y \xrightarrow{\Delta} Y^q \xrightarrow{\times_{i=1}^{q} \phi_i} X^q \xrightarrow{*} X$$

is given by $y \to [q](\phi_{1*}\, y)$ for $y \in H_*Y$.

<u>Proof.</u> Define $i_q: X \to X$ by $i_q x = (1 * \ldots * 1) \# x$, q factors of 1. The distributivity diagram (applied to $(g_2, c_q, (1)^q, 1, \phi_1(y))$, $y \in Y$) implies that $*\circ\phi_1^q\circ\Delta$ is homotopic to $i_q \circ \phi_1$, and the result follows.

When the $\tilde{\xi}(S_i)_*$ are distinct, the following observation will allow computation of Δ_* on G-equivariant homology.

Lemma 2.4. The following diagram is commutative:

$$\begin{array}{ccc}
\mathcal{G}(k) \times (\mathcal{C}(j) \times X^j)^k & \xrightarrow{1 \times \Delta^k} & \mathcal{G}(k) \times ((\mathcal{C}(j) \times X^j)^q)^k \\
\downarrow{\scriptstyle \Delta} & & \downarrow{\scriptstyle \Delta \times 1} \\
(\mathcal{G}(k) \times (\mathcal{C}(j) \times X^j)^k)^q & \xleftarrow{\mu} & \mathcal{G}(k)^q \times ((\mathcal{C}(j) \times X^j)^q)^k
\end{array}$$

(where μ is the evident shuffle homeomorphism).

When $k = p$ and G contains the cyclic group $\pi = \pi \times 1^j \subset \Sigma_p \int \Sigma_j$, the maps $(1 \times \Delta^k)_*$ and $\mu_*(\Delta \times 1)_*$ on G-equivariant homology can be readily evaluated by the naturality of equivariant homology and by use of the explicit coproduct on the standard π-free resolution of Z_p (compare [A, Proposition 2.6]).

The following theorem was first proven, for $X = \widetilde{F}$, by Madsen [15] when $p = 2$ (using Kochman's calculations [13] of the operations in H_*O) and by myself [20] when $p > 2$; the present proof is a simplification of that given by Tsuchiya in a later reformulation of my result [38].

Theorem 2.5 (The mixed Cartan formula). Let $x, y \in H_*X$. Then

(i) $\widetilde{Q}^s(x * y) = \sum_{s_0 + \dots + s_p = s} \sum \widetilde{Q}_0^{s_0}(x^{(0)} \otimes y^{(0)}) * \dots * \widetilde{Q}_p^{s_p}(x^{(p)} \otimes y^{(p)})$,

where $x \otimes y \rightarrow \sum x^{(0)} \otimes y^{(0)} \otimes \dots \otimes x^{(p)} \otimes y^{(p)}$ under the iterated coproduct of $H_*X \otimes H_*X$;

(ii) $\widetilde{Q}_0^s(x \otimes y) = (\varepsilon y)\widetilde{Q}^s x$ and $\widetilde{Q}_p^s(x \otimes y) = (\varepsilon x)\widetilde{Q}^s y$; and

(iii) $\widetilde{Q}_i^s(x \otimes y) = [\frac{1}{p}(i, p-i)]Q^s(\sum\sum x^{(1)} \dots x^{(p-i)} y^{(1)} \dots y^{(i)})$, $0 < i < p$.

(iv) If $p = 2$, $\tilde{Q}^s(x * y) = \sum\limits_{s_0 + s_1 + s_2 = s} \sum\sum \tilde{Q}^{s_0} x' * Q^{s_1}(x''y') * \tilde{Q}^{s_2} y''$.

Proof. Formula (iv) is just the case $p = 2$ of (i) through (iii). We use the distributivity diagram and Proposition 2.2 with $k = p$ and $j_r = j = 2$; we fix $c_2 \in \mathcal{C}(2)$ and omit the coordinates $\mathcal{C}(2)$ from the notation in these results. We must analyze the set $S(2;p)$ of sequences $I = \{i_1, \ldots, i_p\}$, $i_r = 1$ or 2. Let $|I|$ denote the number of indices r such that $i_r = 2$ and note that $|I|$ is invariant under permutations of the entries of I. For $0 \leq i \leq p$, let $S_i = \{I \mid |I| = i\}$. By the cited results, with $G = \Sigma_p \subset \Sigma_p \int \Sigma_2$ in Proposition 2.2, the following diagram is Σ_p-equivariantly homotopy commutative:

$$\begin{array}{ccccc}
\mathcal{G}(p) \times (X^2)^p & \xrightarrow{1 \times *^p} & \mathcal{G}(p) \times X^p & \xrightarrow{\xi_p} & X \\
\Big\downarrow \Delta & & & & \Big\uparrow * \\
(\mathcal{G}(p) \times (X^2)^p)^{p+1} & & \xrightarrow{\quad \times_{i=0}^{p} \tilde{\xi}(S_i) \quad} & & X^{p+1}
\end{array}$$

Define $\tilde{Q}_i^s(x \otimes y)$, $0 \leq i \leq p$, by use of the map $\tilde{\xi}(S_i)$ in precisely the same way that the homology operations were defined for E_∞ spaces (X, θ) by use of the map θ_p in the proof of Theorem I.1.1. Then formula (i) follows from the diagram by use of Lemma 2.4 and the subsequent remarks. The sets S_0 and S_p each have a single element and, for these i, $\tilde{\xi}(S_i)$ is the composite

$$\mathcal{G}(p) \times (X^2)^p \xrightarrow{\xi(S_i)} \mathcal{C}(1) \times X \xrightarrow{\theta_1} X .$$

Since $\mathcal{C}(1)$ is contractible to the point 1, θ_1 is homotopic to the projection π onto the second factor. By Definition 2.1, $\pi \circ \xi(S_i)$ coincides with the composite of ξ_p with either $1 \times \pi_1^p$ (when $i = 0$) or $1 \times \pi_2^p$

(when $i = p$), where $\pi_j: X^2 \to X$ is the projection onto the j-th factor, $j = 1$ or 2. Formula (ii) follows. It remains to prove (iii). Fix i, $0 < i < p$, and let $r_i = \frac{1}{p}(i, p-i)$. Let π be the cyclic group of order p with generator σ and let τ_j, $1 \leq j \leq r_i$, run through a set of double coset representatives for π and $\Sigma_{p-i} \times \Sigma_i$ in Σ_p (under the standard inclusions). Thus $\Sigma_p = \bigcup_j \pi\tau_j(\Sigma_{p-i} \times \Sigma_i)$. Note that, for $\tau \in \Sigma_p$, the group $\tau^{-1}\pi\tau \cap (\Sigma_{p-i} \times \Sigma_i)$ is trivial and there are therefore $p = [\tau^{-1}\pi\tau : \{e\}]$ complete right cosets of $\Sigma_{p-i} \times \Sigma_i$ in the double coset $\pi\tau(\Sigma_{p-i} \times \Sigma_i)$; thus precisely r_i double cosets are indeed required. Let $I_i = \{1, \ldots, 1, 2, \ldots, 2\}$, i twos, observe that $\Sigma_{p-i} \times \Sigma_i$ acts trivially on I_i, and define $S_{ij} = \{\sigma^k \tau_j I_i \mid 1 \leq k \leq p\}$. Clearly S_i is the disjoint union of the sets S_{ij} and, by Proposition 2.2, the following diagram is π-equivariantly homotopy commutative:

$$\begin{array}{ccc}
\mathcal{B}(p) \times (X^2)^p & \xrightarrow{\quad \tilde{\xi}(S_i) \quad} & X \\
\Big\downarrow \Delta & & \Big\uparrow * \\
(\mathcal{B}(p) \times (X^2)^p)^{r_i} & \xrightarrow{\quad \times_{j=1}^{r_i} \tilde{\xi}(S_{ij}) \quad} & X^{r_i}
\end{array}$$

We claim that, on the π-equivariant homology classes relevant to (iii), each $\tilde{\xi}(S_{ij})_*$ agrees with the map induced on homology by the composite

$$\mathcal{C}(p) \times (X^2)^p \xrightarrow{1 \times (\Delta \times \Delta)^p} \mathcal{C}(p) \times (X^{p-i} \times X^i)^p \xrightarrow{1 \times \#^p} \mathcal{C}(p) \times X^p \xrightarrow{\theta_p} X.$$

(The claim makes sense since $C_*\mathcal{B}(p)$ and $C_*\mathcal{C}(p)$ are both π-free resolutions of Z_p.) Formula (iii) will follow by use of Lemma 2.3 . To prove our claim, consider the following diagram:

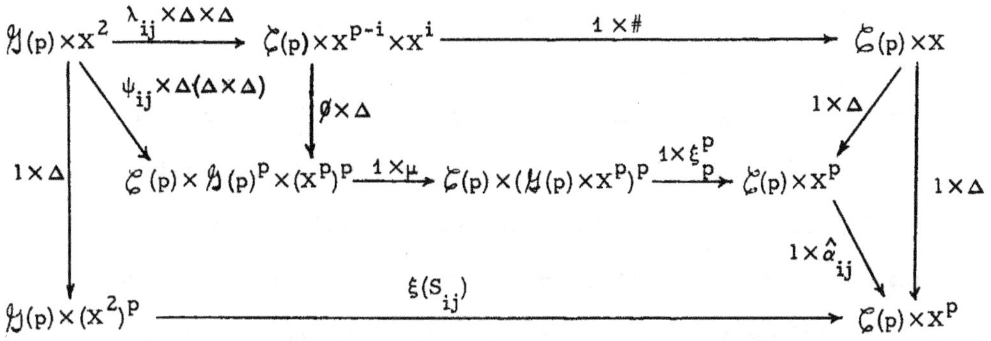

The map $\phi\colon \mathcal{C}(p)\to\mathcal{C}(p)\times\mathcal{G}(p)^p$ is defined by $\phi(c)=(c,g_p^p)$ and the upper right trapezoid commutes because $\#=\xi_p(g_p)\colon X^p\to X$. The maps $\lambda_{ij}\colon\mathcal{G}(p)\to\mathcal{C}(p)$ and $\psi_{ij}\colon\mathcal{G}(p)\to\mathcal{C}(p)\times\mathcal{G}(p)^p$ are defined by

$$\lambda_{ij}(g)=\lambda(S_{ij})(g,c_2^p)\quad\text{and}\quad\psi_{ij}(g)=(\lambda_{ij}(g),\mathop{\times}_{k=1}^{p} g\sigma^k\tau_j)\ .$$

If we could identify S_{ij} with $\{1,\ldots,p\}$ by $\sigma^k\tau_jI_i\to k$, then the homomorphism $\iota(S_{ij})\colon\pi\to\Sigma_p$ given by the action of π on S_{ij} would coincide with the standard inclusion $\iota\colon\pi\to\Sigma_p$. Actually, the given ordering of S_{ij} as a subset of $S(2;p)$ may differ from the standard ordering of $\{k\}$, hence $\iota(S_{ij})$ may differ from ι by an inner automorphism, given by α_{ij} say, of Σ_p. The map $\hat{\alpha}_{ij}\colon X^p\to X^p$ is left action by α_{ij}, and the right-hand triangle commutes trivially. Of course, $\hat{\alpha}_{ij}$ is π-equivariant if π acts on the domain via ι (that is, by cyclic permutations) and on the range via $\iota(S_{ij})$. If π acts on $\mathcal{G}(p)$ via ι and on $\mathcal{C}(p)$ via $\iota(S_{ij})$, then λ_{ij} is π-equivariant. If π acts on $\mathcal{G}(p)^p$ by cyclic permutations and acts diagonally on $\mathcal{C}(p)\times\mathcal{G}(p)^p$, then ψ_{ij} and ϕ are π-equivariant. Thus the upper left triangle is π-equivariantly homotopy commutative since $\mathcal{G}(p)$ is π-free and $\mathcal{C}(p)\times\mathcal{G}(p)^p$ is contractible. The bottom part of the diagram commutes since ξ_p is Σ_p-equivariant and since, for $g\in\mathcal{G}(p)$ and $(x_1,x_2)\in X^2$,

$$\xi(S_{ij})(1\times\Delta)(g,x_1,x_2) = (\lambda_{ij}(g), \underset{I\in S_{ij}}{\times} \xi_p(g,x_{i_1},\ldots,x_{i_p}))$$

while $(g,x_1,x_2) \to (\lambda_{ij}(g), \alpha_{ij}(\overset{p}{\underset{k=1}{\times}} \xi_p(g\sigma^k\tau_j, x_1^{p-i}, x_2^i))$ under the composite through the center of the diagram. If $j: W \to C_*\mathcal{G}(p)$ and $j': W \to C_*\mathcal{E}(p)$ are any two morphisms of π-complexes over Z_p (as used in the proof of Theorem I.1.1), then $(C_*\lambda_{ij})\circ j$ is π-homotopic to j' by elementary homological algebra. Thus, when we pass to π-equivariant homology, we may ignore λ_{ij}. By the diagram and by [A, Proposition 9.1], which evaluates $(1\times\Delta)_*: H_*(\pi; H_*X) \to H_*(\pi; H_*X^p)$ for any space X, we conclude (by induction on the degree of $x\otimes y$) that

$$\tilde{\xi}(S_{ij})_*(e_r\otimes(x\otimes y)^p) = e_r \otimes \sum\sum (x^{(1)}\ldots x^{(p-i)}y^{(1)}\ldots y^{(i)})^p .$$

By [A, Lemma 1.3], $H_*(\pi; H_*(X^2)^p)$ is generated as a Z_p-space by classes of the forms $e_r\otimes(x\otimes y)^p$ and $e_0\otimes x_1\otimes y_1\otimes\ldots\otimes x_p\otimes y_p$. The latter classes are clearly irrelevant to our formulas. The proof of our claim, and of the theorem, are now complete.

In $H_*\tilde{F}$, the computation of the operations $\tilde{Q}^s$ therefore reduces to their computation on generators under the loop product and thus, by the ordinary Cartan formula, to their computation on generators under both products. We have the following lemma (as always, for any E_∞ ring space X).

Lemma 2.6. (i) $\tilde{Q}^s[0] = 0$ and $\tilde{Q}^s[1] = 0$ for all p and all $s > 0$.

(ii) If $\pi_0 X$ is a ring and $p > 2$, then $\tilde{Q}^s[-1] = 0$ for all $s > 0$.

(iii) If $\pi_0 X$ is a ring and $p = 2$, then $\tilde{Q}^s[-1] = Q^s[1] * [-1]$, $s \geq 0$.

Proof. $\tilde{Q}^s[1] = 0$ for $s > 0$ by Theorem I.1.1(5) and $\tilde{Q}^s[0] = 0$ for $s > 0$ since $\xi_p(g, (0)^p) = 0$ for all $g \in \mathcal{G}(p)$. Now assume that

$\pi_0 X$ is a ring. Let $p > 2$ and assume that $\tilde{Q}^k[-1] = 0$ for $0 < k < s$ (a vacuous assumption if $s = 1$). Then, by the Cartan formula and the fact that $\tilde{Q}^0[-1] = [-1]^p = [-1]$, we have

$$0 = \tilde{Q}^s[1] = \tilde{Q}^s([-1].[-1]) = (\tilde{Q}^s[-1])[-1] + [-1](\tilde{Q}^s[-1]) = 2[-1]\tilde{Q}^s[-1].$$

$\tilde{Q}^s[-1] = 0$ follows. Finally, let $p = 2$, observe that $\tilde{Q}^0[-1] = [1]$, and assume that $\tilde{Q}^k[-1] = Q^k[1]*[-1]$ for $0 \leq k < s$. Then, by the mixed Cartan formula, by $Q^j[-1] = \chi Q^j[1]$, and by the defining formula for χ, we have

$$\begin{aligned} 0 &= \tilde{Q}^s[0] = \tilde{Q}^s([-1]*[1]) = \sum_{i=0}^{s} \tilde{Q}^i[-1]*Q^{s-i}[-1]*[1] \\ &= \tilde{Q}^s[-1]*[-1] + \sum_{i=0}^{s-1} Q^i[1]*\chi Q^{s-i}[1] = \tilde{Q}^s[-1]*[-1] + Q^s[1]*[-2]. \end{aligned}$$

$\tilde{Q}^s[-1] = Q^s[1]*[-1]$ follows.

The following implication of Theorem 2.5 is due to Madsen [15].

<u>Lemma 2.7.</u> If $\pi_0 X$ is a ring and $p = 2$, then, for $s > 0$,

$$\tilde{Q}^{2s+1}(Q^s Q^s[1]*[-3]) = 0.$$

<u>Proof.</u> $Q^{2s+1}Q^s = 0$ and $\tilde{Q}^{2s+1}\tilde{Q}^s = 0$ by the Adem relations, and $\tilde{Q}^s Q^s[1] = Q^s Q^s[1]$ by Lemma 1.9. An easy calculation from the mixed Cartan formula and the lemma above shows that $\tilde{Q}^1[-3] = 0$. Each summand of $\tilde{Q}^{2s+1}(Q^s Q^s[1]*[-3])$, as evaluated by the mixed Cartan formula, has a $*$ factor of one of the forms $\tilde{Q}^{2i+1}\tilde{Q}^i Q^i[1]$, $Q^{2i+1}Q^i Q^i[-3]$, or $\tilde{Q}^1[-3]$ (or else is zero by (3) of Theorem I.1.1 The result follows.

Finally, we record the following consequence of Theorem 2.5 for use in [R, VIII §4], where it plays a key role in the proof that SF splits as an infinite loop space when localized at any odd prime .

Lemma 2.8. For any $r > 0$ and $s > 0$, $\tilde{Q}^s[r]$ lies in the subalgebra of H_*X generated under $*$ by $[1]$ and the $Q^t[1]$ and, modulo elements decomposable as linear combinations of $*$-products of positive degree elements,

$$\tilde{Q}^s[r] \equiv \frac{1}{p}(r^p - r)Q^s[1]*[r^p - p].$$

Proof. The result holds trivially when $r = 1$ and we proceed by induction on r. The first part is evident and, inductively, Theorem 2.5 gives that

$$\tilde{Q}^s([r-1]*[1]) \equiv \frac{1}{p}((r-1)^p - (r-1))Q^s[1]*[r^p - p]$$
$$+ \sum_{i=1}^{p-1} [\frac{1}{p}(i, p-i)]Q^s[(r-1)^{p-i}]*[r^p - (r-1)^{p-i}(i, p-i)].$$

By Proposition 1.5(iii), the second term is congruent to

$$(\sum_{i=1}^{p-1} \frac{1}{p}(i, p-i)(r-1)^{p-i})Q^s[1]*[r^p - p].$$

Since $r^p = \sum_{i=0}^{p}(i, p-i)(r-1)^{p-i}$, the coefficient here is equal to $\frac{1}{p}(r^p - (r-1)^p - 1)$ and the conclusion follows.

§3. The mixed Adem relations

We shall first obtain precise (but incredibly complicated) formulas which implicitly determine $\tilde{Q}^r\beta^\varepsilon Q^s x$ by induction on the degree of x, $x \in H_*X$. We shall then derive simpler expressions in the case $x = [1]$. Modulo corrections arrived at in correspondence between us, the latter formulas are due to Tsuchiya [38].

The proofs will again be based on Proposition 2.2, and the following lemma will aid in the homological evaluation of certain of the maps $\tilde{\xi}(S)$.

Lemma 3.1. The following diagram is commutative for any subset S of $S(j;k)$:

$$\begin{array}{ccccc}
\mathcal{B}(k)\times\mathcal{C}(j)\times X & \xrightarrow{1\times 1\times\Delta} & \mathcal{B}(k)\times\mathcal{C}(j)\times X^j & \xrightarrow{1\times\Delta} & \mathcal{B}(k)\times(\mathcal{C}(j)\times X^j)^k \\
\downarrow{\scriptstyle \Delta\times 1\times 1} & & & & \downarrow{\scriptstyle \xi(S)} \\
\mathcal{B}(k)\times\mathcal{B}(k)\times\mathcal{C}(j)\times X & & & & \mathcal{C}(n(S))\times X^{n(S)} \\
\downarrow{\scriptstyle 1\times t\times 1} & & & & \uparrow{\scriptstyle 1\times\Delta} \\
\mathcal{B}(k)\times\mathcal{C}(j)\times\mathcal{B}(k)\times X & \xrightarrow{1\times\Delta\times 1\times\Delta} & \mathcal{B}(k)\times\mathcal{C}(j)^k\times\mathcal{B}(k)\times X^k & \xrightarrow{\lambda(S)\times\xi_k} & \mathcal{C}(n(S))\times X
\end{array}$$

Proof. For $x \in X$, each coordinate of $\delta(S)(x^{jk})$ is just x^k. The result follows trivially by inspection of Definition 2.1.

Let G be a subgroup of $\Sigma_k \times \Sigma_j \subset \Sigma_k \int \Sigma_j$ such that S is fixed under the action of G. The lemma reduces the evaluation of $\tilde{\xi}(S)_*$ on classes coming from the G-equivariant homology of $\mathcal{B}(k)\times\mathcal{C}(j)\times X$ to the analysis of the map $[\lambda(S)(1\times\Delta)]_*$ from the G-equivariant homology of $\mathcal{B}(k)\times\mathcal{C}(j)$ to the $\Sigma_{n(S)}$-equivariant homology of $\mathcal{C}(n(S))$, and this map clearly depends only on the homomorphism $\iota(S): G \to \Sigma_{n(S)}$ determined by the action of G on S.

In order to simplify the statement of the mixed Adem relations, we introduce some notations.

Definition 3.2. Define $\tilde{\mathcal{Q}}^r x = \Sigma\, \tilde{Q}^{r+k} P^k_* x$ for $x \in H_*X$ and $r \geq 0$. Observe that evaluation of the $\tilde{\mathcal{Q}}^r x$ is in principle equivalent to the evaluation of the $\tilde{Q}^r x$ in view of the equations

$$\tilde{Q}^r x = \sum_{n\geq 0}\sum_{j\geq 0} \tilde{Q}^{r+n} P^j_*(\chi P^{n-j})_*(x) = \sum_{k\geq 0} \tilde{\mathcal{Q}}^{r+k}(\chi P^k)_*(x),$$

where χ is the conjugation in the Steenrod algebra. Observe too that, by Proposition 1.6, the analogous operation $\mathcal{Q}^r x = \sum Q^{r+k} P^k_* x$ coincides with the # product $Q^r[1]x$.

Theorem 3.3 (The mixed Adem relations). Let $x \in H_*X$ and fix $r \geq 0$, $\varepsilon = 0$ or 1, and $s \geq \varepsilon$. Then $\tilde{Q}^r \beta^\varepsilon Q^s x$ is implicitly determined by the following formulas ($p > 2$ in (i)-(v)).

(i) $$\mathcal{Q}^r \beta^\varepsilon Q^s x = \sum\sum (-1)^\gamma \tilde{Q}_0^{r_0,\varepsilon_0,s_0} x' * \tilde{Q}_1^{r_1,\varepsilon_1,s_1} x'' * [p^{p-2}-1]\tilde{Q}_2^{r_2,\varepsilon_2,s_2} x''',$$

where $\psi x = \sum x' \times x'' \times x'''$, $\gamma = \varepsilon_1 \deg x' + \varepsilon_2(\deg x' + \deg x'')$, and the $(r_m, \varepsilon_m, s_m)$, $0 \leq m \leq 2$, range over those triples with $r_m \geq 0$, $\varepsilon_m = 0$ or 1, and $s_m \geq \varepsilon_m$ whose termwise sum is (r, ε, s). (Here each operation $\tilde{Q}_m^{r,\varepsilon,s}$ has degree $2(r+s)(p-1)+\varepsilon$.)

Define $\tilde{\mathcal{Q}}_m^{r,\varepsilon,s} x = \sum_{k\geq 0} \tilde{Q}_m^{r,\varepsilon,s+k} P^k_* x$ for $0 \leq m \leq 2$.

(ii) $$\tilde{\mathcal{Q}}_1^{r,\varepsilon,s} x = \sum\sum (-1)^\delta \tilde{Q}_{1,1}^{r_1,\varepsilon_1,s_1} x^{(1)} * \cdots * \tilde{Q}_{1,p-1}^{r_{p-1},\varepsilon_{p-1},s_{p-1}} x^{(p-1)},$$

where $\psi x = \sum x^{(1)} \otimes \cdots \otimes x^{(p-1)}$, $\delta = \sum_{i<j} \varepsilon_j \deg x^{(i)}$, and the $(r_n, \varepsilon_n, s_n)$, $1 \leq n \leq p-1$, range over those triples with $r_n \geq 0$,

$\varepsilon_n = 0$ or 1, $s_n \geq \varepsilon_n$, and $r_n + s_n \equiv 0 \mod (p-1)$ whose termwise sum is $(r(p-1), \varepsilon, s(p-1))$. (Here each operation $\tilde{Q}^{r,\varepsilon,s}_{1,n}$ has degree $2(r+s)+\varepsilon$.)

Define $\tilde{\mathcal{L}}^{r,\varepsilon,s}_{1,n} x = \sum_{k\geq 0} Q^{r,\varepsilon,s+k(p-1)}_{1,n} P^k_* x$ for $1 \leq n \leq p-1$.

(iii) $\tilde{\mathcal{L}}^{r,0,s}_0 x = Q^s[1] \cdot \tilde{\mathcal{L}}^r x$ and

$$\tilde{\mathcal{L}}^{r,1,s}_0 x = \tilde{\mathcal{L}}^{r,0,s}_0 \beta x + \beta Q^s[1] \cdot \tilde{\mathcal{L}}^r x .$$

(iv) $\tilde{\mathcal{L}}^{r,0,s}_{1,n} x = \sum_{i\geq 0} n^r(r-i(p-1), s) Q^{t-i}[1] \cdot \tilde{\mathcal{L}}^i x$ and

$$\tilde{\mathcal{L}}^{r,1,s}_{1,n} x = \tilde{\mathcal{L}}^{r,0,s}_{1,n} \beta x + \sum_{i\geq 0} n^r(r-i(p-1), s-1)\beta Q^{t-i}[1] \cdot \tilde{\mathcal{L}}^i x ,$$

where $r+s = t(p-1)$.

(v) $\tilde{\mathcal{L}}^{r,0,s}_2 x = \sum_{i\geq 0} Q^{r-i}[1] \cdot Q^s[1] \cdot \tilde{\mathcal{L}}^i x$ and

$$\tilde{\mathcal{L}}^{r,1,s}_2 x = \tilde{\mathcal{L}}^{r,0,s}_2 \beta x + \sum_{i\geq 0} Q^{r-i}[1] \cdot \beta Q^s[1] \cdot \tilde{\mathcal{L}}^i x .$$

(vi) If $p = 2$, $\tilde{\mathcal{L}}^r Q^s x = \sum_{(r_0, s_0)+(r_1, s_1) = (r,s)} \sum \tilde{Q}^{r_0, s_0}_0 x' * \tilde{Q}^{r_1, s_1}_1 x''$

where, if $\tilde{\mathcal{L}}^{r,s}_m x = \sum_{k\geq 0} \tilde{Q}^{r,s+k}_m P^k_* x$ for $m = 0$ or 1, then

$$\tilde{\mathcal{L}}^{r,s}_0 x = Q^s[1] \cdot \tilde{\mathcal{L}}^r x \quad \text{and} \quad \tilde{\mathcal{L}}^{r,s}_1 x = \sum_{i\geq 0} (r-i, s) Q^{r+s-i}[1] \cdot \tilde{\mathcal{L}}^i x .$$

<u>Proof.</u> Since $[p^{p-2} - 1] = [0]$ if $p = 2$, (vi) can be viewed as the special case $p = 2$ and $\varepsilon = 0$ of (i) through (iv). We shall use the distributivity diagram and Proposition 2.2, with $k = j_r = j = p$, and we must analyze the set $S(p; p)$ of sequences $I = \{i_1, \ldots, i_p\}$, $1 \leq i_r \leq p$. Let U de-

note the set of all sequences $J = \{j_1, \ldots, j_p\}$ such that $0 \leq j_k \leq p$ and $j_1 + \ldots + j_p = p$. For $I \in S(p;p)$, define $J(I) \in U$ by letting the kth entry of $J(I)$, $1 \leq k \leq p$, be the number of entries of I whose value is k (that is, the number of r such that $i_r = k$). Let π and ν be cyclic groups of order p with generators σ and τ. Embed ν as the diagonal in ν^p and embed π and ν in (copies of) Σ_p in the standard way as cyclic permutations. These embeddings fix inclusions

$$\pi \times \nu \subset \Sigma_p \times \nu \subset \Sigma_p \int \nu \subset \Sigma_p \int \Sigma_p$$

of subgroups of Σ_{p^p}, where Σ_{p^p} acts as permutations of $S(p;p)$. Thus σ acts on sequences $I \in S(p;p)$ by cyclic permutation of the entries and τ acts diagonally, adding one to each entry. Let τ act on sequences $J \in U$ by cyclic permutation of the entries and observe that $J(\tau I) = \tau J(I)$. Let T be a subset of U obtained by choosing one sequence in each orbit under the action of ν ; we insist that T contain the particular sequences

$$J_0 = \{p, 0, \ldots, 0\} \text{ and } J_1 = \{1, \ldots, 1\}.$$

For $J \in T$, define $S_J = \{I \mid J(I) = \tau^k J \text{ for some } k,\ 0 \leq k < p\}$. Since permutations of the entries of I do not change $J(I)$, each S_J is closed under the action of $\Sigma_p \times \nu$. Obviously $S(p;p)$ is the disjoint union of the S_J. For $J = \{j_1, \ldots, j_p\} \in T$, define $I(J) \in S_J$ by

$$I(J) = \{1, \ldots, 1, 2, \ldots, 2, \ldots, p, \ldots, p\}, \text{ where } k \text{ appears } j_k \text{ times.}$$

We next break the S_J into smaller subsets which are still closed under the action of $\pi \times \nu$. First, consider the possibility $\sigma I = \tau^n I$ for $I = \{i_1, \ldots, i_p\} \in S(p;p)$ and some n. Then

$$\{i_p, i_1, \ldots, i_{p-1}\} = \{i_1 + n, \ldots, i_p + n\}, \text{ hence } i_r + n = i_{r-1}.$$

Here $n = 0$ if and only if $J(I) \in \nu \cdot J_0$; we agree to write $S_0 = S_{J_0}$. Thus

$$S_0 = \{\tau^k I(J_0) \mid 1 \leq k \leq p\} .$$

If $1 \leq n \leq p-1$, then all entries of I are distinct and $J(I) = J_1$. Define $\gamma_n \in \Sigma_p$ by $\gamma_n^{-1}(i) = (p+1-i)n$. Then $\gamma_n I(J_1) = \{pn, (p-1)n, \ldots, n\}$. Define

$$S_{1,n} = \{\tau^k \gamma_n I(J_1) \mid 1 \leq k \leq p\} \quad , \quad 1 \leq n \leq p-1 ,$$

and note that

$$S_1 = \bigcup_{n=1}^{p-1} S_{1,n} = \{I \mid J(I) = J_1 \text{ and } \pi I = \nu I\}$$

Each $S_{1,n}$, hence also S_1, is closed under the action of $\pi \times \nu$ since $\sigma\gamma_n I(J_1) = \tau^n \gamma_n I(J_1)$. Clearly the complement S_1' of S_1 in S_{J_1} is also closed under the action of $\pi\times\nu$. Define

$$S_2 = S_1' \cup \Big(\bigcup_{J \in T, J \neq J_0, J \neq J_1} S_J \Big) = S(p;p) - (S_0 \cup S_1).$$

Note that $\pi\times\nu$ acts freely on S_2. Choose a subset $\{\phi_q(J_1)\}$ of Σ_p such that $\{\phi_q(J_1)\cdot I(J_1)\}$ is a $\pi\times\nu$-basis for S_1' (the $\phi_q(J_1)$ may be chosen from among a set of left coset representatives for π in Σ_p). Similarly, for $J \in T$, $J \neq J_0$ and $J \neq J_1$, let $\{\phi_q(J)\}$ be a set of double coset representatives for π and $\Sigma_{j_1} \times \ldots \times \Sigma_{j_p}$ in Σ_p; thus $\Sigma_p = \bigcup_q \pi\phi_q(J)(\Sigma_{j_1} \times \ldots \times \Sigma_{j_p})$. Note that, for any $\phi \in \Sigma_p$, the group $\phi^{-1}\pi\phi \cap (\Sigma_{j_1} \times \ldots \times \Sigma_{j_p})$ is trivial (since $j_k < p$ for all k) and there are $p = [\phi^{-1}\pi\phi : \{e\}]$ complete right cosets of $\Sigma_{j_1} \times \ldots \times \Sigma_{j_p}$ in the double coset $\pi\phi(\Sigma_{j_1} \times \ldots \times \Sigma_{j_p})$; thus precisely $\frac{1}{p}(j_1, \ldots, j_p)$ double cosets are required, where $(j_1, \ldots, j_p)$ denotes the multinomial coefficient. Observe that $\Sigma_{j_1} \times \ldots \times \Sigma_{j_p}$ acts trivially on each $\tau^k I(J)$ and define

$$S_{J,q} = \{\sigma^j \tau^k \phi_q(J)\cdot I(J) \mid 1 \le j \le p,\ 1 \le k \le p\}$$

($J = J_1$ is allowed here). Clearly S_J (or S_1' if $J = J_1$) is the disjoint union of the sets $S_{J,q}$, and each $S_{J,q}$ is closed under the action of $\pi \times \nu$. By Proposition 2.2, we now have the following three $\pi\times\nu$-equivariantly homotopy commutative diagrams:

$$\begin{array}{ccccccc}
\mathcal{B}(p)\times\mathcal{C}(p)\times X^p & \xrightarrow{1\times\Delta} & \mathcal{B}(p)\times(\mathcal{C}(p)\times X^p)^p & \xrightarrow{1\times\theta_p^p} & \mathcal{B}(p)\times X^p & \xrightarrow{\xi_p} & X \\
\downarrow{\scriptstyle\Delta} & & \downarrow{\scriptstyle\Delta} & & & & \uparrow{\scriptstyle *} \\
(\mathcal{B}(p)\times\mathcal{C}(p)\times X^p)^3 & \xrightarrow{(1\times\Delta)^3} & (\mathcal{B}(p)\times(\mathcal{C}(p)\times X^p)^p)^3 & & \xrightarrow{\tilde\xi(S_0)\times\tilde\xi(S_1)\times\tilde\xi(S_2)} & & X^3
\end{array}$$

$$\begin{array}{ccccc}
\mathcal{B}(p)\times\mathcal{C}(p)\times X^p & \xrightarrow{1\times\Delta} & \mathcal{B}(p)\times(\mathcal{C}(p)\times X^p)^p & \xrightarrow{\tilde\xi(S_1)} & X \\
\downarrow{\scriptstyle\Delta} & & \downarrow{\scriptstyle\Delta} & & \uparrow{\scriptstyle *} \\
(\mathcal{B}(p)\times\mathcal{C}(p)\times X^p)^{p-1} & \xrightarrow{(1\times\Delta)^{p-1}} & (\mathcal{B}(p)\times(\mathcal{C}(p)\times X^p)^p)^{p-1} & \xrightarrow{\overset{p-1}{\underset{n=1}{\times}}\tilde\xi(S_{1,n})} & X^{p-1}
\end{array}$$

$$\begin{array}{ccccc}
\mathcal{B}(p)\times\mathcal{C}(p)\times X^p & \xrightarrow{1\times\Delta} & \mathcal{B}(p)\times(\mathcal{C}(p)\times X^p)^p & \xrightarrow{\tilde\xi(S_2)} & X \\
\downarrow{\scriptstyle\Delta} & & \downarrow{\scriptstyle\Delta} & & \uparrow{\scriptstyle *} \\
(\mathcal{B}(p)\times\mathcal{C}(p)\times X^p)^{p^{p-2}-1} & \xrightarrow{(1\times\Delta)^{p^{p-2}-1}} & (\mathcal{B}(p)\times(\mathcal{C}(p)\times X^p)^p)^{p^{p-2}-1} & \xrightarrow{\underset{J,q}{\times}\tilde\xi(S_{J,q})} & X^{p^{p-2}-1}
\end{array}$$

(Here $n(S_2) = p^p - p^2$ since $n(S_0) = p$ and $n(S_1) = p(p-1)$; compare $p^p = \sum_{J\in U} (j_1, \ldots, j_p)$.) In each case, the left-hand square is required in order to obtain explicit formulas since we only have $\pi\times\nu$-equivariance in the right- hand square. The left-hand vertical maps Δ_* on $\pi\times\nu$-equivariant homology are easily computed by use of Lemma 2.4. The map $(1\times\Delta)_*$ is explicitly computed in [A, Proposition 9.1]. For $x \in H_q X$, define classes

$$f_{r,\varepsilon,s}(x) = (-1)^{r+s}\nu(q)e_{2r(p-1)} \otimes e_{(2s-q)(p-1)-\varepsilon} \otimes x^p$$

$$[f_{r,s}(x) = e_r \otimes e_{s-q} \otimes x^2 \quad \text{if } p = 2]$$

and $\quad g_{r,\varepsilon,s}(x) = (-1)^t \nu(q) e_{2r} \otimes e_{2s-q(p-1)-\varepsilon} \otimes x^p \quad$ if $\quad r+s = t(p-1)$

in $H_*(\pi; H_*(\nu; (H_*X)^p))$. Then define

$$\tilde{Q}_0^{r,\varepsilon,s}(x) = \tilde{\xi}(S_0)_*(1\times\Delta)_* f_{r,\varepsilon,s}(x) \qquad [\tilde{Q}_0^{r,s}(x) = \tilde{\xi}(S_0)_*(1\times\Delta)_* f_{r,s}(x)],$$

$$\tilde{Q}_1^{r,\varepsilon,s}(x) = \tilde{\xi}(S_1)_*(1\times\Delta)_* f_{r,\varepsilon,s}(x) \qquad [\tilde{Q}_1^{r,s}(x) = \tilde{\xi}(S_1)_*(1\times\Delta)_* f_{r,s}(x)],$$

$$\tilde{Q}_{1,n}^{r,\varepsilon,s}(x) = \tilde{\xi}(S_{1,n})_*(1\times\Delta)_* g_{r,\varepsilon,s}(x) , \quad \text{and}$$

$$\tilde{Q}_2^{r,\varepsilon,s}(x) = \tilde{\xi}(S_{J,q})_*(1\times\Delta)_* f_{r,\varepsilon,s}(x) .$$

We claim that, for $p > 2$ and $y = e_{2i} \otimes e_j \otimes x^p$, $\tilde{\xi}(S_0)_*(1\times\Delta)_*(y) = 0$ and $\tilde{\xi}(S_{J,q})_*(1\times\Delta)_*(y) = 0$ unless y is a multiple of some $f_{r,\varepsilon,s}(x)$ and $\tilde{\xi}(S_{1,n})_*(y) = 0$ unless y is a multiple of some $g_{r,\varepsilon,s}(x)$. We also claim that the $\tilde{Q}_2^{r,\varepsilon,s}(x)$ are indeed well-defined, in the sense that the same operations are obtained as J and q vary, and that formulas (iii), (iv), and (v) hold. Formulas (i) and (ii) will follow by use of Lemma 2.3 and chases of the three diagrams above. Thus it remains to evaluate the maps $\tilde{\xi}(S_0)_*$, $\tilde{\xi}(S_{1,n})_*$, and $\tilde{\xi}(S_{J,q})_*$. In each case, we shall rely on Lemma 3.1; we shall ignore those classes of $H_*(\mathcal{B}(p)\times_\pi(\mathcal{C}(p)\times_\nu X^p)^p)$ which are not in the image of $H_*(\mathcal{B}(p)\times_\pi(\mathcal{C}(p)\times_\nu X))$ under $(1\times\Delta)_*(1\times 1\times\Delta)_*$ since these classes are clearly irrelevant to our formulas. First, consider S_0; since

σ acts trivially on $I(J_0) = \{1, \dots, 1\}$, the homomorphism $\iota(S_0)\colon \pi \times \nu \to \Sigma_p$ determined by the action of $\pi \times \nu$ on S_0 is just the composite of the projection $\pi \times \nu \to \nu$ and the inclusion of ν in Σ_p. Therefore $\lambda(S_0)(1 \times \Delta)\colon \mathcal{B}(p) \times \mathcal{C}(p) \to \mathcal{C}(p)$ is $\pi \times \nu$-equivariantly homotopic to the projection on the second factor. By Lemma 3.1, we conclude that the following diagram is $\pi \times \nu$-equivariantly homotopy commutative:

$$\begin{array}{ccccccc}
\mathcal{B}(p) \times \mathcal{C}(p) \times X & \xrightarrow{1 \times 1 \times \Delta} & \mathcal{B}(p) \times \mathcal{C}(p) \times X^p & \xrightarrow{1 \times \Delta} & \mathcal{B}(p) \times (\mathcal{C}(p) \times X^p)^p & \xrightarrow{\tilde{\xi}_0(S_0)} & X \\
\Big\downarrow {\scriptstyle t \times 1} & & & & & & \Big\uparrow {\scriptstyle \theta_p} \\
\mathcal{C}(p) \times \mathcal{B}(p) \times X & & & & & & \mathcal{C}(p) \times X^p \\
\Big\downarrow {\scriptstyle 1 \times 1 \times \Delta} & & & & & & \Big\uparrow {\scriptstyle 1 \times \Delta} \\
\mathcal{C}(p) \times \mathcal{B}(p) \times X^p & & & \xrightarrow{1 \times \xi_p} & & & \mathcal{C}(p) \times X
\end{array}$$

Formula (iii) follows by a chase of the resulting diagram on $\pi \times \nu$-equivariant homology, starting with the element $e_{2r(p-1)} \otimes e_{2s(p-1)-\varepsilon} \otimes x$ (or $e_r \otimes e_s \otimes x$ if $p = 2$) and applying [A, Proposition 9.1] to evaluate maps $(1 \times \Delta)_*$. When $p > 2$, the vanishing of $\xi(S_0)_*(1 \times \Delta)_*$ on classes $e_{2i} \otimes e_j \otimes x^p$ which are not multiples of some $f_{r,\varepsilon,s}(x)$ follows from the $\Sigma_p \times \Sigma_p$-equivariance of all maps other than $\tilde{\xi}(S_0)$ in the diagram. Next, consider $S_{1,n}$, $1 \le n \le p-1$; $\iota(S_{1,n})\colon \pi \times \nu \to \Sigma_p$ is the composite

$$\pi \times \nu \xrightarrow{\chi_n \times 1} \nu \times \nu \xrightarrow{\phi} \nu \subset \Sigma_p,$$

where $\chi_n(\sigma^i) = \tau^{ni}$ and ϕ is the multiplication of ν. Therefore $\lambda(S_{1,n})(1 \times \Delta)\colon \mathcal{B}(p) \times \mathcal{C}(p) \to \mathcal{C}(p)$ is $\pi \times \nu$-equivariantly homotopic to the composite

$$\mathcal{B}(p) \times \mathcal{C}(p) \xrightarrow{\hat{\chi}_n \times 1} \mathcal{C}(p) \times \mathcal{C}(p) \xrightarrow{\hat{\phi}} \mathcal{C}(p),$$

where $\hat{\chi}_n$ and $\hat{\phi}$ are any χ_n and ϕ equivariant maps. By Lemma 3.1, we conclude that the following diagram is $\pi \times \nu$-equivariantly homotopy commutative:

$$
\begin{array}{ccccccc}
\mathcal{B}(p) \times \mathcal{C}(p) \times X & \xrightarrow{1\times 1\times \Delta} & \mathcal{B}(p)\times \mathcal{C}(p)\times X^p & \xrightarrow{1\times\Delta} & \mathcal{B}(p)\times(\mathcal{C}(p)\times X^p)^p & \xrightarrow{\tilde{\xi}(S_{1,n})} & X \\
\Big\downarrow {\scriptstyle (1\times t\times 1)(\Delta\times 1\times 1)} & & & & & & \Big\uparrow {\scriptstyle \theta_p} \\
\mathcal{B}(p)\times\mathcal{C}(p)\times\mathcal{B}(p)\times X & & & & & & \mathcal{C}(p)\times X^p \\
\Big\downarrow {\scriptstyle 1\times 1\times 1\times\Delta} & & & & & & \Big\uparrow {\scriptstyle 1\times\Delta} \\
\mathcal{B}(p)\times\mathcal{C}(p)\times\mathcal{B}(p)\times X^p & \xrightarrow{1\times 1\times \xi_p} & \mathcal{B}(p)\times\mathcal{C}(p)\times X & & \xrightarrow{\hat{\phi}(\hat{\chi}_n\times 1)\times 1} & & \mathcal{C}(p)\times X
\end{array}
$$

Formula (iv) and our vanishing claim when $p > 2$ follow by chases of the resulting diagram on $\pi\times\nu$-equivariant homology. When $p > 2$, the key facts are that $\hat{\chi}_{n*}(e_{2i-\varepsilon}) = n^i e_{2i-\varepsilon}$ for $1 \leq n \leq p-1$ (and all $i \geq 0$, $\varepsilon = 0$ or 1) by the proof of [A, Lemma 1.4], and that

$$\hat{\phi}_*(e_{2i} \otimes e_{2j-\varepsilon}) = (i, j-\varepsilon)e_{2i+2j-\varepsilon}$$

since $H_*\nu$ is the tensor product $\Gamma(e_2)\otimes E(e_1)$, with $e_{2i} = \gamma_i(e_2)$ and $e_{2i+1} = e_{2i}e_1$; when $p = 2$, $\chi_{1*} = 1$ and $\hat{\phi}_*(e_i \otimes e_j) = (i,j)e_{i+j}$. Finally, consider any $S_{J,q}$; $\iota(S_{J,q}): \pi\times\nu \to \Sigma_{p^2}$ is determined by the action of $\pi\times\nu$ on $S_{J,q} \subset S(p;p)$. If we could identify $S_{J,q}$ with $\{(j,k)\}$ via $\sigma^j\tau^k\phi_q(J)I(J) \to (j,k)$, then $\iota(S_{J,q})$ would coincide with the standard composite inclusion

$$\pi\times\nu \subset \pi\textstyle\int\nu^p \subset \Sigma_{p^2}$$

(see [A, p. 172-173]). Actually, the given ordering of $S_{J,q}$ as a subset of $S(p;p)$ may differ from the lexicographic ordering of $\{(j,k)\}$, hence $\iota(S_{J,q})$ may differ from the specified composite by an inner automorphism, given by

$\alpha_{J,q}$ say, of Σ_{p^2}. It follows that $\lambda(S_{J,q})(1\times\Delta)\colon \mathcal{B}(p)\times\mathcal{C}(p)\to\mathcal{C}(p^2)$ is $\pi\times\nu$-equivariantly homotopic to the composite

$$\mathcal{B}(p)\times\mathcal{C}(p)\xrightarrow{\hat{\chi}\times 1}\mathcal{C}(p)\times\mathcal{C}(p)\xrightarrow{1\times\Delta}\mathcal{C}(p)\times\mathcal{C}(p)^p\xrightarrow{\gamma}\mathcal{C}(p^2)\xrightarrow{\hat{\alpha}_{J,q}}\mathcal{C}(p^2)$$

where γ is the operad structure map, $\chi = \chi_1\colon \pi\to\nu$ is the isomorphism $\chi\sigma^i = \tau^i$, $\hat{\chi}$ is any χ-equivariant map, and $\hat{\alpha}_{J,q}$ is right action by $\alpha_{J,q}$. By Lemma 3.1, we conclude that the following diagram is $\pi\times\nu$-equivariantly commutative:

$$\begin{array}{ccccccc}
\mathcal{B}(p)\times\mathcal{C}(p)\times X & \xrightarrow{1\times 1\times\Delta} & \mathcal{B}(p)\times\mathcal{C}(p)\times X^p & \xrightarrow{1\times\Delta} & \mathcal{B}(p)\times(\mathcal{C}(p)\times X^p)^p & \xrightarrow{\tilde{\xi}(S_{J,q})} & X \\
\Big\downarrow{\scriptstyle (1\times t\times 1)(\Delta\times 1\times 1)} & & & & & & \Big\uparrow{\scriptstyle \theta_{p^2}} \\
\mathcal{B}(p)\times\mathcal{C}(p)\times\mathcal{B}(p)\times X & & & & & & \mathcal{C}(p^2)\times X^{p^2} \\
\Big\downarrow{\scriptstyle 1\times 1\times 1\times\Delta} & & & & & & \Big\uparrow{\scriptstyle 1\times\Delta} \\
\mathcal{B}(p)\times\mathcal{C}(p)\times\mathcal{B}(p)\times X^p & \xrightarrow{\hat{\chi}\times 1\times\xi_p} & \mathcal{C}(p)\times\mathcal{C}(p)\times X & & \xrightarrow{(\hat{\alpha}_{J,q}\gamma(1\times\Delta))\times 1} & & \mathcal{C}(p^2)\times X
\end{array}$$

Here, on the bottom right, we may pass to $\mathcal{C}(p^2)\times_{\Sigma_{p^2}} X$, where $\alpha_{J,q}\times 1$ is the identity, since $\theta_{p^2}(1\times\Delta)$ is Σ_{p^2}-equivalent. To evaluate the composite

$$\theta_{p^2}(1\times\Delta)(\gamma\times 1)(1\times\Delta\times 1)\colon \mathcal{C}(p)\times\mathcal{C}(p)\times X\to X,$$

observe that, by the definition of an action of an operad on a space (see [G, Lemma 1.4]) and by a trivial diagram chase, this composite coincides with the composite

$$\mathcal{C}(p)\times\mathcal{C}(p)\times X\xrightarrow{1\times 1\times\Delta}\mathcal{C}(p)\times\mathcal{C}(p)\times X^p\xrightarrow{1\times\theta_p}\mathcal{C}(p)\times X\xrightarrow{1\times\Delta}\mathcal{C}(p)\times X^p\xrightarrow{\theta_p}X.$$

Now formula (v) and our vanishing claim follow by simple diagram chases. The proof of the theorem is complete.

Remark 3.4. In [38, 3.14], Tsuchiya stated without proof an explicit, rather than implicit, formula for the evaluation of $\tilde{Q}^r Q^s x$ when $p = 2$. The formula appears to be incorrect, and Tsuchiya's unpublished proof contains an error stemming from a subtle difficulty with indices of summation.

While the full strength of the theorem is necessary (and, with the mixed Cartan formula, sufficient) to determine the $\tilde{Q}^r$ on $H_*Q(X^+)$ for a general E_∞ space X, our results greatly overdetermine the operations $\tilde{Q}^r$ on H_*QS^0: all that was absolutely necessary was a knowledge of $\tilde{Q}^r \beta^\varepsilon Q^s[1]$. The Nishida relations, Theorem I.1.1(9), can be used to derive simpler expressions for these operations.

Corollary 3.5. Let $p = 2$ and fix $r > s \geq 0$. Then

$$\tilde{Q}^r Q^s[1] = \sum_{j=0}^{s} (r-s-1, s-j) Q^j[1] * Q^{r+s-j}[1].$$

Proof. We have $\sum_k (k, s-2k) \tilde{Q}^{r+k} Q^{s-k}[1] = \sum_j (r, s-j) Q^j[1] * Q^{r+s-j}[1]$ since $\tilde{Q}_0^{0,s}[1] = Q^s[1]$, $\tilde{Q}_0^{r,s}[1] = 0$ if $r > 0$, and $\tilde{Q}_1^{r,s}[1] = (r,s) Q^{r+s}[1]$. By induction on s, it follows that if $r \geq s \geq 0$, then

$$(1) \qquad \tilde{Q}^r Q^s[1] = \sum_{j=0}^{s} a_{rsj} Q^j[1] * Q^{r+s-j}[1],$$

where the constants a_{rsj} satisfy the formulas

$$(2) \qquad \sum_{k=0}^{s-j} (k, s-2k) a_{r+k, s-k, j} = (r, s-j), \quad 0 \leq j \leq s.$$

Visibly $\tilde{Q}^r Q^0[1] = [2] * Q^r[1]$, hence $a_{r00} = 1$ for $r \geq 0$. By induction on s, there is a unique solution of the equations (2) for $r > s$ which starts with

$a_{r00} = 1$. By Adem's summation formula [3, Theorem 25.3],

$$\sum_{k=0}^{s-j} (k, s-2k)(r-s-1+2k, s-j-k) = (r, s-j) \quad \text{for } 0 \leq j \leq s \quad \text{and } r > 0,$$

hence $a_{rsj} = (r-s-1, s-j)$ is this solution.

Remark 3.6. By the mixed Cartan formula and Lemma 2.6, we have

$$\tilde{Q}^r(Q^s[1]*[-1]) = \sum \tilde{Q}^{r_1} Q^{s_1}[1] * Q^{r_2} Q^{s_2}[-1] * Q^{r_3}[1]*[-1]$$

when $\pi_0 X$ is a ring. The $\tilde{Q}^r Q^s[1]$ are evaluated by the corollary and, when $r = s$, by Lemma 1.9.

For the case of odd primes, we need a lemma on binomial coefficients.

Lemma 3.7. The following identity holds for all $a \geq 1$ and $b \geq 0$:

$$\sum_{k \geq 0} (-1)^{k+b}(k, b-pk)(b-k(p-1), a-1-b+pk) \equiv (a(p-1), b) \bmod p.$$

Proof. Calculate in Z_p. When $a = 1$, both sides are one if $b \equiv 0$ (p) and are zero if $b \not\equiv 0$ (p). When $b < p$, the identity reads

$$(-1)^b(b, a-1-b) = (a(p-1), b),$$

and this is true since if $a \equiv a'$ (p), $1 \leq a' \leq p$, then both sides are zero if $1 \leq a' \leq b$ and are $(-1)^b(b, a'-1-b) = (p-a', b)$ if $b < a' \leq p$. For $a > 1$ and $b \geq p$, proceed by induction on a. By iterative use of $(i-1, j) + (i, j-1) = (i, j)$, we see that, for $n \geq 0$,

$$(a(p-1), b) = \sum_{m=0}^{n} (m, n-m)(a(p-1)-m, b-n+m).$$

Set $n = p$, where $(m, p-m) = 0$ for $0 < m < p$. By another use of $(i-1, j) + (i, j-1) = (i, j)$ and by the induction hypothesis,

$$(a(p-1), b) = (a(p-1), b-p) + (a(p-1)-p, b)$$

$$= (a(p-1), b-p) - ((a-1)(p-1), b-1) + ((a-1)(p-1), b)$$

$$= \sum_{j\geq 0} (-1)^{j+1+b}(j, b-p-pj)(b-p-j(p-1), a-1-b+p+pj)$$

$$+ \sum_{k\geq 0} (-1)^{k+b}(k, b-1-pk)(b-1-k(p-1), a-1-b+pk)$$

$$+ \sum_{k\geq 0} (-1)^{k+b}(k, b-pk)(b-k(p-1), a-2-b+pk)$$

$$= \sum_{k\geq 0} (-1)^{k+b}(k, b-pk)(b-k(p-1), a-1-b+pk) \quad .$$

Here the last equality is obtained by changing the dummy variable in the first sum to $k = j+1$, so that this sum becomes

$$\sum_{k\geq 1} (-1)^{k+b}(k-1, b-pk)(b-1-k(p-1), a-1-b+pk) ,$$

and then adding the first and second sums and adding the result to the third sum.

Corollary 3.8. Let $p > 2$ and fix $r \geq 1$, $\varepsilon = 0$ or 1, and $s \geq \varepsilon$. Then

$$\chi^r\beta^\varepsilon Q^s[1] = \sum (-1)^k (k, s(p-1) - pk - \varepsilon)\tilde{Q}^{r+k}\beta^\varepsilon Q^{s-k}[1]$$

$$= \sum 1^{r_1} 2^{r_2} \cdots (p-1)^{r_{p-1}} (r_1, s_1 - \varepsilon_1) \cdots (r_{p-1}, s_{p-1} - \varepsilon_{p-1})$$

$$\beta^{\varepsilon_0} Q^{s_0}[1] * \beta^{\varepsilon_1} Q^{t_1}[1] * \cdots * \beta^{\varepsilon_{p-1}} Q^{t_{p-1}} * [p^{p-2}-1] Q^{r_p}[1] \beta^{\varepsilon_p} Q^{s_p}[1]$$

summed over all triples $(0, \varepsilon_0, s_n)$ and $(r_n, \varepsilon_n, s_n)$, $1 \leq n \leq p$, with $r_n + s_n = t_n(p-1)$ for some t_n and with termwise sum

$$(0, \varepsilon_0, s_0(p-1)) + \sum_{n=1}^{p-1} (r_n, \varepsilon_n, s_n) + (r_p(p-1), \varepsilon_p, s_p(p-1)) = (r(p-1), \varepsilon, s(p-1)).$$

Moreover, modulo linear combinations of elements decomposable as $*$ products between positive degree elements of H_*X,

$$\tilde{Q}^r\beta^\varepsilon Q^s[1] \equiv -(-1)^\varepsilon (s(p-1) - \varepsilon, r - s(p-1) + \varepsilon - 1)\beta^\varepsilon Q^{r+s}[1] * [p^p - p]$$

$$- Q^r\beta^\varepsilon Q^s[1] * [p^p - p^2] .$$

Proof. The first statement is a direct consequence of the theorem and the Nishida relations. For the second statement, note that $[p^n-1]x \equiv -x*[(p^n-1)i]$ modulo $*$-decomposables if $x \in H_*X_i$, $n > 0$, and $\deg x > 0$. Therefore inspection of our formulas gives

$$\tilde{\mathcal{Q}}^r\beta^\varepsilon Q^s[1] \equiv -(r(p-1), s(p-1)-\varepsilon)\beta^\varepsilon Q^{r+s}[1]*[p^p-p] - Q^r[1]\beta^\varepsilon Q^s[1]*[p^p-p^2].$$

In view of Proposition 1.6, it follows by induction on s that

$$(1)\qquad \tilde{Q}^r\beta^\varepsilon Q^s[1] \equiv -a_{r\varepsilon s}\beta^\varepsilon Q^{r+s}[1]*[p^p-p] - Q^r\beta^\varepsilon Q^s[1]*[p^p-p^2],$$

where the constants $a_{r\varepsilon s}$ satisfy the formula

$$(2)\qquad \sum_{k=0}^{s}(-1)^k(k, s(p-1)-pk-\varepsilon)a_{r+k,\varepsilon,s-k} = (r(p-1), s(p-1)-\varepsilon).$$

We claim that $a_{r\varepsilon s} = (-1)^\varepsilon(s(p-1)-\varepsilon, r-s(p-1)+\varepsilon-1)$. Visibly

$$\tilde{Q}^r Q^0[1] \equiv -Q^r[1]*[p^p-p] - Q^r Q^0[1]*[p^p-p^2],$$

hence $a_{r00} = 1$. Just as visibly,

$$\tilde{Q}^r\beta Q^1[1] \equiv -(r(p-1), p-2)\beta Q^{r+1}[1]*[p^p-p] - Q^r\beta Q^1[1]*[p^p-p^2],$$

hence $a_{r11} = (r(p-1), p-2)$. Calculating in Z_p, we see that

$$(r(p-1),\ p-2) = \begin{cases} 1 & \text{if } r \equiv 0\ (p) \\ -1 & \text{if } r \equiv -1\ (p) \\ 0 & \text{otherwise} \end{cases},$$

which is in agreement with the claimed value $-(p-2, r-p+1)$. By induction on s, there is a unique solution for the $a_{r\varepsilon s}$ which agrees with the known values for a_{r00} and a_{r11}. By the lemma, with $a = r$ and $b = s(p-1)-\varepsilon$, our claimed values for the $a_{r\varepsilon s}$ do give this solution.

Remark 3.9. The mixed Cartan formula and the corollary imply that, when

$\pi_0 X$ is a ring,

$$\tilde{Q}^r(\beta^\varepsilon Q^s[1]*[1-p]) \equiv -(-1)^\varepsilon (s(p-1)-\varepsilon, r-s(p-1)+\varepsilon-1)\beta^\varepsilon Q^{r+s}[1]*[1-p]$$

modulo elements decomposable under the $\underline{*}$ product. The point is that, in the mixed Cartan formula, the term involving

$$\tilde{Q}^r_{p-1}(\beta^\varepsilon Q^s[1]\otimes[1-p]) = Q^r(\beta^\varepsilon Q^s[1]([1-p])^{p-1})$$

gives rise to a summand $Q^r\beta^\varepsilon Q^s[1]*[1-p^2]$ which cancels with the negative of the same summand which arises from $\tilde{Q}^r\beta^\varepsilon Q^s[1]$. (The terms which involve the $\tilde{Q}^r_i(\beta^\varepsilon Q^s[1]\otimes[1-p])$ for $1 \leq i < p-1$ are $\underline{*}$ decomposable by Proposition 6.5 (i) below.)

§4. The homology of $C(X^+)$ and $Q(X^+)$

Recall that we have assumed given a fixed pair of E_∞ operads $\mathcal{C}$ and $\mathcal{G}$ such that $\mathcal{G}$ acts on $\mathcal{C}$. Let X be a $\mathcal{G}$-space, with basepoint 1, and let X^+ be the union of X and a disjoint basepoint 0. Clearly X^+ is then a $\mathcal{G}_0$-space (Definition 1.3) with X as a sub $\mathcal{G}$-space. As pointed out in [R, VI §2], $C(X^+)$ is the free $(\mathcal{C}, \mathcal{G})$-space generated by the $\mathcal{G}_0$-space X^+. An obvious example is $X = \{1\}$, when $C(X^+) = CS^0$ is the disjoint union $\coprod_{j\geq 0} K(\Sigma_j, 1)$. This example generalizes to $X = G$, a discrete Abelian group, when $C(X^+) = \coprod_{j\geq 0} K(\Sigma_j \int G, 1)$ by I.5.7. Similarly, if $\mathcal{G}$ maps to the linear isometries operad $\mathcal{L}$ (as can always be arranged [R, IV 1.10]), then $Q(X^+)$ is the zeroth space of the free $\mathcal{G}$-spectrum $Q_\infty(X^+)$ [R, IV.1.8]. The example of greatest interest is QS^0, but the Kahn-Priddy theorem [10, 11] and its analog due to Segal [30] make it clear that $Q(RP^{\infty +})$ and $Q(CP^{\infty +})$ are also of considerable interest.

Define an allowable AR-Hopf bialgebra to be an allowable AR-Hopf algebra under two R-algebra structures $(*, Q^s)$ and $(\#, \tilde{Q}^s)$ such that the formulas of Proposition 1.5 (i), (iii), (iv), Proposition 1.6, Lemma 2.6(i), and the mixed Cartan formula and mixed Adem relations are satisfied. Such an object with (additive) conjugation χ is also required to satisfy Proposition 1.5(ii) (namely $\chi(x) = [-1] \# x$), and the formulas of Lemmas 1.9, 2.6(ii) and (iii), and 2.7 then follow.

Let K be an allowable AR-Hopf algebra with product #, unit [1], and operations $\tilde{Q}^s$. Construct $WE(K \oplus Z_p[0])$ and $GWE(K \oplus Z_p[0])$ by I§2 from the component A-coalgebra $K \oplus Z_p[0]$ with unit $\eta(1) = [0]$; the constructed R-algebra structure $(*, Q^s)$ is thought of as additive. The formulas cited above determine a unique extension of the product # and operations $\widetilde{Q}^s$ from K to all of $WE(K \oplus Z_p[0])$ and $GWE(K \oplus Z_p[0])$. Tedious formal verifications (omitted since they are automatic when K can be realized topologically) demonstrate that $WE(K \oplus Z_p[0])$ is a well-defined allowable AR-Hopf bialgebra and that $GWE(K \oplus Z_p[0])$ is a well-defined allowable AR-Hopf bialgebra with conjugation. Moreover, these are the free such structures generated by K.

Returning to our $\mathcal{G}$-space X, we see that freeness gives morphisms $\bar{\eta}_*$ of allowable AR-Hopf bialgebras and $\tilde{\eta}_*$ of allowable AR-Hopf bialgebras with conjugation such that the following diagrams are commutative:

$$\begin{array}{ccc} H_*X & \longrightarrow & WE(H_*X \oplus Z_p[0]) \\ \downarrow & & \downarrow \bar{\eta}_* \\ H_*X^+ & \xrightarrow{\ \eta_*\ } & H_*C(X^+) \end{array} \qquad \text{and} \qquad \begin{array}{ccc} H_*X & \longrightarrow & GWE(H_*X \oplus Z_p[0]) \\ \downarrow & & \downarrow \bar{\eta}_* \\ H_*X^+ & \xrightarrow{\ \eta_*\ } & H_*Q(X^+) \end{array}$$

The following pair of theorems are immediate consequences of I.4.1 and I.4.2. (In the second, we assume that $\mathcal{G}$ maps to $\mathcal{L}$.)

<u>Theorem 4.1.</u> For every $\mathcal{G}$-space $X, \bar{\eta}_*: WE(H_*X \oplus Z_p[0]) \to H_*C(X^+)$ is an isomorphism of AR-Hopf bialgebras.

<u>Theorem 4.2.</u> For every $\mathcal{G}$-space X, $\tilde{\eta}_*: GWE(H_*X \oplus Z_p[0]) \to H_*Q(X^+)$ is an isomorphism of AR-Hopf bialgebras with conjugation.

These results are simply conceptual reformulations of the observations that Propositions 1.5 and 1.6 completely determine # on $H_*C(X^+)$ and $H_*Q(X^+)$ from # on H_*X and that the mixed Cartan formula and mixed Adem relations completely determine the $\widetilde{Q}^s$ on $H_*C(X^+)$ and $H_*Q(X^+)$ from the $\widetilde{Q}^s$ on H_*X. In other words, if we are given a basis for H_*X and if we give $H_*C(X^+)$ and $H_*Q(X^+)$ the evident derived bases of *-monomials in degree zero elements and operations $Q^I x$ on the given basis elements $x \in H_*X$, then, in principle, for any basis elements y and z, our formulas determine $y \# z$ and the $\tilde{Q}^s y$ as linear combinations in the specified basis.

There are two obvious difficulties. The formulas for the computation of $y \# z$ and the $\tilde{Q}^s y$ are appallingly complicated, and the result they give is not a global description of $H_* \Omega(X^+)$ as an R-algebra under $\#$ and the $\tilde{Q}^s$. In practice, one wants to determine the homology of the component $Q_1(X^+)$ of the basepoint of X as an R-algebra, with minimal reference to $*$ and the Q^s. We shall only study this problem for $Q_1 S^0 = SF$, but it will be clear that the methods generalize.

In view of these remarks, the term AR-Hopf bialgebra should be regarded merely as a quick way of referring to the sort of algebraic structure possessed by the homology of E_∞ ring spaces. In the absence of an illuminating description of the free objects, the concept is of limited practical value. (The term Hopf ring has been used by other authors; this would be reasonable only if one were willing to rename Hopf algebras Hopf groups.)

§5. The homology of SF, F/O, and BSF

The results of §1 (or of I§1) completely determine H_*SF as an algebra. In this section, which is independent of §3 and §4, we shall analyze the sequence of Hopf algebras obtained by passage to mod p homology from the sequence of spaces

$$SO \xrightarrow{j} SF \xrightarrow{\tau} F/O \xrightarrow{q} BSO \xrightarrow{Bj} BSF$$

Here $j: SO \to SF$ is the natural inclusion, Bj is its classifying map, and τ and q are the natural maps obtained by letting F/O be the fibre of Bj. As explained in [R, I], these are all maps of $\mathcal{L}$-spaces, where $\mathcal{L}$ is the linear isometries operad, and thus of infinite loop spaces. When $p = 2$, a variant of this exposition has been presented in [5, §8]. We end with a detailed proof of the evaluation of the suspensions of the Stiefel-Whitney and, if $p > 2$, Wu classes on $\tilde{H}_*SF$.

Recall Definition I.2.1. For any admissible sequence with $d(I) > 0$, define

$$x_I = Q^I[1] * [1 - p^{\ell(I)}] \in H_*SF .$$

In particular, $x_s = Q^s[1] * [1-p]$ for $s \geq 1$.

For a graded set S, let AS, PS, and ES denote the free commutative, polynomial, and exterior algebras over Z_p generated by S. If $p = 2$, $AS = PS$. If $p > 2$, $AS = ES^- \otimes PS^+$ where S^- and S^+ are the odd and even degree parts of S.

The following theorem is due to Milgram [22] (except that (ii) and the algebra structure in (iii) are addenda due, respectively, to myself and Madsen [15]).

Theorem 5.1. The following conclusions hold in mod 2 homology.

(i) $H_*SF = E\{x_s\} \otimes AX$ as an algebra under #, where

$$X = \{x_I \mid \ell(I) > 2 \text{ and } e(I) > 0 \text{ or } \ell(I) = 2 \text{ and } e(I) \geq 0\}.$$

(ii) $\operatorname{Im} j_* = E\{x_s\}$ and $H_*F/O \cong H_*SF//j_* \cong AX$.

(iii) $H_*BSF = H_*BSO \otimes E\{\sigma_* x_{(s,s)}\} \otimes ABX$ as a Hopf algebra, where

$$BX = \{\sigma_* x_I \mid \ell(I) > 2 \text{ and } e(I) > 1 \text{ or } \ell(I) = 2 \text{ and } e(I) \geq 1\}.$$

(iv) $H_*BF = H_*BO \otimes E\{\sigma_* x_{(s,s)}\} \otimes ABX$ as a Hopf algebra.

Part (i) of the following theorem is due to Tsuchiya [36] and myself, independently, while (ii) is due to me and the first correct line of argument for (iii) is due to Tsuchiya. Recall from [1, p. 91] that, when localized at an odd prime p, BO splits as $W \times W^{\perp}$ as an infinite loop space, where the non-zero homotopy groups of W are $\pi_{2i(p-1)}W = Z_{(p)}$ and where H_*W is a polynomial algebra on generators of degree $2i(p-1)$, $i \geq 1$.

<u>Theorem 5.2.</u> The following conclusions hold in mod p homology, $p > 2$.

(i) $H_*SF = E\{\beta x_s\} \otimes P\{x_s\} \otimes AX$ has an algebra under #, where

$$X = \{x_I \mid \ell(I) \geq 2 \text{ and } e(I) + b(I) > 0\}.$$

(ii) $\operatorname{Im} j_* = E\{b_s\}$, where b_s is a primitive element of degree $2s(p-1)-1$ to be specified below, and

$$H_*F/O \cong H_*SF//j_* \otimes H_*BSO\backslash\backslash (Bj)_* \cong P\{x_s\} \otimes AX \otimes H_*W^{\perp}.$$

(iii) $H_*BSF = H_*BF = H_*W \otimes E\{\sigma_* x_s\} \otimes ABX$ as a Hopf algebra, where

$$BX = \{\sigma_* x_I \mid \ell(I) > 2 \text{ and } e(I) + b(I) > 1 \text{ or } \ell(I) = 2 \text{ and } e(I) + b(I) \geq 1\}.$$

Of course, the elements $\sigma_* y \in H_*BSF$ are primitive. The map j_* will be computed in Lemmas 5.8, 5.11, and 5.12, and the maps τ_*, q_*, and $(Bj)_*$ are as one would expect from the statements of the theorems. It is instructive to compare these results with those obtained for the * product on H_*QS^0 in I§4. Recall that $H_*SF = AX_0$ as an algebra under $\underline{*}$, where

$$X_0 = \{x_I \mid \ell(I) \geq 1 \quad \text{and} \quad e(I) + b(I) > 0\},$$

and that

$$H_*QS^1 = A\{\sigma_*(x_I * [-1]) \mid \ell(I) \geq 1 \text{ and } e(I) + b(I) > 1\}.$$

The following remarks show how Bj dictates the difference between H_*BSF and H_*QS^1.

<u>Remarks 5.3.</u> Let $p = 2$. Here X is obtained from X_0 by deleting the elements x_s and adjoining their squares, $x_{(s,s)}$, under the $\underline{*}$ product. Thus the appearance of the generators $x_{(s,s)}$ in H_*SF is forced by the relations $x_s^2 = 0$, which in turn are forced by the fact that H_*SO is an exterior algebra. Again, while

$$\sigma_*(x_{(s+1,s)} * [-1]) = \sigma_*(x_s * [-1])^2$$

in H_*QS^1, the squares of the elements $\sigma_* x_s$ in H_*BSF lie in H_*BSO and the elements $\sigma_* x_{(s+1,s)}$ are exceptional generators. This behavior may propagate. If, as could in principle be checked by use of the mixed Adem relations and the lemmas of the next section, $\tilde{Q}^{2s+2i+1} x_{(s+i,s)}$ is decomposable in H_*SF, then, for all $i \geq 0$, $\sigma_*^i x_{(s+i,s)}$ is an element of H_*B^iSF the suspension of which is an element of square zero in $H_*B^{i+1}SF$. The problem of calculating H_*B^iSF for $i \geq 2$ depends on the evaluation of differentials on divided polynomial algebras, which arise as torsion products of exterior algebras, in the E^2-terms of the relevant Eilenberg-Moore spectral sequences.

Remarks 5.4. Let $p > 2$. Here X is obtained from X_0 by deleting the elements $\beta^{\varepsilon} x_s = x_{(\varepsilon, s)}$, and the two Pontryagin algebra structures on H_*SF are abstractly isomorphic. In H_*QS^1,

$$\sigma_*(x_{(0, ps-s, 1, s)} * [-1]) = \sigma_*(\beta x_s * [-1])^p \text{ and } \sigma_*(x_{(1, 1s-s, 1, s)} * [-1]) = 0.$$

In H_*BSF, the pth powers of the elements $\sigma_*(b_s)$ lie in H_*W and the exceptional generators $\sigma_* x_I$ with $\ell(I) = 2$ and $e(I) + b(I) = 1$, namely those with $I = (\varepsilon, ps-s, 1, s)$, appear. Calculation of H_*B^iSF for $i \geq 2$ is less near here than in the case $p = 2$ because of the lesser precision in the results of the next section.

The following remarks may help clarify the structure of H_*BSF.

Remarks 5.5. For $n \geq 2$, let $X_n = \{x_I \mid \ell(I) = n\} \subset X$. Obviously $AX = \bigotimes_{n \geq 2} AX_n$ as an algebra. When $p = 2$, AX_2 is a sub Hopf algebra of AX. When $p > 2$ or when $p = 2$ and $n \geq 3$, AX_n is not a sub Hopf algebra of H_*SF because the coproduct on x_I can have summands $x_J \otimes x_K$ with $e(K) = b(K) = 0$; x_K is then a pth power in the $\underline{*}$ product and therefore, by Proposition 6.4 below, in the # product. (For this reason, [5, 8.12 and 8.16] are incorrect as stated.) Similarly, we can compute the Steenrod operations on elements of X, modulo elements which are p^{th} powers in both products, by use of the Nishida and Adem relations for the operations $Q^I[1]$ (compare I.3.8 and I.3.9). It follows that these relations completely determine the Steenrod operations in H_*BSF. Moreover, if

$$B_n X = \{\sigma_* x_I \mid \ell(I) = n\} \subset BX,$$

then, as a Hopf algebra over the Steenrod algebra,

$$H_*BSF = H_*BSO \otimes (E\{\sigma_* x_{(s,s)}\} \otimes AB_2X) \otimes \bigotimes_{n \geq 3} AB_nX \quad \text{if } p = 2$$

and

$$H_*BSF = H_*W \otimes \bigotimes_{n\geq 2} AB_nX \qquad \text{if } p > 2,$$

the point being that, when $p = 2$, $E\{\sigma_* x_{(s,s)}\} \otimes AB_2X$ and each AB_nX with $n \geq 3$, or, when $p > 2$, each AB_nX with $n \geq 2$ is a sub A-Hopf algebra of H_*BSF.

To begin the proofs, we define a weight function w on $H_*\widetilde{F}$ by

$$w[i] = 0, \quad wQ^I[1] = p^{\ell(I)} \quad \text{if} \quad d(I) > 0,$$

and, if $x \neq 0$ and $y \neq 0$,

$$w(x * y) = wx + wy \quad \text{and} \quad w(x + y) = \min(wx, wy).$$

It is easy to verify that w is well-defined. Clearly wx is divisible by p for all x, and we define a decreasing filtration on $H_*\widetilde{F}$ by

$$F_iH_*\widetilde{F} = \{x \mid wx \geq pi\} \quad \text{for } i \geq 0.$$

Define $E^0_{ij}H_*\widetilde{F} = (F_iH_*\widetilde{F}/F_{i+1}H_*\widetilde{F})_{i+j}$. Since the product $*$ is homogeneous with respect to w, $E^0H_*\widetilde{F}$ may be identified with $H_*\widetilde{F}$ as an algebra under $*$. Clearly ψ and χ are filtration preserving and reduce in $E^0H_*\widetilde{F}$ to

$$\psi Q^I[1] = Q^I[1] \otimes [p^{\ell(I)}] + [p^{\ell(I)}] \otimes Q^I[1]$$

and

$$\chi Q^I[1] = -Q^I[1] * [-2p^{\ell(I)}] .$$

<u>Lemma 5.6</u>. The product $\#$ is filtration preserving. In $E^0H_*\widetilde{F}$

$$(x * [i])(y * [j]) = j^r i^s x * y * [(k+i)(\ell+j) - (k+\ell)]$$

for $x = Q^{I_1}[1] * \ldots * Q^{I_r}[1] \in E^0H_*\widetilde{F}_k$ and $y = Q^{J_1}[1] * \ldots * Q^{J_s}[1] \in E^0H_*\widetilde{F}_\ell$, where $d(I_n) > 0$, $d(J_n) > 0$, and, if $p = 2$, either $\ell(I_n) \geq 2$ or $\ell(J_n) \geq 2$ for all n.

<u>Proof</u>. By Proposition 1.6, $w(Q^I[1]Q^J[1]) = p^{\ell(I) + \ell(J)}$, which is

greater than $p^{\ell(I)} + p^{\ell(J)}$ unless $p = 2$ and $\ell(I) = \ell(J) = 1$. By Proposition 1.5 and the form of ψ, we easily deduce that

$$(x * [i])(y * [j]) = x[j] * y[i] * [ij + k\ell]$$

in $E^0H_*\tilde{F}$. In $H_*\tilde{F}$, we have

$$(x_1 * x_2)[j] = x_1[j] * x_2[j] \text{ and, for } j > 0,$$

$$Q^I[1][j] = Q^I[1]([1] * \dots * [1]) = \sum \pm Q^{I^{(1)}}[1] * \dots * Q^{I^{(j)}}[1]$$

and
$$Q^I[1][-j] = \chi(Q^I[1][j]) .$$

It follows that, in $E^0H_*\tilde{F}$, $Q^I[1][j] = jQ^I[1] * [(j-1)p^{\ell(I)}]$ and therefore $x[j] = j^r x * [(j-1)k]$ for any integer j.

With $i = 1-k$ and $j = 1-\ell$, the lemma gives most of the multiplication table for $\#$ on E^0H_*SF. If $p > 2$, $\underline{*}$ and $\#$ coincide on $E^0H_*\tilde{F}$, and Theorem 5.2(i) follows. If $p = 2$, let $A(X; \underline{*})$ denote the subalgebra of H_*SF under $\underline{*}$ generated by the set X. Propositions 1.5 and 1.6 imply that $A(X; \underline{*})$ is closed under $\#$ and contains the subalgebra of H_*SF generated under $\#$ by X. By the lemma, $\#$ and $\underline{*}$ coincide on $E^0A(X; \underline{*})$. Therefore X generates a free commutative subalgebra of H_*SF under $\#$ and this subalgebra coincides (as a subset of H_*SF) with $A(X; \underline{*})$. We know by Lemma 1.9 that $\{x_s\}$ generates an exterior subalgebra under $\#$. Visibly $E\{x_s\}$ and AX are sub coalgebras of H_*SF, and it follows easily that $H_*SF = E\{x_s\} \otimes AX$ as a Hopf algebra. This proves Theorem 5.1(i).

In order to compute H_*BSF as an algebra and to compute $j_*: H_*SO \to H_*SF$ when $p > 2$, we need information about $\tilde{Q}^r x_I$ when $p = 2$ and $r = d(I) + 1$ and when $p > 2$ and $2r = d(I) + 1$. Together with Lemma 2.7, the following result more than suffices.

Lemma 5.7. Let $\ell(I) = k \geq 2$ and let $r \geq 0$ be such that $e(r, I) < k$. Then $\tilde{Q}^r x_I \equiv x_{(r,I)}$ modulo $F_{p^k+1} H_* SF$.

Proof. $e(r, I) = r - d(I)$ if $p = 2$ and $e(r, I) = 2r - d(I)$ if $p > 2$. By Proposition 1.7, $Q^I[1]$ is a linear combination of monomials $\beta^{\varepsilon_1} Q^{s_1}[1] \ldots \beta^{\varepsilon_k} Q^{s_k}[1]$ (where irrelevant Bocksteins are to be suppressed when $p = 2$). The Cartan formula gives

$$\tilde{Q}^r(\beta^{\varepsilon_1} Q^{s_1}[1] \ldots \beta^{\varepsilon_k} Q^{s_k}[1]) = \sum \tilde{Q}^{r_1} \beta^{\varepsilon_1} Q^{s_1}[1] \ldots \tilde{Q}^{r_k} \beta^{\varepsilon_k} Q^{s_k}[1] .$$

If $2r_i < 2s_i(p-1) - \varepsilon_i$, then $\tilde{Q}^{r_i} \beta^{\varepsilon_i} Q^{s_i}[1] = 0$. Thus $2r_i \geq 2s_i(p-1) - \varepsilon_i$ for all i in each non-zero summand and, since

$$r - d(I) = \sum (r_i - s_i) < k \text{ if } p = 2 \text{ and } 2r - d(I) = \sum (2r_i - 2s_i(p-1) + \varepsilon_i) < k \text{ if } p > 2,$$

$2r_i = 2s_i(p-1)$ and $\varepsilon_i = 0$ for at least one index i. Here $\tilde{Q}^{r_i} Q^{s_i}[1]$ is the $\#$ p^{th} power of $Q^s[1]$, and it follows that each non-zero summand has weight at least $p^{k-1} \cdot p^p = p^{p+k-1}$. Now the mixed Cartan formula gives

$$\tilde{Q}^r(x_I) \equiv x_{(r,I)} + \tilde{Q}^r Q^I[1] * [1 - p^{pk}] \mod F_{p^k+1} H_* SF ,$$

and the conclusion follows.

We first complete the proof of Theorem 5.1 and then that of Theorem 5.2.

Let $p = 2$. Then $H_* SO = E\{a_s \mid s \geq 1\}$ where a_s is the image of the non-zero element of $H_* RP^\infty$ under the standard map $RP^\infty \to SO$. Clearly $\psi(a_s) = \sum_{i=0}^{s} a_i \otimes a_{s-i}$, where $a_0 = 1$. Define Stiefel-Whitney classes $w_s = \Phi^{-1} Sq^s \Phi(1)$ in both $H^* BO$ and $H^* BF$, where Φ is the stable Thom isomorphism. Since $(Bj)^*(w_s) = w_s$, $\sigma^*(Bj)^* = j^* \sigma^*$, and $\langle w_{s+1}, \sigma_* a_s \rangle = 1$,

$$(Bj)^* : H^* BF \to H^* BO, \ (Bj)^* : H^* BSF \to H^* BSO, \text{ and } j^* : H^* SF \to H^* SO$$

are certainly epimorphisms.

Lemma 5.8. Let $p = 2$. Then $j_*(a_s) = x_s$ for all $s \geq 1$.

Proof. Clearly $j_*(a_1) = x_1$. Assume $j_*(a_i) = x_i$ for $i < s$. Then $j_*(a_s) + x_s$ is a primitive element of H_*SF whose square is zero. Since $H_*SF = E\{x_t\} \otimes AX$ as a Hopf algebra, its primitive elements split as $PE\{x_t\} \oplus PAX$. Therefore $j_*(a_s) + x_s$ must be in $E\{x_t\}$. Since j_* is a monomorphism, $j_*(a_s) + x_s$ must be decomposable. Since the natural homomorphism $PE\{x_t\} \to QE\{x_t\}$ is a monomorphism, it follows that $j_*(a_s) + x_s = 0$.

Remark 5.9. The lemma asserts that the two maps

$$RP^\infty \to SO \to SF \quad \text{and} \quad RP^\infty \simeq \mathcal{C}_\infty(2)/\Sigma_2 \xrightarrow{\theta_2} Q_2S^0 \xrightarrow{*[-1]} SF$$

induce the same homomorphism on mod 2 homology, where $\mathcal{C}_\infty$ is the infinite little cubes operad and θ_2 is given by the action map restricted to $\mathcal{C}_\infty(2) \times \{1\} \times \{1\}$. R. Schultz and J. Tornehave have unpublished proofs that these two maps are actually homotopic.

Remark 5.10. Kochman [13, Theorem 56] has proven that, in H_*SO,

$$Q^r a_s = \sum_{i+j+k=r+s} (s-i, r-s-j-1) a_i a_j a_k \qquad (i, j, k \geq 0 \text{ and } a_0 = 1).$$

In view of the lemma and the mixed Cartan formula, this formula implies and is implied by Lemma 1.9 and Corollary 3.5 (compare Remark 3.6). The actual verification of either implication would entail a lengthy and unpleasant algebraic calculation. A proof of Kochman's formula will be given in section 11.

Proof of Theorem 5.1. We have proven (i). For (ii), Lemma 5.8 shows that $\operatorname{Im} j_* = E\{x_s\}$ and H_*SF is a free H_*SO-module. Therefore the E^2-term $\operatorname{Tor}^{H_*SO}(Z_2, H_*SF)$ of the Eilenberg-Moore spectral sequence converging to H_*F/O (e.g. [8, § 3] or [21, 13.10]) reduces to $H_*SF/\!/j_* \cong AX$. For (iii), consider the Eilenberg-Moore spectral sequence which converges from $E^2SF = \mathrm{Tor}^{H_*SF}(Z_2, Z_2)$ to H_*BSF and the analogous spectral sequence $\{E^rSO\}$. In view of (i) and (ii),

$$E^2SF = E^2SO \otimes E\{\sigma x_{(s,s)}\} \otimes E\{\sigma x_I \mid \ell(I) \geq 2 \text{ and } e(I) \geq 1\}.$$

The elements σy have homological degree 1, hence are permanent cycles, and the homomorphism $\sigma: \tilde{H}_iSF \to E^2_{1,i}SF$ induces the homology suspension $\sigma_*: \tilde{H}_iSF \to H_{i+1}BSF$ (e.g. [8.3.12]). $E^2SO = \Gamma\{\sigma a_s\} = E^\infty SO$ and it follows that $E^2SF = E^\infty SF$. If $\deg y = r-1$, $\sigma_*\tilde{Q}^r y = \tilde{Q}^r\sigma_* y = (\sigma_* y)^2$. In particular, $(\sigma_* x_{(s,s)})^2 = 0$ by Lemma 2.7. Recall that if $J = (r, I)$, then $e(J) = r - d(I)$. Thus $e(J) = 1$ implies $r = d(I) + 1$. By Lemma 5.7, we may as well replace $\sigma x_{(r,I)}$ by $\sigma\tilde{Q}^r x_I$ in our description of E^2SF, and (iii) follows by a trivial counting argument. In view of the obvious compatible splittings $BO \simeq BO(1) \times BSO$ and $BF \simeq BO(1) \times BSF$, (iv) follows from (iii).

In order to describe the image of H_*SO in H_*SF when $p > 2$, we shall have to replace x_s and βx_s by the elements y_s and βy_s specified in the following result.

Lemma 5.11. Let $p > 2$ and let $r = r(p)$ be a power of a prime q such that r reduces mod p^2 to a generator of the group of units of Z_{p^2}. There exist unique elements $\beta^\varepsilon y_s$ H_*SF such that

$$(\beta^\varepsilon y_s)[r^p] = \beta^\varepsilon \tilde{Q}^s[r] \in H_*\tilde{F}_{p^r}.$$

$\beta^\varepsilon y_s$ is an element of the subalgebra of H_*SF under the $*$ product generated by $\{\beta^\varepsilon x_s\}$, and $\beta^\varepsilon x_s - k\beta^\varepsilon y_s$ is $*$ decomposable, where $k = r^{-p}\frac{1}{p}(r^p - r)$. Moreover, the subalgebra $E\{\beta y_s\} \otimes P\{y_s\}$ of H_*SF under the # product is a sub AR-Hopf algebra.

Proof By [R, VII. 5. 3], the localization of SF at p is the 1-component of an infinite loop space in which the component r^p is invertible in π_0. This implies the existence and uniqueness of the $\beta^{\varepsilon}y_s$, and the second statement follows from Proposition 1. 5 and Lemma 2. 8. The proof of the splitting of SF as an infinite loop space at p in [R, VIII. 4. 1] gives that $E\{\beta y_s\} \otimes P\{y_s\}$ is precisely the image of $H_*J_p^{\delta}$ in H_*SF under a certain infinite loop map $\alpha_p^{\delta}: J_p^{\delta} \to SF$, where J_p^{δ} is an appropriate discrete model for the fibre of $\psi^r - 1: BU \to BU$ at p (see §10 below), and the last statement follows.

Let $p > 2$. Then $H_*SO = E\{a_s \mid s \geq 1\}$, where $\deg a_s = 4s-1$. The a_s may be specified as the unique primitive elements such that $\langle P_s, \sigma_* a_s \rangle = (-1)^{s+1}$, where P_s is the sth Pontryagin class reduced mod p. (The sign is introduced in order to simplify constants below.) Define Wu classes $w_s = \Phi^{-1}P^s\Phi(1)$ in both H^*BO and H^*BF. There seems to be no generally accepted notation for these classes; our choice emphasizes the analogy with the Stiefel-Whitney classes mod 2. Let $m = \frac{1}{2}(p-1)$. Since $(Bj)^*(w_s) = w_s$, $\sigma^*(Bj)^* = j^*\sigma^*$, and w_s is indecomposable (indeed, $w_s \equiv (-1)^{m+1} mP_s$ modulo decomposable elements), $j_*(a_{ms})$ is certainly a non-zero primitive element of H_*SF. Moreover, since SO has no p-torsion, $\beta\tilde{Q}^t j_*(a_s) = 0$ for all s and t.

Lemma 5. 12. Let $p > 2$. Then $j_*(a_s) = 0$ if $s \not\equiv 0 \bmod m$ and $j_*(a_{ms}) = (-1)^s cb_s$, where $0 \neq c \in Z_p$ and b_s is the unique primitive element of $E\{\beta y_s\} \otimes P\{y_s\}$ such that $b_s - \beta y_s$ is decomposable.

Proof. The Z_p space of odd degree primitive elements of H_*SF has a basis consisting of the b_s and of elements of the form $p_I = x_I + y_I$, where $\ell(I) \geq 2$ and y_I is a linear combination of (decomposable) $\underline{*}$ monomials all of whose positive degree $*$ factors can be written in the form $Q^J[1]$ with $\ell(J) = \ell(I)$. If $d(I) = 2t-1$, then, by Lemma 5.7 and the mixed Cartan formula,

$$\beta \tilde{Q}^t p_I \equiv x_{(1,t,I)} \mod F_{p^{\ell(I)}+1} H_* SF .$$

On the other hand, $\beta \tilde{Q}^t b_s = 0$ for all s and t since β annihilates all odd degree elements of $E\{\beta y_s\} \otimes P\{y_s\}$. Therefore scalar multiples of the b_s are the only odd degree primitives annihilated by all operations $\beta \tilde{Q}^t$. It follows that $j_*(a_s) = 0$ if $s \not\equiv 0 \mod m$ and that $j_*(a_{ms}) = c_s b_s$, $0 \neq c_s \in Z_p$. Let $c = -c_1$. By the known values of the Steenrod operations on the P_s and by the Nishida relations and the previous lemma,

$$P_*^i a_{ms} = (i, s(p-1) - pi - 1) a_{m(s-i)} \quad \text{and} \quad P^i b_s = (-1)^i (i, s(p-1) - pi - 1) b_{s-i}$$

Thus $(-1)^i c_s = c_{s-i}$ if $(i, s(p-1) - pi - 1) \neq 0$. Since $(i, s(p-1) - pi - 1) \neq 0$ for all $i > 0$ implies $s = p^k$ and since, if $s = p^k$,

$$P_*^1 b_{s+1} = b_s \quad \text{and} \quad P_*^p b_{s+1} = -b_{s+1-p} \quad \text{when } k \geq 2$$

and

$$P_*^{p-1} b_{s+p-1} = b_s \quad \text{and} \quad P_*^p b_{s+p-1} = 2b_{s-1} \quad \text{when } k = 1,$$

we see by induction on s that $c_s = (-1)^s c$ for all $s \geq 1$.

<u>Remarks 5.13</u>. In [20], I asserted the previous result with b_s replaced by the unique primitive element b_s' in $E\{\beta x_s\} \otimes P\{x_s\}$ such that $b_s' - \beta x_s$ is decomposable (and an argument for this was later published by Tsuchiya [38]). This assertion would be true if and only if b_s were kb_s', and this would hold if $\beta Q^{s(p-1)} b_s'$ were zero. In principle, this could be checked by direct calculation from the results of the next section, but the details are forbidding. Looked at another way, the point is that $\beta y_s - k\beta x_s$ is $\underline{*}$ decomposable but possibly not # decomposable.

<u>Remarks 5.14.</u> Kochman [13, p. 105] has proven that, in H_*SO,

$$Q^r a_s = (-1)^r (2s-1, r-2s) a_{s+mr} .$$

It follows that $\tilde{Q}^r b_s = (s(p-1)-1, r-s(p-1)) b_{r+s}$, as could also be deduced from Remarks 3.9. Indeed, by use of the cited remarks, Lemma 5.11, the mixed Cartan formula, and the filtration on H_*SF, we easily deduce that

$$\tilde{Q}^r \beta^\varepsilon y^s \equiv -(-1)^\varepsilon (s(p-1)-\varepsilon, r-s(p-1)+\varepsilon-1)\beta^\varepsilon y_{r+s} \quad \text{modulo } \tilde{H}SF \# \tilde{H}SF.$$

<u>Proof of Theorem 5.2.</u> Note first that (i) remains true with $E\{\beta x_s\}$ replaced by $E\{b_s\}$. For (ii), observe that, in the Eilenberg-Moore spectral sequence converging to H_*F/O,

$$E^2 = \mathrm{Tor}^{H_*SO}(Z_p, H_*SF) = H_*SF /\!/ j_* \otimes \Gamma\{\sigma a_s \mid s \not\equiv 0 \mod m\} ,$$

where $H_*SF/\!/j_* \cong P\{x_s\} \otimes AX$. By [R, V§3 and §4], the Adams conjecture yields a map of fibration sequences (localized at p)

$$\begin{array}{ccccccc} SO & \longrightarrow & J_p & \longrightarrow & BO & \xrightarrow{\psi^r - 1} & BO \\ \| & & \downarrow{\scriptstyle \alpha_p} & & \downarrow{\scriptstyle \gamma_p} & & \| \\ SO & \xrightarrow{j} & SF & \xrightarrow{\tau} & F/O & \xrightarrow{q} & BO \end{array}$$

such that α_p and γ_p are equivalent to inclusions of direct factors with common complementary factor C_p. It follows that $E^2 = E^\infty$ and that γ_{p*} maps $H_*W^\perp$ onto a complementary tensor product factor to $H_*SF/\!/j_*$. (Warning: $J_p^\delta \simeq J_p$, but it is not known that $\alpha_p^\delta \simeq \alpha_p$, where α_p^δ is as in Lemma 5.11; compare [R, p. 306].) For (iii), consider the Eilenberg-Moore spectral sequences $\{E^r SF\}$ and $\{E^r SO\}$ converging to H_*BSF and H_*BSO.

$$E^2 SF = \Gamma\{\sigma b_s\} \otimes E\{\sigma x_s\} \otimes E\{\sigma x_I \mid \ell(I) \geq 2,\ e(I) + b(I) > 0,\ d(I) \text{ even}\}$$
$$\otimes\ \Gamma\{\sigma x_I \mid \ell(I) \geq 2,\ e(I) + b(I) > 0,\ d(I) \text{ odd}\} .$$

Here $\Gamma\{\sigma b_s\}$ is the image of E^2SO and consists of permanent cycles. Recall that if $J = (\varepsilon, s, I)$, then $b(J) = \varepsilon$ and $e(J) + b(J) = 2s - d(I)$. Thus $e(J) + b(J) = 1$ implies $2s = d(I) + 1$. If $\ell(I) \geq 2$ and $2s = d(I) + 1$, then $\sigma\tilde{Q}^s x_I$ survives to $(\sigma_* x_I)^p$ and $\sigma\beta\tilde{Q}^s x_I$ survives to $\beta(\sigma_* x_I)^p = 0$. By Lemma 5.7, we may as well replace $\sigma x_{(\varepsilon, s, I)}$ by $\sigma\beta^{\varepsilon}\tilde{Q}^s x_I$ in our description of E^2SF. Then

$$d^{p-1}\gamma_{p+j}(\sigma x_I) = -(\sigma\beta\tilde{Q}^s x_I)\gamma_j(\sigma x_I) \quad \text{for } j \geq 0.$$

This statement is just an application of the appropriate analog for the Eilenberg-Moore spectral sequence of Kudo transgression for the Serre spectral sequence and follows from Kochman's result [12] that $-\beta Q^s y$ is the p-fold symmetric Massey product $\langle y \rangle^p$ if $\deg y = 2s-1$ together with either a direct calculation in the bar construction on the chains of SF or quotation of [8, Theorem 5.6], which codifies the relationship between Massey products and differentials in the Eilenberg-Moore spectral sequence. Clearly all generators of E^pSF not in $\Gamma\{\sigma b_s\}$ have homological degree less than p and are thus permanent cycles. Therefore $E^pSF = E^{\infty}SF$, and Theorem 5.2(iii) follows by a trivial counting argument.

It remains to give the promised evaluation of $\langle \sigma^* w_s, x \rangle$ for $x \in H_*SF$. This depends on the following folklore result, which is usually stated without proof. Since I find the folklore argument based on use of the Hopf construction somewhat misleading and since the precise unstable form of the result will be needed in the study of $H^*BSF(2n)$ at odd primes (see IV), I will give a somewhat different argument (which was known to Milgram).

<u>Lemma 5.15.</u> Let A be a connnected based CW-complex, let $\alpha: A \to SF(n)$ be a based map, let $\bar{\alpha}: \Sigma A \to BSF(n)$ be the composite of

$\Sigma\alpha$ and the adjoint of the standard equivalence $\zeta : SF(n) \to \Omega BSF(n)$, and let $\alpha_0 : \Sigma^n A \to S^n$ be a based map with adjoint $A \to \Omega_0^n S^n$ homotopic to $\alpha * [-1]$. Then the Thom complex of the spherical fibration classified by $\overline{\alpha}$ is homotopy equivalent to the mapping cone of α_0.

Proof. $BSF(n)$ classifies integrally oriented spherical fibrations with a canonical cross-section which is a fibrewise cofibration, and the Thom complex of such a fibration is defined to be the quotient of the total space by the base space (see [21, 5. 2 and 9. 2] and [R,III]). With notations for the two-sided bar construction as in [21, §7], let $\nu : D \to \Sigma A$ be the pull-back in the following diagram:

$$\begin{array}{ccc} D & \longrightarrow & B(*, SF(n), S^n) \\ \nu \big\downarrow \big\uparrow \sigma & & p \big\downarrow \big\uparrow \sigma \\ \Sigma A & \xrightarrow{\ \overline{\alpha}\ } & BSF(n) = B(*, SF(n), *) \end{array}$$

Here $\overline{\alpha}(a \wedge t) = |[\alpha(a)], (t, 1-t)|$. Via the correspondence

$$x \in S^n \longleftrightarrow (*, |[\ ]x, (1)|) \in D$$

and

$$(a \wedge t, x) \in CA \times S^n \longleftrightarrow (a \wedge t, |[\alpha(a)]x, (t, 1-t)|) \in D,$$

D is homeomorphic to the quotient of the disjoint union of S^n and $CA \times S^n$ (where $CA = A \times I / A \times \{0\} \times \{*\} \cup I$) obtained by identifying $(*, x) \in CA \times S^n$ with $x \in S^n$ and $(a \wedge 1, x) \in CA \times S^n$ with $\alpha(a)(x) \in S^n$. Moreover, ν is specified by $\nu(x) = *$ for $x \in S^n$ and $\nu(a \wedge t, x) = a \wedge t$. By the standard Dold-Thom argument (e. g. [G, 7. 1]), ν is a quasi-fibration. The section of p is determined by the basepoint $\infty \in S^n$ and pulls back to the cofibration $\sigma : \Sigma A \to D$ specified by $\sigma(a \wedge t) = (a \wedge t, \infty)$. By comparison with the diagram obtained by replacing p and ν by fibrations with section [21, 5. 3], we conclude that the Thom complex of the fibration classified by

$\overline{\alpha}$ is homotopy equivalent to the Thom complex $T\nu = D/\Sigma A$. Clearly $T\nu$ can be viewed as obtained from $S^n \coprod \mathrm{Cyl}\, A \ltimes S^n$ by identifying $(\alpha, 1) \ltimes x$ to $\alpha(a)(x) \in S^n$ and $(a, 0) \ltimes x$ to $x \in S^n$ (where $\mathrm{Cyl}\, A = A \times I/\{*\} \times I$ and $\mathrm{Cyl}\, A \ltimes S^n = \mathrm{Cyl}\, A \times S^n/\mathrm{cyl}\, A \times \{\infty\}$). Modulo neglect of the basepoint of A ($\alpha * [-1]: A \to \Omega_0^n S^n$ not being basepoint preserving), $C\alpha_0$ can be viewed as obtained from $S^n \coprod C(A \wedge S^n)$ by first using the pinch map $S^n \to S^n \vee S^n$ to collapse the base $A \wedge S^n$ of the cone to $(A \wedge S^n) \vee (A \wedge S^n)$ and then identifying the point (a, x) of the first wedge summand to $\alpha(a)(x) \in S^n$ and the point (a, x) of the second wedge summand to $[-1]x \in S^n$ (where $[-1]: S^n \to S^n$ is any fixed map of degree -1). The conclusion follows by an easy direct comparison of these constructions.

<u>Remarks 5.16</u>. For a spherical fibration $\xi: E \to X$ of the sort classified by $BSF(n)$, define $w_s = \Phi^{-1} P^s \Phi(1) \in H^* X$, where $\Phi: H^* X \to \tilde{H}^* T\xi$ is the mod p Thom isomorphism. For $w_s \in H^* BSF(n)$, $\sigma^* w_s \in \tilde{H}^* SF(n)$ is characterized by $\Sigma \sigma^* w_s = \tilde{\zeta}^* w_s \in H^* \Sigma SF(n)$, $\tilde{\zeta}: \Sigma SF(n) \to BSF(n)$. Thus, for the fibration ξ over ΣA classified by $\overline{\alpha} = \Sigma\alpha \circ \tilde{\zeta}$, $w_s \in H^* \Sigma A$ is the suspension of $\alpha^* \sigma^* w_s \in \tilde{H}^* A$. In terms of the cofibration

$$\Sigma^n A \xrightarrow{\alpha_0} S^n \xrightarrow{i} T\xi \xrightarrow{j} \Sigma^{n+1} A$$

given by the lemma, the Thom class $\mu(\xi) \in H^n T\xi$ is the unique element such that $i^* \mu_\xi$ is the fundamental class in $H^n S^n$ and, for $x \in \tilde{H}^* \Sigma A$, the Thom isomorphism $\Phi(x) = x \cup \mu(\xi)$ can equally well be specified as $\Phi(x) = j^* \Sigma^n x$. In particular,

$$j^* \Sigma^{n+1} \alpha^* \sigma^* w_s = j^* \Sigma^n w_s = \Phi(w_s) = P^s \mu(\xi).$$

Since $\sigma_* x = 0$ if $x \in H_* SF$ is # decomposable, $\langle \sigma^* w_s, x \rangle = 0$ unless x is indecomposable. Tsuchiya [36, 6.3] showed that $\langle \sigma^* w_s, x_I \rangle = 0$

if $\ell(I) \geq 2$. By Corollary 1.8, this assertion is a consequence of the following technically simpler result, which is due to Brumfiel, Madsen, and Milgram [5, 3.5] and gives maximal unstable information.

Lemma 5.17. $<\sigma^*\beta^\varepsilon w_s, x \underline{*} y> = 0$ for $s \geq 1$, $\varepsilon = 0$ or 1, and all $x, y \in \tilde{H}_* SF(n)$.

Proof. Since $\sigma^*\beta = -\beta\sigma^*$, $<\beta\omega, z> = \pm <\omega, \beta z>$, and $\beta(x \underline{*} y) = (\beta x) \underline{*} y \pm x \underline{*} (\beta y)$, the result for $\varepsilon = 1$ will follow immediately from the result for $\varepsilon = 0$. Let $x = \alpha_*(a)$ for $\alpha: A \rightarrow SF(n)$ and $y = \beta_*(b)$ for $\beta: B \rightarrow SF(n)$, where A and B are connected CW-complexes. Then $x \underline{*} y = \gamma_*(a \otimes b)$ where $\gamma = \underline{*} \circ (\alpha \times \beta): A \times B \rightarrow SF(n)$. Let $\alpha_0: \Sigma^n A \rightarrow S^n$, $\beta_0: \Sigma^n B \rightarrow S^n$, and $\gamma_0: \Sigma^n(A \times B) \rightarrow S^n$ have adjoints homotopic to $\alpha * [-1]$, $\beta * [-1]$, and $\gamma * [-1]$. If $p: S^n \rightarrow S^n \vee S^n$ and $f: S^n \vee S^n \rightarrow S^n$ are the pinch and fold maps and if $\lambda_0 = f \circ (\alpha_0 \Sigma^n \pi_1 \vee \beta_0 \Sigma^n \pi_2)$, then standard properties of cofibre sequences give dotted arrows such that the following diagram, whose rows are cofibre sequences, is homotopy commutative:

$$
\begin{array}{ccccccc}
\Sigma^n(A\times B) & \xrightarrow{\gamma_0} & S^n & \xrightarrow{i} & C\gamma_0 & \xrightarrow{j} & \Sigma^{n+1}(A\times B) \\
\downarrow p & & \| & & \downarrow & & \downarrow p \\
\Sigma^n(A\times B)\vee\Sigma^n(A\times B) & \xrightarrow{\lambda_0} & S^n & \xrightarrow{i} & C\lambda_0 & \xrightarrow{j} & \Sigma^{n+1}(AXB)\vee\Sigma^{n+1}(AXB) \\
\| & & \uparrow f & & \uparrow & & \| \\
\Sigma^n(AXB)\vee\Sigma^n(AXB) & \xrightarrow{\alpha_0\Sigma^n\pi_1\vee\beta_0\Sigma^n\pi_2} & S^n\vee S^n & \xrightarrow{i} & C\alpha_0\Sigma^n\pi_1\vee C\beta_0\Sigma^n\pi_2 & \xrightarrow{j} & \Sigma^{n+1}(AXB)\vee\Sigma^{n+1}(AXB) \\
\downarrow \Sigma^n\pi_1\vee\Sigma^n\pi_2 & & \| & & \downarrow & & \downarrow \Sigma^{n+1}\pi_1\vee\Sigma^{n+1}\pi_2 \\
\Sigma^nA\vee\Sigma^nB & \xrightarrow{\alpha_0\vee\beta_0} & S^n\vee S^n & \xrightarrow{i} & C\alpha_0\vee C\beta_0 & \xrightarrow{j} & \Sigma^{n+1}A\vee\Sigma^{n+1}B
\end{array}
$$

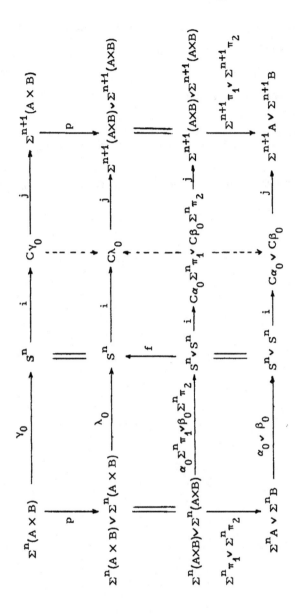

By Lemma 5.16, each of the displayed cofibres is a Thom complex. By the previous remarks, we can use the i^* to read off relationships between the various Thom classes and can then use the j^* to read off relationships between their Steenrod operations. We conclude that

$$j^*\Sigma^{n+1}\gamma^*\sigma^*w_s = j^*\Sigma^{n+1}(\alpha^*\sigma^*w_s \otimes 1 + 1 \otimes \beta^*\sigma^*w_s).$$

Thus $\gamma^*\sigma^*w_s = \alpha^*\sigma^*w_s \otimes 1 + 1 \otimes \beta^*\sigma^*w_s$, and $\langle \sigma^*w_s, \gamma_*(a \otimes b)\rangle = 0$ whenever a and b both have positive degree.

§6. The R-algebra structure of H_*SF

Theorems 5.1 and 5.2 describe generators of H_*SF in terms of the loop operations Q^r in H_*QS^0. For the analysis of infinite loop maps in and out of SF and for the understanding of characteristic classes for spherical fibrations in terms of the infinite loop structure of BF, it would be highly desirable to have a description of generators of H_*SF primarily in terms of the operations $\tilde{Q}^r$. The following conjecture of mine was proven by Madsen [15].

Theorem 6.1. When $p = 2$, $H_*SF = E\{x_s\} \otimes A\tilde{X}$ as an algebra under #, where

$$\tilde{X} = \{\tilde{Q}^J x_K \mid \ell(K) = 2 \text{ and } x_{(J,K)} \in X\}.$$

Both $E\{x_s\} = H_*SO$ and $A\tilde{X}$ are sub AR-Hopf algebras of H_*SF.

The analogous result for $p > 2$ would read as follows

Conjecture 6.2. When $p > 2$, $H_*SF = E\{\beta y_s\} \otimes P\{y_s\} \otimes A\tilde{X}$ as an algebra under #, where

$$\tilde{X} = \{\tilde{Q}^J y_K \mid \ell(K) = 2 \text{ and } x_{(J,K)} \in X\}$$

and $y_K - x_K$ is an appropriately chosen element of $A\{\beta^\varepsilon y_s\}$. Both $A\{\beta^\varepsilon y_s\} = H_*J_p^\delta$ and $A\tilde{X}$ are sub AR-Hopf algebras of H_*SF.

The change of generators from x_K to y_K, $\ell(K) = 2$, serves to ensure that $A\tilde{X}$ is a sub AR-Hopf algebra of H_*SF and will be specified in section 10 below. Since we know that $A\{\beta^\varepsilon y_s\}$ is closed under the operations $\tilde{Q}^r$, the conjecture will be true as a statement about R-algebra generators if and only if it is true with the y_K replaced by the x_K.

Thus $\{x_K \mid \ell(K) = 1 \text{ or } \ell(K) = 2\}$ certainly generates H_*SF as an R-algebra when $p = 2$ and is conjectured to do so when $p > 2$. The operations $\tilde{Q}^r x_K$ with (r, K) inadmissible can be shown to decompose many of the x_K with K admissible and $\ell(K) = 2$. Madsen proved the following theorem; since use of the mixed Adem relations would not appreciably simplify his argument, we refer the reader to [15] for the proof.

Theorem 6.3. Let $p = 2$. Then the following set is a basis for the Z_2-module of R-algebra indecomposable elements of H_*SF:

$$\{x_{2^s} \mid s \geq 0\} \cup \{x_{(2^s, 2^s)} \mid s \geq 0\} \cup \{x_{(2^s n + 2^s, 2^s n)} \mid n \geq 1 \text{ and } s \geq 0\}.$$

Observe that $\tilde{X}$ contains precisely one element of this set in each degree ≥ 2. Even if Conjecture 6.2 is correct, the analog of the previous result when $p > 2$ will have a considerably more complicated statement and proof. Even when $p = 2$, determination of a defining set of R-algebra relations in terms of the displayed minimal set of generators would probably be prohibitively difficult.

We shall give a variant of Madsen's proof of Theorem 6.1. The argument is based on analysis of the # decomposable elements of H_*SF, and we shall carry out this analysis simultaneously for all primes. In the process, we shall see where the gap in Tsuchiya's proof of Conjecture 6.2 occurs and shall make clear what remains to be done in order to prove that statement.

The following three propositions generalize results of Madsen [15] to the case of odd primes, and much of this material was stated without proof by Tsuchiya [38]. The key to these results is our analysis of the dual of the Dyer-Lashof algebra in I §3.

Proposition 6.4. Let ξ and $\tilde{\xi}$ denote the pth power operations on $H_*\tilde{F}$ in the $*$ and # products respectively and let $\tilde{F}_{p*}$ denote the union of the components $\tilde{F}_{pj}$ for $j \in Z$. Then

(i) If $x \in H_*\widetilde{F}_{p*}$, $\widetilde{\xi} x \in \operatorname{Im} \xi$.

(ii) If $y \in H_*SF$, $\widetilde{\xi} y \in (\operatorname{Im} \xi) * [1]$; that is, any # p^{th} power in H_*SF is also a $\underline{*}$ p^{th} power.

(iii) If $p > 2$ and $y \in H_*SF$ or $p = 2$ and $y \in AX$, $\xi(y * [-1]) * [1] \in \operatorname{Im} \widetilde{\xi}$; that is, any $\underline{*}$ p^{th} power in H_*SF (of an element of AX if $p = 2$) is also a # p^{th} power.

Proof. We first prove (i) when $x = Q^I[1]$. To this end, observe that the evaluation map $f: R \to H_*\widetilde{F}$, $f(r) = r[1]$, is a monomorphism of coalgebras with image closed under ξ (obviously) and $\widetilde{\xi}$ (by Proposition 1.6). We may therefore regard ξ and $\widetilde{\xi}$ as morphisms of coalgebras defined on R. We must show that the image of $\widetilde{\xi}: R[k] \to R[pk]$ is contained in the image of $\xi: R[pk-1] \to R[pk]$. Dually, it suffices to show that $\operatorname{Ker} \xi^* \subset \operatorname{Ker} \widetilde{\xi}^*$. Now ξ^* and $\widetilde{\xi}^*$ are morphisms of algebras (which annihilate all odd degree elements if $p > 2$), and it is immediate from Theorem I.3.7 that ξ^* is given on generators (of even degree if $p > 2$) by

$$\xi^*(\xi_{i,pk}) = \begin{cases} \xi_{i,pk-1} & \text{if } i < pk \\ 0 & \text{if } i = pk \end{cases}$$

and, if $p > 2$,

$$\xi^*(\sigma_{i,j,pk-1}) = \begin{cases} \sigma_{i,j,pk-1} & \text{if } j < pk \\ 0 & \text{if } j = pk \end{cases}$$

Since the degrees of $\xi_{pk,pk}$ and the $\sigma_{i,pk,pk}$ are not divisible by p, $\widetilde{\xi}^*$ also annihilates these elements, as required. By the mixed Cartan formula (together with the facts that operations below the pth power are identically zero and that $\operatorname{Im} \xi$ is a subalgebra of $H_*\widetilde{F}$ under $*$),

$\tilde{\xi}(x_1 * x_2) \in \operatorname{Im} \xi$ whenever $\tilde{\xi}(x_1) \in \operatorname{Im} \xi$ and $\tilde{\xi}(x_2) \in \operatorname{Im} \xi$. This proves (i), and (ii) follows from (i) by the mixed Cartan formula applied to the calculation of $\tilde{\xi}(x * [1])$, $x = y * [-1]$. Part (iii) follows by dimensional considerations from (ii), part (i) of Theorems 5.1 and 5.2, and the fact that $AX = A(X; \underline{*})$ when $p = 2$.

Part of the usefulness of the proposition lies in the fact that many linear combinations that arise in the analysis of our basic formulas turn out to be in the image of ξ.

Proposition 6.5. Let $x, y \in H_*\tilde{F}$, with $y \in H_*\tilde{F}_{p*}$ in (ii) and (iii). Then

(i) $\sum_{i=0}^{n} (Q^{n-i}x)(Q^i y) \in \operatorname{Im} \xi$

(ii) $\sum_{i=0}^{n} (Q^{n-i}x)(\tilde{Q}^i y) \in \operatorname{Im} \xi$

(iii) $\sum_{i=0}^{n} (Q^{n-i}x)(\tilde{\mathcal{Q}}^i y) \in \operatorname{Im} \xi$.

Proof. We prove (iii); (i) and (ii) are similar but simpler. By Proposition 1.6, Definition 3.2, the Nishida relations, and the change of dummy variables $m = i-j$, we find that

$$\sum_{i=0}^{n} Q^{n-i}x\, \tilde{\mathcal{Q}}^i y = \sum_{i=0}^{n} \sum_{j} Q^{n-i+j}(xP_*^j \tilde{\mathcal{Q}}^i y)$$

$$= \sum_{i=0}^{n} \sum_{j,k} Q^{n-i+j}(xP_*^j \tilde{Q}^{i+k} P_*^k y)$$

$$= \sum_{i=0}^{n} \sum_{j,k,\ell} (-1)^{j+\ell} (j-p\ell, (i+k)(p-1)-pj+p\ell)\, Q^{n-i+j}(x\tilde{Q}^{i+k-j+\ell} P_*^\ell P_*^k y)$$

$$= \sum_{k,\ell,m} \sum_{i=0}^{n} (-1)^{i+\ell+m} (i-m-p\ell, k(p-1)+pm+p\ell-i)\, Q^{n-m}(x\tilde{Q}^{k+\ell+m} P_*^\ell P_*^k y)$$

Fix k, ℓ, and m. Observe that $P_*^k y = 0$ if $2kp > \deg y$ (or $2k > \deg y$ if $p = 2$).

It follows easily that $Q^{n-m}(x\tilde{Q}^{k+\ell+m}P_*^{\ell}P_*^{k}y) \doteq 0$ unless $m+p\ell \geq 0$ and $k(p-1)+pm+p\ell \leq n$. We may therefore let i run over $m+p\ell \leq i \leq k(p-1)+pm+p\ell$ in the last sum. If we set $q = i-m-p\ell$, the constant becomes

$$\sum_{q=0}^{(k+m)(p-1)} (-1)^q(q,(k+m)(p-1)-q),$$

which is zero unless $k+m = 0$. We therefore have

$$\sum_{i=0}^{n} Q^{n-i}x\chi^{i}y = \sum_{k,\ell} Q^{n+k}(x\tilde{Q}^{\ell}P_*^{\ell}P_*^{k}y).$$

For any z, an easy excess argument gives $\tilde{Q}^{\ell}P_*^{\ell}z = 0$ unless $2\ell p = \deg z$ (or $2\ell = \deg z$ if $p = 2$), when $\tilde{Q}^{\ell}P_*^{\ell}z = \tilde{\xi}P_*^{\ell}z$. By (i) of the previous proposition, each $\tilde{Q}^{\ell}P_*^{\ell}P_*^{k}y$ is in $\operatorname{Im}\xi$. Since $\operatorname{Im}\xi$ is an ideal under # (by Proposition 1.5) and is closed under the Q^s (by the Cartan formula), the result follows.

We can now show that a variety of combinations of the operations *, #, Q^s, and $\tilde{Q}^s$ in $H_*\tilde{F}$ lead to elements which become decomposable under # when translated to H_*SF. It will often be convenient to write $x*[?]$ for translates of $x \in H_*\tilde{F}$; in such expressions, the unspecified number inside square brackets will always be uniquely determined by the context.

<u>Proposition 6.6.</u> Let I_j denote the set of positive degree elements of $H_*\tilde{F}_j$ and let $I_{p*} = \sum I_{pj}$ and $I = \sum I_j$. Define $D_j \subset I_j$ by

$$D_j = \{x \mid x \in I_j \text{ and } x*[1-j] \in I_1 \# I_1\}$$

and let $D = \sum D_j$. Then D satisfies the following properties.

(i) $x*\xi y \in D$ if $x \in I_{pi}$, $y \in I_j$, and j is even if $p = 2$.

(ii) $x_1*\cdots*x_r + (-1)^r(r-1)!\,x_1\cdots x_r*[?] \in D$ if $x_k = Q^{I_k}[1] \in I$ for $1 \leq k \leq r$.

(iii) If $x_k = Q^{I_k}[1] \in I$ for $1 \leq k \leq r$ and $y_\ell = Q^{J_\ell}[1] \in I$ for $1 \leq \ell \leq s$, then

$$(x_1 * \cdots * x_r)(y_1 * \cdots * y_s) \in D \text{ if } 1 \leq r < s \text{ and, if } r = s,$$

$$(x_1 * \cdots * x_r)(y_1 * \cdots * y_r) + (-1)^r r!\,(r-1)!\, x_1 \cdots x_r y_1 \cdots y_r * [?] \in D.$$

(iv) $\sum_{i=0}^{n} Q^{n-i} x * Q^i y \in D$ if $x \in I$ and $y \in I$.

(v) $\sum_{i=0}^{n} Q^{n-i} x * \tilde{Q}^i y \in D$ if $x \in I$ and $y \in I_{p*}$.

Proof. The weight function takes on only finitely many values in any given degree, and the proofs of (i) and (ii) proceed by downwards induction on weight. For (i), Proposition 1.5 gives

$$(x * [1-pi])(\xi y * [1-pj]) = x * \xi y * [1-pi-pj] + x \xi y * [1-p^2 ij]$$

plus terms of the form $u * \xi v * [1-pk-p\ell]$ with $u \in I_{pk}$, $v \in I_\ell$, ℓ even if $p = 2$, and $wu + pwv > wx + pwy$. Either $x\xi y = 0$ or, if $\deg x = pq$, $x \xi y = \xi(P_*^q x \cdot y)$. When $p = 2$, Propositions 1.5 and 1.6 imply that $P_*^q x \cdot y * [1-2ij] \in AX$ (since j is even). By Proposition 6.4(iii), $x\xi y$ is in D, and it follows that $x * \xi y$ is also in D. For (ii) and (iii), note first that $[pk]x \in \operatorname{Im} \xi$ for any $x \in H_*\tilde{F}$ and any k and that if $x = Q^I[1]$ then $\sum x'x'' \in \operatorname{Im} \xi$ by (i) of the previous proposition. If $pi = \sum_{k=1}^{r-1} p^{\ell(I_k)}$ and $pj = p^{\ell(I_r)}$, then, modulo terms known to be #-decomposable by (i),

$$(x_1 * \cdots * x_{r-1} * [1-pi])(x_r * [1-pj])$$

$$\sum \pm x_1' x_r^{(1)} * \cdots * x_{r-1}' x_r^{(r-1)} * x_1'' * \cdots * x_{r-1}'' * x_r^{(r)} * [?].$$

By the induction hypothesis on weight and by (i), all terms which have a * factor $x_i' x_r^{(i)} \in I$ with either $0 \leq \deg x_i' < \deg x_i$ or $0 \leq \deg x_r^{(i)} < \deg x_r$ add up to an element of D. Therefore the right side reduces modulo D to

$$x_1 * \cdots * x_r * [1-pi-pj] + \sum_{k=1}^{r-1} \pm x_k x_r * x_1 * \cdots * x_{k-1} * x_{k+1} * \cdots * x_{r-1} * [?].$$

By induction on weight (and, when $p = 2$, by induction on r for fixed weight) and by the commutativity of #, the sum reduces modulo D to $(-1)^r(r-1)!\, x_1 \cdots x_r * [?]$, as required. Part (iii) follows without difficulty from Proposition 1.5 and (i) and (ii). Parts (iv) and (v) are easy consequences of Proposition 1.5, the previous two propositions, and (i).

Part (ii) implies that all $(p+1)$-fold $\underline{*}$ products in H_*SF are decomposable under # and allows us to express any element decomposable under $\underline{*}$ as a linear combination of elements of X plus terms decomposable under #. Parts (iii) and (iv) imply that $Q^I[1](I_{p*} * I_{p*}) \subset D$ and $Q^n(I_{p*} * I_{p*}) \subset D$. Note that D is obviously closed under the operations β and P_*^s.

<u>Remark 6.7.</u> When $p = 2$, the arguments above are essentially those of Madsen [15], although his details depend on several assertions true only at 2 (and a few of his claims are marginally too strong; e.g. $I_{2*}(I_{2*} * I_{2*}) \subset D$, not $I(I * I) \subset D$). The key effective difference between the cases $p = 2$ and $p > 2$ comes from the factorial coefficients in Proposition 6.6.

<u>Proposition 6.8.</u> The following congruences hold modulo D.

(i) $\tilde{Q}^r(x * [1]) \equiv \tilde{Q}^r x * [?] + Q^r x * [?]$ if $x \in I_{p*}$.

(ii) $\tilde{Q}^r(x * [pj]) \equiv \tilde{Q}^r x * [?]$ if $x \in I_{p*}$ and $j \equiv 0 \bmod p$.

(iii) $Q^r(x * [pj]) \equiv Q^r x * [?]$ if $x \in I_{p*}$ and $p > 2$ or $j \equiv 0 \bmod 2$.

(iv) $\tilde{Q}^r(x_I) \equiv x_{(r,I)} + \tilde{Q}^r Q^I[1] * [?]$ if $d(I) > 0$ and $\ell(I) \geq 2$.

<u>Proof.</u> Since $x_I = Q^I[1] * [1-p^{\ell(I)}]$ (where I is not assumed to be admissible), (iv) will follow immediately from (i), (ii), and (iii). Part (i) holds since Propositions 6.5(i) and (ii) and 6.6 (ii), (iv), and (v) imply

that all $*$ decomposable terms in the mixed Cartan formula for the evaluation of $Q^r(x*[1])$ are in D, that $[j]Q^r y \equiv jQ^r y$ for any $y \in I_{p*}$, and that $Q^r(x^{(1)}\dots x^{(i)}) \in D$ for $i \geq 2$. For (ii), note first that all terms of the mixed Cartan formula for the evaluation of $\tilde{Q}^r(x*[pj])$ with a positive degree $*$ factor $\tilde{Q}_i^s(y \otimes [pj])$, $0 < i < p$, are in D because all such $\tilde{Q}_i^s(y \otimes [pj])$ involve products $z \cdot [pj]$ and are thus in $\operatorname{Im} \xi$. We claim that $\tilde{Q}^r[pj]$ is also in $\operatorname{Im} \xi$, and this will imply (ii). Obviously $[pj] = Q^0[j]$, and the mixed Adem relations reduce to give

$$\tilde{Q}^r Q^0[j] = \sum Q^0[1]\tilde{Q}^{r_0}[j] * Q^{r_1 - i}[1]\tilde{Q}^i[j] \quad \text{if } p = 2$$

and, if $p > 2$,

$$\tilde{Q}^r Q^0[j] = \sum 1^{r_1} 2^{r_2} \cdots [p-1]^{r_{p-1}}$$

$$Q^0[1]\tilde{Q}^{r_0}[j] * Q^{r_1 - i_1}[1]\tilde{Q}^{i_1}[j] * \dots * Q^{r_{p-1} - i_{p-1}}[1]\tilde{Q}^{i_{p-1}}[j] * Q^0[1]Q^{r_p - i_p}[1]\tilde{Q}^{i_p}[j].$$

With $j \equiv 0 \bmod p$, Propositions 6. 5(ii), 6. 4(iii), and 6. 6(i) imply that $\tilde{Q}^r Q^0[j] \in \operatorname{Im} \xi$. Note that Proposition 6. 5(ii) would no longer apply with $j \not\equiv 0 \bmod p$, hence that (ii) may well fail then (a point missed in [15] and [38], both of which neglect to consider possible terms arising from the $\tilde{Q}^r[pj]$). Part (iii) holds by the Cartan formula, the fact that $Q^r[pj] \in \operatorname{Im} \xi$, and Propositions 6. 4(iii) and 6. 6(i).

Unfortunately, the mixed Adem relations appear not to simplify so pleasantly modulo D. We do have that one type of term drops out when $p > 2$ however.

<u>Lemma 6. 9</u>. If $p > 2$ and $x \in I_{p*}$, then

$$\tilde{Q}^r \beta^{\varepsilon} Q^s x \equiv \sum (-1)^{\varepsilon_1 \deg x'} \tilde{Q}_0^{r_0, \varepsilon_0, s_0} x' * \tilde{Q}_1^{r_1, \varepsilon_1, s_1} x'' \quad \text{mod D.}$$

<u>Proof</u>. Comparison of Proposition 6. 5(iii) to Theorem 3. 3(v) shows that

$\mathcal{L}_2^{r,\varepsilon,s} x \in \operatorname{Im} \xi$. It follows by induction on the degree of x that $\tilde{Q}_2^{r,\varepsilon,s} x \in \operatorname{Im} \xi$ and thus that $[p^{p-2}-1]\tilde{Q}_2^{r,\varepsilon,s} x \in \operatorname{Im} \xi$. The conclusion follows from Proposition 6.6(i).

Fortunately, we need only use a special case of the mixed Adem relations. The following result, which is now merely an observation, is the core of the proof of Theorem 6.1.

Proposition 6.10. Let $p = 2$ and let $k = \ell(I) \geq 2$. Then

$$\tilde{Q}^r Q^I[1] \equiv \sum_{\ell(J) = k} a_{I,J} Q^J[1] * [?] \text{ modulo } D.$$

Proof. By Proposition 1.7, we may write $Q^I[1]$ as a linear combination of elements $Q^{s_1}[1] \cdots Q^{s_k}[1]$. By the Cartan formula, Corollary 3.5, and Proposition 6.6(iii),

$$\tilde{Q}^r(Q^{s_1}[1] \cdots Q^{s_k}[1]) \equiv \sum \Big(\mathop{\times}_{i=1}^{k} (r_i - s_i - 1, s_i)\Big) Q^{r_1+s_1}[1] \cdots Q^{r_k+s_k}[1] * [?],$$

the essential point being that the $*$ decomposable summands of the $\tilde{Q}^{r_i} Q^{s_i}[1]$ make no contribution modulo D since $I_{2*}(I_{2*} * I_{2*}) \subset D$. The conclusion follows by Proposition 1.6.

Propositions 6.8(iii) and 6.10 imply the first of the following corollaries, and the second follows from the first by induction on $\ell(I)$. It should be noted that neither of these corollaries requires restriction to admissible sequences I.

Corollary 6.11. Let $p = 2$ and let $k = \ell(I) \geq 2$. Then

$$\tilde{Q}^r x_I \equiv x_{(r,I)} + \sum_{\ell(J) = k} a_{I,J} x_J \quad \text{modulo } I_1 \# I_1 .$$

Corollary 6.12. Let $p = 2$ and let $I = (J, K)$, $\ell(K) = 2$. Then

$$\tilde{Q}^J x_K \equiv x_I + \sum_{2 \leq \ell(L) < \ell(I)} c_{I,L} x_L \quad \text{modulo } I_1 \# I_1 .$$

In view of Theorem 5.1, Corollary 6.12 implies Theorem 6.1.

We try to complete the proof of Conjecture 6.2 in the same way.

Proposition 6.13. Let $p > 2$ and let $k = \ell(I) \geq 2$. Then

$$\tilde{Q}^r Q^I[1] \equiv \sum_{\ell(J) = k} a_{I,J} Q^J[1] * [?] + \sum_{\ell(K) \geq 2k} b_{I,K} Q^K[1] * [?] \quad \text{modulo } D.$$

Proof. By Proposition 1.7, we may write $Q^I[1]$ as a linear combination of elements $\beta^{\varepsilon_1} Q^{s_1}[1] \ldots \beta^{\varepsilon_k} Q^{s_k}[1]$. We evaluate $\tilde{Q}^r(\beta^{\varepsilon_1} Q^{s_1}[1] \ldots \beta^{\varepsilon_k} Q^{s_k}[1])$ by the Cartan formula and Corollary 3.8. By Proposition 6.7(iii), the only possible contributions modulo D from $*$-decomposable summands of the $\tilde{Q}^{r_i} \beta^{\varepsilon_i} Q^{s_i}[1]$ come from products of such summands with each other (and not with $*$ indecomposable summands); such products lead to the sum written $\sum b_{I,K} Q^K[1] * [?]$ with $\ell(K) \geq 2k$. By Propositions 6.5(ii) and 6.4(iii), all terms which involve the $*$ indecomposable summands $-Q^{r_i} \beta^{\varepsilon_i} Q^{s_i}[1] * [?]$ of the $\tilde{Q}^{r_i} \beta^{\varepsilon_i} Q^{s_i}[1]$ add up to an element of the image of ξ and thus to an element of D. We are left with the products of the $*$ indecomposable summands which are multiples of the $\beta^{\varepsilon_i} Q^{r_i + a_i}[1] * [?]$, and these lead to the sum written $\sum a_{I,J} Q^J[1] * [?]$ with $\ell(J) = k$.

Corollary 6.14. Let $p > 2$ and let $k = \ell(I) \geq 2$. Then

$$\tilde{Q}^r x_I \equiv x_{(r,I)} + \sum_{\ell(J) = k} a_{I,J} x_J + \sum_{\ell(K) \geq 2k} b_{I,K} x_K \quad \text{modulo } I_1 \# I_1 .$$

Unfortunately, we cannot go on to obtain an analog of Corollary 6.12 since, upon application of iterated operations, the higher length terms can give rise to successively lower length terms which might cancel with the desired dominant terms.

Tsuchiya [38, 4.6(3)] asserts that the $b_{I,K}$ are all zero, but the core of his argument (namely the last two sentences on page 308 and the next to last sentence on page 310) is stated without any indication of proof.[1] He may conceivably be right, but a proof will surely require many pages of very careful computation. In view of Proposition 6.7(ii) and (iii) and Corollary 3.8, it would suffice (for example) to show that, for all $k \geq 2$,

$$\sum i_1^{r_{i_1}} \cdots i_k^{r_{i_k}} (r_{i_1}, s_{i_1} - \varepsilon_{i_1}) \cdots (r_{i_k}, s_{i_k} - \varepsilon_{i_k}) \beta^{\varepsilon_{i_1}} Q^{t_{i_1}}[1] \cdots \beta^{\varepsilon_{i_k}} Q^{t_{i_k}}[1] \in \operatorname{Im} \xi ,$$

where $0 \leq i_1 < \cdots < i_k \leq p-1$, $r_0 = 0$ and $s_0 = t_0 > 0$ when $i_1 = 0$, $r_{i_j} + s_{i_j} = t_{i_j}(p-1) > 0$ when $i_j > 0$, and the sum ranges over all such terms with corresponding $*$-monomial a summand of $\tilde{\mathcal{Z}}^r \beta^\varepsilon Q^s[1]$.

Note that this is definitely not implied by the much simpler statements of Proposition 6.5 (or by analogous proofs). Finally, it should be observed that Conjecture 6.2 could well be true even if some of the $b_{I,K}$ were actually non-zero.

---

[1] My letters to Tsuchiya pointing out the difficulty went unanswered.

§7. Homology operations for matrix groups

Let A be a topological ring and let G(n) be a topological subgroup of GL(n, A) such that $G(i)\oplus G(j) \subset G(i+j)$ and $\Sigma_n \subset G(n)$. Let $\mathcal{G}$ denote the category with objects the non-negative integers and with morphisms from n to n the elements of G(n). Then $\mathcal{G}$ is a sub permutative category of the category $\mathcal{G}\mathcal{L}A$ [R, VI §3 and 5.2]; its classifying space $B\mathcal{G} = \coprod BG(n)$ is a $\mathcal{D}$-space over a certain E_∞ operad $\mathcal{D}$ [R, VI. 4.1]. The homology operations Q^r on $B\mathcal{G}$ are induced from the action maps

(1) $$\mathcal{D}(p) \times_{\Sigma_p} BG(n)^p \cong B(\Sigma_p \int G(n)) \xrightarrow{Bc_p} BG(pn),$$

where $c_p : \Sigma_p \int G(n) \rightarrow G(pn)$ is the homomorphism specified by

(2) $$c_p(\sigma; g_1, \ldots, g_p) = (g_{\sigma^{-1}(1)} \oplus \cdots \oplus g_{\sigma^{-1}(p)})\sigma(n, \ldots, n)$$

for $\sigma \in \Sigma_p$ and $g_i \in G(n)$. (See I. 5.3 for our conventions on wreath products and [G. 1.1 or R. VI. 1.1] for the notations on the right-hand side.)

Let $\Gamma B\mathcal{G}$ denote the zeroth space of the spectrum derived from $B\mathcal{G}$ in [G §14 or R VII §3]. $\pi_0 \Gamma B\mathcal{G} = Z$ and $\Gamma_n B\mathcal{G}$ denotes the nth component. There is a natural map $\iota : B\mathcal{G} \rightarrow \Gamma B\mathcal{G}$ which sends BG(n) to $\Gamma_n B\mathcal{G}$ and which preserves the E_∞ structure [R, VII 3.1 and VIII 1.1]. Moreover, ι is a "group completion", so that $H_*\Gamma B\mathcal{G} = GH_*B\mathcal{G}$ as an AR-Hopf algebra with conjugation, G being the functor specified at the end of I §2. Less formally, $H_*\Gamma B\mathcal{G}$ is generated as an algebra under $*$ by [-1] and $\coprod_{n\geq 1} \iota_* H_* BG(n)$, hence the Q^r in $H_*\Gamma B\mathcal{G}$ are entirely determined by those in $H_*B\mathcal{G}$ via Lemma I. 1.2, which gives $Q^r[-1] = \chi Q^r[1]$, and the Cartan formula. Of course, the operations in $H_*B\mathcal{G}$ can be computed by purely group (or representation) theoretic techniques in view of (1) and (2).

Now suppose that A is commutative and that $G(i)\otimes G(j) \subset G(ij)$. Then $\mathcal{G}$ is a sub bipermutative category of $\mathcal{GL}A$ [R, VI §3 and 5.2]. The operad $\mathcal{P}$ acts on itself [R, VI. 2.6], and $B\mathcal{G}$ is a $(\mathcal{P},\mathcal{P})$-space [R, VI. 4.4]. The operations $\tilde{Q}^s$ on $B\mathcal{G}$ are induced from the action maps

(3) $$\mathcal{P}(p) \times_{\Sigma_p} BG(n)^p \cong B(\Sigma_p \int G(n)) \xrightarrow{B\tilde{c}_p} BG(n^p),$$

where $\tilde{c}_p : \Sigma_p \int G(n) \to G(n^p)$ is the homomorphism specified by

(4) $$\tilde{c}_p(\sigma; g_1, \ldots, g_p) = (g_{\sigma^{-1}(1)} \otimes \ldots \otimes g_{\sigma^{-1}(p)})\sigma\langle n, \ldots, n\rangle$$

for $\sigma \in \Sigma_p$ and $g_i \in G(n)$. (See Definition 1.1 for the notations on the right.) Moreover, $\Gamma B\mathcal{G}$ is an E_∞ ring space and $\iota : B\mathcal{G} \to \Gamma B\mathcal{G}$ respects both E_∞ structures [R, VII. 2.4, 4.1, and 4.2]. The homology operations $\tilde{Q}^r$ in $H_*\Gamma B\mathcal{G}$ are entirely determined by those in $H_*B\mathcal{G}$ via Lemma 2.6, which specifies the operations $\tilde{Q}^r[-1]$, and the mixed Cartan formula. Again, the operations $\tilde{Q}^r$ in $H_*B\mathcal{G}$ can be computed by group theoretic techniques in view of (3) and (4).

We shall give a uniform general discussion of procedures for the explicit calculation of the Q^r and $\tilde{Q}^r$ in a number of special cases. We assume that A is a (commutative) field which contains a primitive p^i th root of unity τ for some $i \geq 1$ and we assume that $\mathcal{G}$ is a sub bipermutative category of $\mathcal{GL}A$ such that $\tau \in G(1)$. Let π_i be the cyclic group of order p^i with generator σ and let $\eta: \pi_i \to G(1)$ be the injection specified by $\eta(\sigma) = \tau$. Recall that $H_*B\pi_i$ has a basis consisting of standard elements e_s of degree s, $s \geq 0$, and define $f_s = \eta_*(e_s) \in H_*BG(1)$. Here we agree to write ζ_* for the map on homology induced by the classifying map $B\zeta$ of a homomorphism ζ.

Define elements (some of which may be zero)

$$v_s = \iota_*(f_s) \in H_*\Gamma_1 B\mathfrak{G} \quad \text{and} \quad \bar{v}_s = v_s * [-1] \in H_*\Gamma_0 B\mathfrak{G} \;.$$

In particular, $v_0 = [1]$ and $\bar{v}_0 = [0]$. In many interesting cases, $H_*\Gamma B\mathfrak{G}$ is generated as an algebra under $*$ by $[-1]$ and the elements v_s, or, equivalently, $H_*\Gamma_0 B\mathfrak{G}$ is generated by the $\bar{v}_s$. Priddy [27] discovered a remarkably simple way of exploiting (1) and (2) to compute the operations $Q^r f_s$ and thus the operations $Q^r v_s$. Indeed, consider the following diagram of groups and homomorphisms, where π is cyclic of order p with generator ρ, μ is the evident shuffle isomorphism, $\chi_n : \pi \to \pi$ is specified by $\chi_n(\rho) = \rho^n$, $\zeta : \pi \to \pi_i$ is the injection $\zeta(\rho) = \sigma^{p^{i-1}}$, ϕ is the multiplication of π_i, and γ denotes conjugation by a suitably chosen matrix (also denoted γ) in the normalizer $NG(p)$ of $G(p)$ in $GL(p, A)$.

$$\begin{array}{lcccccccc}
 & \pi \times \pi_i & \xrightarrow{1\times\Delta} & \pi \int \pi_i \subset \Sigma_p \int \pi_i & \xrightarrow{1 \int \eta} & \Sigma_p \int G(1) & \xrightarrow{c_p} & & G(p) \\
 & \downarrow{\scriptstyle \Delta\times\Delta} & & & & & & & \downarrow{\scriptstyle \gamma} \\
(*) & \pi^p \times (\pi_i)^p & & & & & & & G(p) \\
 & \downarrow{\scriptstyle \mu} & & & & & & & \uparrow{\scriptstyle \oplus} \\
 & (\pi\times\pi_i) & \xrightarrow{\times_{n=0}^{p-1}(\chi_n\times 1)} & (\pi\times\pi_i)^p & \xrightarrow{(\zeta\times 1)^p} & (\pi_i\times\pi_i)^p & \xrightarrow{\phi^p} & (\pi_i)^p \xrightarrow{\eta^p} & G(1)^p
\end{array}$$

Thinking of $\tau^s \in \mathfrak{G}(1) \subset GL(1, A)$ as a scalar in A and thinking of $\sigma^r(1, \ldots, 1)$ as a permutation matrix, we see that (ρ^r, σ^s) maps to $\tau^s \gamma \sigma^r(1, \ldots, 1)\gamma^{-1}$ under $\gamma c_p(1 \int \eta)(1 \times \Delta)$ and to $\tau^s \mathrm{diag}(1, \tau^{p^{i-1}r}, \ldots, \tau^{p^{i-1}r(p-1)})$ under the lower path. The characteristic polynomial of $\sigma(1, \ldots, 1)$ is $x^p - 1 = \times_{n=0}^{p-1} (x - \tau^{p^{i-1}n})$. It follows that we can choose $\gamma \in GL(p, A)$ such that the diagram commutes when $\mathcal{G} = \mathcal{GL}A$, and the diagram will remain commutative for $\mathcal{G} \subset \mathcal{GL}A$ if γ can be chosen in $NG(p)$. If γ is actually in $G(p)$, then $B\gamma \simeq 1: BG(p) \to BG(p)$ and $\gamma_* = 1$ on $H_*BG(p)$. The following result is due to Priddy [27] when $p = 2$ and to Moore [24] when $p > 2$ and $i = 1$.

Theorem 7.1. Assume that the matrix γ can be chosen in $NG(p)$. Then the following formulas evaluate the operations $Q^r f_s$.

(i) Let $p = 2$ and $i = 1$. Then, for $r > s \geq 0$,

$$Q^r f_s = \sum_j (r-s-1, s-j)\gamma_*^{-1}(f_j * f_{r+s-j})$$

(ii) Let $p = 2$ and $i > 1$. Then $Q^{2r+1} f_s = 0$ for $r, s \geq 0$ and, for $r > s \geq 0$,

$$Q^{2r} f_{2s} = \sum_j (r-s-1, s-j)\gamma_*^{-1}(f_{2j} * f_{2r+2s-2j})$$

and

$$Q^{2r} f_{2s+1} = \sum_j (r-s-1, s-j)\gamma_*^{-1}(f_{2j} * f_{2r+2s+1-2j} + f_{2j+1} * f_{2r+2s-2j}).$$

(iii) Let $p > 2$ and $i \geq 1$. Then, for $r \geq 0$, $s \geq 0$, and $\varepsilon = 0$ or 1,

$$\sum_k (k, s-pk) Q^{r+k} f_{2s+\varepsilon-2k(p-1)}$$

$$= (-1)^r \sum \Big(\mathop{\times}_{n=1}^{p-1} n^{r_n}(r_n, s_n) \Big) \gamma_*^{-1}(f_{2s_0+\varepsilon_0} * f_{2r_1+2s_1+\varepsilon_1} * \cdots * f_{2r_{p-1}+2s_{p-1}+\varepsilon_{p-1}}),$$

where the right-hand sum ranges over all sets of triples $(r_n, \varepsilon_n, s_n)$, $0 \leq n < p$, with $r_0 = 0$, $r_n \geq 0$, $s_n \geq 0$, $\varepsilon_n = 0$ or 1, and with termwise sum $(r(p-1), \varepsilon, s)$; moreover, if $\gamma_* = 1$,

$$Q^r f_{2s+\varepsilon} \equiv -(-1)^{r+s}(s, r-s-1) f_{2r(p-1)+2s+\varepsilon}^{*}[p-1]$$

modulo elements decomposable under $*$.

Proof. Apply the classifying space functor to the diagram (*) and pass to homology. As Hopf algebras, $H^*B\pi_i = P\{e_1^*\}$ if $p = 2$ and $i = 1$ and $H^*B\pi_i = P\{\beta_i e_1^*\} \otimes E\{e_1^*\}$ if $p > 2$ or if $p = 2$ and $i > 1$ (e.g., by [8, p. 85-86]); $\{e_s\}$ is the evident dual basis, and we read off formulas for the product ϕ_*, coproduct Δ_*, and Steenrod operations P_*^k on $H_*B\pi_i$ by dualization. When $p > 2$, the resulting formulas are clearly independent of i. In (*), $(1\times\Delta)_*$ is evaluated in terms of the $P_*^k e_s$ in [A, Proposition 9.1], and $(1 \int \eta)_*$ can be read off from η_* by [A, Lemma 1.3]. $\phi(\zeta\chi_0 \times 1) : \pi \times \pi_i \to \pi_i$ is just the projection on the second factor, while $\chi_{n*}(e_{2r-\varepsilon}) = n^r e_{2r-\varepsilon}$ for $0 < n < p$ by the proof of [A, Lemma 1.4]. For $i = 1$, $\zeta = 1$ on $\pi = \pi_1$; for $i > 1$, $\zeta_* : H_*B\pi \to H_*B\pi_i$ is given by $\zeta_*(e_{2r}) = e_{2r}$ (e.g., by direct comparison of the standard resolutions or by a Chern class argument) and $\zeta_*(e_{2r-1}) = 0$ (because $\beta e_{2r} = 0$ in $H_*B\pi_i$). In particular, comparison of the diagrams (*) for $i = 1$ and $i > 1$ shows that our formulas for $i = 1$ and s even imply our formulas for $i > 1$ and s even. When $p = 2$, note for consistency that $(r-s-1, s-j) \equiv (2r-2s-1, 2s-2j)$ mod 2. To prove (i), chase $e_r \otimes e_s$ around the diagram. The resulting formula is

$$\sum_k (k, s-2k) Q^{r+k} f_{s-k} = \sum_j (r, s-j) \gamma_*^{-1}(f_j * f_{r+s-j}) .$$

This formula is precisely analogous to that obtained with $x = [1]$ in the mixed Adem relations, and the derivation of formula (i) is formally identical to the proof of Corollary 3.5. (Priddy [27] reversed this observation, deriving our Corollary 3.5 from his proof of (i).) In (ii), $Q^{2r+1} f_s = 0$ holds by induction on s since, inductively, $e_{2r+1} \otimes e_s$ maps to $Q^{2r+1} f_s$ along the top of (*) and to zero along the bottom (because $\zeta_*(e_{2r+1}) = 0$). The formula for $Q^{2r} f_{2s}$ follows from (i) and that for $Q^{2r} f_{2s+1}$ is proven by chasing $e_{2r} \otimes e_{2s+1}$ around the diagram to obtain the formula

$$\sum_k (k, s-2k) Q^{2r+2k} f_{2s+1-2k} = \sum_j (r, s-j) \gamma_*^{-1} (f_{2j} * f_{2r+2s+1-2j} + f_{2j+1} * f_{2r+2s-2j})$$

and again formally repeating the argument used to prove Corollary 3.5. Finally, the first part of (iii) is proven by chasing $(-1)^r e_{2r(p-1)} \otimes e_{2s+\varepsilon}$ around the diagram and the second part follows from the first by an application of Lemma 3.7 along precisely the same lines as the proof of Corollary 3.8.

Moore [24] goes further and derives closed formulas for the $Q^r f_{2s+\varepsilon}$, keeping track of the $*$-decomposable summands.

Determination of the operations $\tilde{Q}^r f_s$ is much simpler. Since $\eta : \pi_i \to G(1)$ is a homomorphism of topological Abelian groups, it and $B\eta : B\pi_i \to BG(1)$ are infinite loop maps. By [R, VI.4.5], $BG(1)$ is a sub E_∞ space of $B\mathcal{J}$ with its tensor product E_∞ structure. By Lemma I.6.1, we therefore have the following result.

Lemma 7.2. $\tilde{Q}^r f_s = 0$ unless both $r = 0$ and $s = 0$ (when $\tilde{Q}^0[1] = [1]$).

In summary, when $H_* \Gamma B\mathcal{J}$ is generated as an algebra under $*$ by $[-1]$ and the elements $v_s = \iota_*(f_s)$, its operations Q^r are determined by $Q^r[-1] = \chi Q^r[1]$, by Theorem 7.1, and by the Cartan formula while its operations $\tilde{Q}^r$ are determined by $\tilde{Q}^r[-1] = Q^r[1] * [-1]$ if $p = 2$, by $\tilde{Q}^0[-1] = [-1]$ and $\tilde{Q}^r[-1] = 0$ when $r > 0$ if $p > 2$, by Lemma 7.2, and by the mixed Cartan formula.

§8. The orthogonal, unitary, and symplectic groups

We turn to examples to which the procedures of the preceding section apply. Consider first the bipermutative categories $\mathcal{O} \subset \mathcal{GL}\mathbb{R}$ and $\mathcal{U} \subset \mathcal{GL}\mathbb{C}$ of orthogonal and unitary groups and the additive permutative category $\mathcal{S}p$ of symplectic groups [R, VI. 5.4]. Of course, $\Gamma_0 B\mathcal{O} \simeq BO$ and $\Gamma_0 B\mathcal{U} \simeq BU$, but the essential fact is that the machine-built spectra of [R, VII§3], which we denote by kO and kU, are equivalent to the connective ring spectra associated to the periodic Bott spectra and so represent real and complex connective K-theory [R, VIII. 2.1]. The weaker assertion that the maps (7.1) are compatible up to homotopy with the E_∞ actions on $BO \times Z$ and $BU \times Z$ regarded as the zeroth spaces of the Bott spectra is an unpublished theorem of Boardman.

To apply the theory of the previous section, we define $\eta: \pi \rightarrow U(1)$ by means of a primitive complex p^{th} root of unity for each prime p. For $p = 2$, we define $\eta: \pi \rightarrow O(1)$ to be the obvious identification. It is trivial to check that the matrix γ needed to make the diagram (*) commute can be chosen in $U(p)$ in the former case and in $O(2)$ in the latter case. We will thus have $\gamma_* = 1$ in the formulas of Theorem 7.1. Recall that $f_s = \eta_*(e_s)$, $v_s = \iota_*(f_s)$, and $\bar{v}_s = v_s * [-1]$. We recollect a few standard facts about the homologies of BO, BU, and BSp in the following theorem. It will be clear from these facts that the result of the previous section in principle determine all operations Q^r for $\Gamma B\mathcal{O}$, $\Gamma B\mathcal{U}$, and $\Gamma B\mathcal{S}p$ and all operations $\tilde{Q}^r$ for $\Gamma B\mathcal{O}$ and $\Gamma B\mathcal{U}$. Note that the standard functors

$$\mu: \mathcal{U} \rightarrow \mathcal{O}, \quad \nu: \mathcal{O} \rightarrow \mathcal{U}, \quad \mu: \mathcal{S}p \rightarrow \mathcal{U}, \text{ and } \nu: \mathcal{U} \rightarrow \mathcal{S}p$$

are all morphisms of additive permutative categories, and complexification $\nu: \mathcal{O} \to \mathcal{U}$ is a morphism of bipermutative categories.

Theorem 8.1. The following statements hold in mod p homology.

(i) In $H_*BU(1)$, $f_{2s+1} = 0$ and $\{f_{2s}\}$ is the standard basis; $H_*BU = P\{\overline{v}_{2s} \mid s \geq 1\}$ as an algebra under $*$.

(ii) Let $p = 2$. In $H_*BO(1)$, $\{f_s\}$ is the standard basis; $H_*BO = P\{\overline{v}_s \mid s \geq 1\}$ as an algebra under $*$. Moreover, $\nu_*: H_*BO(1) \to H_*BU(1)$ sends f_s to f_s and $\mu_*: H_*BU(1) \to H_*BO(2)$ sends f_{2s} to $f_s * f_s$.

(iii) Let $p > 2$. Define $\overline{u}_{4s} = (-1)^s \mu_*(\overline{v}_{4s}) \in H_*BO$. Then $H_*BO = P\{\overline{u}_{4s} \mid s \geq 1\}$ and $\nu_*(\overline{u}_{4s}) = \sum_{i+j=2s} (-1)^i \overline{v}_{2i} * \overline{v}_{2j} \in H_*BU$.

(iv) Define $\overline{z}_{4s} = (-1)^s \nu_*(\overline{v}_{4s}) \in H_*BSp$. Then $H_*BSp = P\{\overline{z}_{4s} \mid s \geq 1\}$ and $\mu_*(\overline{z}_{4s}) = \sum_{i+j=2s} (-1)^i \overline{v}_{2i} * \overline{v}_{2j} \in H_*BU$.

To illustrate the use of Theorem 7.1, we consider H_*BO when $p = 2$. We think of BO and $\Gamma_0 B\mathcal{O}$ and we have $Q^r \overline{v}_s = \sum_{i+j=r} Q^i v_s * Q^j[-1]$ by the Cartan formula. $Q^0[-1] = [-2]$ and, for $j > 0$,

$$Q^j[-1] = \chi Q^j[1] = Q^j[1]*[-4] + \sum_{i=1}^{j-1} Q^i[1]*\chi Q^{j-i}[1]*[-2].$$

Clearly this formula can be solved recursively for $Q^j[-1]$ in terms of the $Q^i[1]$ for $i \leq j$. Finally, Theorem 7.1 gives

$$Q^r v_s = \sum_j (r-s-1, s-j) v_j * v_{r+s-j}, \text{ hence } Q^r[1] = v_r * [1].$$

In particular, $Q^r \overline{v}_s \equiv (r-s-1, s)\overline{v}_{r+s}$ modulo $*$ decomposable elements. Kochman [13, p. 133] and Priddy [27, § 2] give tables of explicit low dimensional calculations of the $Q^r\overline{v}_s$.

<u>Remarks 8.2.</u> The first calculations of the homology operations Q^r of the classifying spaces of classical groups were due to Kochman [13]. Actually, although Theorem 7.1 is a considerable improvement on his result [13, Theorem 6], there is otherwise very little overlap between the material above and his work. His deepest results are most naturally expressed in terms of the dual operations Q^r_* in cohomology. In particular, he proved that, on the Chern and Stiefel-Whitney classes,

$$Q^r_* c_s = (-1)^{r+s}(s-r(p-1)-1, pr-s)c_{s-r(p-1)}$$

and

$$Q^r_* w_s = (s-r-1, 2r-s)w_{s-r} \; .$$

Due to the awkward change of basis required to relate the bases for H_*BO and H_*BU given in Theorem 8.1 with those given by the duals to monomials in the Chern and Stiefel-Whitney classes, it is quite difficult to pass back and forth algebraically between the formulas of Theorem 7.1 and those just stated. To illustrate the point, we note that Theorem 7.1 gives

$$Q^r[1] = (-1)^r \sum_{r_1+\dots+r_{p-1} = r(p-1)} 1^{r_1}2^{r_2}\cdots(p-1)^{r_{p-1}} v_{2r_1} * \cdots * v_{2r_{p-1}} * [1],$$

whereas Kochman's result [13, Theorem 22] gives $Q^r[1] = (c^r_{p-1})^* * [p]$.

<u>Remarks 8.3</u> (i) Well before the theory of E_∞ ring spaces was invented, Herrero [8] determined the operations $\tilde{Q}^r$ in $H_*(BO \times Z)$ and $H_*(BU \times Z)$ by proving all of the formulas of sections 1 and 2 relating the products $*$ and $\#$ and the operations Q^r and $\tilde{Q}^r$ and proving Lemma 7.2. She worked homotopically, using models for $BO \times Z$ and $BU \times Z$

defined in terms of Fredholm operators (and it does not seem likely that any such models are actually E_∞ ring spaces).

(ii) While the mixed Adem relations are not required for the evaluation of the $\tilde{Q}^r$ on $H_*\Gamma B\mathcal{O}$ and $H_*\Gamma B\mathcal{U}$, they are nevertheless available. Tsuchiya [38, 4.17] asserted what amounts to a simplification of these relations for $\tilde{Q}^r Q^s[1]$ in $H_*\Gamma B\mathcal{U}$, $p > 2$, but his argument appears to be incorrect (as it appears to require the composites $\pi \times \pi \xrightarrow{\phi} \pi \xrightarrow{\varepsilon} U(p)$ and $\pi \times \pi \xrightarrow{\eta \times \varepsilon} U(1) \times U(p) \xrightarrow{\otimes} U(p)$ to be equivalent representations, where ε is the regular representation).

Let $BO_\otimes$ and $BU_\otimes$ denote $\Gamma_1 B\mathcal{O}$ and $\Gamma_1 B\mathcal{U}$ regarded as E_∞ spaces and thus infinite loop spaces under $\otimes$. By [R, IV.3.1], $BO_\otimes \simeq BO(1) \times BSO_\otimes$ and $BU_\otimes \simeq BU(1) \times BSU_\otimes$ as infinite loop spaces. We shall need to know $H_*BSO_\otimes$ as an algebra in section 10. Adams and Priddy [2] have proven that the localizations of BSO and $BSO_\otimes$ and of BSU and $BSU_\otimes$ at any given prime p are equivalent as infinite loop spaces, and we could of course obtain the desired information from this fact. However, to illustrate the present techniques, we prefer to give a quick elementary calculation. We first recall some standard facts about Hopf algebras.

Lemma 8.4. Let A be a connected commutative Hopf algebra of finite type over Z_p which is concentrated in even degrees if $p > 2$. If all primitive elements of A have infinite height, then A is a polynomial algebra.

Proof. By Borel's theorem, $A \cong \bigotimes_{i \geq 1} A_i$ as an algebra, where A_i has a single generator a_i and $\deg a_i \leq \deg a_j$ if $i < j$. By induction on i, each a_i has infinite height since either a_i is primitive or $\psi(a_i)$

has infinite height (by application of the induction hypothesis to the calculation of its $p^{n\underline{th}}$ powers).

Note that the hypothesis on primitives certainly holds if A is a sub Hopf algebra of a polynomial Hopf algebra. We shall later want to use this lemma in conjunction with the following result of Milnor and Moore [23, 7.21].

Lemma 8.5. Let A be a connected commutative and cocommutative Hopf algebra of finite type over Z_p, $p > 2$. Then $A \cong E \otimes B$ as a Hopf algebra, where E is an exterior algebra and B is concentrated in even degrees.

We also need the following useful general observations.

Lemma 8.6. Let X be an E_∞ ring space and let $x \in \tilde{H}_*X_0$ be a primitive element. Then $\tilde{Q}^r(x*[1]) = \tilde{Q}^r x * [1] + Q^r x * [1]$ for all r. In particular, $\tilde{\xi}(x*[1]) = \tilde{\xi}(x)*[1] + \xi(x)*[1]$, where $\tilde{\xi}$ and ξ are the $p^{\underline{th}}$ power operations in the # and * products. Moreover, the # product of two primitive elements of H_*X_0 is again primitive, and the # product of a primitive and a *-decomposable element of $\tilde{H}_*X_0$ is zero.

Proof. For $0 < i < p-1$, the terms $\tilde{Q}_i^r(x \otimes [1])$ of the mixed Cartan formula are zero since $x \cdot [0] = 0$. The first part follows, and the last part is also immediate from $x \cdot [0] = 0$.

Proposition 8.7. $H_*BSO_\otimes$ and $H_*BSU_\otimes$ are polynomial algebras.

Proof. For $p > 2$, $\nu_*: H_*BSO_\otimes \rightarrow H_*BSU_\otimes$ is a monomorphism of Hopf algebras (by translation of Theorem 8.1 (iii) to the 1-components), hence it suffices to consider $BSU_\otimes$ for all p and $BSO_\otimes$ for $p = 2$. Since the two arguments are precisely the same, we consider only $BSU_\otimes$. Let $p_n = \sum_{j=1}^{n-1} (-1)^{j+1} \bar{v}_{2j} * p_{n-j} + (-1)^{n+1} n \bar{v}_{2n}$ be the basic primitive element in $H_{2n}BU$. We shall prove that $\tilde{\xi}(p_n) = 0$ for $n > 1$. Since $\{p_n * [1] | n > 1\}$ is a basis for $PH_*BSU_\otimes$ and since the p_n certainly have infinite height under $*$, the result will follow from Lemmas 8.4 and 8.6. Since $p_1 = v_1 * [-1]$ and $\tilde{\xi}(v_1) = 0$, $\tilde{\xi}(p_1) = -\xi(p_1)$ by Lemma 8.6. Since $p_i p_j$ is primitive, $p_1^j = c_j p_j$ with $c_j \neq 0$ for $2 \leq j \leq p$. Since

$$p_1^{p+1} = p_1^p p_1 = \tilde{\xi}(p_1)p_1 = -\xi(p_1)p_1 = 0,$$

by Proposition 1.5 (iii), it follows that $\tilde{\xi}(p_j) = 0$ for $2 \leq j \leq p$. Again, since $p_i p_j$ is primitive, we necessarily have $\tilde{\xi}(p_n) = k_n p_{pn} = k_n \xi(p_n)$ for some $k_n \in Z_p$, and then

$$\tilde{\xi}(P_*^i p_n) = P_*^{pi}\tilde{\xi}(p_n) = k_n P_*^{pi}\xi(p_n) = k_n \xi(P_*^i p_n).$$

$P_*^i p_n = (i, n-pi-1)p_{n-i(p-1)}$, and a standard calculation shows that for $n > p$, either $P_*^i p_n \neq 0$ for some $i > 0$ or else $n = mp^r$ with $1 \leq m \leq p-1$ and $r \geq 1$, in which case $p_n = aP_*^k p_{n+k(p-1)}$ and $P_*^\ell p_{n+k(p-1)} \neq 0$ for some $k < \ell$ and $a \neq 0$. It follows by induction on n that all $k_n = 0$.

In the case $p = 2$, explicit algebra generators in terms of the standard basis for $H_*BO_\otimes$ (obtained by translation to the 1-component from Theorem 8.1) will be given in Remarks 12.7.

§9. General linear groups of finite fields

As explained in [R, VIII §1], when A is a discrete ring the zero component $\Gamma_0 B\mathcal{GL}A$ of the infinite loop space $\Gamma B\mathcal{GL}A$ is equivalent to Quillen's plus construction on BGLA and its homotopy groups are therefore Quillen's algebraic K-groups of A.

Let k_r be the field with $r = q^a$ elements, $q \neq p$. We shall recall (and give an addendum to) Quillen's calculation of $H_*\Gamma B\mathcal{GL}k_r$ [29] and shall show that the procedures of section 7 again suffice for the computation of both types of homology operations. We shall also compute $H_*\Gamma B\mathcal{GL}k_r$ as an algebra under $\#$.

Via Brauer lifting and the Frobenius automorphism, these calculations translate to give information about spaces of topological interest. We shall utilize this translation to study the odd primary homology of Coker J, B(SF; kO), and BTop in the next section.

Let d be the smallest positive number such that $r^d \equiv 1 \mod p$ and let $r^d - 1 = p^i t$ with t prime to p. Let μ_p be the group of pth roots of unity in the algebraic closure $\overline{k}_q$ of k_r and let $k_r(\mu_p)$ be the extension over k_r generated by μ_p. Clearly $k_r(\mu_p)$ has degree d over k_r, hence its multiplicative group is cyclic of order r^d-1 and contains a primitive $p^{i\,th}$ root of unity τ. Define $\eta: \pi_i \to GL(1, k_r(\mu_p))$ by $\eta(\sigma) = \tau$. As in section 7, set $f_s = \eta_*(e_s)$, $v_s = \iota_*(f_s) \in H_*\Gamma_1 B\mathcal{GL}k_r(\mu_p)$, and $\overline{v}_s = v_s * [-1]$. Define a morphism of additive permutative categories

$$\mu: \mathcal{GL}k_r(\mu_p) \to \mathcal{GL}k_r$$

by $\mu(n) = dn$ on objects, with $\mu: GL(n, k_r(\mu_p)) \to GL(d_n, k_r)$ specified by fixing a basis for $k_r(\mu_p)$ over k_r, using it to identify $k_r(\mu_p)$ with k_r^d as a k_r-space, and then identifying $k_r(\mu_p)^n$ with $k_r^{dn} = (k_r^d)^n$. The Galois group of $k_r(\mu_p)$ over k_r is cyclic of order d with generator the Frobenius automorphism ϕ^r, $\phi^r(z) = z^r$, and it is easy to see that $\mu \circ \phi^r \circ \eta = \alpha \circ \mu \circ \eta$, where $\alpha: GL(d, k_r) \to GL(d, k_r)$ is a suitably chosen inner automorphism. It follows that $\mu_*(f_s) = 0$ unless $s \equiv 0$ or $s \equiv -1$ mod 2d. Finally, note that the inclusion of k_r in $k_r(\mu_p)$ induces a morphism of bipermutative categories

$$\nu: \mathscr{G}\mathscr{L} k_r \to \mathscr{G}\mathscr{L} k_r(\mu_p).$$

With these notations we have the following theorem, all but the last statement of which is due to Quillen [29].

<u>Theorem 9.1.</u> $H_*\Gamma_0 B\mathscr{G}\mathscr{L} k_r = P\{\mu_*\overline{v}_{2ds} \mid s \geq 1\} \otimes E\{\mu_*\overline{v}_{2ds-1} \mid s \geq 1\}$ as an algebra under $*$. Moreover, for $s \geq 1$ and $\varepsilon = 0$ or 1,

$$\nu_*\mu_*(f_{2ds-\varepsilon}) = \sum_{\substack{s_0+\dots+s_{d-1} = 2ds \\ \varepsilon_0+\dots+\varepsilon_{d-1} = \varepsilon}} r^{s_1} r^{2s_2} \cdots r^{(d-1)s_{d-1}} f_{2s_0-\varepsilon_0} * f_{2s_1-\varepsilon_1} * \dots * f_{2s_{d-1}-\varepsilon_{d-1}}.$$

<u>Proof.</u> We must analyze $\nu\mu: GL(1, k_r(\mu_p)) \to GL(d, k_r(\mu_p))$. We claim that the following diagram commutes, where δ denotes conjugation by a suitably chosen matrix:

$$\begin{array}{ccccc}
GL(1, k_r(\mu_p)) & \xrightarrow{\mu} & GL(d, k_r) & \xrightarrow{\nu} & GL(d, k_r(\mu_p)) \\
\Big\downarrow \Delta & & & & \Big\downarrow \delta \\
GL(1, k_r(\mu_p))^d & \xrightarrow{\times_{b=0}^{d-1} (\phi^r)^b} & GL(1, k_r(\mu_p))^d & \xrightarrow{\oplus} & GL(d, k_r(\mu_p))
\end{array}$$

Let $k_r(\tau) \subset k_r(\mu_p)$ be the subfield generated by τ. Since d is minimal such that $(\phi^r)^d(\tau) = \tau$, the degree of $k_r(\tau)$ over k_r is d and thus $k_r(\tau) = k_r(\mu_p)$. We may therefore choose $\{1, \tau, \dots, \tau^{d-1}\}$ as our basis for $k_r(\mu_p)$ over k_r. If $g(x) = \sum_{j=0}^{d} c_j x^j$, $c_d = 1$, is the minimal polynomial over k_r satisfied by τ, then the matrix $\mu(\tau)$ is obviously the companion matrix of $g(x)$. This remains true for $\nu\mu(\tau)$, but here, as a polynomial with entries in $k_r(\mu_p)$,

$$g(x) = \underset{b=0}{\overset{d-1}{\times}} (x - \tau^{r^b}) = \underset{b=0}{\overset{d-1}{\times}} (x - (\phi^r)^b(\tau)) .$$

Our claim follows. We may restrict ϕ^r to π_i, and we see that $\phi^r_*(e_{2s-\varepsilon}) = r^s e_{2s-\varepsilon}$ by a trivial calculation with the standard resolutions [8, p. 86] when $s = 1$ and by use of $\Delta_* \phi^r_* = (\phi^r_* \times \phi^r_*)\Delta_*$ for $s > 1$ and $\varepsilon = 0$ and of $\beta_i \phi^r_* = \phi^r_* \beta_i$ for $s > 1$ and $\varepsilon = 1$. The formula for $\nu_*\mu_*(f_{2ds-\varepsilon})$ follows by a diagram chase. In interpreting it, it is useful to remember that

$$r^e \equiv 1 \bmod p \Longleftrightarrow r^e \equiv 1 \bmod p^i \Longleftrightarrow d \text{ divides } e$$

and that, since $r^{dj} - 1 = (r^j - 1)(\sum_{b=0}^{d-1} r^{bj})$,

$$\sum_{b=0}^{d-1} r^{bj} \equiv 0 \bmod p \text{ if } j \not\equiv 0 \bmod d.$$

Of course, when $d = 1$ (which holds automatically if $p = 2$), μ and ν are the identity functors. In this case, the procedures of section 7, with $\gamma_* = 1$ in Theorem 7.1, apply directly to allow computation of the operations Q^s and $\tilde{Q}^s$ in $H_*\Gamma B\mathcal{G}\mathcal{L}k_r$ (compare the discussion following Theorem 8.1). In the case $d > 1$, the operations Q^s are determined by commutation with μ_* and the operations $\tilde{Q}^s$ are determined (not very efficiently) by commutation with the monomorphism ν_*. In the key case $d = p-1$ and $i = 1$, a more efficient procedure will be given in the next section.

<u>Remark 9.2.</u> The Bockstein spectral sequence of $\Gamma_0 B\mathcal{G}\mathcal{L}k_r$ can be

read off from Theorem 9.1 and Lemma I.4.11 since $\beta_i \overline{v}_{2ds} = \overline{v}_{2ds-1}$. Explicitly, if $p > 2$ or if $p = 2$ and $i > 1$ we have

$$E^{r+i} = P\{(\mu_* \overline{v}_{2ds})^{p^r}\} \otimes E\{(\mu_* \overline{v}_{2ds})^{p^r - 1} \overline{v}_{2ds-1}\}$$

with $\beta_{r+i}(\mu_* \overline{v}_{2ds})^{p^r} = (\mu_* \overline{v}_{2ds})^{p^r-1} \overline{v}_{2ds-1}$, while if $p = 2$ and $i = 1$ we have

$$E^{r+1} = P\{\overline{v}_{2s}^{2^r}\} \otimes E\{\overline{v}_{2s}^{2^r-2}(\overline{v}_{2s}\overline{v}_{2s-1} + Q^{2s}\overline{v}_{2s-1})\}$$

with $\beta_{r+1}\overline{v}_{2s}^{2^r} = \overline{v}_{2s}^{2^r-2}(\overline{v}_{2s}\overline{v}_{2s-1} + Q^{2s}\overline{v}_{2s-1})$. Here $Q^{2s}\overline{v}_{2s-1} = \sum_{j=0}^{2s-1} \overline{v}_j * \overline{v}_{4s-1-j}$ by Theorem 7.1 and the Cartan formula.

We next consider the homology algebra of the multiplicative infinite loop space $\Gamma_1 B\mathscr{G}\mathscr{L}k_r$. $\pi_1\Gamma_1 B\mathscr{G}\mathscr{L}k_r = Z_{r-1}$, and we write $\widetilde{\Gamma}_1 B\mathscr{G}\mathscr{L}k_r$ for the fibre of the infinite loop map $\Gamma_1 B\mathscr{G}\mathscr{L}k_r \to K(Z_{r-1}, 1)$ which represents the identity element of $\mathrm{Hom}(Z_{r-1}, Z_{r-1}) = H^1(\Gamma_1 B\mathscr{G}\mathscr{L}k_r; Z_{r-1})$. As a space, $\widetilde{\Gamma}_1 B\mathscr{G}\mathscr{L}k_r$ is equivalent to $(BSGLk_r)^+$.

<u>Lemma 9.3.</u> $\Gamma_1 B\mathscr{G}\mathscr{L}k_r$ is equivalent as an infinite loop space to the product $BGL(1, k_r) \times \widetilde{\Gamma}_1 B\mathscr{G}\mathscr{L}k_r$.

<u>Proof.</u> The inclusion of $BGL(1, k_r)$ in $B\mathscr{G}\mathscr{L}k_r$ is an E_∞ map with respect to $\otimes$ by [R, VI.4.5], hence the evident composite

$$BGL(1, k_r) \times \widetilde{\Gamma}_1 B\mathscr{G}\mathscr{L}k_r \longrightarrow \Gamma_1 B\mathscr{G}\mathscr{L}k_r \times \Gamma_1 B\mathscr{G}\mathscr{L}k_r \xrightarrow{\#} \Gamma_1 B\mathscr{G}\mathscr{L}k_r$$

is an infinite loop map. It clearly induces an isomorphism on homotopy groups, and the conclusion follows.

We have the following analog to Proposition 8.7.

<u>Proposition 9.4.</u> $H_*\widetilde{\Gamma}_1 B\mathscr{G}\mathscr{L}k_r$ is the tensor product of an exterior algebra on primitive generators of degrees $2ds-1$ and a polynomial algebra on generators of degrees $2ds$, where $s \geq 2$ if $d = 1$ and $s \geq 1$ if $d > 1$.

Proof. If $d = 1$ and $p > 2$, the result follows from Lemmas 8.4-8.6 and an argument precisely analogous to the proof of Proposition 8.7. If $d > 1$ (hence $p > 2$), then $r-1$ is prime to p and $H_*\tilde{\Gamma}_1 B\mathcal{GL}k_r = H_*\Gamma_1 B\mathcal{GL}k_r$ maps monomorphically to $H_*\tilde{\Gamma}_1 B\mathcal{GL}k_r(\mu_p)$ under ν_*, hence the conclusion follows from Lemmas 8.4 and 8.5. Finally, let $p = 2$. The following elements comprise a basis for the primitive elements in $H_*\Gamma_0 B\mathcal{GL}k_r$:

$$p_{2s-1} = \bar{v}_{2s-1} + \sum_{j=1}^{s-1} \bar{v}_{2j} * p_{2s-2j-1}, \quad s \geq 1,$$

$$p_{4s} = s\bar{v}_{2s} * \bar{v}_{2s} + \sum_{j=1}^{s-1} \bar{v}_{2j} * \bar{v}_{2j} * p_{4s-4j}, \quad s \geq 1, \text{ when } i = 1,$$

and

$$p_{2s} = s\bar{v}_{2s} + \sum_{j=1}^{s-1} \bar{v}_{2j} * p_{2s-2j}, \quad s \geq 1, \text{ when } i > 1.$$

Both $\xi(p_{2s-1}) = 0$ and $\tilde{\xi}(p_{2s-1}) = 0$, the latter being trivial when $i = 1$ and requiring a calculation from Proposition 1.5 and Lemma 8.6 when $i > 1$, hence $\tilde{\xi}(p_{2s-1} * [1]) = 0$. Thus

$$E\{p_{2s-1} * [1] \mid s \geq 2\} \otimes \Gamma\{v_1\} \text{ if } i = 1 \quad \text{or} \quad E\{p_{2s-1} * [1] \mid s \geq 1\} \otimes \Gamma\{v_2\} \text{ if } i > 1$$

is a sub Hopf algebra of $H_*\Gamma_1 B\mathcal{GL}k_r$, where $\Gamma\{v_1\}$ if $i = 1$ or $E\{v_1\} \otimes \Gamma\{v_2\}$ if $i > 1$ is the image of $H_*B\pi_i$. It follows easily from the last sentence of Lemma 8.6 that the primitive elements in the quotient of $H_*\Gamma_1 B\mathcal{GL}k_r$ by the Hopf ideal generated by this sub Hopf algebra have infinite height. The conclusion follows from Lemma 8.4 and an obvious lifting of generators argument.

Quillen's calculations in [28 and 29] yield an equivalence of fibration sequences completed away from q (where $\bar{k}_q$ is the algebraic closure of k_q):

$$
\text{(A)}\qquad
\begin{array}{ccccccc}
\Omega\Gamma_0 B\mathscr{GL}\,\bar{k}_q & \xrightarrow{\ \zeta\ } & \Gamma_0 B\mathscr{GL}\,k_r & \xrightarrow{\ \kappa\ } & \Gamma_0 B\mathscr{GL}\,\bar{k}_q & \xrightarrow{\ \phi^r - 1\ } & \Gamma_0 B\mathscr{GL}\,\bar{k}_q \\
\downarrow{\scriptstyle \Omega\hat{\lambda}} & & \downarrow{\scriptstyle \hat{\lambda}} & & \downarrow{\scriptstyle \hat{\lambda}} & & \downarrow{\scriptstyle \hat{\lambda}} \\
\Omega BU & \longrightarrow & F(\psi^r - 1) & \longrightarrow & BU & \xrightarrow{\ \psi^r - 1\ } & BU
\end{array}
$$

Here $F(\psi^r-1)$ denotes the homotopy theoretic fibre of ψ^r-1, ϕ^r is induced from the Frobenius automorphism, and the maps $\hat{\lambda}$ are derived from Brauer lifting. There is an analogous equivalence of fibration sequences completed away from q:

$$
\text{(B)}\qquad
\begin{array}{ccccccc}
\Omega\Gamma_1 B\mathscr{GL}\,\bar{k}_q & \xrightarrow{\ \zeta\ } & \Gamma_1 B\mathscr{GL}\,k_r & \xrightarrow{\ \kappa\ } & \Gamma_1 B\mathscr{GL}\,\bar{k}_q & \xrightarrow{\ \phi^r/1\ } & \Gamma_1 B\mathscr{GL}\,\bar{k}_q \\
\downarrow{\scriptstyle \Omega\hat{\lambda}} & & \downarrow{\scriptstyle \hat{\lambda}} & & \downarrow{\scriptstyle \hat{\lambda}} & & \downarrow{\scriptstyle \hat{\lambda}} \\
\Omega BU_{\otimes} & \longrightarrow & F(\psi^r/1) & \longrightarrow & BU_{\otimes} & \xrightarrow{\ \psi^r/1\ } & BU_{\otimes}
\end{array}
$$

A detailed discussion of these diagrams may be found in [R, VIII §2 and §3]. As explained there, results originally due to Tornehave [34, unpublished] imply that both (A) and (B) are commutative diagrams of infinite loop spaces and maps.

In Proposition 9.4, the exterior subalgebra of $H_*\tilde{\Gamma}B\mathscr{GL}\,k_r$ is the image of $H_*\Omega\Gamma_1 B\mathscr{GL}\,\bar{k}_q$, as can be verified by an easy spectral sequence argument. There is a general conceptual statement which can be used to obtain an alternative proof of part of that proposition and which will later be used in an algebraically more complicated situation

For any E_∞ ring space X, let $\rho: X_0 \to X_1$ denote the translation map, $\rho(x) = x*1$.

<u>Lemma 9.5</u>. Let $X \xrightarrow{\kappa} Y \overset{\phi,\psi}{\rightrightarrows} Z$ be maps of E_∞ ring spaces such that $\phi\kappa = \psi\kappa$. Then there are infinite loop maps $\mu_\oplus : X_0 \to F(\phi - \psi)$

and $\mu_{\otimes}: X_1 \to F(\phi/\psi)$ such that the following triangles of infinite loop spaces and maps homotopy commute

$$\begin{array}{ccc} & X_0 & \\ {\scriptstyle\mu_\oplus}\swarrow & & \searrow{\scriptstyle\kappa} \\ F(\phi-\psi) & \xrightarrow{\pi} & Y_0 \end{array} \qquad \text{and} \qquad \begin{array}{ccc} & X_1 & \\ {\scriptstyle\mu_\otimes}\swarrow & & \searrow{\scriptstyle\kappa} \\ F(\phi/\psi) & \xrightarrow{\pi} & Y_1 \end{array}$$

and there is a map $\overline{\rho}: F(\phi - \psi) \to F(\phi/\psi)$ such that the following diagram homotopy commutes

$$\begin{array}{ccccc} \Omega Z_0 & \xrightarrow{\iota} & F(\phi - \psi) & \xleftarrow{\mu_\oplus} & X_0 \\ {\scriptstyle\Omega\rho}\downarrow & & \downarrow{\scriptstyle\overline{\rho}} & & \downarrow{\scriptstyle\rho} \\ \Omega Z_1 & \xrightarrow{\iota} & F(\phi/\psi) & \xleftarrow{\mu_\otimes} & X_1 \end{array}$$

Here the maps π and ι are natural maps of fibration sequences.

<u>Proof.</u> $\phi - \psi = *(\phi, \chi\psi)\Delta : Y_0 \to Z_0$ and $\phi/\psi = \#(\phi, \chi\psi)\Delta : Y_1 \to Z_1$. Since ρ is not an H-map, $(\phi/\psi)\rho$ is not homotopic to $\rho(\phi - \psi)$. It is therefore convenient to replace the fibres $F(\phi - \psi)$ and $F(\phi/\psi)$ by the homotopy equalizers $E_0(\phi, \psi)$ and $E_1(\phi, \psi)$ of $\phi, \psi: Y_0 \to Z_0$ and $\phi, \psi: Y_1 \to Z_1$. To justify this, recall that the homotopy equalizer $E(\alpha, \beta)$ of maps $\alpha, \beta: C \to D$ of spaces or spectra is the pullback of the endpoints fibration $(p_0, p_1): F(I^+, D) \to D \times D$ along the map $(\alpha, \beta): C \to D \times D$, where $F(I^+, D)$ is the function space or spectra of unbased maps $I \to D$, and that, in the case of spectra, there is a map $E(\alpha, \beta) \to F(\alpha - \beta)$ which makes the following an equivalence of fibration sequences:

$$\begin{array}{ccccc} \Omega D & \longrightarrow & E(\alpha, \beta) & \longrightarrow & C \\ \| & & \downarrow & & \| \\ \Omega D & \xrightarrow{\iota} & F(\alpha - \beta) & \xrightarrow{\pi} & C \end{array}$$

This diagram yields a diagram of the same form on passage to zeroth

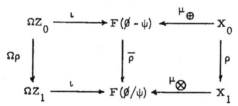

spaces. (These statements are immediate verifications from the definitions of spectra and function spectra in [R, II].) Since ϕ and ψ are maps of E_∞ ring spectra, $\phi\rho = \rho\phi$ and $\psi\rho = \rho\psi$ (with no homotopies required), and we are given that $\phi\kappa = \psi\kappa$. We can therefore write down explicit maps

$$\mu_\oplus : X_0 \to E_0(\phi,\psi)\,,\quad \mu_\otimes : X_1 \to E_1(\phi,\psi)\,,\quad \text{and}\quad \bar{\rho} : E_0(\phi,\psi) \to E_1(\phi,\psi)$$

by $\mu_\oplus(x) = (\kappa x, \omega_{\phi\kappa x})$ for $x \in X_0$, $\mu_\otimes(x) = (\kappa x, \omega_{\phi\kappa x})$ for $x \in X_1$, and $\bar{\rho}(y,\omega) = (\rho y, \rho\circ\omega)$ for $y \in Y_0$ and $\omega \in F(I^+, Z_0)$ with $(p_0, p_1)(\omega) = (\phi y, \psi y)$, where ω_z denotes the constant path at z. The requisite diagrams commute trivially. $\mu_\oplus$ and $\mu_\otimes$ are infinite loop maps because the passage from E_∞ spaces to spectra is functorial and the formulas for $\mu_\oplus$ and $\mu_\otimes$ make perfect sense on the spectrum level (applied on the spaces which make up the spectra determined by X_0, X_1 etc.) where they yield maps equivalent to the given $\mu_\oplus$ and $\mu_\otimes$ on the zeroth space level.

In diagrams (A) and (B) above, the identification of the top rows as fibration sequences proceeds by construction of maps $\mu_\oplus$ and $\mu_\otimes$ as in the lemma, with $(\phi,\psi) = (\phi^r, 1)$. (Completion away from q was only needed to establish the equivalence with the bottom rows.) Here $\mu_\oplus$ and $\mu_\otimes$ are equivalences, and the maps ζ of (A) and (B) are $\mu_\oplus^{-1}\iota$ and $\mu_\otimes^{-1}\iota$.

<u>Corollary 9.6</u> The following diagram is homotopy commutative:

$$\begin{array}{ccc}
\Omega\Gamma_0 B\mathcal{GL}\bar{k}_q & \xrightarrow{\ \zeta\ } & \Gamma_0 B\mathcal{GL}k_r \\
\Big\downarrow{\scriptstyle \Omega\rho} & & \Big\downarrow{\scriptstyle \rho} \\
\Omega\Gamma_1 B\mathcal{GL}\bar{k}_q & \xrightarrow{\ \zeta\ } & \Gamma_1 B\mathcal{GL}k_r
\end{array}$$

Therefore $\rho\zeta$ is an H-map (since $\Omega\rho$ and the ζ are) and

$$(x * [1])(y * [1]) = x * y * [1] \quad \text{for} \quad x, y \in \zeta_* H_* \Omega\Gamma_0 B\mathcal{GL}\bar{k}_q\,.$$

§10. The homology of BCoker J, B(SF; kO), and BTop at $p > 2$

When E is a commutative ring spectrum, we have a fibration sequence, natural in E:

$$\text{(1)} \qquad SF \xrightarrow{e} SFE \xrightarrow{\tau} B(SF;E) \xrightarrow{q} BSF$$

Here SFE denotes the component of the identity element of the zeroth space of E, e is obtained by restriction to the 1-components of zeroth spaces from the unit $e: Q_\infty S^0 \to E$ of E (where $Q_\infty S^0$ denotes the sphere spectrum), B(SF; E) is the classifying space for E-oriented stable spherical fibrations, and q corresponds to neglect of orientation. See [R, III §2] for details. When E is an E_∞ ring spactrum, (1) is naturally a fibration sequence of infinite loop spaces, by [R, IV §3]. Thus to calculate characteristic classes for E-oriented stable spherical fibrations, we need only compute $e_*: H_*SF \to H_*SFE$ and use the Eilenberg-Moore spectral sequence converging from

$$\text{(2)} \qquad E^2 = \mathrm{Tor}^{H_*SF}(H_*SFE, Z_p) \quad \text{to} \quad H_*B(SF;E).$$

Explicitly, B(SF; E) is the two-sided bar construction B(SFE, SF, *), and the spectral sequence is obtained from the obvious filtration of this space (e.g. [G, 11.14 and 21, 13.10]). One way of analyzing e_* is to note that, if X denotes the zeroth space of E (which is an E_∞ ring space), then we have the homotopy commutative diagram

$$\begin{array}{ccc} Q_0S^0 & \xrightarrow{\rho} & Q_1S^0 = SF \\ \downarrow{\scriptstyle e} & & \downarrow{\scriptstyle e} \\ X_0 & \xrightarrow{\rho} & X_1 \end{array}$$

Thus we can use the additive infinite loop structures to compute

$e_*: H_*Q_0S^0 \to H_*X_0$ and can then translate to the 1-components. The advantage of this procedure is that, as our earlier work makes amply clear, analysis of the additive homology operations tends to be considerably simpler than analysis of the multiplicative homology operations.

As explained in [R, IV § 6], results of Sullivan [32, 33] give an equivalence of fibration sequences localized away from 2:

$$(*)\qquad \begin{array}{ccccccc} SF & \longrightarrow & F/Top & \longrightarrow & BTop & \longrightarrow & BSF, \\ \downarrow{\scriptstyle \chi} & & \downarrow{\scriptstyle \bar{f}} & & \downarrow{\scriptstyle \bar{g}} & & \| \\ SF & \xrightarrow{e} & BO_{\otimes} & \xrightarrow{\tau} & B(SF;kO) & \xrightarrow{q} & BSF \end{array}$$

where $BO_{\otimes} = SFkO$. By [R, VIII §1], kO is an E_{∞} ring spectrum and the rows are fibration sequences of infinite loop spaces. Recent results of Madsen, Snaith, and Tornehave [19], which build on earlier results of Adams and Priddy [2], imply that (*) is a commutative diagram of infinite loop spaces and maps. (See [R, V § 7].) Thus analysis of characteristic classes for stable topological (or PL) bundles away from 2 is equivalent to analysis of characteristic classes for kO-oriented stable spherical fibrations.

Henceforward, complete all spaces and spectra at a fixed odd prime p. Here analysis of B(SF; kO) in turn reduces to analysis of another special case of (1). To see this, let $r = r(p)$ be a power of a prime $q \neq p$ such that r reduces mod p^2 to a generator of the group of units of Z_{p^2}. Equivalently, p-1 is the smallest positive number d such that $r^d \equiv 1 \bmod p$ and $r^{p-1} - 1 = pt$ with t prime to p. Define j_p^{δ} to be (the completion at p of) the E_{∞} ring spectrum derived from the bipermutative category $\mathscr{G}\mathscr{L}k_r$. The superscript δ stands for "discrete model", and j_p^δ is equivalent to the fibre j_p of $\psi^r - 1 : kO \to bo$ (bo being the 0-connected cover of kO) by [R, VIII, 3.2]. As shown in [R, VIII. 3.4], we have a commutative diagram of infinite loop spaces and maps (completed at p):

$$(**)\qquad \begin{array}{ccccccc}
SF & \xrightarrow{e} & J^\delta_{\otimes p} & \xrightarrow{\tau} & B(SF; j_p^\delta) & \xrightarrow{q} & BSF \\
\| & & \downarrow{\scriptstyle \hat{\lambda}\circ\kappa} & & \downarrow{\scriptstyle B\hat{\lambda}\circ B\kappa} & & \| \\
SF & \xrightarrow{e} & BO_\otimes & \xrightarrow{\tau} & B(SF; kO) & \xrightarrow{q} & BSF
\end{array}$$

where $J^\delta_{\otimes p} = SFj_p^\delta = \Gamma_1 B\mathscr{GL}k_r$, $\kappa : J^\delta_{\otimes p} \to BO^\delta_\otimes = \Gamma_1 B\mathscr{O}\bar{k}_r$ is such that its composite with $\Gamma_1 B\mathscr{O}\bar{k}_r \to \Gamma_1 B\mathscr{GL}\bar{k}_r$ is induced from the inclusion of $\mathscr{GL}k_r$ in $\mathscr{GL}\bar{k}_r$, and $\hat{\lambda} : BO^\delta_\otimes \to BO_\otimes$ is the equivalence obtained by Brauer lifting. As explained in [R, VIII §3 and V §4 and §5], $B(SF; j_p^\delta)$ is equivalent to the infinite loop space usually called BCoker J_p, abbreviated BC_p and defined as the fibre of the universal cannibalistic class $c(\psi^r): B(SF; kO) \to BO_\otimes$.

By [1 and 2], any infinite loop space of the homotopy type of BO (completed at p) splits as $W \times W^\perp$ as an infinite loop space, where $\pi_{2i(p-1)}W = \hat{Z}_{(p)}$ and $\pi_j W = 0$ if $j \not\equiv 0 \bmod 2(p-1)$. By [R, V.4.8 and VIII. 3.4], the composite

$$(3)\qquad B(SF; j_p^\delta) \times W \times W^\perp \to B(SF; j_p^\delta) \times BO \times BO_\otimes \longrightarrow$$

$$\xrightarrow{B\Lambda\circ B\kappa \times g \times \tau} B(SF; kO)^3 \xrightarrow{\phi} B(SF; kO)$$

is an equivalence, where $g: BO \to B(SF; kO)$ can be taken to be either the Atiyah-Bott-Shapiro orientation or the restriction to BO of the Sullivan orientation $\bar{g}$. The cited results of Madsen, Snaith, and

Tornehave [19] and of Adams and Priddy [2] imply that both choices are infinite loop maps (here at an odd prime; see [R, V §7]). Thus the specified composite is an equivalence of infinite loop spaces, and analysis of $B(SF;kO)$ reduces to analysis of $B(SF;j_p^\delta) \simeq BC_p$.

Write $J_p^\delta = \Gamma_0 B\mathcal{GL}k_r$ (in conformity with $J_{\otimes p}^\delta = \Gamma_1 B\mathcal{GL}k_r$). By diagrams (A) and (B) of the previous section, these are discrete models for the additive and multiplicative infinite loop spaces usually called $\mathrm{Im}\, J_p$ and $\mathrm{Im}\, J_{\otimes p}$, abbreviated J_p and $J_{\otimes p}$.

We have reduced the computation of H_*BTop and $H_*B(SF;kO)$ to the study of $e_*: H_*SF \to H_*J_{\otimes p}^\delta$, and we have analyzed H_*SF in sections 5 and 6 and $H_*J_{\otimes p}^\delta$ in section 9. As explained in [R, VIII §4], there is a commutative diagram of infinite loop spaces and maps

$$\begin{array}{ccc} Q_0S^0 & \xrightarrow{\alpha_p^\delta e} & SF \\ {\scriptstyle e}\downarrow & \overset{\alpha_p^\delta}{\nearrow} & \downarrow{\scriptstyle e} \\ J_p^\delta & \xrightarrow{e\alpha_p^\delta} & J_{\otimes p}^\delta \end{array}$$

such that $e\alpha_p^\delta : J_p^\delta \to J_{\otimes p}^\delta$ is an (exponential) equivalence. The diagram produces a splitting of SF as $J_p^\delta \times (SF;j_p^\delta)$ as an infinite loop space, where $(SF;j_p^\delta) = \Omega B(SF;j_p^\delta)$. Moreover, the proof given in [R, VIII §4] shows that

$$H_*J_p^\delta = P\{Q^s[1]*[-p] \mid s \geq 1\} \otimes E\{\beta Q^s[1]*[-p] \mid s \geq 1\}$$

as an algebra under $*$. These generators are not the same as the generators $\mu_*(\bar{v}_{2(p-1)s-\varepsilon})$ of Theorem 9.1. Indeed, Theorem 7.1 implies that

$$(4) \qquad Q^s[1]*[-p] = (-1)^s \sum_{s_1+\cdots+s_{p-1}=s} \mu_*(\bar{v}_{2(p-1)s_1} * \cdots * \bar{v}_{2(p-1)s_{p-1}}),$$

and the operations $Q^r(Q^s[1]*[-p])$ are now determined by that result. The elements $y_s = (\alpha_p^\delta)_*(Q^s[1]*[-p])$ in H_*SF were discussed in Lemma 5.11. Their images $e_*(y_s)$ in $H_*J_{\otimes p}^\delta$ give explicit algebra generators on which the operations $\tilde{Q}^r$ are determined by commutation with $(e\alpha_p^\delta)_*$. In particular, note that Theorem 7.1 implies

(5) $$\tilde{Q}^r\beta^\varepsilon y_s \equiv -(-1)^\varepsilon(s(p-1)-\varepsilon, r-s(p-1)+\varepsilon-1)\beta^\varepsilon y_{r+s}$$

modulo #-decomposable elements, in agreement with Remarks 5.14. Looking at diagram (**), we see that $\{e_*y_s = (\hat{\lambda}\circ\kappa\circ e)_*(y_s)\}$ is a set of polynomial generators for $H_*W \subset \operatorname{Ker}\tau_* \subset H_*BO_\otimes$.

Of course, the splitting of SF gives $H_*SF = H_*J_p^\delta \otimes H_*(SF; j_p^\delta)$ and $E^2 = \operatorname{Tor}^{H_*(SF;j_p^\delta)}(Z_p, Z_p)$ in the spectral sequence (2) for $E = j_p^\delta$. We require a set of generators for $H_*(SF; j_p^\delta) \subset H_*SF$. Certainly $\tilde{H}_*(SF; j_p^\delta)$ is contained in the kernel of $e_*: H_*SF \to H_*J_{\otimes p}^\delta$, and standard Hopf algebra arguments [23, § 4] show that $H_*(SF; j_p^\delta)$ is in fact exactly the set of all elements x such that if $\psi x = \sum x' \otimes x''$, then $e_*x' = 0$ when $\deg x' > 0$. We shall content ourselves with the specification of generators in $\operatorname{Ker} e_*$. Their suspensions will be primitive elements in $\operatorname{Ker}(Be)_*$ and thus, by simpler Hopf algebra arguments, will necessarily be elements of $H_*B(SF; j_p^\delta) \subset H_*BSF$. For K admissible of length 2, choose $z_K \in A\{\beta^\varepsilon y_s\}$ such that $e_*(x_K + z_K) = 0$ and set $y_K = x_K + z_K$. These are the elements y_K referred to in Conjecture 6.2. If we knew that conjecture to be true, we could take the set $\tilde{X}$ of elements $\tilde{Q}^j y_K$ specified there as our set of generators in $\operatorname{Ker} e_*$. As we don't know this, we instead choose $z_I \in A\{\beta^\varepsilon y_s\}$ such that $y_I = x_I + z_I$ is in the kernel of e_* for each $x_I \in X$ and set $Y = \{y_I \mid x_I \in X\}$. It follows immediately from Theorem 5.2 that

$$H_*SF = A\{\beta^{\varepsilon} y_s\} \otimes AY$$

as an algebra under # . We have the following complementary pair of results.

Theorem 10.1. As a sub Hopf algebra of H_*BF (via q_*), $H_*B(SF; j_p^{\delta}) = ABY$ where

$BY = \{\sigma_* y_I \mid \ell(I) > 2$ and $e(I) + b(I) > 1$ or $\ell(I) = 2$ and $e(I) + b(I) \geq 1\}$.

Proof. By (5), $\tilde{Q}^{s(p-1)} \beta y_s \equiv \beta y_{ps}$ modulo #-decomposable elements. The requisite calculation of the spectral sequence (2) is virtually identical to the proof of Theorem 5.2(ii) (which follows 5.14).

Theorem 10.2. As a sub Hopf algebra of H_*BF (via $B\alpha^{\delta}_{p*}$),

$$H_*BJ_p^{\delta} = (Bj)_* H_*W \otimes E\{\sigma_* y_s\}$$

Proof. This follows from Lemma 5.12 and Theorem 5.2. Note that $\beta\sigma_* y_s \in (Bj)_* H_*W$ is indecomposable if $s \not\equiv 0 \bmod p$ but that $\beta\sigma_* y_{ps} = -\sigma_* \beta y_{ps} = -\tilde{Q}^{s(p-1)} \sigma_* \beta y_s = -(\sigma_* \beta y_s)^p$.

As discussed in [R, V §7 and VIII §4], I conjecture that $Bj: BO \to BSF$ actually factors through $B\alpha_p^{\delta}$. Of course, Bj coincides with the composite $BO \xrightarrow{g} B(SF; kO) \xrightarrow{q} BF$, whereas $q\tau \simeq * : BO_{\otimes} \to BF$. By the splitting (3) and diagrams (*) and (**), we obtain the following corollary.

Theorem 10.3. $H_*BTop \cong H_*B(SF; kO) \cong H_*W \otimes H_*W^{\perp} \otimes H_*B(SF; j_p^{\delta})$. Under the natural map to H_*BF, H_*W maps isomorphically onto $(Bj)_* H_*W$, $H_*W^{\perp}$ maps trivially, and $H_*B(SF; j_p^{\delta})$ maps isomorphically to ABY.

The correction summands $\sigma_* z_I$ by which $\sigma_* y_I$ differs from $\sigma_* x_I$ are significant. They explain Peterson's observation [25] that, while βw_s maps to zero under $(Bj)^*: H^*BF \to H^*BTop$ for $s \leq p$ (for dimensional reasons, $B(SF; j_p^\delta)$ being $2p(p-1)-3$ connected and $H^{2p(p-1)+1}B(SF; j_p^\delta)$ being zero), βw_s maps non-trivially for all $s > p$. As pointed out by Peterson, $\psi w_s = \sum w_i \otimes w_{s-i}$ implies $(Bj)^*(\beta w_s) \neq 0$ for all $s > p$ if $(Bj)^*(\beta w_{p+1}) \neq 0$. The following result is considerably stronger.

<u>Theorem 10.4.</u> Let $p(k) = 1 + p + \ldots + p^{k-1}$, $k \geq 2$. Then $(Bj)^*(\beta w_{p(k)})$ is an indecomposable element of $H^*B(SF; j_p^\delta) \subset H^*BTop$.

<u>Proof.</u> Let $I_{kk} = (0, p^{k-1}, 0, p^{k-2}, \ldots, 0, 1)$, as in formula (II) of I§3. $d(I_{kk}) = 2(p^k - 1) = \deg w_{p(k)}$ and $e(I_{kk}) = 2$. We claim that

$$\langle w_{p(k)}, \beta\sigma_* y_{I_{kk}} \rangle \neq 0, \text{ hence } \langle \beta w_{p(k)}, \sigma_* y_{I_{kk}} \rangle \neq 0,$$

and the theorem will follow immediately from the claim. By Corollary 1.8 and Lemma 5.17, $\langle \beta^\varepsilon w_s, \sigma_* x \rangle = 0$ if x is $\underline{*}$ or #-decomposable or if $x = x_I$ with $\ell(I) \geq 2$. By Lemma 5.11, it suffices to verify that, in $H_* J_p^\delta$,

$$Q^{I_{kk}}[1] * [-p^k] \equiv \pm Q^{p(k)}[1] * [-p]$$

modulo * decomposable elements, since then $z_{I_{kk}}$ will differ from $x_{p(k)}$ by summands annihilated by $w_{p(k)}$ and we will have

$$\langle w_{p(k)}, \beta\sigma_* y_{I_{kk}} \rangle = \langle w_{p(k)}, \beta\sigma_* z_{I_{kk}} \rangle = \pm \langle w_{p(k)}, \beta x_{p(k)} \rangle \neq 0.$$

By Theorem 7.1, $Q^{p^{k-1}} f_{2(p^{k-1}-1)} \equiv f_{2(p^k-1)}$, and this remains true with the f's replaced successively by v', $\mu_* v$'s, and $\mu_* \bar{v}$'s. By (4),

$Q^s[1]*[-p] \equiv (-1)^{s+1}\mu_*\bar{v}_{2s(p-1)}$, and the conclusion follows.

The following consequence of this theorem was first conjectured by Peterson [25] and first proven by Tsuchiya [37, § 4], who used quite different techniques.

Theorem 10.5. Let $\Phi: H^*BSTop \to H^*MStop$ be the Thom isomorphism and define $\phi: A \to H^*MSTop$ by $\phi(a) = a\Phi(1)$. Then $\phi(Q_i) \neq 0$ for $i \geq 2$, and $\operatorname{Ker}\phi$ is the left ideal $A(Q_0, Q_1)$ generated by Q_0 and Q_1.

Proof. Q_0 is the cohomology Bockstein and $Q_i = [P^{p^i}, Q_{i-1}]$ for $i \geq 1$. As pointed out by Peterson and Toda [26, 3.1], the definition $P^s\Phi(1) = \Phi(w_s) = \Phi(1) \cup w_s$ and the Wu formulas

$$P^r\beta^\varepsilon w_s \equiv (-1)^r(r, s(p-1)-r+\varepsilon-1)\beta^\varepsilon w_{r+s} \text{ mod decomposable elements}$$

in H^*BSF formally imply the relation

$$Q_i\Phi(1) = \lambda_i\Phi(1) \cup (\beta w_{p(i)} + \text{decomposable terms}), \quad 0 \neq \lambda_i \in Z_p,$$

in H^*MSF. Application of $(Mj)^*$ and $(Bj)^*$ and use of the previous theorem shows that $Q_i\Phi(1) \neq 0$ for $i \geq 2$. $Q_0\Phi(1) = 0$ in H^*MSF, $Q_1\Phi(1) = \Phi(1) \cup \beta w_1$ in H^*MSF, and $(Bj)^*(\beta w_1) = 0$. Thus $\operatorname{Ker}\phi$ contains $A(Q_0, Q_1)$ and ϕ induces a morphism of coalgebras $\bar{\phi}: A/A(Q_0, Q_1) \to H^*MSTop$. $\{Q_k \mid k \geq 2\} \cup \{P_j^0\}$ is a basis for the primitive elements of $A/A(Q_0, Q_1)$, where P_j^0 is the Milnor basis element given by the sequence with 1 in the jth position and zero in all other positions, and P_j^0 maps nontrivially to H^*MSO. Thus $\bar{\phi}$ is a monomorphism because it is a monomorphism on primitive elements.

Remarks 10.6. Although I have made no attempt to do so, it should not be unreasonably difficult to push on and obtain sufficient information on the structure of H^*MSTop as an A-module to compute E_2 of the Adams spectral sequence converging to π_*MSTop. This requires calculation of certain Steenrod operations in H^*BSTop, and these are accessible from the Nishida relations and the theorems above (compare Remarks 5.5). Indeed, considerable relevant information on this dualization problem has already been tabulated in our calculation of the dual of the Dyer-Lashof algebra in I §3.

Remarks 10.7. It is straightforward to read off the mod p Bockstein spectral sequences of BJ_p^δ, $B(SF; j_p^\delta)$, BF, and BTop from Lemma I.4.11 (and its analog in cohomology). For BJ_p^δ, it is most convenient to work in cohomology where $H^*BJ_p^\delta = P\{w_s\} \otimes E\{\beta w_s\}$ as a quotient of H^*BF and thus

$$E_{r+1} = P\{w_s^{p^r}\} \otimes E\{w_s^{p^r-1}\beta w_s\} \text{ with } \beta_{r+1} w_s^{p^r} = w_s^{p^r-1}\beta w_s .$$

The homology Bockstein spectral sequence of $B(SF; j_p^\delta)$ is specified by

$$E^{r+1} = P\{y^{p^r}\} \otimes E\{y^{p^r-1}\beta y\} \text{ with } \beta_{r+1} y^{p^r} = y^{p^r-1}\beta y ,$$

where y runs through $\{\sigma_* y_I \mid b(I) = 0, d(I) \text{ is odd}\} \subset BY$. The Bockstein spectral sequence of J_p^δ is specified in Remarks 9.2, and that of $(SF; j_p^\delta)$ can again be read off from Lemma I.4.11.

Remarks 10.8. In [4, §5], Brumfiel conjectured that the image of $H^*(BTop; Z[1/2])$ in $H^*(BTop; Q)$ was a polynomial algebra on classes $R_i \in H^{4i}(BTop; Q)$ such that $\psi R_n = \sum_{i+j=n} R_i \otimes R_j$ and $R_n \equiv (2^{2n-1} - 1)\, \mathrm{num}(B_n/4n) P_n$ modulo decomposable elements, where B_n is the nth Bernoulli number and $\mathrm{num}(B_n/4n)$ is the numerator of the fraction $B_n/4n$ in lowest terms. As observed by Tsuchiya [37, §3], up to an undetermined factor 2^{a_n} in the congruence, the conjecture is an easy consequence of the fact that the primitive elements of $H_*(BTop; Z[1/2])$/ torsion are generated by the images of the basic primitive elements elements in $H_*(BO; Z[1/2])$ and in $H_*(F/Top; Z[1/2])$. Madsen and Milgram [18] have recently proven the conjecture in its original form.

§11. Orthogonal groups of finite fields

When A is a discrete commutative ring, the zero component $\Gamma_0 B\mathcal{O}A$ is equivalent to Quillen's plus construction on BOA and its homotopy groups can reasonably be called KO_*A.

Let k_q be the field with $q = p^a$ elements, where p is an odd prime. Fiedorowicz and Priddy [6] have made an exhaustive study of the homologies of the orthogonal groups and of various related families of matrix groups of k_q. We shall recall their calculations of $H_*\Gamma BOk_q$ and shall show that a slight elaboration of the procedures of section 7 suffices for the computation of both types of homology operations. We shall also compute $H_*\Gamma_1 B\mathcal{O}k_q$ as an algebra under $\#$.

Via Brauer lifting and the Frobenius automorphism, these calculations translate to give information about spaces of topological interest. We shall utilize this translation to study the 2-primary homology of BCoker J and B(SF; kO) in the following sections.

We repeat that the homological calculations in this section are due to Fiedorowicz and Priddy [6]. All homology groups are to be taken with Z_2 coefficients.

$O(1, k_q) = Z_2$ and we let $\eta: Z_2 \to O(1, k_q)$ be the identification. As usual, this fixes elements $f_s = \eta_* e_s$, $v_s = \iota_* f_s \in H_*\Gamma_1 B\mathcal{O}k_q$, and $\bar{v}_s = v_s * [-1] \in H_*\Gamma_0 B\mathcal{O}k_q$.

Let $\mathcal{O}^{ev}k_q$ be the full subcategory of $\mathcal{O}k_q$ whose objects are the even non-negative integers. Let

$$\delta = \begin{pmatrix} -a & b \\ b & a \end{pmatrix} \in GL(2, k_q),$$

where $a^2 + b^2$ is a non-square. Let $\delta_n = \delta \oplus \ldots \oplus \delta \in GL(2n, k_q)$. Define $\Phi: \mathcal{O}^{ev}k_q \to \mathcal{O}^{ev}k_q$ by $\Phi(2n) = 2n$ on objects and $\Phi(A) = \delta_n A \delta_n^{-1}$ on matrices $A \in O(2n, k_q)$. Then Φ is a morphism of (additive) permutative categories such that $\Phi^2 = 1$.

With $p = 2$ and $G(n) = O(n, k_q)$, the diagram (*) above Theorem 7.1 is made commutative by use of conjugation by the matrix

$$\gamma = \begin{pmatrix} -1 & 1 \\ 1 & 1 \end{pmatrix} \in GL(2, k_q).$$

If $q \equiv \pm 1 \bmod 8$, then 2 is a square in k_q and conjugation by γ is equal to conjugation by $(1/\sqrt{2})\gamma \in O(2, k_q)$. Thus $\gamma_* = 1$ in Theorem 7.1 in this case. If $q \equiv \pm 3 \bmod 8$, then 2 is a non-square and we may take $\delta = \gamma$ in the definition of Φ. Thus $\gamma_*^{-1} = \Phi_*$ in Theorem 7.1 in this case.

With these notations, we have the following theorem.

Theorem 11.1. $H_*\Gamma_0 B\mathcal{O}k_q = P\{\overline{v}_s \mid s \geq 1\} \otimes E\{\overline{u}_s \mid s \geq 1\}$ as an algebra under $*$, where $(\Phi - 1)_*(\overline{v}_s) = \overline{u}_s$. Therefore $\Phi_*\overline{v}_s = \sum_{i+j=s} \overline{v}_i * \overline{u}_j$, $\Phi_*\overline{u}_s = \overline{u}_s$, and the following formulas are satisfied.

(i) $Q^r v_s = \sum_j (r-s-1, s-j) v_j * v_{r+s-j}$ if $q \equiv \pm 1 \bmod 8$,

(ii) $Q^r v_s = \sum_{i,j,k} (r-s-1, s-j) v_i * v_k * \overline{u}_{j-i} * \overline{u}_{r+s-j-k}$ if $q \equiv \pm 3 \bmod 8$,

(iii) $Q^r \overline{u}_s = \sum_{i,j} (r-s-i-1, s-j) \overline{u}_i * \overline{u}_j * \overline{u}_{r+s-i-j}$ for all q.

Proof. We refer to [6] for the first sentence and show how the rest follows. $\Phi - 1 = *(\Phi, \chi)\Delta$, hence

$$\overline{u}_s = (\Phi-1)_*(\overline{v}_s) = \sum_{m+n=s} \Phi_*\overline{v}_m * \chi\overline{v}_n .$$

By induction on s, these formulas admit a unique solution for the $\Phi_*\overline{v}_s$. That solution is $\Phi_*\overline{v}_s = \sum_{i+j=s} \overline{v}_i * \overline{u}_j$ since

$$\sum_{m+n=s}\Big(\sum_{i+j=m}\bar{v}_i*\bar{u}_j\Big)*\chi\bar{v}_n = \sum_{i+j+n=s}(\bar{v}_i*\chi\bar{v}_n)*\bar{u}_j = \bar{u}_s .$$

Since $\Phi(\Phi-1) = 1-\Phi = \chi(\Phi-1)$ and $\chi\bar{u}_s = \bar{u}_s$ (by induction on s), $\Phi_*\bar{u}_s = \bar{u}_s$. Now formulas (i) and (ii) are immediate consequences of Theorem 7.1(i). Formula (iii) is proven by writing $\bar{v}_s*[2] = v_s*[1]$, computing $Q^r(\bar{v}_s*[2])$ by the Cartan formula, noting that $(\Phi-1)_*$ commutes with $*$ and the Q^s and takes $[2]$ to $[0]$, and explicitly calculating $Q^r\bar{u}_s = (\Phi-1)_*Q^r(\bar{v}_s*[2])$. An essential point is that $(\Phi-1)_*(\bar{u}_s) = 0$ for $s > 0$, and it is this which allows (i) and (ii) to yield the same formula (iii).

Of course, these formulas completely determine the operations Q^r in $H_*\Gamma B\mathcal{O}k_q$. The operations $\tilde{Q}^r v_s$ are trivial, and we shall compute the operations $\tilde{Q}^r u_s$ in Proposition 11.7 below, $u_s = \bar{u}_s*[1]$. Formula (ii) is extremely illuminating, as our later work will make clear. For example, it yields the following simple observation. (Compare Madsen [16, § 2].)

<u>Remarks 11.2.</u> Consider the Bockstein spectral sequence of $\Gamma_0 B\mathcal{O}k_q$. Obviously $\beta\bar{v}_{2r} = \bar{v}_{2r-1}$, hence $\beta\bar{u}_{2r} = \bar{u}_{2r-1}$. Thus

$$E^2 = P\{\bar{v}_{2s}^2\} \otimes E\{\bar{u}_{2s-1}*\bar{u}_{2s}\}.$$

By I.4.11, $\beta_2\bar{v}_{2s}^2 = \bar{v}_{2s}*\bar{v}_{2s-1}+Q^{2s}\bar{v}_{2s-1}$. Let $q \equiv \pm 3 \bmod 8$. Then

$$\begin{aligned} Q^{2s}\bar{v}_{2s-1} &= Q^{2s}(v_{2s-1}*[-1]) = Q^{2s}v_{2s-1}*[-2]+\bar{v}_{2s-1}^2*\bar{v}_1 \\ &= \sum_{i,j\leq 2s-1,k} \bar{v}_i*\bar{v}_k*\bar{u}_{j-i}*\bar{u}_{4s-1-j-k}+\bar{v}_{2s-1}^2*\bar{v}_1 . \end{aligned}$$

Adding in $\bar{v}_{2s}*\bar{v}_{2s-1}$ and reducing modulo $\operatorname{Im}\beta$, we find that

$$\beta_2\bar{v}_{2s}^2 = \sum_{i+j=s,\,j\geq 1} \bar{v}_{2i}^2*\bar{u}_{2j-1}*\bar{u}_{2j} .$$

We conclude from Lemma I.4.11 that, for $r \geq 2$,

$$E^{r+1} = P\{\bar{v}_{2s}^{2^r}\} \otimes E\{\bar{v}_{2s}^{2^r-2}\beta_2 v_{2s}^2\} \text{ with } \beta_{r+1}\bar{v}_{2s}^{2^r} = \bar{v}_{2s}^{2^r-2}\beta_2\bar{v}_{2s}^2 .$$

When $q \equiv \pm 1 \bmod 8$, $\beta_2\bar{v}_{2s}^2 = 0$ and a rather complicated calculation of Fiedorowicz and Priddy [6] shows that if r is maximal such that 2^r divides $\frac{1}{2}(q^2-1)$, then $\beta_r v_{2s}^2 \neq 0$ for all s.

The homology algebra of the multiplicative infinite loop space $\Gamma_1 B\mathcal{O}k_q$ can be computed the same way that of $\Gamma_1 B\mathcal{GL}k_r$ was computed in section 8. $\pi_1\Gamma_1 B\mathcal{O}k_q = Z_2 \oplus Z_2$ and the non-zero elements of $H^1\Gamma_1 B\mathcal{O}k_q$ correspond to the families of homomorphisms $O(n,k_q) \to Z_2$ given by the determinant, the spinor norm, and their product. The first and last of these restrict non-trivially to $O(1,k_q)$. Let $\tilde{\Gamma}_1 B\mathcal{O}k_q$ be the fibre of the infinite loop map $\det: \Gamma_1 B\mathcal{O}k_q \to K(Z_2,1)$. As a space, $\tilde{\Gamma}_1 B\mathcal{O}k_q$ is equivalent to $(BSOk_q)^+$. The proof of the following result is the same as that of Lemma 9.3.

<u>Lemma 11.3</u>. $\Gamma_1 B\mathcal{O}k_q$ is equivalent as an infinite loop space to the product $BO(1,k_q) \times \tilde{\Gamma}_1 B\mathcal{O}k_q$.

<u>Proposition 11.4.</u> $H_*\tilde{\Gamma}_1 B\mathcal{O}k_q$ is the tensor product of the exterior algebra on the generators $u_s = \bar{u}_s * [1]$ and a polynomial algebra on one generator in each degree ≥ 2.

While an elementary proof along the lines of those of Propositions 8.7 and 9.4 is possible, we prefer to rely on application of the Serre spectral sequence to diagram B below (together with Proposition 8.7) for identification of the polynomial algebra and on Corollary 11.5 below for identification of the exterior algebra.

As explained in detail in [R, VIII §2 and §3], the calculations of Fiedorowicz and Priddy [6] together with the machinery of [R] and, for diagram B, results of [2] and [19] imply that, when completed away from p, both of the following are equivalences of fibration sequences and commutative diagrams of infinite loop spaces and maps. Here the infinite loop spaces $\tilde{\Gamma}_0 B\mathcal{O}\bar{k}_p$ and $\tilde{\Gamma}_1 B\mathcal{O}\bar{k}_p$ are the fibres of the nontrivial maps from $\Gamma_0 B\mathcal{O}\bar{k}_p$ and $\Gamma_1 B\mathcal{O}\bar{k}_p$ to $K(Z_2, 1)$.

(A)
$$\begin{array}{ccccccc}
\Omega\tilde{\Gamma}_0 B\mathcal{O}\bar{k}_p & \xrightarrow{\zeta} & \Gamma_0 B\mathcal{O}k_q & \xrightarrow{\kappa} & \Gamma_0 B\mathcal{O}\bar{k}_p & \xrightarrow{\phi^q-1} & \tilde{\Gamma}_0 B\mathcal{O}\bar{k}_p \\
\downarrow{\scriptstyle \Omega\hat{\lambda}} & & \downarrow{\scriptstyle \hat{\lambda}} & & \downarrow{\scriptstyle \hat{\lambda}} & & \downarrow{\scriptstyle \hat{\lambda}} \\
\Omega BSO & \longrightarrow & F(\psi^q-1) & \longrightarrow & BO & \xrightarrow{\psi^q-1} & BSO
\end{array}$$

(B)
$$\begin{array}{ccccccc}
\Omega\tilde{\Gamma}_1 B\mathcal{O}\bar{k}_p & \xrightarrow{\zeta} & \Gamma_1 B\mathcal{O}k_q & \xrightarrow{\kappa} & \Gamma_1 B\mathcal{O}\bar{k}_p & \xrightarrow{\phi^q/1} & \tilde{\Gamma}_1 B\mathcal{O}\bar{k}_p \\
\downarrow{\scriptstyle \Omega\hat{\lambda}} & & \downarrow{\scriptstyle \hat{\lambda}} & & \downarrow{\scriptstyle \hat{\lambda}} & & \downarrow{\scriptstyle \hat{\lambda}} \\
\Omega BSO_{\otimes} & \longrightarrow & F(\psi^q/1) & \longrightarrow & BO_{\otimes} & \xrightarrow{\psi^q/1} & BSO_{\otimes}
\end{array}$$

$\hat{\lambda}\kappa: \Gamma_1 B\mathcal{O}k_q \to BO_{\otimes}$ lifts to an infinite loop map $\tilde{\Gamma}_1 B\mathcal{O}k_q \to BSO_{\otimes}$ since $(\hat{\lambda}\kappa)^*(w_1) = \det$. Since the Brauer lift of $\eta: Z_2 \to O(1, k_q)$ is $\eta: Z_2 \to O(1, R)$, $(\hat{\lambda}\kappa)_*(v_i) = v_i$.

Since $\bar{k}_p$ contains a square root of $a^2 + b^2$, conjugation by δ_n is an inner automorphism of $O(n, \bar{k}_p)$ for all n. It follows that $\kappa\Phi \simeq \kappa: \Gamma_0 B\mathcal{O}k_q \to \Gamma_0 B\mathcal{O}\bar{k}_p$. Therefore $\Phi - 1$ factors as $\zeta\omega$ for some map $\omega: \Gamma_0 B\mathcal{O}k_q \to \Omega\Gamma_0 B\mathcal{O}\bar{k}_p$. By Theorem 11.1, we see that $E\{\bar{u}_s\}$ coincides with $\zeta_* H_* \Omega\tilde{\Gamma}_0 B\mathcal{O}\bar{k}_p$. Of course, the composite

$$RP^{\infty} = BZ_2 \xrightarrow{\eta} BO(1, k_q) \xrightarrow{\iota} \Gamma_1 B\mathcal{O}k_q \xrightarrow{\rho^{-1}} \Gamma_0 B\mathcal{O}k_q \xrightarrow{\omega} \Omega\Gamma_0 B\mathcal{O}\bar{k}_p \longrightarrow$$
$$\xrightarrow{\Omega\hat{\lambda}} \Omega BSO \simeq SO$$

is homologically non-trivial. There is only one non-trivial A- algebra homomorphism $H^*SO \to H^*RP^{\infty}$, hence this composite must coincide homologically with the standard map $RP^{\infty} \to SO$. This proves that $(\Omega\hat{\lambda})_*(\overline{u}_s) = a_s \in H_*SO$. In particular, Theorem 11.1(iii) implies Kochman's calculation of $Q^r a_s$; compare Remarks 5.10. Now the following analog of Corollary 9.6, which is again an immediate consequence of Lemma 9.5, implies the identification of exterior algebra generators specified in Proposition 11.4.

Corollary 11.5. The following diagram is homotopy commutative

$$\begin{array}{ccc} \Omega\tilde{\Gamma}_0 B\mathcal{O}\overline{k}_p & \xrightarrow{\zeta} & \Gamma_0 B\mathcal{O}k_q \\ \Omega\rho \downarrow & & \downarrow \rho \\ \Omega\tilde{\Gamma}_1 B\mathcal{O}\overline{k}_p & \xrightarrow{\zeta} & \Gamma_1 B\mathcal{O}k_q \end{array}$$

Therefore $\rho\zeta$ is an H-map and

$$(x*[1])(y*[1]) = x*y*[1] \quad \text{for } x, y \in \zeta_* H_* \Omega\tilde{\Gamma}_0 B\mathcal{O}\overline{k}_p .$$

In a rather roundabout way, quite explicit generators for the polynomial part of $H_*\tilde{\Gamma}_1 B\mathcal{O}k_q$ will appear in the next section. It is also useful to have, in addition to the global statement Proposition 11.4, particular formulas which determine the #-product on $H_*\Gamma_1 B\mathcal{O}k_q$ in terms of its basis in the $\underline{*}$-product. Lemma 11.3 implies that $v_r v_s = (r,s)v_{r+s}$, the previous corollary gives $u_r u_s = u_r \underline{*} u_s$, and the remaining formula required is given in the following result.

Proposition 11.6. $u_r v_s = \sum_{i+j+k=s} (i,r) u_{r+i} \underline{*} u_j \underline{*} v_k$ for all r and s.

Proof. By Proposition 1.5 (iv), the specified formula is equivalent to

$$\overline{u}_r \overline{v}_s = \sum_{i,j} (i,j)\overline{u}_{i+j} * \overline{u}_{r-i} * \overline{u}_{s-j} .$$

We claim that the following diagram is homotopy commutative, where we have abbreviated $X = \Gamma_0 B\mathcal{O}k_q$:

$$
\begin{array}{ccccccc}
X\times X & \xrightarrow{\Delta\times 1} & X\times X\times X & \xrightarrow{\phi\times\chi\times 1} & X\times X\times X & \xrightarrow{*\times 1} & X\times X \\
 & \searrow^{\Delta} & \downarrow{\scriptstyle (1\times t\times 1)(1\times 1\times\Delta)} & & \downarrow{\scriptstyle (1\times t\times 1)(1\times 1\times\Delta)} & & \\
\downarrow{\scriptstyle \#} & & X\times X\times X\times X & \xrightarrow{\phi\times 1\times\chi\times 1} & X\times X\times X\times X & & \Big|{\scriptstyle \#} \\
 & & \downarrow{\scriptstyle \#\times\#} & & \downarrow{\scriptstyle \#\times\#} & & \\
X & \xrightarrow{\Delta} & X\times X & \xrightarrow{\phi\times\chi} & X\times X & \xrightarrow{*} & X
\end{array}
$$

The crucial relation $\Phi\circ\# \simeq \#\circ(\Phi\times 1)$ follows from the corresponding commutative diagram on the level of categories and functors, and the analogous relation $\chi\circ\# \simeq \#\circ(\chi\times 1)$ follows from the fact that $\chi\, x = x\,\#\,(-1)$ (see I.1.5). Chasing $\overline{v}_r\otimes\overline{v}_s$ around the diagram, we obtain

$$\overline{u}_r\overline{v}_s = (\Phi-1)_*(\overline{v}_r)\overline{v}_s = (\Phi-1)_*(\overline{v}_r\overline{v}_s).$$

By Proposition 1.5(iii) and the formula $v_r v_s = (r,s)v_{r+s}$, we have $\overline{v}_r\overline{v}_s = \sum_{i,j}\overline{v}_{i+j} * \chi\overline{v}_{r-i} * \chi\overline{v}_{s-j}$. Since $(\Phi-1)_*$ commutes with $*$ and χ and since $\chi\overline{u}_s = \overline{u}_s$, the conclusion follows.

Finally, we prove the analog of Corollary 11.5 for the operations $\widetilde{Q}^r$ and thus complete the theoretical determination of these operations in $H_*\Gamma_1 B\mathcal{O}k_q$.

<u>Proposition 11.7.</u> $\widetilde{Q}^r(x*[1]) = Q^r x * [1]$ for $x\in \zeta_* H_*\Omega\widetilde{\Gamma}_0 B\mathcal{O}\overline{k}_p$.

<u>Proof.</u> By the Adams-Priddy theorem [2], there exists an equivalence of infinite loop spaces completed at 2, $\xi:\widetilde{\Gamma}_0 B\mathcal{O}\overline{k}_p\rightarrow\widetilde{\Gamma}_1 B\mathcal{O}\overline{k}_p$. Any two H-equivalences $\Omega\widetilde{\Gamma}_0 B\mathcal{O}\overline{k}_p\rightarrow\Omega\widetilde{\Gamma}_1 B\mathcal{O}\overline{k}_p$ necessarily induce the same homomorphism on homology since they necessarily restrict to the same homomorphism $H_*RP^\infty\rightarrow H_*\widetilde{\Gamma}_1 B\mathcal{O}\overline{k}_p$ and since the image of H_*RP^∞ generates $H_*\Omega\widetilde{\Gamma}_0 B\mathcal{O}\overline{k}_p$ as an algebra. Therefore, in the diagram of Corollary 11.5, $\Omega\rho$ behaves homologically as if it were $\Omega\xi$ and, since the ζ are infinite loop maps, $\rho\zeta$ behaves homologically as if it were

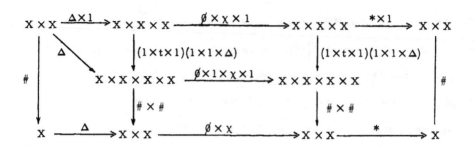

an infinite loop map.

The reader is referred to [6, App. B] for alternative group theoretical proofs of the previous results and for further details on the algorithm they imply for the computation of the operations $\tilde{Q}^r$ on $H_*\Gamma_1 B\mathcal{O}k_q$.

12. Orientation sequences at $p = 2$; analysis of $e: SF \to J_{\otimes}$

We return to the study of orientation sequences of the form (10.1), but here all homology groups are to be taken with Z_2 coefficients and all spaces and spectra are to be completed at 2.

Although apparently unrelated to BTop at 2, B(SF;kO) does play a central role, explained in [R, V §3-5], in Adams' study of the groups J(X) both at and away from 2. It is thus also of interest to understand its mod 2 homology. We begin by using the spaces studied in the previous section, with $q = 3$, to reduce the analysis of B(SF;kO) to that of $B(SF; j_2^{\delta})$ and then give a detailed analysis of the behavior of the relevant unit map $e: SF \to J^{\delta}_{\otimes 2}$ on homology.

Define jO_2^{δ} to be the (completion at 2 of) the E_{∞} ring spectrum derived from the bipermutative category $\mathcal{O}k_3$. As explained in [R, VI. 5.7], $\mathcal{O}k_3$ contains the bipermutative subcategory $\mathcal{N}k_3$ whose morphisms are those elements $\tau \in O(n, k_3)$, $n \geq 0$, such that $\nu(\tau)\det(\tau) = 1$, where ν is the spinor norm. Define j_2^{δ} to be the (completion at 2 of) the E_{∞} ring spectrum derived from $\mathcal{N}k_3$. By [R, VIII. 3.2], jO_2^{δ} is equivalent to the fibre jO_2 of $\psi^3 - 1: kO \to bso$ and j_2^{δ} is equivalent to the fibre j_2 of $\psi^3 - 1: kO \to bspin$. Let JO_2^{δ}, J_2^{δ}, $JO^{\delta}_{\otimes 2}$, and $J^{\delta}_{\otimes 2}$ be the 0-components and 1-components of the zeroth spaces of jO_2^{δ} and j_2^{δ}, and similarly without the superscript δ. Diagrams (A) and (B) of the previous section give infinite loop equivalences $JO_2^{\delta} \to JO_2$ and $JO^{\delta}_{\otimes 2} \to JO_{\otimes 2}$. By [R, VIII. 3.2 and 3.4], we have analogous equivalences of fibration sequences of infinite loop spaces

$$
\begin{array}{ccccccc}
\mathrm{Spin}^{\delta} & \xrightarrow{\zeta} & J_2^{\delta} & \xrightarrow{\kappa} & BO^{\delta} & \xrightarrow{\phi^3 - 1} & BSpin^{\delta} \\
\Omega\hat{\lambda}\downarrow & & \downarrow\hat{\lambda} & & \downarrow\hat{\lambda} & & \downarrow\hat{\lambda} \\
\mathrm{Spin} & \longrightarrow & J_2 & \longrightarrow & BO & \xrightarrow{\psi^3 - 1} & BSpin
\end{array} \tag{A'}
$$

$$
\begin{array}{ccccccc}
\mathrm{Spin}_{\otimes}^{\delta} & \xrightarrow{\zeta} & J_{\otimes 2}^{\delta} & \xrightarrow{\kappa} & BO_{\otimes}^{\delta} & \xrightarrow{\phi^3/1} & BSpin_{\otimes}^{\delta} \\
\Omega\hat{\lambda}\downarrow & & \downarrow\hat{\lambda} & & \downarrow\hat{\lambda} & & \downarrow\hat{\lambda} \\
\mathrm{Spin}_{\otimes} & \longrightarrow & J_{\otimes 2} & \longrightarrow & BO_{\otimes} & \xrightarrow{\psi^3/1} & BSpin_{\otimes}
\end{array} \tag{B'}
$$

Here $BO^{\delta} = \Gamma_0 B\mathcal{O}\bar{k}_3$, $BO_{\otimes}^{\delta} = \Gamma_1 B\mathcal{O}\bar{k}_3$, and $BSpin^{\delta}$ and $BSpin_{\otimes}^{\delta}$ are their 2-connected covers. J_2 is our choice for the space Im J at 2. Many authors instead use the fibre J_2' of $\psi^3 - 1 : BSO \to BSO$, which has the same homotopy type as J_2 but a different H-space structure. Various reasons for preferring J_2 were given in [R, V §4], and the calculations below give further evidence that this is the correct choice.

By [R, VIII. 3.4], we have a commutative diagram of infinite loop spaces and maps

$$
\begin{array}{ccccccc}
SF & \xrightarrow{e} & J_{\otimes 2}^{\delta} & \xrightarrow{\tau} & B(SF; j_2^{\delta}) & \xrightarrow{q} & BSF \\
\| & & \downarrow \hat{\lambda}\circ\kappa & & \downarrow B\hat{\lambda}\circ B\kappa & & \| \\
SF & \xrightarrow{e} & BO_{\otimes} & \xrightarrow{\tau} & B(SF; kO) & \xrightarrow{q} & BSF
\end{array} \tag{*}
$$

As explained in [R, VIII §3 and V §4 and §5], $B(SF; j_2^{\delta})$ is equivalent to the infinite loop space BCoker J_2, abbreviated BC_2, where the latter can and should be defined as the fibre of the universal cannibalistic class $c(\psi^3): B(SF; kO) \to BSpin_{\otimes}$. By [R, V. 4.8 and VIII. 3.4], the composite

$$
B(SF; j_2^{\delta}) \times BSpin \xrightarrow{B\hat{\lambda}\circ B\kappa \times g} B(SF; kO) \times B(SF; kO) \xrightarrow{\phi} B(SF; kO) \tag{1}
$$

is an equivalence, where g is the Atiyah-Bott-Shapiro orientation. Using work of Madsen, Snaith, and Tornehave [19] and of Adams and Priddy [2], Ligaard recently proved that g is an infinite loop map (see [R, V § 7]). Thus the specified composite is an equivalence of infinite loop spaces, and analysis of B(SF; kO) reduces to analysis of $B(SF; j_2^\delta) \simeq BC_2$.

As explained in [R, VIII §4], there is a commutative diagram of infinite loop spaces and maps

$$\begin{array}{ccc} Q_0S^0 & \xrightarrow{\alpha_2^\delta e} & SF \\ {\scriptstyle e}\downarrow & \nearrow{\scriptstyle \alpha_2^\delta} & \downarrow{\scriptstyle e} \\ J_2^\delta & \xrightarrow{e\alpha_2^\delta} & J_{\otimes 2}^\delta \end{array}$$

Here α_2^δ is emphatically not an equivalence. In homology, the proof of [R, VIII. 4. 1] gives the formula

$$(\alpha_2^\delta)_*\,(Q^I[1] * [-2^{\ell(I)}])\cdot[3^{2^{\ell(I)}}] = \tilde{Q}^I[3].$$

On the other hand, the Adams conjecture yields a homotopy commutative diagram

$$\begin{array}{ccccc} J_2 & \longrightarrow & BO & \xrightarrow{\psi^3-1} & BSpin \\ {\scriptstyle \alpha_2}\downarrow & & \downarrow{\scriptstyle \gamma_2} & & \| \\ SF & \xrightarrow{\tau} & SF/Spin & \xrightarrow{q} & BSpin \end{array}$$

Here $SF/Spin \simeq BO(1) \times F/O$ as an infinite loop space by [R, V. 3. 4], and the following composites are equivalences by [R, V. 4. 7 and VIII§ 3], where $(SF; j_2^\delta) = \Omega B(SF; j_2^\delta) \simeq C_2$.

(2) $J_2 \xrightarrow{\alpha_2} SF \xrightarrow{e} J_{\otimes 2}^\delta$

(2) $J_2 \times (SF, j_2^\delta) \xrightarrow{\alpha_2 \times \Omega(B\hat{\lambda} \circ B\kappa)} SF \times SF \xrightarrow{\phi} SF$

(3) $BO \times (SF; j_2^\delta) \xrightarrow{\gamma_2 \times \tau\Omega(B\hat{\lambda} \circ B\kappa)} SF/Spin \times SF/Spin \xrightarrow{\phi} SF/Spin.$

With these facts in mind, we return to homology. Comparison of diagram A' to diagram A of the previous section yields the following addendum to Theorem 11.1. Let $\bar{u}_0 = \bar{v}_0 = [0]$.

Proposition 12.1. $H_*J_2^\delta = P\{\sum_{i+j=s} \bar{u}_i * \bar{v}_j \mid s \geq 1\} \otimes E\{\bar{u}'_s \mid s \neq 2^i\} \subset H_*JO_2^\delta$, where $\bar{u}'_s \in E\{\bar{u}_s\}$ and $\bar{u}_s + \bar{u}'_s$ is decomposable (under $*$).

Proof. The following evaluation formulas hold for JO_2^δ:

$$<\det, \bar{v}_1> = 1 \quad , \quad <\nu \cdot \det, \bar{v}_1> = 1, \quad , \quad <\nu, \bar{v}_1> = 0$$

$$<\det, \bar{u}_1> = 0 \quad , \quad <\nu \cdot \det, \bar{u}_1> = 1 \quad , \quad <\nu, \bar{u}_1> = 1 \, .$$

Indeed, for $\bar{v}_1$, these hold by consideration of the composite $BO(1, k_3) \rightarrow JO^\delta_{\otimes 2} \xrightarrow{\det} K(Z_2, 1)$, while $<\det, \bar{u}_1> = 0$ since $\det = (\det) \circ \Phi$; the remaining formulas are forced by the fact that $\det$, $\nu . \det$, and ν are distinct cohomology classes. Therefore $\bar{u}_1 + \bar{v}_1$ is the image of $H_1 J_2^\delta$ in $H_1 JO_2^\delta$. Since $\tilde{H}_* J_2^\delta$ consists of

$$\{x \mid \psi x = \sum x' \otimes x'', (\nu \cdot \det)_*(x') = 0 \text{ if } \deg x' > 0\} \subset H_*JO_2^\delta ,$$

it follows by induction and the fact that $H_s K(Z_2, 1)$ contains no non-zero primitive elements if $s > 1$ that $\sum \bar{u}_i * \bar{v}_j \in H_* J_2^\delta$. The rest is clear.

Recall that $H^*SO = P\{\sigma^* w_{2s} \mid s \geq 1\}$ and $H^*Spin = H^*SO/(\sigma^* w_2)$. In the dual basis, with $<\sigma^* w_{2s}, a'_{2s-1}> = 1$,

$$H_*Spin = \Gamma\{a'_{2s-1} \mid s \geq 2\} \subset \Gamma\{a'_{2s-1} \mid s \geq 1\} = H_*SO \, .$$

The change of basis implicit in the proposition corresponds to that comparing this description of H_*SO to that given by $H_*SO = E\{a_s\}$, $a_s \in \operatorname{Im} H_*RP^\infty$.

The following corollary, a less obvious analog of which for J'_2 was first proven by Madsen [15], explains why $e\alpha_2^\delta$ could not possibly be an equivalence and demonstrates that no choice of $\alpha_2 : J_2 \to SF$ can be an H-map.

Corollary 12.2. No H-map $J_2 \to SF$ can induce a monomorphism in (mod 2) homology in degree 2.

Proof. $\{\bar{v}_2, \bar{v}_1^2 = (\bar{u}_1 + \bar{v}_1)^2\}$ and $\{x_2, x_{(1,1)}\}$ are bases for $H_2 J_2^\delta \cong H_2 J_2$ and $H_2 SF$, and $\bar{v}_1^2$ maps to zero under any H-map.

We wish to study $e: SF \to J^\delta_{\otimes 2}$, and it is convenient to first use Proposition 12.1 and Theorem 11.1 to study $e: Q_0 S^0 \to J_2^\delta$.

Theorem 12.3. The restriction of $e_* : H_* Q_0 S^0 \to H_* J_2^\delta$ to the $*$-subalgebra

$$P\{Q^s[1]*[-2] \mid s \geq 1\} \otimes P\{Q^{2^s n + 2^s} Q^{2^s n}[1]*[-4] \mid s \geq 0 \text{ and } n \geq 1\}$$

of $H_* Q_0 S^0$ is an epimorphism.

Proof. By Theorem 11.1, the following congruences hold modulo $*$-decomposable elements of $H_* \Gamma B \mathcal{O} k_3$:

$$Q^s[1] = \sum v_k * u_{s-k} \equiv v_s * [1] + u_s * [1]$$

and

$$Q^r Q^s[1] \equiv Q^r(v_s * [1]) + Q^r(u_s * [1]) \equiv (r-s-1, s) v_{r+s} * [3].$$

Since $(r-s-1, s) = 0$ for all $r > s$ such that $r+s = t$ if and only if t is a power of 2, the coefficient prevents decomposition of u_{2^q} in terms of $*$ and the operations Q^r (as is consistent with $\operatorname{Im} e_* \subset H_* J_2^\delta \subset H_* JO_2^\delta$). If $t = 2^s(2n+1)$, then $(2^s - 1, 2^s n) = 1$ and therefore

$$Q^{2^s n + 2^s} Q^{2^s n}[1] \equiv v_t * [3].$$

The conclusion follows immediately from Proposition 12.1.

Turning to multiplicative structures, we note first that comparison of diagram B' to diagram B of the previous section yields the following addendum to Propositions 11.4 and 11.7 and Corollary 11.5. Let $\tilde{J}^{\delta}_{\otimes 2}$ denote the universal cover of $J^{\delta}_{\otimes 2}$, namely the fibre of $\det : J^{\delta}_{\otimes 2} \to K(Z_2, 1)$, and observe that we have the following commutative diagram of fibration sequences of infinite loop spaces:

$$\begin{array}{ccccccc}
\mathrm{Spin}^{\delta}_{\otimes} & \xrightarrow{\zeta} & \tilde{J}^{\delta}_{\otimes 2} & \longrightarrow & BSO^{\delta}_{\otimes} & \xrightarrow{\phi^3/1} & B\mathrm{Spin}^{\delta}_{\otimes} \\
\| & & \downarrow & & \downarrow & & \| \\
\mathrm{Spin}^{\delta}_{\otimes} & \xrightarrow{\zeta} & J^{\delta}_{\otimes 2} & \longrightarrow & BO^{\delta}_{\otimes} & \xrightarrow{\phi^3/1} & B\mathrm{Spin}^{\delta}_{\otimes} \\
\downarrow & & \downarrow \det & & \downarrow \det & & \downarrow \\
* & \longrightarrow & K(Z_2, 1) & = & K(Z_2, 1) & \longrightarrow & *
\end{array}$$

<u>Proposition 12.4.</u> $H_* \tilde{J}_{\otimes 2}$ is the tensor product of the exterior algebra on the generators $u'_s = \bar{u}'_s * [1]$ and a polynomial algebra on one generator in each degree ≥ 2. For $x, y \in E\{u'_s\} = \zeta_* H_* \mathrm{Spin}_{\otimes}$,

$$xy = x \,\underline{*}\, y \quad \text{and} \quad \tilde{Q}^r x = Q^r(x * [-1]) * [1].$$

Recall that $\underline{*}$ is the translate of the $*$ product from the 0-component to the 1-component.

$\tilde{J}^{\delta}_{\otimes 2}$ has been constructed by first taking the fibre of $\nu \cdot \det : JO^{\delta}_{\otimes 2} \to K(Z_2, 1)$ and then that of det. Since we could equally well reverse the order, $\tilde{J}^{\delta}_{\otimes 2}$ is also the universal cover of $\tilde{\Gamma}_1 B\mathcal{O}k_3$. Since $\nu \cdot \det$ restricts non-trivially to $O(1, k_3)$, $JO^{\delta}_{\otimes 2} \simeq BO(1, k_3) \times J^{\delta}_{\otimes 2}$ as an infinite loop space, this splitting being distinct from the splitting $JO^{\delta}_{\otimes 2} \simeq BO(1, k_3) \times \tilde{\Gamma}_1 B\mathcal{O}k_3$ of Lemma 11.3.

<u>Theorem 12.5</u> The restriction of $e_* : H_* SF \to H_* J^{\delta}_{\otimes 2}$ to the #-subalgebra

$$E\{x_r \mid r \geq 1\} \otimes P\{x_{(2^s, 2^s)} \mid s \geq 0\} \otimes P\{x_{(2^s n+2^s, 2^s n)} \mid s \geq 0 \text{ and } n \geq 1\}$$

of H_*SF is an isomorphism.

Proof. Note first that $e_*x_r = \sum v_k \mathbin{\underline{*}} u_{r-k}$ maps non-trivially to $H_*K(Z_2, 1)$ under $\det: J^\delta_{\otimes 2} \to K(Z_2, 1)$ since this assertion holds for x_1 by Theorem 12.3. Of course, we must not confuse

$$H_*SO \xrightarrow{j_*} H_*SF \xrightarrow{e_*} H_*J^\delta_{\otimes 2} \quad \text{with} \quad H_*Spin_\otimes \xrightarrow{\zeta_*} H_*J^\delta_{\otimes 2} \ .$$

Indeed, $u'_r \in \operatorname{Im} \zeta_*$ and the inverse image of u'_r in the #-subalgebra of H_*SF specified in the statement is most unobvious. Let $p_{2s+1} = x_{2s+1} + \sum_{j=1}^{s} x_j x_{2s+1-j}$ be the non-zero primitive element of degree $2s+1$ in $E\{x_r\}$ and let $p'_s = sx_s + \sum_{j=1}^{s-1} x_j \mathbin{\underline{*}} p'_{s-j}$ be the non-zero primitive element of degree s in $P(\{x_r\}; \underline{*})$. Obviously $p_1 = p'_1$, and Propositions 1.5 and 1.6 give

$$x_1x_2 = x_{(2,1)} + x_2 \mathbin{\underline{*}} x_1 + x_1 \mathbin{\underline{*}} x_1 \mathbin{\underline{*}} x_1 \ , \text{ hence } p_3 = p'_3 + x_{(2,1)} \ .$$

Therefore $e_*(p_1) \neq 0$ and $e_*(p_3) \neq 0$ by Theorem 12.3. Since $P^r_* Q^s[1] = (r, s-2r)Q^{s-r}[1]$, we have

$$P^{2r}_* p_{2s+1} = (2r, 2s+1-4r)p_{2s-2r+1} = (r, s-2r)p_{2s-2r+1} \ .$$

In particular,

$$P^{2s}_* p_{4s+1} = p_{2s+1} \quad \text{and} \quad P^{2^s}_* p_{2^s(t+1)+1} = p_{2^s t+1} \quad \text{if } s \geq 1 \text{ and } t \text{ is odd.}$$

The second of these shows that $p_{2^s t+1}$ can be hit by iterated Steenrod operations acting on some p_{2^q+1}, and the first of these shows that p_{2^q+1} hits p_3 under some iterated Steenrod operation. The same formulas also hold for the Steenrod operations on the odd degree primitive elements of

$P\{\bar{u}_s + \bar{v}_s\}$ and $E\{\bar{u}_s'\}$ hence, by translation to the 1-component, on the two basic families of odd degree primitive elements of $H_*J^\delta_{\otimes 2}$. Therefore $e_* p_{2s+1} \neq 0$ for all $s \geq 0$, and the restriction of e_* to $E\{x_r\}$ is a monomorphism. Let f denote the composite

$$P\{x_{(2^s, 2^s)}, x_{(2^s n+2^s, 2^s n)}\} \subset H_*SF \xrightarrow{e_*} H_*J^\delta_{\otimes 2} \xrightarrow{\kappa_*} H_*BO^\delta_{\otimes} \longrightarrow H_*BSO^\delta_{\otimes},$$

where $BO^\delta_{\otimes} \to BSO^\delta_{\otimes}$ is induced by the evident splitting of infinite loop spaces $BO^\delta_{\otimes} \simeq BO(1, \bar{k}_3) \times BSO^\delta_{\otimes}$. By an easy comparison of dimensions argument, it suffices to prove that f is an epimorphism and thus an isomorphism. By the Adams-Priddy theorem [2], $BSO^\delta_{\otimes} \simeq BSO$ as an infinite loop space. In particular, the known formulas for the Steenrod operations on the indecomposable elements of H_*BSO apply equally well to $H_*BSO^\delta_{\otimes}$. Explicitly, if y_s is the non-zero element of degree s in $QH_*BSO^\delta_{\otimes}$, then

$$P^s_* y_{2s} = y_s \quad\text{and}\quad P^{2r}_* y_{2s+1} = (r, s-2r) y_{2s-2r+1} .$$

By the Nishida relations I.1.1(9), the analogous formulas

$$P^{r+s}_* x_{(2r, 2s)} = x_{(r, s)} \quad\text{and}\quad P^{2r}_* x_{(s+1, s)} = (r, s-2r) x_{(s-r+1, s-r)}$$

hold in H_*SF. By use of appropriate special cases (exactly like those in the first half of the proof), it follows that f induces an epimorphism on indecomposable elements since, by Theorem 12.3, $fx_{(1,1)}$ and $fx_{(2,1)}$ are non-zero indecomposable elements.

An obvious comparison of dimensions argument gives the following corollary, which complements Proposition 12.4. For notational convenience, define $t_r \in H_r J^\delta_{\otimes 2}$ for $r \geq 2$ by

$$t_r = \begin{cases} e_* x_{(2^s, 2^s)} & \text{if } r = 2^{s+1} \\ e_* x_{(2^s n + 2^s, 2^s n)} & \text{if } r = 2^s(2n+1) \end{cases}$$

Corollary 12.6. As an algebra under $\#$,

$$H_* J^{\delta}_{\otimes 2} = E\{u'_r \mid r \neq 2^q\} \otimes E\{e_* x_{2^q} \mid q \geq 0\} \otimes P\{t_r \mid r \geq 2\}.$$

Comparison of this result to Theorem 6.3 is illuminating: The R-algebra generators of H_*SF map onto algebra generators for the complement of $\zeta_* H_* Spin^{\delta}_{\otimes}$ in $H_* J^{\delta}_{\otimes 2}$.

Remark 12.7. By (*), the composite $SF \xrightarrow{e} J_{\otimes 2} \xrightarrow{\kappa} BO_{\otimes} \xrightarrow{\hat{\lambda}} BO_{\otimes}$ is the unit infinite loop map $e: SF \to BO_{\otimes}$. Since $x_{(r,s)} = Q^r Q^s[1] * [-3]$, $e_* x_{(r,s)}$ can be read off in terms of the basis for $H_* BO_{\otimes}$ specified by $H_* BO_{\otimes} = P(\{v_s\}, \underline{*})$ by application of Theorem 7.1(i). Thus the results above yield explicit polynomial generators for $H_* BSO_{\otimes} \subset H_* BO_{\otimes}$; that the images of the t_r do actually lie in $H_* BSO_{\otimes}$ can be checked by verifying that they come from $H_* BSO(2)$.

We require two further technical results in order to obtain complete information about the various algebra structures on the classifying space level. The following result was first noted by Fiedorowicz.

Lemma 12.8. For $r \neq 2^q$, $e_* x_{(r,r)}$ is a $\#$-decomposable element of $H_* J^{\delta}_{\otimes 2}$.

Proof. Working in $H_* \Gamma B \mathcal{O} k_3$, we find that

$$Q^r Q^r[1] = Q^r[1] * Q^r[1] = (\textstyle\sum v_j * u_{r-j}) * (\textstyle\sum v_k * u_{r-k})$$

$$= \textstyle\sum v_j * v_j * u_{r-j} * u_{r-j} = v_r * v_r * [2]$$

by Theorem 11.1, symmetry, and the fact that $u_i * u_i = 0$. Thus $e_*x_{(r,r)} = v_r \underline{*} v_r$. Since $\psi(v_r \underline{*} v_r) = \sum_{i+j=r} (v_i \underline{*} v_i) \otimes (v_j \underline{*} v_j)$, we can form the (Newton) primitive elements

$$b_s = s(v_s \underline{*} v_s) + \sum_{j=1}^{s-1} (v_j \underline{*} v_j) b_{s-j} .$$

If s is odd, $b_s \equiv v_s \underline{*} v_s$ modulo #-decomposable elements and b_s has even degree. In degrees > 2, all even degree primitive elements of $H_*J^{\delta}_{\otimes 2}$ are squares. Therefore $v_r \underline{*} v_r$ is #-decomposable if $r \geq 3$ is odd. Since $H^*J^{\delta}_{\otimes 2}$ is a polynomial algebra, the squaring homomorphism on its primitive elements is a monomorphism. Dually, the halving homomorphism $P^n_*: H_{2n}J^{\delta}_{\otimes 2} \to H_nJ^{\delta}_{\otimes 2}$ induces an epimorphism, and thus an isomorphism if $n \geq 2$, on indecomposable elements. Since $P^{2r}_*(v_{2r} \underline{*} v_{2r}) = v_r \underline{*} v_r$, the conclusion follows.

Lemma 12.9. For $r \neq 2^q$, $\tilde{Q}^{r+1} t_r \equiv t_{2r+1} + e_*x_{2r+1}$ modulo #-decomposable elements of $H_*J^{\delta}_{\otimes 2}$.

Proof. By the Nishida relations and the proof of Theorem 12.5,

$$P^{2r}_* \tilde{Q}^{2r+1} t_{2r} = \tilde{Q}^{r+1} t_r , \quad P^{2r}_* t_{4r+1} = t_{2r+1} , \quad \text{and} \quad P^{2r}_* x_{4r+1} = x_{2r+1} .$$

Thus, by the argument of the previous proof, it suffices to prove the result when r is odd. Similarly, if $2q \leq r+1$,

$$P^{4q}_* \tilde{Q}^{2r+2} t_{2r+1} = (q, r-2q) \tilde{Q}^{2r-2q+2} t_{2r-2q+1} ,$$

$$P^{4q}_* t_{4r+3} = (q, r-2q) t_{4r-4q+3} \quad \text{and} \quad P^{4q}_* x_{4r+3} = (q, r-2q) t_{4r-4q+3} .$$

By the special cases cited in the proof of Theorem 12.5, it suffices to prove the result when $r = 3$. Since $t_3 = e_*x_{(2,1)}$ is primitive, $\tilde{Q}^4 t_3$ is also primitive and is therefore of the form $aq_7 + br_7$, where

$$q_k = kv_k + \sum_{j=1}^{k-1} v_j \underline{*}\, q_{k-j} \quad \text{and} \quad r_{2k+1} = u_{2k+1} + \sum_{j=1}^{k} u_j \underline{*}\, u_{2k+1-j} .$$

Clearly, the coefficients a and b can be read off from a calculation of $\tilde{Q}^4 t_3$ modulo $\underline{*}$-decomposable elements. Theorem 11.1 implies that

$$e_* x_{(2,1)} = q_3 = v_3 + v_2 \underline{*}\, v_1 + v_1 \underline{*}\, v_1 \underline{*}\, v_1 .$$

The mixed Cartan formula, Proposition 1.5 (in particular $x\cdot[0] = 0$ if $\deg x > 0$), Theorem 11.1, and the fact that $\tilde{Q}^r v_s = 0$ if $r > 0$ and $s > 0$ imply that, modulo $\underline{*}$-decomposable elements, $\tilde{Q}^4(v_1 \underline{*}\, v_1 \underline{*}\, v_1) \equiv 0$, $\tilde{Q}^4 v_3 = 0$, and

$$\tilde{Q}^4(v_2 \underline{*}\, v_1) \equiv Q^4(v_2 \bar{v}_1) * [1] \equiv Q^4 v_3 * [-1] \equiv u_7 + v_7 .$$

Therefore $a = b = 1$ and $\tilde{Q}^4 t_3 = q_7 + r_7$. We must still calculate $q_7 + r_7$ modulo #-decomposable elements. Recall from the proof of Theorem 12.5 that $p_3 = p_3' + x_{(2,1)}$. Since $p_{2s+1} \equiv x_{2s+1}$ modulo #-decomposable elements, the formulas for Steenrod operations in the cited proof imply that

$$e_* p'_{2s+1} \equiv e_* x_{2s+1} + e_* x_{(s+1,s)} \quad \text{modulo \#-decomposable elements.}$$

Since $p_s' = sx_s + \sum_{j=1}^{s-1} x_j \underline{*}\, p'_{s-j}$, Theorem 11.1 implies that

$$e_* p'_{2s+1} \equiv e_* x_{2s+1} \equiv v_{2s+1} + u_{2s+1} \quad \text{modulo } \underline{*}\text{-decomposable elements}$$

Therefore $e_* p'_{2s+1} = q_{2s+1} + r_{2s+1}$ and

$$q_{2s+1} + r_{2s+1} \equiv e_* x_{2s+1} + e_* x_{(s+1,s)} \quad \text{modulo \#-decomposable elements.}$$

We shall also need the following consequence of the previous lemma.

Corollary 12.10. For $r \neq 2^q$, $\tilde{Q}^{2r+2} t_{2r} \equiv t_{4r+2} + e_* x_{4r+2}$ modulo #-decomposable elements of $H_* J^{\delta}_{\otimes 2}$.

Proof. Certainly $\tilde{Q}^{2r+2} t_{2r} \equiv at_{4r+2} + be_* x_{4r+2}$ for some constants a and b, and we see that $a = b = 1$ by applying β to both sides.

§13. The homology of BCoker J, BSF, and $BJ_{\otimes}$ at $p = 2$.

Until otherwise specified, all homology and cohomology groups are to be taken with Z_2 coefficients. Again, all spaces and spectra are to be completed at 2.

Precisely as in Section 10, we can now exploit our understanding of $e_*: H_*SF \to H_*J^{\delta}_{\otimes 2}$ to compute $H_*B(SF; j^{\delta}_2)$ as a sub Hopf algebra of H_*BSF. We first specify certain elements of H_*SF which lie in the kernel of e_*. Here there are two different choices available according to whether we choose to use the description of H_*SF given in Theorem 5.1 or in Theorem 6.1. For each $I = (J, K)$, $\ell(K) = 2$, such that $x_I \in X$ (as in 5.1), write $\tilde{x}_I = \tilde{Q}^J x_K$ for the corresponding element of $\tilde{X}$ (as in 6.1). There are unique elements

$$z_I, \tilde{z}_I \in E\{x_r\} \otimes P\{x_{(2^s,2^s)}\} \otimes P\{x_{(2^s n+2^s, 2^s n)}\}$$

such that $e_* z_I = e_* x_I$ and $e_* \tilde{z}_I = e_* \tilde{x}_I$, and we define

$$y_I = x_I + z_I \quad \text{and} \quad \tilde{y}_I = \tilde{x}_I + \tilde{z}_I \;.$$

The following sequence of results gives a complete analysis of the behavior on mod 2 homology of the diagram

$$(*) \qquad \begin{array}{ccccc}
 & & BSO & & BSpin^{\delta}_{\otimes} \\
 & & \downarrow Bj & & \downarrow \\
B(SF; j^{\delta}_2) & \xrightarrow{q} & BSF & \xrightarrow{Be} & BJ^{\delta}_{\otimes 2} \\
 & & \downarrow & & \downarrow \\
 & & B(F/O) & & BO^{\delta}_{\otimes} \simeq K(Z_2, 2) \times BBSO^{\delta}_{\otimes}
\end{array}$$

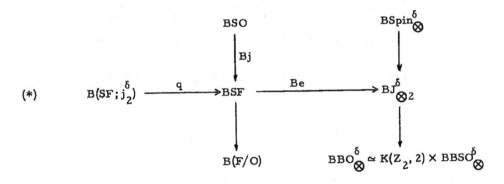

Our first result is an immediate consequence of Theorems 5.1 and 6.1 and the observation that $E^2 = E^\infty$ in the Eilenberg-Moore spectral sequence converging from $\mathrm{Tor}^{H_*(F/O)}(Z_2, Z_2)$ to $H_*B(F/O)$. It is recorded in order to clarify the counting arguments needed to prove the following two results, which implicitly contain alternative, more geometrically based, descriptions of H_*BSF and $H_*B(F/O)$.

Theorem 13.1. As Hopf algebras,

$$H_*BSF = H_*BSO \otimes E\{\sigma_* x_{(s,s)} \mid s \geq 1\} \otimes AB\widetilde{X}$$

and

$$H_*B(F/O) = H_*BSF /\!/ H_*BSO = E\{\sigma_* x_{(s,s)} \mid s \geq 1\} \otimes AB\widetilde{X},$$

where $B\widetilde{X} = \{\sigma_* \widetilde{x}_I \mid \ell(I) > 2 \text{ and } e(I) > 1 \text{ or } \ell(I) = 2 \text{ and } e(I) \geq 1\}$.

Theorem 13.2. The image of $H_*B(SF; j_2^\delta)$ in H_*BSF (under q_*) is the tensor product of the following three sub Hopf algebras:

$$E\{\sigma_* \widetilde{y}_{(r,r)} \mid r \geq 3 \text{ and } r \neq 2^q\}$$

$$P\{\sigma_* \widetilde{y}_{(r+1,r)} \mid r \geq 3 \text{ and } r \neq 2^q\}$$

$$P\{\sigma_* \widetilde{y}_I \mid \ell(I) \geq 2 \quad \text{and} \quad e(I) \geq 2\,;\ I \neq (2^s n + 2^s, 2^s n)\}.$$

Moreover, $\sigma_* \widetilde{y}_{(r,r)} = \sigma_* x_{(r,r)}$ and, if $r = 2^s(2n+1)$,

$$\sigma_* \widetilde{y}_{(r+1,r)} = \sigma_* x_{(r+1,r)} + \sigma_* x_{2r+1} + (\sigma_* x_{(2^s n+2^s, 2^s n)})^2 .$$

The result remains true with $\sigma_* \widetilde{y}_I$ replaced by $\sigma_* y_I$ in the third algebra.

Theorem 13.3. The restriction of $(Be)_*: H_*BSF \to H_*BJ^\delta_{\otimes 2}$ to the sub Hopf algebra

$$H_*BSO \otimes E\{\sigma_* x_{(2^s, 2^s)} \mid s \geq 0\} \otimes P\{\sigma_* x_{(2^s+1, 2^s)} \mid s \geq 0\}$$

$$\otimes\, P\{\sigma_* x_{(2^s n+2^s, 2^s n)} \mid s \geq 1 \text{ and } n \geq 1\}$$

of H_*BSF is an isomorphism.

Corollary 13.4. $(B\zeta)_*: H_*BSpin^{\delta}_{\otimes} \to H_*BJ^{\delta}_{\otimes 2}$ is a monomorphism and

$$H_*BJ^{\delta}_{\otimes 2} = H_*BSpin^{\delta}_{\otimes} \otimes P\{\sigma_* e_* x_{2^s} \mid s \geq 0\} \otimes E\{\sigma_* t_{2^s} \mid s \geq 1\} \otimes P\{\sigma_* t_{2^s+1} \mid s \geq 1\}$$

$$\otimes P\{\sigma_* t_{2^s(2n+1)} \mid s \geq 1 \text{ and } n \geq 1\}$$

Corollary 13.5. The sub coalgebra $\Gamma\{\sigma_* a_{2^s} \mid s \geq 0\}$ of H_*BSO maps isomorphically onto $H_*K(Z_2, 2)$ under the composite

$$H_*BSO \xrightarrow{(Bj)_*} H_*BSF \xrightarrow{(Be)_*} H_*BJ^{\delta}_{\otimes} \longrightarrow H_*BBO^{\delta}_{\otimes} \xrightarrow{(B\det)_*} H_*K(Z_2, 2).$$

Corollary 13.6. The sub Hopf algebra

$$E\{\sigma_* t_{2^s} \mid s \geq 1\} \otimes P\{\sigma_* t_{2^s+1} \mid s \geq 1\} \otimes P\{\sigma_* t_{2^s(2n+1)} \mid s \geq 1 \text{ and } n \geq 1\}$$

of $H_*BJ^{\delta}_{\otimes 2}$ maps isomorphically onto $H_*BSO^{\delta}_{\otimes}$ under the natural map.

Proofs. Write $\{E^r X\}$ for the Eilenberg-Moore spectral sequence converging from $E^2X = \mathrm{Tor}^{H_*X}(Z_2, Z_2)$ to H_*BX. By Theorem 12.5, the composite

$$E^2SO \otimes E\{\sigma x_{(2^s, 2^s)}\} \otimes E\{\sigma x_{(2^s n + 2^s, 2^s n)}\} \subset E^2SF \to E^2J^{\delta}_{\otimes 2}$$

is an isomorphism. Thus $E^2SO \otimes E\{\sigma t_r \mid r \geq 2\} \cong E^2J^{\delta}_{\otimes 2} = E^{\infty}J^{\delta}_{\otimes 2}$. By Lemma 12.9, $\sigma_* t_{2r+1} = (\sigma_* t_r)^2 + (Be)_* \sigma_* x_{2r+1}$ if $r \neq 2^s$, and this implies Theorem 13.3. Corollary 13.4 follows in view of Corollary 12.6. Theorem 13.3 implies that $E^2 = E^{\infty}$ in the Serre spectral sequence of $B(SF; j^{\delta}_2) \to BSF \to BJ^{\delta}_{\otimes 2}$ and thus that q_* is a monomorphism. The sub Hopf algebra of H_*BSF specified in Theorem 13.2 certainly lies in the image of q_* and is all of this image by an obvious counting argument. The formulas for $\sigma_* y_{(r,r)}$ and $\sigma_* \tilde{y}_{(r+1,r)}$ are immediate from Lemmas 12.8

and 12.9. Recall that $E^2SO = \Gamma\{\sigma_* a_r\} = E^\infty SO$ and $E^\infty SO = H_*SO$ as a coalgebra. This makes sense of Corollary 13.5, which now follows easily from the first sentence in the proof of Theorem 12.5. Note that only the odd degree generators of H_*BSO are in the image of σ_* since $\sigma_* a_{2r+1} = (\sigma_* a_r)^2$ for $r \geq 2$. Finally, Corollary 13.6 holds since the analogous assertion for $E^2 J^\delta_{\otimes 2} \rightarrow E^2 BSO^\delta_\otimes$ holds by the proof of Theorem 12.5.

In the remainder of this section, we shall analyze the behavior of the diagram (*) with respect to higher torsion. Henceforward, $\{E^r X\}$ will denote the mod 2 homology Bockstein spectral sequence of a space X. We begin by identifying those parts of $\{E^r BSF\}$ which are already determined by the formula $\beta Q^{s+1} = sQ^s$, the fact that $\sigma_*\beta = \beta\sigma_*$, and the general formulas for higher Bocksteins on squares given in I.4.11. We need a lemma.

<u>Lemma 13.7.</u> $\tilde{Q}^{2i+2} x_{(i,i)}$ is #-decomposable in H_*SF for all i.

<u>Proof.</u> $x_{(i,i)} = Q^i Q^i[1] * [-3]$, and $Q^i Q^i[1] = \tilde{Q}^i Q^i[1]$ by Lemma 1.9. As pointed out in the proof of Lemma 2.7, $\tilde{Q}^1[-3] = 0$, $\tilde{Q}^{2i+1}\tilde{Q}^i = 0$, and $Q^{2i+1} Q^i = 0$. In the evaluation of $\tilde{Q}^{2i+2} x_{(i,i)}$ by the mixed Cartan formula, all terms not zero by these facts or by I.1.1(iii) have either $\tilde{Q}^{2i} Q^i Q^i[1]$ or $Q^{2i} Q^i Q^i[-3]$ as a *-factor and are therefore #-decomposable by Propositions 6.4 and 6.6.

The following two propositions should be regarded as establishing notation and identifying certain differentials in $\{E^r BSF\}$. From this point of view, the proofs are immediate from I.4.11 and the lemma. That the spectral sequences $\{{}_0E^r\}$ and $\{{}_1E^r\}$ do actually embed as stated in the various spectral sequences $\{E^r X\}$ (with no relations and no interference from other differentials) will emerge from later counting arguments.

Proposition 13.8. Define a spectral sequence $\{{}_0E^r\}$ by

$${}_0E^{r+1} = \bigotimes_y P\{(\sigma_* y)^{2^r}\} \otimes E\{\beta_{r+1}(\sigma_* y)^{2^r}\} \quad \text{for } r \geq 1,$$

with $\beta_{r+1}(\sigma_* y)^{2^r} = (\sigma_* y)^{2^r - 2}[(\sigma_* y)(\beta\sigma_* y) + \tilde{Q}^{2q}\beta\sigma_* y]$ if $\deg y = 2q-1$, where y runs through the union of the following two sets:

$$\{\tilde{y}_I \mid I = (2s, J), d(I) \text{ odd}, e(I) \geq 3\} \quad \text{and} \quad \{\tilde{y}_{(2i,\, 2i-1)} \mid i \geq 2\},$$

the error term $\tilde{Q}^{2q}\beta\sigma_* y$ being zero for all y in the second set. Then $\{{}_0E^r\}$ is a sub spectral sequence of $\{E^r BSF\}$ which is the image of an isomorphic copy of $\{{}_0E^r\}$ in $\{E^r B(SF; j_2^\delta)\}$ and which maps onto an isomorphic copy of $\{{}_0E^r\}$ in $\{E^r B(F/O)\}$.

Proposition 13.9. Define a spectral sequence $\{{}_1E^r\}$ by

$${}_1E_{r+1} = P\{(\sigma_* x_{(2,1)})^{2^r}\} \otimes E\{\beta_{r+1}(\sigma_* x_{(2,1)})^{2^r}\} \quad \text{for } r \geq 1,$$

with $\beta_{r+1}(\sigma_* x_{(2,1)})^{2^r} = (\sigma_* x_{(2,1)})^{2^r - 1}(\sigma_* x_{(1,1)})$. Then $\{{}_1E^r\}$ is a sub-spectral sequence of $\{E^r BSF\}$ which maps onto an isomorphic copy of $\{{}_1E^r\}$ in $\{E^r X\}$ for $X = BJ^\delta_{\otimes 2}$, $BBSO^\delta_{\otimes 2}$, and $B(F/O)$.

To calculate the portions of the various Bockstein spectral sequences not determined by the results above, we require information about the first Bocksteins in $H_* BJ^\delta_{\otimes 2}$ and about the higher Bocksteins on the elements $\sigma_* x_{(2i,\, 2i)}$ in $H_* BSF$. It is immediate from the definition of the elements $t_i \in H\, BJ^\delta_{\otimes 2}$ (above Corollary 12.6) that

$$\beta\sigma_* t_{2^j} = 0 \text{ and } \beta\sigma_* t_{2^j+1} = 0 \text{ for } j \geq 2 \text{ and } \beta t_{4i+2} = t_{4i+2} * t_{4i+1} \text{ for } i \geq 1.$$

The remaining Bocksteins $\beta\sigma_* t_{2^s(2n+1)} = \sigma_* e_* x_{(2^s n + 2^s - 1,\, 2^s n)}$ for $s \geq 2$ and $n \geq 1$ could in principle be determined by direct calculation of

$e_* x_{(2^s n + 2^s - 1, 2^s n)}$ modulo #-decomposable elements. We prefer to obtain partial information by means of the following theorem, the proof of which gives a new derivation of Stasheff's results on the torsion in BBSO [31]. Write $\bar{t}_r$ for the image of t_r in $H_* BBSO_{\otimes}^{\delta}$.

Theorem 13.10. If $i \neq 2^j$, then $\beta\sigma_* \bar{t}_{2i} = \sigma_* \bar{t}_{2i-1}$. For $2 \leq r \leq \infty$,

$$E^r BBSO_{\otimes}^{\delta} = {}_1E^r \otimes E\{\sigma_* \bar{t}_{4i} \mid i = 2^j, j \geq 0\} \otimes E\{\bar{f}_{4i+1} \mid i \geq 3, i \neq 2^j\},$$

where $\bar{f}_{4i+1} = \sigma_* \bar{t}_{4i} + (\sigma_* \bar{t}_{4n+2})(\sigma_* \bar{t}_{4n+1})^{2^s - 1}$ if $4i = 2^{s+1}(2n+1)$, $s \geq 1$ and $n \geq 1$.

Proof. By the Adams-Priddy theorem [2], $BBSO_{\otimes}^{\delta}$ is homotopy equivalent to BBSO. By Bott periodicity, there is a fibration sequence $BBSO \xrightarrow{\iota} BSpin \xrightarrow{\pi} BSU$, where π is the natural map. Since $\pi^*(c_i) = w_i^2$, a standard calculation shows that $H^* BBSO = E\{e_i \mid i \geq 3\}$, where $e_i = \iota^* w_i$ if $i \neq 2^j + 1$ and where $e_{2^j+1} = Sq^I e_3$, $I = (2^{j-1}, 2^{j-2}, \ldots, 2)$, restricts to an indecomposable element of $H^* SU$. This specification of the e_i implies that $\beta e_{2i} = e_{2i+1}$ if $i \neq 2^j$ and that $\beta e_{2^{j+1}+1} = e_{2^j+1}^2 = 0$ if $j \geq 1$, while $\beta e_3 = e_4$ since $\beta\sigma_* \bar{t}_3 = \sigma_* \bar{t}_2$ by Proposition 13.9. Therefore

$$E_2 BBSO = E\{e_3 e_4, e_{2^j+2}, e_{2i} e_{2i+1}, e_{2^j+1} \mid i \neq 2^j, i \geq 3, j \geq 1\}.$$

Obviously $\beta\sigma_* \bar{t}_{4i}$ is either 0 or $\sigma_* \bar{t}_{4i-1} = (\sigma_* t_{4n+1})^{2^s}$ if $4i = 2^{s+1}(2n+1)$ ($\sigma_* \bar{t}_{4n+1}$ being indecomposable if $n = 2^q$ and being $(\sigma_* \bar{t}_{2n})^2$ otherwise). We have just determined $E^2 BBSO_{\otimes}^{\delta} \cong E^2 BBSO$ additively, and an easy counting argument shows that we must have $\beta\sigma_* \bar{t}_{4i} \neq 0$ for $i \neq 2^j$. Therefore $E^2 BBSO_{\otimes}^{\delta}$ has the stated form. Since $H^*(BBSO; Q)$ is clearly an exterior algebra on one generator in each degree $4i+1$, $i \geq 1$, the rest is immediate from the differentials in Proposition 13.9 and counting arguments.

The determination of $\beta_r \sigma_* x_{(2i, 2i)}$ is the central calculation of Madsen's paper [16], and we shall content ourselves with a sketch of his proof. Recall that $E^2 BSO = E^\infty BSO$ is a polynomial algebra on generators d_{4i} (of degree 4i) such that $\psi d_{4i} = \sum d_{4j} \otimes d_{4i-4j}$. Let

$$p_{4i} = i d_{4i} + \sum_{j=1}^{i-1} d_{4j} p_{4i-4j}$$ be the ith non-zero primitive element of $E^2 BSO$ (which is dual to w_{2i}^2 in $E_2 BSO$).

<u>Theorem 13.11.</u> In $\{E^r BSF\}$, $\beta_2 \sigma_* x_{(2i, 2i)} = 0$ and $\beta_3 \sigma_* x_{(2i, 2i)} = (Bj)_*(p_{4i})$.

<u>Proof.</u> Intuitively, the idea is that the differentials in $\{E^r BSF\}$ are specified in Propositions 13.8 and 13.9, except for determination of the $\beta_r \sigma_* x_{(2i, 2i)}$, hence $\beta_r \sigma_* x_{(2i, 2i)}$ must be $(Bj)_*(p_{4i})$ for some r (perhaps depending on i) since there is no other way that $E^\infty BSF$ can be trivial. In $H_* Q_0 S^0$,

$$\beta_2(Q^{2i}[1] * Q^{2i}[1]) = Q^{2i-1}[1] * Q^{2i}[1] + Q^{2i} Q^{2i-1}[1].$$

Therefore $\beta_2 x_{(2i, 2i)}$ can be calculated directly (modulo #-decomposable elements and the image of β). The resulting computation, which Madsen carried out but did not publish, yields $\beta_2 \sigma_* x_{(2i, 2i)} = 0$. Madsen's published proof of this fact relies instead on analysis of $\alpha_{2*}: H_* J \to H_* SF$, for a suitable choice of α_2, and use of $\{E^r J\}$ (see Remarks 11.2). Thus $r \geq 3$. Madsen proves that $r = 3$ by a direct chain level calculation. Alternatively, an obvious dualization argument yields classes in $H^*(BSF; Z_{2^r})$ which pull back to the mod 2^r reductions of the Pontryagin classes, and the equivalent claim (to $r = 3$) that the Z_{16} Pontryagin classes of vector bundles are not fibre homotopy invariants can be verified by geometrical example.

The previous results, together with the cohomological analog of I.4.11

$(\beta_2 z^2 = z\beta z + P^{2q}\beta z$ if $\deg z = 2q$ and $\beta_r z = z\beta_{r-1} z$ if $r > 2)$ and our mod 2 calculations, suffice to determine the Bockstein spectral sequences of all spaces in sight and the natural maps between them. Certain of the relevant spectral sequences are most naturallly described in cohomology, and we shall not introduce the extra notation necessary to state the results obtained in homology by double dualization; appropriate formulations may be found in [16]. We collect results before proceeding to the proofs.

Write $q_*^{-1} y \in H_*B(SF; j_2)$ for the inverse image of an element $y \in \mathrm{Im}\, q_* \subset H_*BSF$.

Theorem 13.12. $E^r B(SF; j_2^\delta) = {}_0E^r$ for all $r > 2$, while

$$E^2 B(SF; j_2) = {}_0E^2 \otimes [\bigotimes_{i \neq 2^j} E\{q_*^{-1}\sigma_* y_{(2i,2i)}\} \otimes P\{\beta_2 q_*^{-1}\sigma_* \tilde{y}_{(2i,2i)}\}],$$

where $\beta_2 q_*^{-1}\sigma_*\tilde{y}_{(2i,2i)}$ is represented in $H_*B(SF; j_2^\delta)$ by $q_*^{-1}\beta g_{4i+1}$ for a certain element $g_{4i+1} \in H_*BSF$.

Theorem 13.13. $\{E^r BSF\} = \{{}_0E^r\} \otimes \{{}_1E^r\} \otimes \{{}_2E^r\}$, where ${}_2E^2 = {}_2E^3 = E^2 BSO \otimes E\{\sigma_* x_{(2i,2i)} \mid i \geq 1\}$ and $\beta_3 \sigma_* x_{(2i,2i)} = (Bj)_*(p_{4i})$.

The dual spectral sequence $\{{}_2E_r\}$ is specified by

$${}_2E_{r+2} = \bigotimes_{i \geq 1} P\{w_{2i}^{2^r}\} \otimes E\{\beta_{r+2} w_{2i}^{2^r}\},\ \beta_{r+1} w_{2i}^{2^r} = w_{2i}^{2^r - 2}(\sigma_* x_{(2i,2i)})^* \quad (r \geq 1).$$

The elements $\sigma_* x_{(2i,2i)}$ map to permanent cycles in $H_*B(F/O)$ and, for $r \geq 2$,

$$E^r B(F/O) = {}_0E^r \otimes {}_1E^r \otimes E\{\sigma_* x_{(2i,2i)} \mid i \geq 1\}.$$

Theorem 13.14. $\{E^r BJ^\delta_{\otimes 2}\} = \{{}_1E^r\} \otimes \{{}_3E^r\} \otimes \{{}_4E^r\}$, where

$${}_3E^2 \otimes {}_4E^2 = E^2 BSO \otimes E\{\sigma_* t_{4i} \mid i = 2^j, j \geq 0\} \otimes E\{f_{4i+1} \mid i \geq 3, i \neq 2^j\},$$

$\beta_2 f_{4i+1} = (BeBj)_*(p_{4i})$ if $i \neq 2^j$, and $\beta_3 \sigma_* t_{4i} = (BeBj)_*(p_{4i})$ if $i = 2^j$.

Here $f_{4i+1} = (Be)_*(g_{4i+1})$ maps to $\overline{f}_{4i+1} \in H_*BBSO^{\delta}_{\otimes}$ and is primitive in $E^2BJ_{\otimes 2}$. The dual spectral sequences $\{{}_3E_r\}$ and $\{{}_4E_r\}$ are specified by

$$ {}_3E_{r+1} = \bigotimes_{i \neq 2^j} P\{w_{2i}^{2^r}\} \otimes E\{\beta_{r+1}w_{2i}^{2^r}\},\ \beta_{r+1}w_{2i}^{2^r} = w_{2i}^{2^r-2}f^*_{4i+1} \quad (r \geq 1) $$

and

$$ {}_4E_{r+2} = \bigotimes_{i = 2^j} P\{w_{2i}^{2^r}\} \otimes E\{\beta_{r+2}w_{2i}^{2^r}\},\ \beta_{r+2}w_{2i}^{2^r} = w_{2i}^{2^r-2}(\sigma_*t_{4i})^* \quad (r \geq 1), $$

where $w_{2i}^2 \in H^*BJ^{\delta}_{\otimes 2}$ survives to $(BeBj)_*(p_{4i})$ in $E_2BJ^{\delta}_{\otimes 2}$.

<u>Proofs.</u> Theorem 13.10 implies that if $4i = 2^{s+1}(2n+1)$, $s \geq 1$ and $n \geq 1$, then

$$ \sigma_*e_*x_{(2^{s+1}n+2^{s+1}-1,\, 2^{s+1}n)} = \beta\sigma_*t_{4i} = \sigma_*t_{4i-1} + a_{4i}\sigma_*e_*x_{4i-1} $$

for some constants a_{4i} (which could be, but have not been, computed explicitly). By the general definition of the $\widetilde{y}_I$, it follows that

$$ \sigma_*\widetilde{y}_{(2^{s+1}n+2^{s+1}-1,\, 2^{s+1}n)} = \sigma_*\widetilde{x}_{(2^{s+1}n+2^{s+1}-1,\, 2^{s+1}n)} + \sigma_*x_{(2i,\, 2i-1)} + a_{4i}\sigma_*x_{4i-1}. $$

Theorem 13.2 and the fact that $(\sigma_*x_{2i-1})^2 = \sigma_*x_{4i-1}$ imply that if $1 \leq k \leq s$, then, with $m = 2n+1$,

$$ (\sigma_*\widetilde{y}_{(2^km,2^km-1)})^{2^{s-k}} = (\sigma_*x_{(2^km,2^km-1)})^{2^{s-k}} + \sigma_*x_{4i-1} + (\sigma_*x_{(2^{k-1}m,\, 2^{k-1}m-1)})^{2^{s+1-k}}. $$

An obvious cancellation argument then shows that

$$ \sigma_*\widetilde{y}_{(2^{s+1}n+2^{s+1}-1,\, 2^{s+1}n)} + \sum_{k=1}^{s} (\sigma_*\widetilde{y}_{2^km,\, 2^km-1})^{2^{s-k}} $$

$$ = \sigma_*x_{(2^{s+1}n+2^{s+1}-1,\, 2^{s+1}n)} + (\sigma_*x_{(2n+1,\, 2n)})^{2^s} + (s+a_{4i})\sigma_*x_{4i-1} $$

$$ = \beta g_{4i+1}, $$

where g_{4i+1} is defined by

$$g_{4i+1} = \sigma_* x_{(2^{s+1}n+2^{s+1}, 2^{s+1}n)} + (\sigma_* x_{(2n+2, 2n)})(\sigma_* x_{(2n+1, 2n)})^{2^s - 1}$$

$$+ (s + a_{4i})\sigma_* x_{4i} .$$

It is now a simple matter to prove Theorem 13.14. We have

$$f_{4i+1} = (Be)_*(g_{4i+1}) = \sigma_* t_{4i} + (\sigma_* t_{4n+2})(\sigma_* t_{4n+1})^{2^s - 1} + (s + a_{4i})\sigma_* e_* x_{4i} .$$

Obviously $\beta f_{4i+1} = 0$ (as could also be verified directly, by use of Lemma 12.9). Therefore $E^2 BJ^\delta_{\otimes 2}$ is as specified. Since $\sigma_* e_* x_{(2i, 2i)} = 0$ for $i \neq 2^j$, by Lemma 12.8, $(BeBj)_*(p_{4i})$ must be zero in $E^3 BJ^\delta_{\otimes 2}$ by Theorem 13.11. The only way this can happen is if $\beta_2 f_{4i+1} = (BeBj)_*(p_{4i})$. On the other hand, $\beta_3 \sigma_* t_{4i} = (BeBj)_*(p_{4i})$ if $i = 2^j$ since then $t_{4i} = \sigma_* e_* x_{(2i, 2i)}$. In view of Proposition 13.9 (and the cohomological formulas cited above), Theorem 13.14 follows by a counting argument.

Theorems 13.12 and 13.13 also follow by counting arguments, but the details are considerably less obvious. We claim first that H_*BSF can be written as the tensor product of the following algebras, each of which is a sub differential algebra under β. Algebras written in terms of elements $\tilde{y}_I$ come from subalgebras of $H_*B(SF; j_2^\delta)$ and algebras written in terms of elements x_I map isomorphically onto subalgebras of $H_*BJ^\delta_{\otimes 2}$.

(i) $P\{\sigma_*\tilde{y}_I, \beta\sigma_*\tilde{y}_I, \tilde{Q}^{2q}\beta\sigma_*\tilde{y}_I \mid d(I) = 2q-1, I = (2s, J), e(I) \geq 3\}$

(ii) $P\{\sigma_*\tilde{y}_{(2i, 2i-1)} \mid i \geq 2\} \otimes E\{\sigma_*\tilde{y}_{(2i-1, 2i-1)} \mid i \geq 2\}$

(iii) $P\{\sigma_* x_{(2, 1)}\} \otimes E\{\sigma_* x_{(1, 1)}\}$

(iv) $H_*BSO \otimes E\{\sigma_*\tilde{y}_{(2i, 2i)} \mid i \neq 2^j\} \otimes E\{\sigma_* x_{(2i, 2i)} \mid i = 2^j\}$

(v) $\bigotimes_z P\{z, \beta z, \tilde{Q}^{H_k} z \mid k \geq 1, H_k = (2^k q + 2^k, 2^{k-1} q + 2^{k-1}, \ldots, 2q+2) \text{ if } \deg z = 2q+1\}$,

where z ranges through the union of the following four sets:

(a) $\{\sigma_* \tilde{y}_I \mid d(I) \text{ even}, I = (2s, J), e(I) \geq 4, I \neq (2^s n + 2^s, 2^s n)\}$

(b) $\{\sigma_* \tilde{y}_I \mid I = (2^{s+1} n + 2^s + 2, 2^s n + 2^s, 2^s n), s \geq 1 \text{ and } n \geq 1\}$

(c) $\{\sigma_* x_I \mid I = (2^s + 2, 2^s), s \geq 1\}$

(d) $\{g_{4i+1} \mid i \neq 2^j, i \geq 1\}$

In (ii), note that $\beta\sigma_* \tilde{y}_{(2i, 2i-1)} = \sigma_* \tilde{y}_{(2i-1, 2i-1)}$ by Theorem 13.2. Clearly (i) and (ii) together have homology ${}_0E^2$ under β, (iii) has homology ${}_1E^2$, (iv) has homology ${}_2E^2$, and (v) is acyclic (because $\beta \tilde{Q}^{H_k} z = (\tilde{Q}^{H_{k-1}} z)^2$). In (v.b), with $m = 2n+1$,

$$\sigma_* \tilde{y}_I = \sigma_* \tilde{x}_I + \sigma_* x_{(2^s m+2, 2^s m)} + \sigma_* x_{2^{s+1} m+2}$$

by Corollary 12.10, hence $\beta\sigma_* \tilde{y}_I = \sigma_* \tilde{y}_{(2^s m+1, 2^s m)}$ by Theorem 13.2. To check the claim, consider H_k as a generic notation (varying with q), let H_0 be empty (so that $Q^{H_0} z = z$) and observe that, modulo decomposable elements and $H_* BSO$,

$$\tilde{Q}^{H_k} g_{4i+1} \equiv \tilde{Q}^{H_k} x_{(2^{s+1} n + 2^{s+1}, 2^{s+1} n)}$$

and

$$\tilde{Q}^{H_k} \sigma_* \tilde{y}_I \equiv \tilde{Q}^{H_{k+1}} \sigma_* x_{(2^s n + 2^s, 2^s n)} + \tilde{Q}^{H_k} \sigma_* x_{(2^s m+2, 2^s m)}$$

where $4i = 2^{s+1}(2n+1)$, $I = (2^{s+1} n + 2^s + 2, 2^s n + 2^s, 2^s n)$, and $m = 2n+1$. Together with the elements $\tilde{Q}^{H_k} \sigma_* x_{(2^s+2, 2^s)}$ from (v.c), these elements account for all of the generators of the form $\tilde{Q}^{H_k} x_{(2^s n + 2^s, 2^s n)}$ in Theorem 13.1 (by an amusing and, in the case $s = 1$, unobvious counting argument). The rest of the verification of the claim is straightforward linear algebra based on Theorem 13.1 and the definition of the $\tilde{y}_I$. The claim clearly implies that the image of $H_*B(SF;j_2^\delta)$ in H_*BSF is the tensor product of the algebras listed in (i), (ii), (v.a), and (v.b) with the algebra

$$\bigotimes_{i \neq 2^j} E\{\sigma_*\tilde{y}_{(2i,2i)}\} \otimes P\{\beta g_{4i+1}\} .$$

Theorems 13.12 and 13.13 now follow from the observation that, for $r \geq 2$, ${}_0E^r$ and ${}_1E^r$ contain no non-zero primitive cycles in degrees 4i (by inspection of Propositions 3.8 and 3.9). Certainly $\beta_2 q_*^{-1}\sigma_*\tilde{y}_{(2i,2i)}$ is a non-zero primitive cycle in $E^2B(SF;j_2^\delta)$, since $\beta_2 q_*^{-1}\sigma_*\tilde{y}_{(2i,2i)} = 0$ would be incompatible with Theorem 13.11, and $q_*^{-1}\beta g_{4i+1}$ is the only candidate. This proves Theorem 13.12. The description of $\{E^rBSF\}$ given in Theorem 13.13 is correct since the observation implies the required splitting of spectral sequences (compare [16, p. 72]). Finally, the description of $\{E^rB(F/O)\}$ is correct since the observation implies that the $\sigma_*x_{(2i,2i)}$ map to permanent cycles in $H_*B(F/O)$.

The following consequence of the theorems explains what is going on integrally in the crucial dimensions.

<u>Corollary 13.15.</u> For $i \neq 2^j$, the sequence $H_{4i}B(SF;j_2^\delta) \to H_{4i}BSF \to H_{4i}BJ_{\otimes 2}^\delta$ of integral homology groups contains a short exact sequence

$$0 \to Z_4 \to Z_2 \oplus Z_8 \to Z_4 \to 0 ,$$

where the Z_4 in $H_{4i}B(SF;j_2^\delta)$ is generated by an element r_{4i} which reduces mod 4 to $\beta_2 q_*^{-1}\tilde{y}_{(2i,2i)}$, the Z_8 in $H_{4i}BSF$ is generated by $(Bj)_*(p_{4i})$ (p_{4i} being the canonical generator of the group of primitive elements of $H_{4i}BSO$/torsion), the Z_2 in $H_{4i}BSF$ is generated by

$q_*(r_{4i}) + 2(Bj)_*(p_{4i})$, and the Z_4 in $H_{4i}BJ^{\delta}_{\otimes 2}$ is generated by $(BeBj)_*(p_{4i})$.

Proof. Choose an integral chain $x \in C_{4i+1}B(SF; j_2)$ such that $dx = 4y$ and the mod 2 reduction of x represents $q_*^{-1}\sigma_*\tilde{y}_{(2i, 2i)}$. The mod 4 reduction of y represents $\beta_2 q_*^{-1}\tilde{y}_{(2i, 2i)}$ (by an abuse of the notation β_2). Of course, $dq_*x = 4q_*y$. Since $\beta_3\sigma_*\tilde{y}_{(2i, 2i)} = (Bj)_*(p_{4i})$, there is an integral chain $z \in C_{4i+1}BSF$ such that $dz = -2q_*y + 4p$, so that $d(q_*x + 2z) = 8p$, and the mod 2 reduction of p represents $(Bj)_*(p_{4i})$. The conclusion follows.

Remark 13.16. In cohomology with Z_8 coefficients, there are classes in $H^{4i}BSF$ which restrict in $H^{4i}BSO$ to the mod 8 reduction of the ith Pontryagin class. Our results clearly imply that there exists such a class in the image of $(Be)^*: H^{4i}BJ^{\delta}_{\otimes 2} \to H^{4i}BSF$ if and only if $i = 2^j$, although there is such a class whose mod 4 reduction is in the image of $(Be)^*$ for all i. This fact makes explicit analysis (e.g., of the coproduct) of such Z_8 Pontryagin classes for spherical fibrations rather intractable.

Bibliography

1. J.F. Adams. Lectures on generalized homology. Lecture Notes in Mathematics Vol. 99, 1-138. Springer, 1969.

2. J.F. Adams and S. Priddy. Uniqueness of BSO. Preprint.

3. J. Adem. The relations on Steenrod powers of cohomology classes. Algebraic geometry and topology. A symposium in honor of S. Lefschetz, 1957.

4. G. Brumfiel. On integral PL characteristic classes. Topology 8, 1969, 39-46.

5. G. Brumfiel, I. Madsen, and R. J. Milgram. PL characteristic classes and cobordism. Annals of Math. 97 (1973), 82-159.

6. Z. Fiedorowicz and S. Priddy. Homology of the orthogonal groups over finite fields and their associated infinite loop spaces. To appear.

7. E. Friedlander. Computations of K-theories of finite fields. Preprint.

8. V. K. A. M. Gugenheim and J. P. May. On the theory and applications of differential torsion products. Memoirs Amer. Math. Soc. 142, 1974.

9. M. Herrero. Thesis. Univ. of Chicago, 1972.

10. D. S. Kahn. Homology of the Barratt-Eccles decomposition maps. Notas de Matematicas y Simposia No.1. Soc. Mat. Mexicana 1975.

11. D. S. Kahn and S. Priddy. Applications of the transfer to stable homotopy theory. Bull. Amer. Math. Soc. 78(1972), 981-987.

12. S. Kochman. Symmetric Massey products and a Hirsch formula in homology. Trans. Amer. Math. Soc. 163 (1972), 245-260.

13. S. Kochman. Homology of the classical groups over the Dyer-Lashof algebra. Trans. Amer. Math. Soc. 185 (1973), 83-136.

14. H. J. Ligaard and I. Madsen. Homology operations in the Eilenberg-Moore spectral sequence. Math. Zeitschrift. 143(1975), 45-54.

15. I. Madsen. On the action of the Dyer-Lashof algebra in H_*G. Pacific J. Math. 60 (1975), 235-275.

16. I. Madsen. Higher torsion in SG and BSG. Math. Zeitschrift. 143 (1975), 55-80.

17. I. Madsen and R. J. Milgram. The universal smooth surgery class. Comm. Math. Helv. To appear.

18. I. Madsen and R. J. Milgram. The oriented bordism of topological manifolds and integrability theorems. Preprint.

19. I. Madsen, V. P. Snaith, and J. Tornehave. Infinite loop maps in geometric topology. Preprint.

20. J. P. May. Homology operations on infinite loop spaces. Proc. Symp. Pure Math. Vol. 22, 171-185. Amer. Math. Soc. 1971.

21. J. P. May. Classifying spaces and fibrations. Memoirs Amer. Math. Soc. 155, 1975.

22. R. J. Milgram. The mod 2 spherical characteristic classes. Annals Math. 92 (1970), 238-261.

23. J. Milnor and J. C. Moore. On the structure of Hopf algebras. Annals Math. 81(1965), 211-264.

24. D. Moore. Thesis. Northwestern Univ. 1974.

25. F. P. Peterson. Some results on PL-cobordism. J. Math. Kyoto Univ. 9 (1969), 189-194.

26. F. P. Peterson and H. Toda. On the structure of $H^*(BSF; Z_p)$. J. Math. Kyoto Univ. 7 (1967), 113-121.

27. S. Priddy. Dyer-Lashof operations for the classifying spaces of certain matrix groups. Quart. J. Math. 26 (1975), 179-193.

28. D. Quillen. The Adams conjecture. Topology 10 (1971), 67-80.

29. D. Quillen. On the cohomology and K-theory of the general linear groups over a finite field. Annals of Math. 96(1972), 552-586.

30. G. Segal. The stable homotopy of complex projective space. Quart. J. Math. 24 (1973), 1-5.

31. J. D. Stasheff. Torsion in BBSO. Pacific J. Math. 28(1969), 677-680.

32. D. Sullivan. Geometric topology seminar. Mimeographed notes, Princeton.

33. D. Sullivan. Geometric topology, part I. Localization, periodicity, and Galois symmetry. Mimeographed notes. M.I.T.

34. J. Tornehave. Thesis. M.I.T. 1971.

35. J. Tornehave. The splitting of spherical fibration theory at odd primes. Preprint.

36. A. Tsuchiya. Characteristic classes for spherical fibre spaces. Nagoya Math. J. 43 (1971), 1-39.

37. A. Tsuchiya. Characteristic classes for PL micro bundles. Nagoya Math. J. 43 (1971), 169-198.

38. A. Tsuchiya. Homology operations on ring spectrum of H^{∞} type and their applications. J. Math. Soc. Japan 25 (1973), 277-316.

THE HOMOLOGY OF $\mathcal{C}_{n+1}$-SPACES, $n \geq 0$

Fred Cohen

The construction of homology operations defined for the homology of finite loop spaces parallels the construction of homology operations defined for the homology of infinite loop spaces, with some major differences. To recall, the operations for infinite loop spaces are defined via classes in the homology of the symmetric group, Σ_j. Working at the prime 2, Browder [24], generalizing and extending the operations of Araki and Kudo [1], found that an appropriate skeleton of $B\Sigma_2$ may be used to describe natural operations which allow computation of $H_*(\Omega^{n+1}\Sigma^{n+1}X; \mathbf{Z}_2)$ as an algebra. Dyer and Lashof [8], using similar methods, subsequently obtained partial analogous results at odd primes. However, comparison of the results of Dyer and Lashof to Milgram's computation [20] of $H_*(\Omega^{n+1}\Sigma^{n+1}X)$ as an algebra made it apparent that the skeleton of $B\Sigma_p$ intrinsic to the geometry of the finite loop space failed to give sufficient operations to compute $H_*(\Omega^{n+1}\Sigma^{n+1}X; \mathbf{Z}_p)$. To be precise, only $1/p-1$ times the requisite number of operations (defined in this paper) may be described using the methods of Dyer and Lashof.

In addition, there is a non-trivial unstable operation in two variables, λ_n, which was invented by Browder; the method of using finite skeleta of $B\Sigma_p$ does not lend itself to finding the relationships between λ_n and the other operations. These relationships are especially important in determination of fine structure and our later work in IV.

An alternative method for defining operations seemed to be provided by the composition pairing and the possibility that $H_*(\Omega^{n+1}S^{n+1}; \mathbf{Z}_p)$ is universal for Dyer-Lashof operations. If $n = \infty$, this method works in principle, but it fails almost completely if $n < \infty$: It is shown in IV §6 that there are finite loop spaces with many non-trivial Dyer-Lashof operations for which the composition pairing is trivial.

The observation of Boardman and Vogt that the space of little (n+1)-cubes acts on (n+1)-fold loop spaces, together with May's theory of iterated loop spaces [G], led one to expect that the equivariant homology of the little cubes ought to enable one to define all requisite homology operations in a natural setting analogous to that provided for an infinite loop spaces by $B\Sigma_p$. This is the case. In addition, one can describe easily understood constructions with the little cubes which, when linked with May's theory of operads, enable one to determine the commutation relations between all of the operations, and between the operations and the product, coproduct, and Steenrod operations on the homology of iterated loop spaces.

Knowledge of this fine structure is essential, for example, in the analysis of the composition pairing and the Pontrjagin ring $H_*(SF(n+1);\mathbb{Z}_p)$ for all n and p in IV. Indeed, all of the formulas in III. 1.1-1.5 are explicitly used there. A further application of the fine structure is an improvement [28] of Snaith's stable decomposition for $\Omega^{n+1}\Sigma^{n+1}X$ [25].

We have tried to parallel the format of I as closely as possible, pointing out essential differences. Sections 1-4, which are analogous to I. 1,2,4, and 5, contain the computations of $H_*\Omega^{n+1}\Sigma^{n+1}X$ and $H_*C_{n+1}X$, $n > 0$, together with a catalogue of the relations amongst the operations.

In more detail, Section 1 gives a list of the commutation relations between all of the operations, coproduct, product, and between them and the Steenrod operations, conjugation, and homology suspension. The relationship between Whitehead products and the λ_n is also described.

Section 2 contains the definition of certain algebraic structures naturally suggested by the preceding section; the free versions of these algebraic structures are constructed.

We compute $H_*\Omega^{n+1}\Sigma^{n+1}X$ in section 3, using the results of section 1 and 2. The associated Bockstein spectral sequences are also computed, and the interesting corollary that $H_*(\Omega^2\Sigma^2X;\mathbb{Z})$ has p-torsion precisely of order p if H_*X has no p-torsion is proved. Using the results of section 3 together with May's approximation theorem [G], we compute $H_*C_{n+1}X$ in section 4. We use these computations to prove the group completion theorem in section 3.

In sections 5-11, the equivariant cohomology of the space of little (n+1)-cubes as an algebra over the Steenrod algebra is computed in order to set up the theory of operations described in the previous sections. Here we replace the little (n+1)-cubes by the configuration space $F(\mathbb{R}^{n+1},p)$ which has the same equivariant homotopy type of the little cubes [G]. The crux of our method of computation lies in the analysis of the (non-trivial) local coefficient system in the Leray spectral sequence $\{E_r^{**}\}$ for the covering $F(\mathbb{R}^{n+1},p) \longrightarrow \frac{F(\mathbb{R}^{n+1},p)}{\Sigma_p}$.

After summarizing our results and giving the definitions of the operations in section 5, we compute the unequivariant cohomology of $F(\mathbb{R}^{n+1},p)$ in section 6.

We analyze the action of the symmetric group on the indecomposables in cohomology in section 7. To obtain complete understanding of the local coefficient system, we also completely analyze the relations in the cohomology algebra of $F(\mathbb{R}^{n+1},p)$ as an algebra over the Steenrod algebra.

We completely describe E_2^{**} of the spectral sequence together with all of the differentials in sections 8 and 9. The "extra" classes present in $E_2^{0,*}$ are essentially the obstructions to the construction of all requisite homology operations via the method of Dyer and Lashof. One of the main tools for computation here is a vanishing theorem for E_2^{**} which is proven in section 10.

An automorphism of $F(\mathbb{R}^{n+1},p)$ which commutes with the Σ_p-action is described in section 11 and is used to compute the precise algebra structure of $H^*(\frac{F(\mathbb{R}^{n+1},p)}{\Sigma_p};\mathbb{Z}_p)$. Of course, the spectral sequence only provides such information up to filtration. The methods used here generalize: We shall give a description of $H_*F(M,j)$ and $H_* \frac{F(M,j)}{\Sigma_j}$ for more general manifolds, M, in [30].

The last 6 sections are occupied with the derivation of the fine structure. First we must obtain information concerning the structure maps γ of the little cubes [G] in unequivariant homology. Using the methods of section 7, we are able to compute γ_* on primitives in section 12. This calculation is crucial to later sections.

In section 13, we prove our statements about the obstruction to the construction of the homology operations using the joins of the symmetric group. The homological properties of the Browder operation, together with its relation to the Whitehead product, are also derived here.

Because of certain recalcitrant behavior of the space of little (n+1)-cubes, $n < \infty$, one must find slightly more geometric methods to compute the rest of the fine structure described in Theorems 1.1, 1.2, and 1.3. Section 14 contains a sketch of the methods and the crucial algebraic lemma.

The commutation of the operations with homology suspension is derived in section 15. The proof is non-standard in the sense that we do not construct an equivariant chain approximation for the space of little (n+1)-cubes, but rather use the methods described in section 14.

The remaining properties of the operations, except for the unstable analogues of the Nishida relations, are derived in section 16; the Nishida relations are derived in section 17.

We also include an appendix giving the description of the homology of the classical braid groups, this information being implicit in sections 3-4. Here, we describe the homology of these groups with $\mathbb{Z}_p$, $\mathbb{Q}$, and $\mathbb{Z}$-coefficients (with trivial action). In the case of $\mathbb{Z}_p$, the action of the Steenrod algebra is also completely described.

Several crucial papers of Peter May are referred to as [A], [G], and I in the text and bibliography. A discussion of these papers is contained in the preface to this volume.

The results announced in [26 and 27] are contained in sections 1 through 5.

I am indebted to my mentors while at Chicago: Richard Lashof, Arunas Liulevicius, Saunders MacLane, Robert Wells, and Michael Barratt. It is a pleasure to extend warm thanks to my ext-tor-mentor Peter May; his patience, interest, and enthusiasm were central to my introduction into a beautiful area of mathematics.

I wish to thank Sara Clayton for typing the manuscript.

Special gratitude is due several close friends: Fred Flowers, Larry Taylor (who should also appear in the above list) and Tim Zwerdling; even more so to Kathleen Whalen, my father Harry Cohen, and my grandmother Bertha Malman.

Contents

1. Homology operations on $\mathcal{C}_{n+1}$-spaces, $n \geq 0$

All spaces are assumed to be compactly generated and Hausdorff with non-degenerate base point. All homology is taken with $\mathbb{Z}_p$-coefficients unless otherwise stated. Modifications required for the case $p = 2$ are stated in brackets.

Recall from [G] that a $\mathcal{C}_{n+1}$-space (X,θ) is a space X together with an action of the little cubes operad, $\mathcal{C}_{n+1}$, on X; $\mathcal{C}_{n+1}[\tau]$ denotes the category of $\mathcal{C}_{n+1}$-spaces. In the following theorems, we assume that all spaces are in $\mathcal{C}_{n+1}[\tau]$. Proofs will be given in sections 12 through 17.

Theorem 1.1. There exist homomorphisms $Q^s\colon H_qX \to H_{q+2s(p-1)}X$ $[H_qX \to H_{q+s}X]$, $s \geq 0$, for $2s - q < n$ $[s-q<n]$ which are natural with respect to maps of $\mathcal{C}_{n+1}$-spaces and satisfy the following properties:

(1) $Q^s x = 0$ if $2s <$ degree (x) $[s <$ degree $(x)]$, $x \in H_*X$.

(2) $Q^s x = x^p$ if $2s =$ degree (x) $[s =$ degree $(x)]$, $x \in H_*X$.

(3) $Q^s\phi = 0$ if $s > 0$, where $\phi \in H_0X$ is the identity element.

(4) The external, internal, and diagonal Cartan formulas hold:

$$Q^s(x\otimes y) = \sum_{i+j=s} Q^i x \otimes Q^j y, \qquad x\otimes y \in H_*(X\times Y)$$

$$Q^s(xy) = \sum_{i+j=s} (Q^i x)(Q^j y), \quad x,\ y \in H_*X; \text{ and}$$

$$\psi Q^s(x) = \sum_{i+j=s} Q^i x' \otimes Q^j x'' \quad \text{if} \quad \psi x = \Sigma x' \otimes x'',\ x \in H_*X.$$

(5) The Adem relations hold: if $p \geq 2$ and $r > ps$, then

$$Q^r Q^s = \sum (-1)^{r+i}(pi-r,\ r-(p-1)s-i-1)\ Q^{r+s-i}Q^i;$$

if $p > 2$, $r \geq ps$ and β is the mod p Bockstein, then

$$Q^r\beta Q^s = \sum_i (-1)^{r+i}(pi-r, r-(p-1)s-i)\beta\ Q^{r+s-i}Q^i$$

$$- \sum_i (-1)^{r+i}(pi-r-1,\ r-(p-1)s-i)\ Q^{r+s-i}\beta Q^i.$$

(6) The Nishida relations hold: Let $P^r_*:H_*X \to H_*X$ be dual to P^r where $P^r = Sq^r$ if $p = 2$. Then

$$P^r_* Q^s = \sum_i (-1)^{r+i}\ (r-pi,\ s(p-1)-pr+pi)\ Q^{s-r+i}P^i_*\ ;$$

if $p > 2$,

$$P^r_*\beta Q^s = \sum_i (-1)^{r+i}(r-pi,\ s(p-1)-pr+pi-1)\beta Q^{s-r+i}P^i_*$$

$$+ \sum_i (-1)^{r+i}(r-pi-1, s(p-1)-pr+pi)Q^{s-r+i}P^i_*\beta.$$

(The coefficients are $(i,j) = \frac{(i+j)!}{i!\ j!}$ if $i > 0$ and $j > 0$, $(i,0) = 1 = (0,i)$ if $i \geq 0$, and $(i,j) = 0$ if $i < 0$ or $j < 0$.)

Compare Theorem 1.1 with Theorem 1.1 of [I].

Remark 1.2. When $X = \Omega^{n+1}Y$, the $Q^s x$ were defined, for $p = 2$, by Araki and Kudo [1], and in the range $2s - q \leq n/p-1$, $x \in H_qX$, for $p > 2$, by Dyer and Lashof [8]. Milgram's calculations [20] indicated that there were operations defined in the range $2s - q \leq n$ for $p > 2$. The "top" operation and its Bockstein, for $2s - q = n$ $[s-q=n]$ has exceptional properties and will be discussed below.

We note that Dyer and Lashof used the (n+1)-fold join of Σ_p, denoted $J^{n+1}\Sigma_p$, in their construction of the Q^s. However, any Σ_p-equivariant map $J^{n+1}\Sigma_p \to \mathcal{C}_{n+1}(p)$ is essential (see section 13). Consequently, there is an obstruction which prevents the construction of all the Q^s of Theorem 1.1 by use of the iterated joins of Σ_p. This obstruction arises from the presence of Browder operations, λ_n, which are related by the following commutative diagram to the Whitehead product:

$$\begin{array}{ccc}
\pi_{p+n+1}Y \otimes \pi_{q+n+1}Y & \xrightarrow{\quad [\ ,\ ]\quad} & \pi_{p+q+2n+1}Y \\
\Big\downarrow \sigma_* \otimes \sigma_* & & \Big\downarrow \sigma_* \\
\pi_p \Omega^{n+1}Y \otimes \pi_q \Omega^{n+1}Y & & \pi_{p+q+n}\Omega^{n+1}Y \\
\Big\downarrow h_* \otimes h_* & & \Big\downarrow h_* \\
H_p \Omega^{n+1}Y \otimes H_q \Omega^{n+1}Y & \xrightarrow{\quad \lambda_n \quad} & H_{p+q+n}\Omega^{n+1}Y
\end{array}$$

[,] denotes the Whitehead product, σ_* the natural isomorphism and h_* the Hurewicz homomorphism. (We are using integral coefficients in this diagram.) We hold the proof of commutativity in abeyance until section 13.

<u>Theorem 1.2</u>. There exist homomorphisms $\lambda_n\colon\ H_qX \otimes H_rX \to H_{q+r+n}X$ which are natural with respect to maps of $\mathcal{C}_{n+1}$-spaces and satisfy the following properties:

(1) If X is a $\mathcal{C}_{n+2}$-space, $\lambda_n(x,y) = 0$ for $x, y \in H_*X$.

(2) $\lambda_0(x,y) = xy - (-1)^{qr}yx$ for $x \in H_qX$, $y \in H_rX$.

(3) $\lambda_n(x,y) = (-1)^{qr+1+n(q+r+1)}\lambda_n(y,x)$ for $x \in H_qX$ and $y \in H_rX$; $\lambda_n(x,x) = 0$ if $p = 2$.

(4) $\lambda_n(\phi,x) = 0 = \lambda_n(x,\phi)$ where $\phi \in H_0X$ is the identity element of H_*X and $x \in H_*X$.

(5) The analogues of the external, internal, and diagonal Cartan formulas hold:

$$\lambda_n(x \otimes y,\ x' \otimes y') = (-1)^{|x'|(|y|+n)}xx' \otimes \lambda_n(y,y') + (-1)^{|y|(|x'|+|y'|+n)}\lambda_n(x,x') \otimes yy'$$

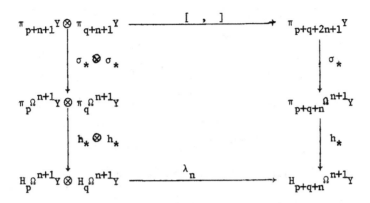

where $|z|$ denotes the degree of z;

$$\lambda_n(xy,x'y') = x\,\lambda_n(y,x')y' + (-1)^{|y|(n+|u|)}\lambda_n(x,x')yy' + (-1)^{|u|(n+|x|+|y|)}x'x\,\lambda_n(y,y') + (-1)^{|y|(n+|v|+|u|)+|u|(n+|x|)}x'\,\lambda_n(x',y')y$$

$$\psi_n(x,y) = \sum(-1)^{n|x'|+|x''||y'|}\lambda_n(x',y') \otimes x''y'' + (-1)^{n|y'|+|x''||y'|}x'y' \otimes \lambda_n(x'',y'')$$

if $\psi x = \Sigma x' \otimes x''$ and $\psi y = \Sigma y' \otimes y''$.

(6) The Jacobi Identity (which is the analogue of the Adem Relations) holds:

$$(-1)^{(q+n)(s+n)}\lambda_n[x,\lambda_n(y,z)] + (-1)^{(r+n)(q+n)}\lambda_n[y,\lambda_n(z,x)] + (-1)^{(s+n)(r+n)}\lambda_n[z,\lambda_n(x,y)] = 0$$

for $x \in H_qX$, $y \in H_rX$, and $z \in H_sX$; $\lambda_n[x,\lambda_n[x,x]] = 0$ for all x if $p = 3$.

(7) The analogues of the Nishida relations hold:

$$P^s_*\lambda_n(x,y) = \sum_{i+j=s} \lambda_n[P^i_*x,P^j_*y]; \text{ and}$$

$$\beta\,\lambda_n(x,y) = \lambda_n(\beta x,y) + (-1)^{n+|x|}\lambda_n(x,\beta y) \quad \text{where } x, y \in H_*X.$$

(8) $\lambda_n[x,Q^sy] = 0 = \lambda_n[Q^sx,y]$ where $x, y \in H_*X$.

We next discuss the "top" operation, ξ_n, and its Bockstein. The operation λ_n is analogous to the bracket operation in a Lie Algebra; ξ_n is analogous to the restriction in a restricted Lie algebra.

Theorem 1.3. There exists a function $\xi_n : H_q X \to H_{pq+n(p-1)} X$ $[H_q X \to H_{2q+n} X]$ defined when $n + q$ is even [for all q] which is natural with respect to maps of $\mathcal{C}_{n+1}$-spaces, and satisfies the following formulas, in which $ad_n(x)(y) = \lambda_n(y,x)$, $ad_n^i(x)(y) = ad_n(x)(ad_n^{i-1}(x)(y))$, and $\zeta_n(x)$ is defined, for $p > 2$, by the formula $\zeta_n(x) = \beta\xi_n(x) - ad_n^{p-1}(x)(\beta x)$:

(1) If X is a $\mathcal{C}_{n+2}$-space, then $\xi_n(x) = Q^{\frac{n+q}{2}}(x)$ $[\xi_n(x) = Q^{n+q}(x)]$, hence $\zeta_n(x) = \beta Q^{\frac{n+q}{2}} x$, for $x \in H_q X$.

(2) If we let $Q^{\frac{n+q}{2}} x$ $[Q^{n+q}x]$ denote $\xi_n(x)$, then $\xi_n(x)$ satisfies formulas (1)-(3), (5) of Theorem 1.1, the external and diagonal Cartan Formulas of Theorem 1.1(4), and the following analog of the internal Cartan formula:

$$\xi_n(xy) = \sum_{\substack{i+j = \frac{n+|xy|}{2} \\ [i+j = n+|xy|]}} Q^i x . Q^j y \quad + \sum_{\substack{0 \le i,\, j \le p \\ 0 \le i+j \le p}} x^i y^j \Gamma_{ij}, \; n > 0,$$

where the Γ_{ij} are functions of x and y specified in section 13; in particular, if $p = 2$,

$$\xi_n(xy) = \sum_{i+j=n+|xy|} Q^i x \cdot Q^j y + x\lambda_n(x,y)y.$$

(The internal Cartan formula for ζ_n follows from those for ξ_n and λ_n.)

(3) The Nishida relations hold:

$$P^r_* \xi_n(x) = \sum (-1)^{r+i}\left(r-pi, \tfrac{n+q}{2}(p-1)-pr+pi\right) Q^{m-r+i} P^i_* x$$
$$+ \sum \frac{1}{i_1} ad_n(P_*^{i_{\sigma(p-1)}} x) ad_n(P_*^{i_{\sigma(p-2)}} x) \ldots ad_n(P_*^{i_{\sigma(1)}} x)(P_*^{i_1} x)$$

where $m = \frac{n+|x|}{2}$, $[m = n + |x|]$ and the second sum runs over all sequences $(i_1,\ldots,i_p)$ such that $i_1 + \ldots + i_p = r$,

$i_1 = \ldots = i_{k_1}, \ldots, i_{k_{\ell-1}+1} = \ldots = i_{k_\ell} = i_p$,

$i_{k_1} < i_{k_2} < \ldots < i_{k_\ell}$, $\ell > 1$, and σ runs over a complete set of distinct coset representatives for

$\Sigma_{k_1-1} \times \Sigma_{k_2-k_1} \times \ldots \times \Sigma_{k_\ell-k_{\ell-1}}$ in Σ_{p-1}.

If $p > 2$, $P^r_*\zeta_n(x) = P^r_*\beta Q^{\frac{n+q}{2}}(x)$ (which is given by Theorem 1.1(7)).

If $p = 2$, $\beta\xi_n(x) = (|x| + n - 1)Q^{|x|+n-1}x + \lambda_n(\beta x,x)$.

(4) $\lambda_n(x,\xi_n y) = ad^p_n(y)(x)$ and $\lambda_n(x,\zeta_n y) = 0$ for $x, y \in H_*X$.

(5) $\xi_n(x+y) = \xi_n(x) + \xi_n(y) + \sum_{i=1}^{p-1} d^i_n(x)(y)$ where

$id^i_n(x)(y) = \sum ad_n^{j_1}(x)ad_n^{k_1}(y)\ldots ad_n^{j_r}(x)ad_n^{k_r}(y)(x)$ for $1 \leq i \leq p-1$

where the second sum runs over all sequences 1 $(j_1,k_1,\ldots,j_r,k_r)$ such that $j_1 \geq 0$, $k_\ell \geq 1$, and $j_\ell \geq 1$ if $\ell > 1$, and $\Sigma j_\ell = i - 1$, $\Sigma k_\ell = p - i$. (Compare Jacobson's Formula [12,p.187] for restricted Lie Algebras.)

(6) $\xi_n(kx) = k^p\xi_n(x)$ for $k \in \mathbf{Z}_p$ whenever $\xi_n x$ is defined.

If $X = \Omega^{n+1}Y$, we have the following theorem relating the previously described operations to the homology suspension $\tilde{\sigma}_*: H_*\Omega^{n+1}Y \to H_*\Omega^n Y$.

<u>Theorem 1.4.</u> If $X = \Omega^{n+1}Y$, then

(1) $\sigma_*Q^s(x) = Q^s\sigma_*(x)$, $x \in H_*X$.

(2) $\sigma_*\xi_n(x) = \xi_{n-1}(\sigma_* x)$, $x \in H_*X$.

(3) $\sigma_*\lambda_n(x,y) = \lambda_{n-1}(\sigma_*x, \sigma_*y)$, $x, y \in H_*X$.

(4) If Ω^nY is simply connected and $x \in H_q\Omega^nY$ transgresses to $y \in H_{q-1}\Omega^{n+1}Y$ in the Serre spectral sequence of the path fibration, then Q^sx, βQ^sx, $\xi_{n-1}(x)$, and $\zeta_{n-1}(x)$ transgress to Q^sy, $-\beta Q^sy$, $\xi_n(y)$, and $-\zeta_n(y)$; if $p > 2$, $n = 1$, and q is even, then $x^{p-1} \otimes y$ "transgresses" to $-\zeta_1(y)$, $d^{q(p-1)}(x^{p-1} \otimes y) = -\zeta_1(y)$; and if $p > 2$, $n > 1$, and $q = 2s$, $x^{p-1} \otimes y$ "transgresses" to $-\beta Q^s(y)$, $d^{q(p-1)}(x^{p-1} \otimes y) = -\beta Q^sy$.

The Hopf algebras $H_*\Omega^{n+1}Y$ admit the conjugation $\chi = C_*$, where C is the standard inverse map, and we have the following formulas.

<u>Proposition 1.5</u>. On $H_*\Omega^{n+1}Y$, $Q^s\chi = \chi Q^s$, $\xi_n\chi = \chi\xi_n$, $\zeta_n\chi = \chi\zeta_n$ if $p > 2$, and $\chi\lambda_n(y,z) = -\lambda_n(\chi y,\chi z)$.

<u>Remark 1.6</u>. In the sequel, we shall often use the notation $*$ to denote the Pontrjagin product in homology.

2. Allowable R_n-structures, $n>0$

We describe some algebraic structures which are naturally suggested by the results above. We restrict attention to the cases $n > 0$ here. We shall consider $\mathbb{Z}_p$-modules and will usually assume that they are unstable A-modules, in the sense of homology, where A is the Steenrod algebra.

Recall the definition of admissible monomials in the Dyer-Lashof algebra, R, [I, §2] and let $R_n(q)$ be the $\mathbb{Z}_p$-subspace of R having additive basis

$$\{Q^I \mid I \text{ admissible, } e(I) \geq q,\ 2\, s_k < n+q\ [s_k < n+q]\}.$$

We do not give $R_n(q)$ any additional structure yet.

Definition 2.1. A $\mathbb{Z}_p$-module L is a restricted λ_n-algebra if there is a homomorphism $\lambda_n\colon L_q \otimes L_r \to L_{q+r+n}$ and functions $\xi_n\colon L_q \to L_{pq+n(p-1)}$ $[\xi_n\colon L_q \to L_{2q+n}]$, and, if $p > 2$, $\zeta_n\colon L_q \to L_{pq+n(p-1)-1}$, for $n+q$ even such that the Lie analogs (3) and (6) of Theorem 1.2 and the restriction analogues (4), (5), and (6) of Theorem 1.3 are satisfied.

Remark 2.2. A restricted λ_n-algebra is a generalization of a restricted Lie algebra in the presence of an additional operation ζ_n for odd primes.

Definition 2.2. A $\mathbb{Z}_p$-module M is an allowable R_n-module if there are homomorphisms

$$Q^s\colon M_q \to M_{q+2s(p-1)} \quad [Q^s\colon M_q \to M_{q+s}]$$

for $0 \leq 2s < q + n$ [$s < q + n$], such that $Q^s = 0$ for $2s < q$[$s < q$] and the composition of the Q^s satisfies the Adem relations. M is an allowable AR_n-module if M is an allowable R_n-module with an A-action which satisfies the Nishida relations. M is an allowable AR_n-algebra if M is an allowable AR_n-module and a commutative algebra which satisfies the internal Cartan formula and (2) and (3) of Theorem 1.1. M is an allowable $AR_n\Lambda_n$-Hopf algebra (with conjugation) if M is a monoidal Hopf algebra (with conjugation) which satisfies the properties of Theorems 1.1, 1.2, and 1.3. If M has the conjugation, χ, then M is required to satisfy Propostion 1.5.

Remark 2.3. Since the coproduct applied to λ_n requires the presence of products (formula 1.2 (5)), we have chosen not to define separate notions of allowable $AR_n\Lambda_n$-algebras or coalgebras. Also, because of the mixing of the ξ_n, ζ_n, Q^s, and λ_n in the presense of products and coproducts (formulas 1.3(2)), we must build the desired properties of our structures in five separate stages.

To exploit the global structure suggested by our definitions, we describe five free functors (left adjoints to the evident forgetful functors) L_n, D_n, V_n, W_n, and G. Note that G has been defined in [I, §2]. The other functors are defined on objects; the morphisms are evident.

L_n: $\mathbb{Z}_p$-modules to restricted λ_n-algebras: Given M, let L_0M denote the free restricted Lie algebra generated by M. (Explicitly, L_0M is the sub-Lie algebra of T(M) generated by M where T(M) is the tensor algebra of M.) If $p > 2$, define $L_1M = s^{-1}L_0sM \oplus (\zeta_1)(s^{-1}L_0sM)$ where sM is a copy of M with all elements raised one higher degree, $s^{-1}L_0sM$

is a copy of L_0sM with all elements lowered one degree, and $(\zeta_1)(s^{-1}L_0sM)$ has $\mathbb{Z}_p$-basis consisting of elements $(\zeta_1)(x)$ of degree $2pq-2$ for each x a basis element of $s^{-1}L_0sM$ of degree $2q-1$. If $p = 2$, set $L_1M = s^{-1}L_0sM$. Inductively, define $L_nM = s^{-1}L_{n-1}sM$ for $n > 1$ and make L_nM into a restricted λ_n-algebra by setting $\lambda_n(x,y) = s^{-1}\lambda_{n-1}(sx,sy)$, $\xi_n(x) = s^{-1}\xi_{n-1}(sx)$, $\zeta_n(x) = -s^{-1}\zeta_{n-1}(sx)$ and $\lambda_n(x,\zeta_n y) = 0$.

We further describe certain elements in L_nM. $x \in M$ is a λ_n-product of weight 1. Assume that λ_n-products of weight j have been defined for $j < k$. Then a λ_n-product of weight k is any $\lambda_n(a,b)$ where a and b are λ_n-products such that weight (a) + weight (b) = k. $x \in M$ is a basic λ_n-product of weight 1. Assume that the basic λ_n-products of weight j have been defined and totally ordered amongst themselves, $j < k$. Then define a basic λ_n-product of weight k to be any $\lambda_n(a,b)$ such that

(1) $\lambda_n(a,b)$ is of weight k, and

(2) $a < b$ where a and b are basic λ_n-products and if $b = \lambda_n(c,d)$ for c and d basic then $c \leq a$.

We include additional basic products not defined by the above inductive procedure.

(2') $a = b$ if $p > 2$ where a is a basic λ_n-product of weight one and n + degree (a) is odd.

Remark 2.3. Compare the above notion of basic λ_n-product to Hilton's [12] or Hall's [11] notion of basic product. Note that (2') is not contained in Hilton's list of criteria. In Hilton's calculations, $\lambda_0(a,a) = 2a^2$ if degree (a) is odd is "seen" as a^2 up to non-zero coefficient (for $p > 2$). Since $\lambda_0(a,a)$ transgresses to $\lambda_1(\tau a, \tau a)$ in the Serre spectral sequence of the path-space fibration (τ denotes the transgression), we find it more convenient to count $\lambda_0(a,a)$ rather than a^2. (See [8; page 80].)

D_n: $\mathbb{Z}_p$-modules to allowable R_n-modules: Given L, define $D_nL = \sum_{q\geq 0} R_n(q) \otimes L_q$. Let $(D_nL)_j$ be the subspace of D_nL spanned by $\{Q^I \otimes \ell \mid \text{degree}(I) + \text{degree}(\ell) = j\}$. Then D_nL is an allowable R_n-module with the action of the Q^s determined by the Adem relations and Q^s: $(D_nL)_j \to (D_nL)_{j+2s(p-1)}$ $[(D_nL)_j \to (D_nL)_{j+s}]$ given by $Q^s(Q^I \otimes \ell) = Q^sQ^I \otimes \ell$. If $e(I) < q$, set $Q^I \otimes \ell = 0$ if the degree of ℓ is q. The inclusion of L in D_nL is given by $\ell \longrightarrow 1 \otimes \ell$.

V_n: Allowable R_n-modules to allowable R_n-algebras: Given D, define $V_nD = \frac{AD}{K}$ where AD is the free commutative algebra generated by D and K is the two-sided ideal generated by $\{x^p - Q^sx \mid 2s = \text{degree}(x)$ $[s = \text{degree}(x)]\}$. The R_n-action is determined by the R_n-action on D, the internal Cartan formula and the formulas for $Q^s\emptyset$.

W_n: Cocommutative component coalgebras over A to allowable $AR_n\Lambda_n$-Hopf algebras: Given M, let η denote the composite $\mathbb{Z}_p \to M \to L_nM \to D_n(L_nM)$. Define W_nM as an allowable AR_n-algebra by $W_n(M) = V_n(JD_nL_nM)$ where JD_nL_nM = cokernel η. The product is determined by the product in V_n, the internal Cartan formulas and the formulas for $\emptyset$ in Theorems 1.2 and 1.3. The coproduct and augmentation are determined by the diagonal Cartan formulas (for Q^s, λ_n, ξ_n, ζ_n), the augmentation of $D_n L_n M = \mathbf{Z}_p \oplus \operatorname{coker}\eta$ and the requirement that $W_n M$ be a Hopf algebra (one must check that K is a Hopf ideal to ensure that $W_n M$ is a well-defined Hopf algebra.)

The action of λ_n is given by the action of λ_n on $L_n M$ and the formula $\lambda_n[x, Q^s y] = 0$. The actions of ξ_n and ζ_n are given by the actions on $L_n M$ and the Adem relations. The A-action on M together with the Nishida relations determine the A-action on $W_n M$.

For convenience, we define $W_0 H_* X$ to be $T(H_* X)$, the tensor algebra of $H_* X$. Here we restrict attention to spaces X which are connected. By Hilton's calculations we may write $T(H_* X)$ additively as a tensor product of polynomial and exterior algebras whose generators are basic λ_0-products [12].

Remark 2.5. By section 1, $H_* C_{n+1} Y$ is an allowable $AR_n\Lambda_n$-Hopf algebra and $H_* \Omega^{n+1} Y$ is an allowable $AR_n\Lambda_n$-Hopf algebra with conjugation.

3. The homology of $\Omega^{n+1}\Sigma^{n+1}X$, $n > 0$; the Bockstein spectral sequence

Recall that $\Omega^{n+1}\Sigma^{n+1}X$ is the free (n+1)-fold loop space generated by X in the sense that if $f\colon X \to \Omega^{n+1}Z$ is any map, then there exists a unique map of (n+1)-fold loop spaces, $g\colon \Omega^{n+1}\Sigma^{n+1}X \to \Omega^{n+1}Z$ such that the following diagram is commutative:

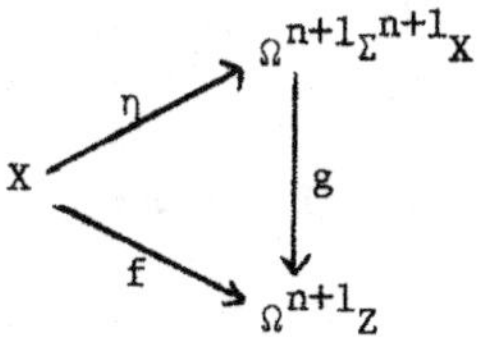

Here η is the standard inclusion of X in $\Omega^{n+1}\Sigma^{n+1}X$ [G;p.43]. Since $\eta_*\colon H_*X \to H_*\Omega^{n+1}\Sigma^{n+1}X$ is a monomorphism, $H_*\Omega^{n+1}\Sigma^{n+1}X$ ought to be an appropriate free functor of H_*X. That is, the classes in H_*X should play an analogous role to that of the fundamental classes in the calculation of the cohomology of $K(\pi,n)$'s.

By the freeness of the functors W_n and GW_n, there are unique morphisms $\bar{\eta}_*$ of allowable $AR_n\Lambda_n$-Hopf algebras and $\tilde{\eta}_*$ of allowable $AR_n\Lambda_n$-Hopf algebras with conjugation such that the following diagram is commutative, where $C_{n+1}X$ and α_{n+1} are as defined in [G; §2 and §5]:

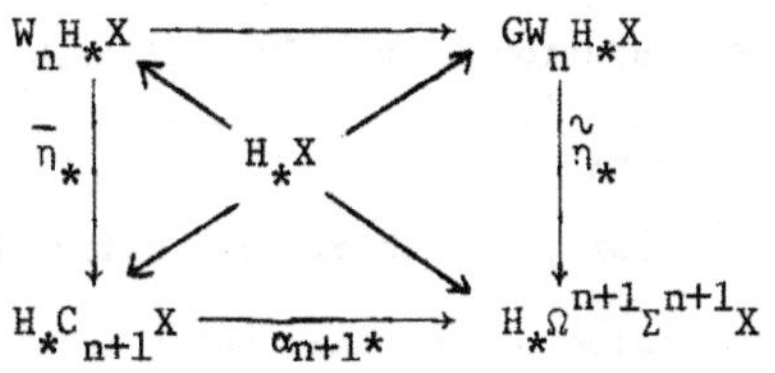

<u>Theorem 3.1.</u> For every space X, $\bar{\eta}_*\colon W_nH_*X \to H_*C_{n+1}X$ is an isomorphism of allowable $AR_n\Lambda_n$-Hopf algebras.

Theorem 3.2. For every space X, $\tilde{\eta}_*: GW_nH_*X \to H_*\Omega^{n+1}\Sigma^{n+1}X$ is an isomorphism of allowable $AR_n\Lambda_n$-Hopf algebras with conjugation.

Corollary 3.3. $\alpha_{n+1}: C_{n+1}X \to \Omega^{n+1}\Sigma^{n+1}X$ is a group completion.

Corollary 3.3 was first proven by Graeme Segal [22] but without a calculation of the homology of $C_{n+1}X$ or $\Omega^{n+1}\Sigma^{n+1}X$.

By [G; Lemma 8.11], $C_{n+1}S^0 = \coprod_{j \geq 0} \mathcal{C}_{n+1}(j)/\Sigma_j$. Since [G; Theorem 4.8] and [9] imply that $\mathcal{C}_2(j)/\Sigma_j$ is a $K(B_j, 1)$ where B_j is Artin's braid group, Theorem 3.1 provides an amusing calculation of the homology of all the braid groups. More complete descriptions of $H_*(B_j; M)$, $M = \mathbb{Z}_p, \mathbb{Z}$, and $\mathbb{Q}$ will appear in the appendix.

We have two obvious corollaries of Theorems 3.1 and 3.2.

Corollary 3.4. If (X, θ) is a $\mathcal{C}_{n+1}$-space, then $\theta_*: H_*C_{n+1}X \to H_*X$ represents H_*X as a quotient allowable $AR_n\Lambda_n$-Hopf algebra of the free allowable $AR_n\Lambda_n$-Hopf algebra W_nH_*X.

Corollary 3.5. If Y is an $(n+1)$-fold loop space, then $\xi_{n+1*}: H_*\Omega^{n+1}\Sigma^{n+1}Y \to H_*Y$ represents H_*Y as a quotient allowable $AR_n\Lambda_n$-Hopf algebra with conjugation of the free allowable $AR_n\Lambda_n$-Hopf algebra with conjugation $GW H_*Y$.

Before proceeding to proofs, we exhibit bases for W_nH_*X and GW_nH_*X. Let $\eta: * \to X$ be the inclusion of the base point in X and let $JH_*X = \operatorname{coker} \eta_*$. Let tX be a totally ordered basis for JH_*X. We define AT_nX to be the free commutative algebra generated by the set

$$T_nX = \left\{ Q^I y \;\middle|\; \begin{array}{l} y \text{ a basic } \lambda_n\text{-product, I admissible, } e(I) + b(I) > |y|,\ |Q^I y| > 0; \text{ if} \\ |y| = q, \text{ then } e(I) \geq q; \text{ and if } p > 2, I = (\varepsilon_1, s_1, \ldots, \varepsilon_k, s_k), \text{ then} \\ \frac{n+q}{2} \leq s_k. \text{ [If } I = (s_1, \ldots, s_k), \text{ then } n + q \leq s_k]. \end{array} \right\}$$

Notice that in our definition of T_nX we denote $\xi_n x$ by $Q^{\frac{n+q}{2}} x [Q^{n+q}x]$ and $\zeta_n x$ by $\beta Q^{\frac{n+q}{2}} x$ for $x \in H_qX$.

Let $\mathbb{Z}_p N\pi_0 X$ and $\mathbb{Z}_p \tilde{N}\pi_0 X$ be as defined in [G;p.80].

Lemma 3.8. $W_nH_*X \cong AT_nX \otimes \mathbb{Z}_p N\pi_0 X$ and $GW_nH_*X \cong AT_nX \otimes \mathbb{Z}_p \tilde{N}\pi_0 X$ as algebras.

Proof: By the definition of W_n and GW_n, it suffices to check that the basic λ_n-products of weight k span the subspace generated by all λ_n-products of weight k. By Hilton's results [12], the basic λ_0-products of weight k span the subspace of L_0H_*X generated by all λ_0-products of weight k. By the inductive definition of L_nH_*X and the definition of basic λ_n-products, the result follows.

For our final preliminary, we recall the calculation of $H_*\Omega\Sigma X$ for connected X.

Lemma 3.9 [5]. If X is connected, $H_*\Omega\Sigma X$ is the free associative algebra on the transgressive elements of H_*X in the Serre spectral sequence of the path fibration.

Evidently, $H_*\Omega\Sigma X = W_0H_*X$ as an algebra.

Alternatively, we may use [G; Proposition 2.6(a)] which states that $(F_jC_1X, F_{j-1}C_1X)$ is an NDR pair for $j \geq 1$. Here the result

$H_*\Omega\Sigma X = H_*C_1X = W_0H_*X$ follows.

Proof of Theorem 3.2: By [I; Lemma 4.6] there is a homotopy equivalence $f\colon \Omega^{n+1}\Sigma^{n+1}X \to \Omega^{n+1}_{\phi}\Sigma^{n+1}X \times \pi_0\Omega^{n+1}\Sigma^{n+1}X$ where $\Omega^{n+1}_{\phi}\Sigma^{n+1}X$ is the component of the basepoint of $\Omega^{n+1}\Sigma^{n+1}X$. Furthermore f is a map of H-spaces if $n > 0$. By [G;8.14],

$\tilde{N}\pi_0X \cong \tilde{H}_0X \cong H_{n+1}\Sigma^{n+1}X \cong \pi_{n+1}\Sigma^{n+1}X \cong \pi_0\Omega^{n+1}\Sigma^{n+1}X$. Since

$f_*\colon H_*\Omega^{n+1}\Sigma^{n+1}X \to H_*\Omega^{n+1}_{\phi}\Sigma^{n+1}X \otimes H_*\pi_0\Omega^{n+1}\Sigma^{n+1}X = H_*\Omega^{n+1}_{\phi}\Sigma^{n+1}X \otimes \mathbb{Z}_p\tilde{N}\pi_0X$,

it clearly suffices to show that $H_*\Omega^{n+1}_{\phi}\Sigma^{n+1}X \cong AT_nX$, as an algebra.

We show that $H_*\Omega^{n+1}_{\phi}\Sigma^{n+1}X \cong AT_nX$ by induction on n. To avoid repetition we state the general step and note the minor modifications required in case $n = 1$.

Obviously $\Sigma_*\colon \tilde{H}_*X \to H_*\Sigma X$ is an isomorphism and we may choose $t\Sigma X = \{\Sigma_*x' \mid x \in tX\}$ as a basis for $H_*\Sigma X$ where $x' = x - (\varepsilon x)\phi$. Define $\tilde{W}_{n-1}H_*\Sigma X$, $n \geq 1$, to be the subalgebra of $W_{n-1}H_*\Sigma X$ generated by all operations on the elements of $t\Sigma X$ of degree greater than one and, in addition, the elements derived from non-trivial applications of the operations, $\beta^{\varepsilon}Q^s, \beta^{\varepsilon}\xi_n$, and λ_n on the elements of $t\Sigma X$ of degree greater than zero. (Compare our definition of $\tilde{W}_{n-1}$ to the definition of $\tilde{W}$ in [I; §4], especially in the case $p = 2$.). Observe that [I; 4.7 and 4.8] together with Lemma 3.9 shows that $H_*U\Omega\Sigma^2X \cong \tilde{W}_0H_*\Sigma X$ as an algebra. Also observe that if our calculations of $H_*\Omega^n\Sigma^nX$ are correct, then $H_*U\Omega^n\Sigma^{n+1}X \cong \tilde{W}_{n-1}H_*\Sigma X$.

We now describe a model spectral sequence $\{'E^r\}$. Define $'E^2$, as an algebra, by the equation

$$'E^2 = \tilde{W}_{n-1}H_*\Sigma X \otimes (GW_nH_*X)_\phi ,$$

(Observe that $'E^2 = W_{n-1}H_*\Sigma X \otimes W_nH_*X$ if X is connected.) Specify the differentials by requiring $\{'E^r\}$ to be a spectral sequence of differential algebras such that the transgression, τ, is given by

$\tau\{ad_{n-1}(\Sigma_*x_1')\dots ad_{n-1}(\Sigma_*x_{k-1}')(\Sigma_*x_k')\} = ad_n(x_1*[-ax_1])\dots ad_n(x_{k-1}*[-ax_{k-1}])(x_k*[-ax_k])$,

$\tau Q^I\Sigma_*x' = (-1)^{d(I)}Q^Ix*[-p^{\ell(I)}ax]$, and if $p > 2$ and the degree of Q^Ix is $2s-1$, $\tau((Q^I\Sigma_*x')^{p-1} \otimes Q^Ix*[-p^{\ell(I)}ax]) = (-1)^{d(I)+1}\beta Q^sQ^Ix*[-p^{\ell(I)}ax]$.

(Recall that $[ax]$ denotes the component of the element x and that if $[ax] \in H_0\Omega X$, $x \in H_*\Omega X$, then the loop product $x * [ax]$ suspends to $\varepsilon[ax](\sigma_*x)$ by [I; Lemma 4.9].) It is easy to see that $\{'E^r\}$ is isomorphic to a tensor product of elementary spectral sequences of the forms $E\{y\} \otimes P\{\tau y\}$ and, if $p > 2$,

$P\{z\}/(z^p) \otimes [E\{\tau z\} \otimes P\{\tau(z^{p-1} \otimes \tau z)\}]$, where E and P

denote exterior and polynomial algebras. The elements y run over

$$\left\{ Q^I\Sigma_*x' \;\middle|\; \begin{array}{l} I \text{ admissible, } e(I) > \text{degree } (x), \text{ degree } (Q^I\Sigma_*x') > 1, \\ s_k \le \dfrac{n-1+|\Sigma_*x'|}{2} \quad [s_k \le n-1+|\Sigma_*x'|], \text{ and of odd degree in} \\ \text{case } p > 2 \end{array} \right\}$$

and if $p > 2$, the elements z run through

$$\{Q^I\Sigma_*x' \mid I \text{ admissible, } e(I) > \text{degree } (x), \text{ degree } Q^I\Sigma_*x' \text{ is even and } s_k \le \frac{n-1+|\Sigma_*x'|}{2}\}.$$

Note that the Serre spectral sequence behaves as if the even degree generators were generators of truncated polynomial algebras since $d^r(x^p) = 0$ if $d^r(x) = \tau x$. Clearly $'E_\infty = \mathbb{Z}_p$.

Evidently, there is a unique morphism of algebras $f\colon {'E^2} \to E^2$ such that the following diagram is commutative:

$$\begin{array}{ccc}
\tilde{W}_{n-1}H_*\Sigma X \otimes (GW_nH_*X)_\phi & \xrightarrow{f\otimes f} & H_*U\Omega^n\Sigma^{n+1}X \otimes H_*\Omega U\Omega^n\Sigma^{n+1}X \\
\downarrow{\scriptstyle i\otimes i} & & \downarrow{\scriptstyle \pi_*\otimes(\Omega\pi)_*} \\
W_{n-1}H_*\Sigma X \otimes GW_nH_*X & \xrightarrow{\tilde{\eta}_*\otimes\tilde{\eta}_*} & H_*\Omega^n\Sigma^{n+1}X \otimes H_*\Omega^{n+1}\Sigma^{n+1}X.
\end{array}$$

Since $\lambda_n(x,y) = \lambda_n(x',y')$ and $Q^Ix = Q^Ix'$ if $d(I) > 0$, [I_j Lemma 4.9] implies that

$$\sigma_*\,\lambda_n(x * [-ax],\ y * [-ay]) = \lambda_{n-1}(\Sigma_*x',\ {}_*y') \text{ and}$$

$$\sigma_*(Q^Ix*[-p^{\ell(I)}ax]) = (-1)^{d(I)}Q^I\Sigma_*x'.$$

By the naturality of σ_*, the same formula holds for

$$\sigma_*\colon\ H_*\Omega U\Omega^n\Sigma^{n+1}X \to H_*U\Omega^n\Sigma^{n+1}X.$$

Note that the classes Σ'_*x' are not present in $H_*U\Omega^n\Sigma^{n+1}X$ if x is a zero dimensional basis element for H_*X, but the transgressive classes $\beta^\varepsilon Q^s\Sigma_*x'$ and $\lambda_n(\Sigma_*x',\ \Sigma_*y')$ are present in $H_*U\Omega^n\Sigma^{n+1}X$ (although not as operations).

By Theorem 1.4 (commutation with suspension), $f\otimes f$ induces a morphism of spectral sequences. Since $f \otimes f$ is an isomorphism on E^{∞}, it suffices to show that f(base) is an isomorphism to show f(fibre) is an isomorphism [15; chap. 12].

Now we show that our results are correct for the case $n = 1$. Recall that by previous remarks $H_* U\Omega\Sigma^2 X \cong \tilde{W}_0 H_* \Sigma X$ as an algebra. By a slight modification of [8; p.80], we may write $\tilde{W}_0 H_* \Sigma X$ additively as a tensor product of polynomial and exterior algebras generated by basic λ_0-products of weight greater than one and basic λ_0-products of weight one on the elements of $t\Sigma X$ of degree greater than 1. It follows from previous remarks that our results are correct for the case $n = 1$.

The remaining details follow directly by induction on n and the above methods. Compare our proof to that of [I; 4.2] and [8; p.80].

The Bockstein spectral sequences for $H_*C_{n+1}X$ and $H_*\Omega^{n+1}\Sigma^{n+1}X$

We require a preliminary lemma concerning the higher Bocksteins on $\xi_n(x)$ mod 2.

Lemma 3.10. Let $X \in \mathcal{C}_{n+1}[\tau]$ and $x \in H_qX$, $p = 2$. Then

(1) $\beta\xi_n(x) = Q^{n+q-1}x + \lambda_n(x,\beta x)$ if $n + q$ is even, and

(2) if $\beta_r x$ is defined, then $\beta_r\xi_n x$ is defined if $n + q$ is odd and $\beta_r\xi_n x = \lambda_n(x,\beta_r x)$ modulo indeterminacy.

Proof: Let a be a chain which represents x and $\partial(a) = 2^r b$. Then $\partial(e_n \otimes a^2) = [\alpha + (-1)^n]e_{n-1} \otimes a^2 + (-1)^n e_n \otimes \partial a \otimes a + (-1)^{n+q} e_n \otimes a \otimes \partial a$ where e_n and α have been defined in [A; section 1 and 6]. Clearly

$$\partial(e_n \otimes a^2) = [(-1)^n + (-1)^q]e_{n-1} \otimes a^2 + 2^r[(-1)^n + (-1)^{n+q}\alpha]e_n \otimes b \otimes a.$$

Since $-2^r \equiv 2^r(2^{r+1})$, we observe that

(a) $$\partial(e_n \otimes a^2) \equiv 2e_{n-1} \otimes a^2 + 2^r[\alpha + (-1)^{n+1}]e_n \otimes b \otimes a \quad (4)$$

if $n + q$ is even and

(b) $$\partial(e_n \otimes a^2) \equiv 2^r[\alpha + (-1)^{n+1}]e_n \otimes b \otimes a \quad (2^{r+1})$$

if $n + q$ is odd.

We recall the definition of λ_n in [A; section 6] and observe that (a) implies (1) and (b) implies (2).

Let $\{E^r x\}$ denote the mod p Bockstein spectral sequence of a space X. If A is an algebra equipped with higher Bocksteins, let $\{E^r A\}$ denote the obvious Bockstein spectral sequence associated to A. By Theorems 3.1, 3.2 and Lemma 3.8,

$$E^rY = E^rAT_nX \otimes H_0Y \quad \text{where} \quad Y = C_{n+1}X \quad \text{or} \quad \Omega^{n+1}\Sigma^{n+1}X.$$

We decompose AT_nX into a tensor product of algebras which are closed under the Bocksteins. The anomaly for the mod 2 Bockstein implied by Lemma 3.10 requires some additional attention.

Definition 3.11. Let the Bockstein spectral sequence of X be given by C_r and D_r as in [I; section 4]. Further, let C_∞ be a set of basis elements which projects to a generating set for $E^\infty X$. Let $x_i \in D_{r_i}$ or $x_i \in C_\infty$. Define $L_n(x_1,\ldots,x_k)$ to be the free commutative algebra generated by all λ_n-products of weight k which are constructed from $y_1,\ldots,y_k$ where $y_{\sigma(i)} = x_i$ or $y_{\sigma(i)} = \beta_{r_i}x_i$ for some fixed $\sigma \in \Sigma_k$. Observe that by [A; 6.7], the algebras $L(x_1,\ldots,x_k)$ are closed (modulo indeterminacy) under the higher Bocksteins. If $p = 2$, further define $G_n(x)$ to be the free commutative algebra generated by $\xi_n(x)$ and $\lambda_n(x,\beta_rx)$ for $n + |x|$ odd and $x \in D_r$ and define $B_n(x)$ to be the free strictly commutative algebra generated by ξ_nx.

We observe that AT_nX may be written as a tensor product of algebras of the following forms:

(A) If $p = 2$,

(i) $E[Q^Ix] \otimes P[\beta Q^Ix]$

(ii) $P[Q^Ix] \otimes E[\beta Q^Ix]$

where in (i) and (ii), I is admissible, $\ell(I) > 1$, or $\ell(I) = 1$, and either $Q^I x \neq \xi_n x$ or in case $Q^I x = \xi_n x$ we then require $n + |x|$ to be even. (The complications here are introduced by the irregular higher Bockstein in Lemma 3.10; the case $Q^J x = \xi_n x$ and $n + |x|$ is odd is accounted for in (iii) below. Here note that $n + |\xi_n x|$ is even.)

(iii) $G_n(x)$ if $n + |x|$ is odd, and

(iv) $L_n(x_1,\dots,x_k)$ where $\{x_1,\dots,x_k\}$ runs over distinct subsets of elements $x_i \in D_{r_i}$ or C_∞, $k \geq 1$.

(B) If $p > 2$,

(i) $E[Q^I x] \otimes P[\beta Q^I x]$,

(ii) $P[Q^I x] \otimes E[\beta Q^I x]$, and

(iii) $L_n(x_1,\dots,x_k)$ where $\{x_1,\dots,x_k\}$ is described above.

With the above preliminaries, we have the following Theorem. The proof is similar to that of [A; section 10] and is deleted.

<u>Theorem 3.12</u>. Define a subset $S_n X$ of $T_n X$ as follows:

(a) $p = 2$:

$$S_n X = \left\{ Q^I x \;\middle|\; \begin{array}{l} Q^I x \in T_n X,\ Q^I x \neq \xi_n x \text{ if } n + |x| \text{ odd, } I = (2s,J), \\ |Q^I x| \text{ even, } \ell(I) > 0 \end{array} \right\}$$

(b) $p > 2$: $S_n X = \{Q^I x \mid Q^I x \in T_n X,\ b(I) = 0,\ |Q^I x| \text{ even, } \ell(I) > 0\}$.

Then if $p = 2$,

$$E^{r+1} AT_n X = P\{y^{2^r} \mid y \in S_n X\} \otimes E\{\beta_{r+1} y^{2^r} \mid y \in S_n X\} \otimes E^{r+1} L_n(x_1,\dots,x_k) \otimes E^{r+1} G_n(x)$$

and if $p > 2$,

$$E^{r+1} AT_n X = P\{y^{p^r} \mid y \in S_n X\} \otimes E\{\beta_{r+1} y^{p^r} \mid y \in S_n X\} \otimes E^{r+1} L_n(x_1,\dots,x_k), \text{ for}$$

all $r > 1$ where

$$\beta_{r+1}y^{p^r} = \begin{cases} y^{2^r-1}(y\beta y + Q^{2q}\beta y) & \text{if } p = 2 \text{ and } |y| = 2q, \\ y^{p^r-1}\beta y & \text{if } p > 2, \text{ and} \end{cases}$$

$$\beta_{r+1}\lambda_n(x,y) = \lambda_n(\beta_{r+1}x,y) + (-1)^{n+|x|}\lambda_n(x,\beta_{r+1}y)$$

if $\beta_{r+1}x$ and $\beta_{r+1}y$ are defined. Therefore if $p = 2$

$$E^{\infty}AT_nX = \otimes L(x_1,\ldots,x_k) \otimes B_n(x)$$

where $\{x_1,\ldots,x_k\}$ runs over distinct subsets, $x_i \in C_\infty$ $\{x_1,x_2\} \neq \{x,x\}$ for $n + |x|$ odd, and $\{x\}$ runs over C_∞ where $n + |x|$ odd. If $p > 2$,

$$E^{\infty}AT_nX = \otimes L(x_1,\ldots,x_k)$$

where $\{x_1,\ldots,x_k\}$ runs over distinct subsets of elements $x_i \in C_\infty$

Since there are only 3 non-trivial operations, ξ_1, ζ_1, and λ_1 defined in $H_*(\Omega^2\Sigma^2X; \mathbb{Z}_p)$, where $\deg(\xi_1 x)$ is odd if $p > 2$, Theorem 3.12 immediately implies the following corollary.

<u>Corollary 3.13</u>. If X has no p-torsion $p \geq 2$, then $E^2AT_1X = E^{\infty}AT_1X$. Hence the p-torsion of $H_*(\Omega^2\Sigma^2X; \mathbb{Z})$ is all of order p.

Observe that 3.13 is obviously false for AT_nX, $n > 1$.

Another immediate corollary of 3.12 is

<u>Corollary 3.14</u>. Let ι be the fundamental class of S^k, $k \geq 0$. Then

(a) $E^{\infty}AT_nS^k$ is the free strictly commutative algebra generated by ι and if $n + k$ is odd, $\lambda_n(\iota,\iota)$ when $p > 2$.

(b) $E^{\infty}AT_nS^k$ is the free strictly commutative algebra generated by ι and if $n + k$ is odd, $\xi_n\iota$ when $p = 2$.

Remarks 3.15. It is amusing to observe how the complications in the Bockstein spectral sequence introduced by Lemma 3.10 give rise to the infinite cycles which must appear. It is, on the face of it, surprising that $\xi_n x$ when $p = 2$ accounts for the infinite cycle corresponding to the class $\lambda_n(x,x)$ when $p > 2$. Since $\lambda_n(x,x) = 0 \bmod 2$, $\xi_n(x)$ is "trying" to be half of the Browder operation.

A remark concerning 3.14 seems approrpiate: $\pi_i\Omega^{n+1}_{\phi}\Sigma^{n+1}S^0 = \pi_{i+n+1}S^{n+1}$ if $i > 0$. Clearly $\pi_i\Omega^{n+1}{}_{\phi}\Sigma^{n+1}S^0$ has a $\mathbf{Z}$ summand if $n + 1$ is even and $i = n$. In this case, the obvious map

$$f: \Omega^{n+1}{}_{\phi}\Sigma^{n+1}S^0 \rightarrow K(\mathbb{Z}, n)$$

is a rational homotopy equivalence with the fundamental class of $K(\mathbf{Z},n)$ corresponding to the Whitehead product $[\iota_{n+1}, \iota_{n+1}] \in \pi_{2n+1}S^{n+1}$. Hence our calculations (3.14) and the remarks in section 1 about Whitehead products are at least reasonable.

4. The homology of $C_{n+1}X$, $n > 0$

We outline the proof of Theorem 3.1. Consider the commutative diagram at the beginning of section 3. Since the composite $\tilde{\eta}_* \circ G$ is a monomorphism, it is immediate that $\bar{\eta}_*$ is a monomorphism. The crux of the proof of 3.1 is to show that $\bar{\eta}_*$ is an epimorphism. To do this, we require a technical lemma (4.3). Conceptually, this lemma states that all homology operations on $\mathcal{C}_{n+1}$-spaces derived from the spaces $\mathcal{C}_{n+1}(k)$ can be expressed in terms of the operations of Theorems 1.1, 1.2, and 1.3. Succinctly, the homology of $C_{n+1}X$ is built up from H_*X, $H_*\mathcal{C}_{n+1}(2)$, $H_*(B(\mathbb{R}^{n+1},p);\ \mathbb{Z}_p(q))$, and homological iterations of the structure map of operads, γ_*.

Before proving Theorem 3.1, we require some preliminary information. Observe that $H_*X \otimes \dots \otimes H_*X = (H_*X)^k$ has a basis given by $A \cup B$ where

$$A = \{x \otimes \dots \otimes x \mid x \text{ a homogeneous basis element for } H_*X\},$$

and

$$B = \{x_1 \otimes \dots \otimes x_k \mid x_i \text{ a homogeneous basis element for } H_*X,$$

$$x_i \neq x_j \text{ if } i \neq j \text{ for some } i \text{ and } j\}.$$

(We do not assume that k is prime.) Clearly, the Σ_k-action on $(H_*X)^k$ induces a permutation action on the set $A \cup B$. Let C be the subset of $A \cup B$ which consists of one element from each Σ_k-orbit in $A \cup B$. If $x \in C$, we let B_x denote the $\mathbb{Z}_p$-submodule of $(H_*X)^k$ spanned by the Σ_k-orbit of x in $A \cup B$.

<u>Lemma 4.1.</u> For any space X, $H_*(\mathcal{C}_{n+1}(k) \times_{\Sigma_k} X^k) = \bigoplus_{x \in C} H_*(C_*\mathcal{C}_{n+1}(k) \otimes_{\Sigma_k} B_x)$.

Let $X^{[k]}$ denote the k-fold smash product of X.

<u>Lemma 4.2</u>. The quotient map $\pi': \mathcal{C}_{n+1}(k)\times_{\Sigma_k} X^k \to \dfrac{\mathcal{C}_{n+1}(k)\times_{\Sigma_k} X^{[k]}}{\mathcal{C}_{n+1}(k)\times_{\Sigma_k} *}$

is an epimorphism in $\mathbb{Z}_p$-homology. Furthermore, the kernel of π'_* has a $\mathbb{Z}_p$-basis consisting of all classes in $H_*(\mathcal{C}_{n+1}(k)\times_{\Sigma_k} X^k)$ of the form $c \otimes x_1 \otimes \ldots \otimes x_k$, $c \in C_*\mathcal{C}_{n+1}(k)$ and $x_i = [0]$ for some i where $[0]$ denotes the class of the base-point.

The conclusions of the following lemma indicate that any operation derived from $\mathcal{C}_{n+1}(k)$ on the variable $x_1 \otimes \ldots \otimes k_k$ can be decomposed into (a) a product or Browder operation on classes which involve operations on fewer than k variables or (b) a Dyer-Lashof operation $(\beta^\varepsilon Q^s, \beta^\varepsilon \xi_n)$ on classes which involve operations on fewer than k variables.

<u>Lemma 4.3</u>. Let $\Delta \otimes x \in H_*(C_*\mathcal{C}_{n+1}(k)\otimes_{\Sigma_k} B_x)$, $k > 2$. Then there exists some $\Gamma \otimes x$ such that $\Delta \otimes x = \gamma_*(\Gamma \otimes x)$, where either

(a) $\Gamma \otimes x \in H_*[C_*(\mathcal{C}_{n+1}(2) \times \mathcal{C}_{n+1}(i_1)\times \mathcal{C}_{n+1}(i_2)) \otimes_{\Sigma_1 \times \Sigma_2} B_x]$ or

(b) $\Gamma \otimes x \in H_*[C_*(\mathcal{C}_{n+1}(p)\times \mathcal{C}_{n+1}(k/p)^p)\otimes_{\Sigma_p \int \Sigma_{k/p}} B_x]$

We prove Lemmas 4.1, 4.2, and 4.3 after the proof of 3.1.

<u>Proof of 3.1:</u>

Consider the monad $(C_{n+1}, \mu_{n+1}, \eta_{n+1})$ associated to the little cubes operad. We write μ for the $\mathcal{C}_{n+1}$-action on $C_{n+1}X$. By [G; § 13], $C_{n+1}X$ is a filtered space such that the product $*$ and the $\mathcal{C}_{n+1}$-action μ_{n+1} restrict to

$$*: F_jC_{n+1}X \times F_kC_{n+1}X \to F_{j+k}C_{n+1}X, \text{ and}$$

$$\mu_k: \mathcal{C}_{n+1}(k) \times F_{j_1}C_{n+1}X \times \dots \times F_{j_k}C_{n+1}X \to F_jC_{n+1}X$$

$$j = j_1 + \dots + j_k.$$

We define an algebraic filtration of W_nH_*X which corresponds to the given filtration of $C_{n+1}X$ by giving the image of an element $Q^I \otimes \lambda \in R_n(q) \otimes L_q$ filtration $p^{\ell(I)}w(\lambda)$ [$w(\lambda)$ denotes the weight of λ] and requiring W_nH_*X to be a filtered algebra. Loosely speaking, this filtration is given by the number of variables (not necessarily distinct) required to define the operation $Q^I\lambda$. Transparently $F_oW_nH_*X$ is spanned by the class of the base-point and $F_1W_nH_*X = H_*X$. We observe that $\overline{\eta}_* : F_kW_nH_*X \to H_*C_{n+1}X$ factors through $H_*F_kC_{n+1}X$ since every operation involving Q^s, ξ_n, λ_n, and Pontrjagin products in $F_kW_nH_*X$ has already occurred geometrically in $H_*F_kC_{n+1}X$. Clearly $H_*X = F_1W_nH_*X \to H_*F_1C_{n+1}X = H_*X$ is an isomorphism. We assume, inductively, that $\overline{\eta}_*: F_jW_nH_*X \to H_*F_jC_{n+1}X$ is an isomorphism for $j < k$. Define

$$E_k^0C_{n+1}X = \frac{F_kC_{n+1}X}{F_{k-1}C_{n+1}X} \quad \text{and} \quad E_k^0W_nH_*X = \frac{F_kW_nH_*X}{F_{k-1}W_n\ H_*X}.$$

Note that $F_{k-1}C_{n+1}X \xrightarrow{i} F_kC_{n+1}X \xrightarrow{\pi} E_k^0C_{n+1}X$ is a cofibration by [G].

We consider the following commutative diagram with exact rows and columns:

$$\begin{array}{ccccccccc}
 & & 0 & & 0 & & & & \\
 & & \downarrow & & \downarrow & & & & \\
 & 0 \to & F_{k-1}W_nH_*X & \longrightarrow & F_kW_nH_*X & \longrightarrow & E_k^0W_nH_*X & \longrightarrow & 0 \\
 & & \Big\downarrow \bar{\eta}_* & & \Big\downarrow \eta_* & & \Big\downarrow \Phi & & \\
 \cdots \longrightarrow & & H_*F_{k-1}C_{n+1}X & \xrightarrow{i_*} & H_*F_kC_{n+1}X & \xrightarrow{\pi_*} & H_*E_k^0C_{n+1}X & \longrightarrow & \cdots \\
 & & \downarrow & & & & & & \\
 & & 0 & & & & & &
\end{array}$$

$\tilde{\eta}_*$ is a monomorphism since (1) the diagram below commutes, (2) $\bar{\eta}_*$ is an isomorphism, and (3) $W_nM \to GW_nM$ is a monomorphism.

$$\begin{array}{ccc}
W_nH_*X & \longrightarrow & GW_nH_*X \\
\Big\downarrow \bar{\eta}_* & & \Big\downarrow \tilde{\eta}_* \\
H_*C_{n+1}X & \xrightarrow{\alpha_{n+1*}} & H\,\Omega^{n+1}\Sigma^{n+1}X
\end{array}$$

We define Φ by commutativity of the right hand square and observe that to show the middle $\bar{\eta}_*$ is an isomorphism it suffices, by the five lemma, to show Φ is an epimorphism.

By lemma 4.2, π'_* is an epimorphism. So we consider the arbitrary class $\Delta \otimes x \in H_*(C_*\mathcal{C}_{n+1}(k) \otimes_{\Sigma_k} B_x)$ and the following commutative diagrams:

$$\text{(i)}\quad \begin{array}{ccc} \mathcal{C}_{n+1}(2)\times\mathcal{C}_{n+1}(i_1)\times\mathcal{C}_{n+1}(i_2)\times X^{i_1+i_2} & \xrightarrow{\mu_2} & F_k C_{n+1}X \\ \downarrow{\scriptstyle \gamma\times 1} & \nearrow{\scriptstyle \gamma_k} & \downarrow{\scriptstyle \pi} \\ \mathcal{C}_{n+1}(k)\times X^k & \xrightarrow{\pi'} & E^0_k C_{n+1}X, \end{array}$$

for $i_1 + i_2 = k$ and

$$\text{(ii)}\quad \begin{array}{ccc} \mathcal{C}_{n+1}(p)\times\mathcal{C}_{n+1}(k/p)^p\times X^k & \xrightarrow{\mu_p} & F_k C_{n+1}X \\ \downarrow{\scriptstyle \gamma\times 1} & \nearrow{\scriptstyle \gamma_k} & \downarrow{\scriptstyle \pi} \\ \mathcal{C}_{n+1}(k)\times X^k & \xrightarrow{\pi'} & E^0_k C_{n+1}X \end{array}$$

for k such that p divides k.

By Lemma 4.3 $\Delta \otimes x = \gamma_*(\Gamma \otimes x)$ for

(a) $\Gamma \otimes x \in H_*(C_*(\mathcal{C}_{n+1}(2)\times\mathcal{C}_{n+1}(i_1)\times\mathcal{C}_{n+1}(i_2))\otimes_{\Sigma_{i_1}\times\Sigma_{i_2}} B_x)$ or

(b) $\Gamma \otimes x \in H_*(C_*(\mathcal{C}_{n+1}(p)\times\mathcal{C}_{n+1}(k/p)^p)\otimes_{\Sigma_p\int\Sigma_{k/p}} B_x)$. In case (a), the diagram (i) shows that $\mu_{2*}(\Delta\otimes x)$ is either $X_1 * X_2$ or $\lambda_n(X_1,X_2)$ where X_i are classes derived from operations on fewer than k variables. In case (b), the diagram (ii) shows that $\mu_{p*}(\Delta\otimes x)$ is given by the operations $\beta^\varepsilon Q^s$, $\beta^\varepsilon\xi_n$, λ_n and $*$ on a class X_3 derived from operations on fewer than k-variables. (Observe that non-equivariant operations from $\mathcal{C}_{n+1}(p)$ are giving products of iterated Browder operations by Theorem 12.1.) By our induction hypothesis, $X_i \in F_{k_i}W_nH_*X$ for $k_1 + k_2 = k$ or $pk_3 = k$. By definition of W_nH_*X and the filtration of W_nH_*X, it is apparent that $X_1 * X_2$, $\lambda_n(X_1,X_2)$, $\beta^\varepsilon Q^s X_3$, and

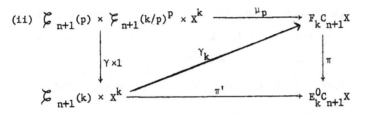

$\beta^{\varepsilon}\xi_n x_3$ are present in $E_k^0 W_n H_* X$. Hence Φ is an epimorphism and we are done.

Proof of Lemma 4.1: By [15; Chap.XI] it is easy to see that $H_*(\mathcal{C}_{n+1}(k)\times_{\Sigma_k} X^k) = H_*(C_*\mathcal{C}_{n+1}(k)\otimes_{\Sigma_k}(C_*X)^k))$. Since we are working over a field, there is a chain homotopy equivalence $i: H_*X \to C_*X$ given by mapping a basis element in H_*X to a cycle which represents it. Obviously $H_*(M\otimes_{\Sigma_k}(H_*X)^k) \to H_*(M\otimes_{\Sigma_k}(C_*X)^k)$ is a homology isomorphism where M is a Σ_k-module. If we let $M = C_*\mathcal{C}_{n+1}(k)$, an easy argument using the spectral sequence of a covering [15] shows that $H_*(C_*\mathcal{C}_{n+1}(k)\otimes_{\Sigma_k}(C_*X)^k) = H_*(C_*\mathcal{C}_{n+1}(k)\otimes_{\Sigma_k}(H_*X)^k)$. We observe that $(H_*X)^k = \bigoplus_{x\in C} B_x$ as Σ_k-modules and we are done.

Proof of Lemma 4.2: By [G; §A.4] (X^k, F) is an equivariant NDR-pair where F is the subspace of X^k given by $\{\langle x_1,\ldots,x_k\rangle \mid x_i = *$ for some $i\}$. Consequently, the inclusion $\mathcal{C}_{n+1}(k)\times_{\Sigma_k} F \to \mathcal{C}_{n+1}(k)\times_{\Sigma_k} X^k$ is a cofibration with cofibre $E_k^0 C_{n+1} X$. Clearly $H_*(\mathcal{C}_{n+1}(k)\times_{\Sigma_k} F) = H_*(C_*\mathcal{C}_{n+1}(k)\otimes_{\Sigma_k} C_*F)$ and $H_*(C_*\mathcal{C}_{n+1}(k)\otimes_{\Sigma_k} C_*F) = H_*(C_*\mathcal{C}_{n+1}(k)\otimes_{\Sigma_k} H_*F)$. Visibly, H_*F is a Σ_k-submodule of $\bigoplus_{x\in C} B_x$ where each basis element in H_*F can be written as $x_1\otimes\cdots\otimes x_k$ for some $x_i = [0]$. We write $\bigoplus_{x\in C} B_x$ as $H_*F \oplus D$ as Σ_k-module where D has additive basis given by $x_1\otimes\cdots\otimes x_k$, $x_i \neq [0]$. The map

$$H_*(C_*\mathcal{C}_{n+1}(k)\otimes_{\Sigma_k} H_*F) \to H_*(C_*\mathcal{C}_{n+1}(k)\otimes_{\Sigma_k} H_*F) \oplus (H_*C_*\mathcal{C}_{n+1}(k)\otimes_{\Sigma_k} D)$$
$$= H_*(C_*\mathcal{C}_{n+1}\otimes_{\Sigma_k}(H_*X)^k)$$

is a monomorphism. Application of the long exact homology sequence for a cofibration finishes the proof.

Proof of Lemma 4.3: Recall that $\alpha_{n+1}: C_{n+1}X \to \Omega^{n+1}\Sigma^{n+1}X$ is a weak homotopy equivalence if X is connected [G]. In this case $\mu_{k*}(\Delta \otimes x_1 \otimes \ldots \otimes x_k)$ is certainly given in terms of the operations $\beta^{\varepsilon}Q^s$, λ_n, and $*$ on the classes x_i. By the definition of these operations (see section 5),

$\mu_{k*}(\Delta \otimes x_1 \otimes \ldots \otimes x_k) = \sum_r Q^{I_1}\lambda_{I_1} * \ldots * Q^{I_s}\lambda_{I_r}$ where the λ_{I_i} are Browder operations on the variables $x_1, \ldots, x_k$ and $\sum_{j=1}^{r} p^{\ell(I_j)}w(\lambda_{I_j}) = k$ ($w(\lambda_{I_j})$ is the weight of λ_{I_j}). In particular, we may express $Q^{I_1}\lambda_{I_1} * \ldots * Q^{I_r}\lambda_{I_r}$ by $\mu_{2*}(\Gamma \otimes x_1 \otimes \ldots \otimes x_k)$ or $\mu_{p*}(\Gamma \otimes x_1 \otimes \ldots \otimes x_k)$, where $\Gamma \otimes x_1 \otimes \ldots \otimes x_r$ has been described in 4.3. By Lemma 4.2, $H_*(C_*\mathcal{C}_{n+1}(k) \otimes_{\Sigma_k} D) \to H_*E_k^0C_{n+1}X$ is a monomorphism. (D has a $\mathbb{Z}_p$-basis given by $x_1 \otimes \ldots \otimes x_k$, $x_i \neq [0]$. See proof of 4.2). By letting $X = S^{q_1} \vee \ldots \vee S^{q_k}$, $q_i > 0$, for appropriate q_i, it is clear that $\gamma_*(\Gamma \otimes x_1 \otimes \ldots \otimes x_k) = \Delta \otimes x_1 \otimes \ldots \otimes x_k$ by the obvious vector space considerations.

A direct geometrical proof of Lemma 4.3, without reference to the approximation theorem of [G], should be possible but would be formidably complicated.

5. The cohomology of braid spaces; the definitions of the operations

In the next 7 sections, we calculate

$$H^*(\mathrm{Hom}_G(C_*F(\mathbb{R}^{n+1},p);\ \mathbb{Z}_p(q)),\quad G = \pi_p \text{ or } \Sigma_p$$

where (1) $F(\mathbb{R}^{n+1},k)$ is the classical configuration space of k-tuples of distinct points in $\mathbb{R}^{n+1}$,

(2) Σ_k is the permutation group on k letters,

(3) π_k is the cyclic group of order k,

(4) $C_*F(\mathbb{R}^{n+1}, k)$ denotes the singular chains of $F(\mathbb{R}^{n+1},k)$,

(5) p is prime, and

(6) $\mathbb{Z}_p(q)$ is $\mathbb{Z}_p$ considered as a Σ_p-module with Σ_p-action defined by $\sigma \cdot x = (-1)^{qs(\sigma)}x$, where $(-1)^{s(\sigma)}$ is the sign of $\sigma \in \Sigma_p$. Since the Σ_p-action on $F(\mathbb{R}^{n+1},p)$ is proper, we may identify $H^*(\mathrm{Hom}_{\Sigma_p}(C_*F(R^{n+1},p);\ Z_p(2q)))$ with $H^*(B(R^{n+1},p);\ Z_p)$ where $B(R^{n+1},p)$ denotes $\dfrac{F(R^{n+1},p)}{\Sigma_p}$ [15; Chap IV]. By an abuse of notation we denote $H^*(\mathrm{Hom}_{\Sigma_p}(C_*F(R^{n+1},p);\ Z_p(q)))$ as $H^*(B(R^{n+1},p);\ Z_p(q))$.

Since $\mathscr{C}_{n+1}(j)$ has the equivariant homotopy type of $F(\mathbb{R}^{n+1},j)$ [G; §4], each class in $H_*(B(R^{n+1},p);\ Z_p(q))$ determines a homology operation on all classes of degree q in the homology of any (n+1)-fold loop space. We summarize the calculations and define the operations in this section.

The main tool used for calculating $H^*(B(R^{n+1},p);\ Z_p(q))$ is the map of fibrations

$$\begin{array}{ccc}
\Sigma_p & \xrightarrow{=} & \Sigma_p \\
\downarrow & & \downarrow \\
F(R^{n+1},p) & \xrightarrow{\hat{f}} & F(R^{\infty},p) \\
\downarrow & & \downarrow \\
B(R^{n+1},p) & \xrightarrow{f} & B(R^{\infty},p)
\end{array}$$

where $F(R^{\infty},p) = \underset{\vec{n}}{\mathrm{Lim}}\, F(R^{n+1},p)$, $B(R^{\infty},p) = \dfrac{F(R^{\infty},p)}{\Sigma_p}$ and $\hat{f}$ and f are the evident inclusions. Since $F(R^{\infty},p)$ is contractible with free Σ_p-action, $B(R^{\infty},p)$ is a $K(\Sigma_p,1)$. The spectral sequence for a covering allows calculation of the desired cohomology classes. Since $H^*(\Sigma_p;\ Z_p(q))$ plays an important role in our calculations, we now recall the following result.

Proposition 5.1 [A; p.158].

Let $j\colon \pi_p \to \Sigma_p$ be the inclusion given by a cyclic permutation and consider $j_*\colon\ H_*(\pi_p;\ Z_p(q)) \to H_*(\Sigma_p;\ Z_p(q))$, p odd. Then

(i) if q is even, $j_*(e_i) = 0$ unless $i = 2t(p-1)-\varepsilon$, $\varepsilon = 0,1$;

(ii) if q is odd, $j_*(e_i) = 0$ unless $i = (2t+1)(p-1)-\varepsilon$, $\varepsilon = 0,1$;

(iii) if q is even, $H^*(\Sigma_p;\ Z_p(q)) = E[v] \otimes P[\beta v]$ as an algebra, where v is a class of degree $2(p-1)-1$; and

(iv) if q is odd, $H^*(\Sigma_p;\ Z_p(q))$ has the additive basis $\{(\beta v)^s \beta^{\varepsilon} v'\}$ where v' is a class of degree $p-2$, $\varepsilon = 0,1$ and $s \geq 0$.

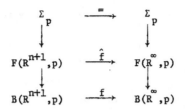

To facilitate the statement of our results, we recall the definition of "product" in the category of connected Z_p-algebras. If A and B are connected Z_p-algebras, their product, denoted $A \pi B$, is defined by $(A\pi B)_q = A_q \pi B_q$ if $q > 0$ and $(A \pi B)_0 = Z_p$, with multiplication specified by $A_q \cdot B_r = 0$ if q and $r > 0$, and by requiring the projections $A\pi B \to A$ and $A\pi B \to B$ to be morphisms of connected Z_p-algebras.

The following two theorems summarize our results:

Theorem 5.2. For p an odd prime and $n \geq 1$,

$$H^*(B(R^{n+1},p);\, Z_p) = A_{n+1}\pi \ \operatorname{Im} f^* \quad \text{as a connected } Z_p\text{-algebra,}$$

where $\operatorname{Ker} f^*$ is the ideal of $H^*(\Sigma_p;\, \mathbb{Z}_p)$ which consists of all elements of degree greater than $n(p-1)$ and, where

$$A_{n+1} = \begin{cases} E[\alpha] & \text{if } n+1 \text{ is even} \\ Z_p & \text{if } n+1 \text{ is odd} \end{cases}$$

for a certain element α of degree n. Further, the Steenrod operations are trivial on α and α restricts to an element $\bar{\alpha} \in H^nF(\mathbb{R}^{n+1},p)$ which is dual to a spherical element in the homology of $F(R^{n+1},p)$.

We remark that by proposition 5.1, $\operatorname{Im} f^*$ is completely known as an algebra over the Steenrod algebra.

Theorem 5.3. For p an odd prime and $n \geq 1$,

$$H^*(B(R^{n+1},p);\, Z_p(2q+1)) = M_n \oplus \operatorname{Im} f^* \quad \text{as a module over } H^*(\Sigma_p; Z_p)$$

where $\operatorname{Ker} f^*$ is the $H^*(\Sigma_p;Z_p)$-submodule of $H^*(\Sigma_p;\, \mathbb{Z}_p(q))$ generated by all elements of degree greater than $n(p-1)$ and where

$$M_{n+1} = \begin{cases} 0 & \text{if } n+1 \text{ is even} \\ Z_p \cdot \lambda & \text{if } n+1 \text{ is odd} \end{cases}$$

for a certain element λ of degree $(\frac{p-1}{2})$ which restricts to an element $\bar{\lambda} \in H^{n(\frac{p-1}{2})}F(\mathbb{R}^{n+1},p)$.

We remark that the statement implies that λ is annihilated by all elements of positive degree in $H^*(\Sigma_p; Z_p)$.

Theorem 5.4. For p an odd prime and $n \geq 1$,

$$H^*(\frac{F(\mathbb{R}^{n+1},p)}{\pi_p}; \mathbb{Z}_p) = \mathrm{Imf}^* \,\pi C \quad \text{additively}$$

where Imf^* is a subalgebra over the Steenrod algebra and is given by the image of the classifying map $f^* : H^*(B\pi_p; \mathbb{Z}_p) \to H^*(\frac{F(\mathbb{R}^{n+1},p)}{\pi_p}; \mathbb{Z}_p)$, Ker f^* is the ideal of $H^*(B\pi_p; \mathbb{Z}_p)$ which consists of all elements of degree greater than $n(p-1)$, and C is a subalgebra of classes in $H^*(F(\mathbb{R}^{n+1}); \mathbb{Z}_p)$ which are fixed under π_p. Furthermore $\alpha \cdot \mathrm{Imf}^* = \lambda \cdot \mathrm{Imf}^* = 0$ where α and λ are the images in $H^*(\frac{F(\mathbb{R}^{n+1},p)}{\pi_p}; \mathbb{Z}_p)$ of the classes specified in Theorems 5.2 and 5.3.

We are deliberately incomplete in our description of C because there are classes in $H^*(F(\mathbb{R}^{n+1},p); \mathbb{Z}_p)$ which are fixed by π_p but are not in C.

For the case $p = 2$, we shall prove the following result in The next section

Proposition 5.5. $F(R^{n+1},2)$ has the π_2-equivariant homotopy type of S^n. Consequently $B(R^{n+1},2)$ has the homotopy type of RP^n.

In passing, we note that $B(R^2,k)$ is a $K(B_k,1)$ where B_k is the braid group on k strings as defined by Artin [4] and considered by Fox and Neuwirth [10]. This fact led Fadell and Neuwirth to the name "braid space" for $B(M,k)$.

We dualize the cohomology of the braid space and let e_i be the homology basis element dual to the i-dimensional generator in the image of $H^*(B\Sigma_p; \mathbb{Z}_p(q))$; α_* and λ_* are basis elements dual to α and λ.

With this notation, we define the operations Q^s in the homology of $\mathcal{C}_{n+1}$-spaces precisely as in [A;2.2] (see [I; 1.1]). In addition, we have two more definitions.

Definition 5.6.

1. $\xi_n(x) = \theta_*(e_n \otimes x \otimes x)$ if $p = 2$, and

2) $\xi_n(x) = (-1)^{\frac{n+q}{2}} \gamma(q)\ \theta_*(e_{n(p-1)} \otimes x^p)$ and

$\zeta_n(x) = (-1)^{\frac{n+q}{2}} \gamma(q)\ \theta_*(e_{n(p-1)-1} \otimes x^p)$ if $n + q$ is even

$x \in H_qX$, and $p > 2$, where $\gamma(2j + \varepsilon) = (-1)^j (m!)^\varepsilon$ for j an integer and $\varepsilon = 0$ or 1, $m = \frac{1}{2}(p-1)$.

(The consistency of this definition of $\zeta_n(x)$ with that given in Theorem 1.3 will be proven in section 17).

We recall that $\mathcal{C}_{n+1}(2)$ has the homotopy type of S^n[Prop. 5.5].

Definition 5.7.

$$\lambda_n(x,y) = (-1)^{nq+1}\ \theta_*(\iota \otimes x \otimes y) \quad \text{for } x \in H_qX \text{ and } y \in H_rX,$$

where ι is the fundamental class of $\mathcal{C}_{n+1}(2)$.

Compare 5.7 with [A; 6.2].

The reader interested in the properties of the operations and willing to accept the results of this section on faith may skip directly to section 12.

We note that Arnold [2 and 3] has obtained information on $H^*F(R^2,j)$ and $H^*B(R^2,j)$.

6. The homology of $F(\mathbb{R}^{n+1},k)$

The definition and Theorem 6.1 of this section are due to Fadell and Neuwirth [9]. Let M be a manifold and define $F(M,k)$ to be the subspace $\{\langle x_1,\ldots,x_k\rangle \mid x_i \in M,\ x_i \neq x_j \text{ if } i \neq j\}$ of M^k. There is a proper left action of Σ_k on $F(M,k)$ given by $\rho \cdot \langle x_1,\ldots,x_k\rangle = \langle x_{\rho^{-1}(1)},\ldots,x_{\rho^{-1}(k)}\rangle$ for $\rho \in \Sigma_k$. Let $B(M,k)$ denote $\frac{F(M,k)}{\Sigma_k}$.

Theorem 6.1. [Fadell and Neuwirth]

Let M be an n-dimensional manifold, $n \geq 2$. Let $Q_0 = \phi$ and let $Q_r = \{q_1,\ldots,q_r\}$, $1 \leq r < j$, be a fixed set of distinct points in M. Define $\pi_k\colon F(M-Q_r,k) \to M-Q_r$ by $\pi_k \langle x_1,\ldots,x_k\rangle = x_1$. Then π_k is a fibration with fibre $F(M-Q_{r+1}, k-1)$ over the point q_{r+1}, and, if $k \geq 1$, π_k admits a cross-section σ_k.

We now specialize by letting $M = R^{n+1}$ and compute the integral cohomology of $F(\mathbb{R}^{n+1},k)$.

Lemma 6.2. Additively, $H^*F(\mathbb{R}^{n+1}-Q_r,k-r) = \bigotimes_{j=r}^{k-1} H^*({}^{j}S^n)$ where ${}^{j}S^n$ denotes the wedge of j copies of S^n.

Proof: For the moment, assume that π_i has trivial local coefficients for $i \geq 1$. Proceed by downward induction on r. If $r = k-1$, $F(R^{n+1}-Q_{k-1},1) = R^{n+1}-Q_{k-1}$; thus assume the result for r and consider the fibration

$$
\begin{array}{c}
F(R^{n+1}-Q_r,k-r) \\
\downarrow \\
F(R^{n+1}-Q_{r-1},k-r+1) \\
\pi_{r-1} \Big\downarrow \Big\uparrow \sigma_{r-1} \\
R^{n+1}-Q_{r-1}.
\end{array}
$$

In the Serre spectral sequence,

$E_2^{**} = H^*(R^{n+1}-Q_{r-1};\ H^*F(R^{n+1}-Q_r,k-r)) = H^*({}^{r-1}S^n) \otimes H^*F(R^{n+1}-Q_r,k-r)$.

By the induction hypothesis, $H^*F(R^{n+1}-Q_r,k-r) = \bigotimes_{j=r}^{n-1} H^*({}^jS^n)$, hence all differentials are zero and $E_2^{**} = E_\infty^{**}$. Since E_2 is free Abelian, the conclusion follows.

Next, we check the triviality of the local coefficients. For $n > 1$, the result is clear. For $n = 1$, we need the following lemma.

__Lemma 6.3__. The fibration $\pi_r\colon F(R^2-Q_r,k-r) \to R^2-Q_r$ has trivial local coefficients.

__Proof__: Again the proof is by downwards induction on r. The result is clear for $r = k - 1$ and, by Propositions 6.4 and 6.5, only the cases $r \geq 2$ require checking. Fix r, $2 \leq r \leq k - 2$, and assume the result for $r + 1$. Consider the fibration $\pi_r\colon F(R^2-Q_r,\ k-r) \to R^2-Q_r$ with fibre $F(R^2-Q_{r+1},\ k-r-1)$. Define a function $\rho_i\colon\ I \times R^2 - Q_r \to R^2-Q_r$

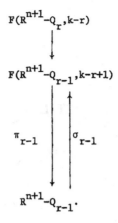

in terms of the following picture where $q_i = 4(i-1)e_1$ and e_1 is the canonical unit vector (1,0):

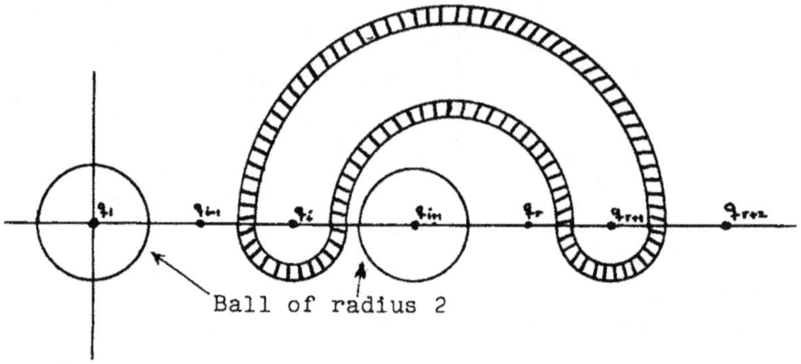

Figure 1. The function ρ_i

The function ρ_i rotates the 2-disc with center q_i contained within the shaded 2-annulus by an angle $2\pi t$ at time t, fixes the unbounded region outside of the shaded 2-annulus, and appropriately "twists" the shaded 2 annulus, at time t, to insure the continuity of ρ_i.

Define $h_i(t) = \rho_i(t, q_{r+1})$; then $h_i\colon I \to R^2 - Q_r$ is a typical generator of $\pi_1(R^2-Q_r, q_{r+1})$. Define a "lift"

$H_i\colon I \times F(R^2-Q_{r+1}, k-r-1) \to F(R^2-Q_r, k-r)$, of h_i by

$H_i(t, \langle z_1,\ldots,z_{k-r-1}\rangle) = \langle\rho_i(t,q_{r+1}), \rho_i(t,z_1),\ldots,\rho_i(t,z_{k-r-1}\rangle$.

Obviously the diagram

$$\begin{array}{ccc}
I \times F(R^2-Q_{r+1}, k-r-1) & \xrightarrow{H_i} & F(R^2-Q_r, k-r) \\
\downarrow \pi_I & & \downarrow \pi_{k-r} \\
I & \xrightarrow{h_i} & R^2-Q_r
\end{array}$$

commutes on the nose.

H_i induces a map of fibrations, for $j \geq 0$,

$$\begin{array}{ccc}
F(R^2-Q_{r+j}, k-r-j) & \xrightarrow{\hat{H}_{ij}} & F(R^2-Q_{r+j}, k-r-j) \\
\downarrow \pi_{k-r-j} & & \downarrow \pi_{k-r-j} \\
R^2-Q_{r+j} & \xrightarrow{\overline{H}_{ij}} & R^2-Q_{r+j}
\end{array}$$

where $\hat{H}_{ij} \langle z_1,\ldots,z_{k-r-j}\rangle = \langle \rho_i(1,z_1),\ldots,\rho_i(1,z_{k-r-j})\rangle$ and $\overline{H}_{ij} \langle z\rangle = \rho_i(1,z)$.

By definition, the local coefficient system for π_r is trivial if $(H_{i,1})_* = 1_*$ for all i or, equivalently, $(H_{i,1})^* = 1^*$. We verify that $(H_{i,1})^* = 1^*$ by an argument involving the maps $\overline{H}_{ij}$ and the existence of appropriate cross-sections. First, appealing to the appropriate picture, we see that $(\overline{H}_{ij})_* = 1_*$: Let α, β, γ, δ be the depicted generators of $\pi_1(R^2-Q_{r+1}, *)$ for $*$ outside of the shaded 2-annulus; then the following picture

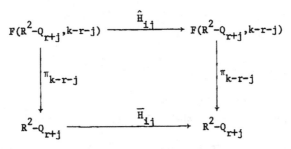

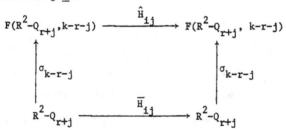

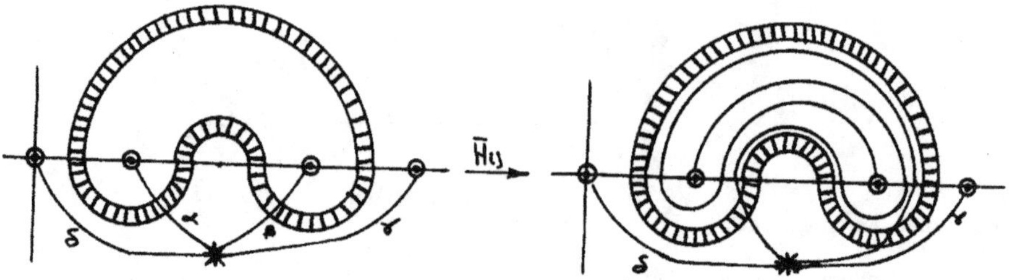

Figure 2. $\overline{H}_{ij}$

represents H_{ij} and shows that (1) $(\overline{H}_{ij})_{\#}(\alpha) = \beta\alpha\beta^{-1}$, (2) $(\overline{H}_{ij})_{\#}(\beta) = \beta^{-1}\alpha^{-1}\beta\alpha\beta$, and (3) $(\overline{H}_{ij})_{\#}(\gamma) = \gamma$. Hence $(\overline{H}_{ij})_{*} = 1_{*}$.

Next define cross-sections, σ_{k-r-j}, such that the following diagram commutes for $j \geq 1$:

$$\begin{array}{ccc}
F(R^2-Q_{r+j}, k-r-j) & \xrightarrow{\hat{H}_{ij}} & F(R^2-Q_{r+j}, k-r-j) \\
\uparrow \sigma_{k-r-j} & & \uparrow \sigma_{k-r-j} \\
R^2-Q_{r+j} & \xrightarrow{\overline{H}_{ij}} & R^2-Q_{r+j}
\end{array}$$

Commutativity of this diagram implies the validity of Lemma 6.3 via iterated applications of Lemma 6.2 on the cohomology algebra of $F(R^2-Q_{r+j}, k-r-j)$.

If $i > 1$, the cross-section defined by Fadell and Neuwirth suffices. Indeed, let $y_1,\ldots,y_{k-r-j-1}$ be $k-r-1-j$ distinct points on the boundary of a ball of radius $\frac{1}{2}$ with center at the origin; then Fadell and Neuwirth define

$$\sigma_{k-r-j}\langle z\rangle = \begin{cases} \langle z, \|z\|\ y_1,\ldots, \|z\|\ y_{k-r-j-1}\rangle & \text{if } \|z\| \geq 1 \\ \langle z, y_1,\ldots, y_{k-r-j-1}\rangle & \text{if } \|z\| \leq 1. \end{cases}$$

We check that $\hat{H}_{ij}\sigma_{k-r-j} = \sigma_{k-r-j}\overline{H}_{ij}$ as follows: if $z \leq 1$, then

$$\begin{aligned} \sigma_{k-r-j}\overline{H}_{ij}\langle z\rangle &= \sigma_{k-r-j}\langle\rho_i(1,z)\rangle = \sigma_{k-r-j}\langle z\rangle \\ &= \langle z, \|z\|\ y_1,\ldots, \|z\|\ y_{k-r-j-1}\rangle \\ &= \hat{H}_{ij}\sigma_{k-r-j}\langle z\rangle \end{aligned}$$

and if $\|z\| \geq 1$, then

$$\begin{aligned} \sigma_{k-r-j}\overline{H}_{ij}\langle z\rangle &= \sigma_{k-r-j}\langle\rho_i(1,z)\rangle \\ &= \langle\rho_i(1,z), y_1,\ldots,y_{k-r-j-1}\rangle \\ &= \hat{H}_{ij}\sigma_{k-r-j}\langle z\rangle. \end{aligned}$$

To check the case $i = 1$, let p_m denote some deleted point outside of the shaded 2-annulus (which exists since $r \geq 2$). Then define a cross-section, σ'_{k-r-j}, in a manner similar to the above. Let $y_1,\ldots,y_{k-r-j-1}$ be $k-r-j-1$ distinct points on the ball of radius $\frac{1}{2}$

center at p_m, and define

$$\sigma'_{k-r-j}\langle z\rangle = \begin{cases} \langle z,\ \|z-p_m\|\ (y_1-p_m)+p_m,\dots,\ \|z-p_m\|\ (y_{k-r-j-1}-p_m)+p_m\rangle & \\ & \text{if } \|z-p_m\| \le 1, \\ \langle z,\ y_1,\dots,\ y_{k-r-j-1}\rangle & \text{if } \|z-p_m\| \ge 1. \end{cases}$$

$\sigma'_{k-r-j}\bar{H}_{ij} = \hat{H}_{ij}\sigma'_{k-r-j}$ is checked as above and the lemma is proved.

Next we demonstrate some interesting geometric properties of $F(R^{n+1},p)$ and $B(R^{n+1},p)$.

Proposition 5.5. $F(\mathbb{R}^{n+1},2)$ has the π_2-equivariant homotopy type of S^n. Consequently $B(R^{n+1},2)$ has the homotopy type of RP^n.

Proof. Define

(1) ϕ: $S^n \to F(\mathbb{R}^{n+1},2)$ by $\phi\langle\xi\rangle = \langle\xi,-\xi\rangle$,

(2) $\hat{\phi}$: $R^{n+1}-\{0\} \to F(R^{n+1},2)$ by $\hat{\phi}\langle z\rangle = \langle z,-z\rangle$, and

(3) ψ: $F(R^{n+1},2) \to R^{n+1}-\{0\}$ by $\psi\langle x,y\rangle = x-y$.

Clearly $\psi\hat{\phi}\langle x\rangle = 2x$ so $\psi\hat{\phi} \simeq 1$.

Define a homotopy $G\colon I \times F(\mathbb{R}^{n+1},2) \to F(\mathbb{R}^{n+1},2)$ from $\phi\psi$ to 1 by the formula $G(t, \langle x,y\rangle) = \langle tx + (1-t)(x-y), ty + (1-t)(y-x)\rangle$.

ϕ induces a map of fibrations

$$\begin{array}{ccc} \Sigma_2 & \xrightarrow{=} & \Sigma_2 \\ \downarrow & & \downarrow \\ S^n & \xrightarrow{\phi} & F(R^{n+1},2) \\ \downarrow & & \downarrow \\ RP^n & \xrightarrow{\bar{\phi}} & B(R^{n+1},2) \end{array}$$

with ϕ given by $\overline{\phi}\{\xi\} = \{<\xi,-\xi>\}$. By the long exact homotopy sequence for a fibration, $(\overline{\phi})_{\#}$ is an isomorphism.

Henceforth, we assume that p is an odd prime in our homological calculations. The following two results are parenthetical.

Proposition 6.4. If M is a topological group, F(M,k) is homeomorphic to $M \times F(M-e, k-1)$, where e is the identity of M.

Proof. This situation is covered by Fadell and Neuwirth's notion of a "suitable" space, but the proof is amusing:

$$<z_1,\ldots,z_k> \longmapsto (z_1, <z_2z_1^{-1},\ldots,z_kz_1^{-1}>)$$

$$<z_1,y_2z_1,\ldots,y_kz_1> \longleftarrow (z_1, <y_2,\ldots, y_2,\ldots,y_k>)$$

Proposition 6.5. $F(R^2-Q_1,k)$ has the homotopy type of $S^1 \times F(R^2-Q_2,k-1)$. (This fact can be generalized to $\mathbb{R}^n$ for $n = 4,8$, but seems to be irrelevant.)

Proof. Define ρ: $I \times R^2-Q_1 \to R^2-Q_1$ by rotating R^2-Q_1 about q_1 through an angle $2\pi t$ at time t. ρ induces maps R: $S^1 \times F(R^2-Q_2,k-1) \to F(R^2-Q_1,k)$ and $\hat{R}$: $S^1 \to R^2-Q_2$ given by $R(\xi,<x_1,\ldots,x_{k-1}>) = <\rho(\xi,q_2), \rho(\xi,x_1),\ldots,\rho(\xi,x_{k-1})>$ and $\hat{R}(\xi) = \rho(\xi,q_2)$. Hence R defines a fibre-wise homotopy equivalence:

$$
\begin{array}{ccc}
F(R^2-Q_2,k-1) & \longrightarrow & F(R^2-Q_2', k-1) \\
\downarrow & & \downarrow \\
S^1 \times F(R^2-Q_2,k-1) & \xrightarrow{\;R\;} & F(R^2-Q_1, k) \\
\downarrow & & \downarrow \\
S^1 & \xrightarrow{\;\hat{R}\;} & R^2-Q_1
\end{array}
$$

7. Action of Σ_k on $H^*F(\mathbb{R}^{n+1},k)$

The geometric action $\Sigma_k \times F(R^{n+1},k) \to F(R^{n+1},k)$ induces a Σ_k-module structure on the (integral) cohomology algebra $H^*F(R^{n+1},k)$. Evaluation of this action allows explicit calculation of the spectral sequence mentioned in section 5.

To determine this action of Σ_k on $H^*(F(R^{n+1},k))$, we first calculate the action of a transposition $\tau_r = (r,r+1)$ on a basis, $\{\alpha_{ij}|k-1\geq i\geq j\geq 1\}$, for $H_nF(R^{n+1},k)$. Since $H^* = (H_*)^*$ here, one then passes to the induced action on the dual basis $\{\alpha^*_{ij}|k-1\geq i\geq j\geq 1\}$ for $H^nF(\mathbb{R}^{n+1},k)$. Since each permutation is induced by a map of spaces the action of a permutation induces an algebra morphism on cohomology and the action of a permutation, τ, on a product of indecomposables is given by the diagonal map, $\tau(\alpha^*_{ij}\ \alpha^*_{km}) = (\tau\alpha^*_{ij})(\tau\alpha^*_{km})$. This information, together with a few technical facts about the product structure in $F(R^{n+1},k)$, is enough to carry out the calculation of the cohomology of the "braid" space.

Fadell and Neuwirth's work allows us to define a Z-basis for $H_nF(R^{n+1},k)$, but a "geometric" change of basis is needed to arrive at the classes, α_{ij}, for which the action is easily computed.

We first give the basis arising from Fadell and Neuwirth's work via the cross-sections of Theorem 6.1. We have the map σ_r: $R^{n+1}-Q_r \to F(R^{n+1}-Q_r,k-r)\subseteq F(R^{n+1},k)$ which induces an isomorphism of $H_n(R^{n+1}-Q_r)$ onto a direct summand of $H_nF(R^{n+1},k)$. Let $S^n = \{\xi|\xi \in R^{n+1},\ \ \|\xi\| = 1\}$ and $Q_r = \{q_1,\ldots,q_r\}$ where $q_i = 4(i-1)e_1$, with e_1 the canonical unit vector, and define $\beta_{r,i}$: $S^n \to F(R^{n+1},k)$ by $\beta_{r,i}<\xi> = <q_1,\ldots,q_r,\ \sigma_r<\xi+q_i>>$. $\beta_{r,i}<\xi> = <q_1,\ldots,q_r,\ \ \xi+q_i,y_1,\ldots,y_{k-r-1}>$. By an abuse of notation, label $(\beta_{r,i})_*\ \ (\iota)$ as $\beta_{r,i}$ where ι is the fundamental class of S^n. Clearly $\{\beta_{r,i}|k-1\geq r\geq i\geq 1\}$ is a Z-basis for

$H_nF(R^{n+1},k)$.

We next define the promised "geometric" change of basis. Let $r \geq i \geq 1$ and let $\alpha_{r,i}\colon S^n \to F(R^{n+1},k)$ be given by $\alpha_{r,i}\langle\xi\rangle = \langle q_1,\ldots,q_r,$ $\xi+q_i, q_{r+1},\ldots,q_{k-1}\rangle$. By further abuse of notation, let $\alpha_{r,i}$ denote $(\alpha_{r,i})_*(\iota)$.

<u>Lemma 7.1</u>. For $F(R^{n+1},k)$,

(1) if $i > 1$, then $(\beta_{r,i})_* = (\alpha_{r,i})_*$,

(2) if $i = 1$, then $(\beta_{r,i})_* = (\alpha_{r,1})_* + \sum_{j=r+1}^{k-1} (-1)^{n+1}(\alpha_{j,r+1})_*$, and

(3) $\{\alpha_{r,i} \mid k-1 \geq r \geq i \geq 1\}$ is a Z-basis for $H_nF(R^{n+1},k)$.

<u>Proof</u>: (3) follows from (1) and (2).

(1): There are paths $\gamma_j\colon I \to R^{n+1}$ such that $\gamma_j(0) = y_j, \gamma_j(1) = q_{r+j}$, $\|\gamma_j(t)-q_\ell\| > 1$ if $1<\ell\leq r$ and image $\gamma_i \cap$ image $\gamma_j = \phi$ if $i \neq j$. Let $H\colon I\times S^n \to F(R^{n+1},k)$ be given by $H\langle t,\xi\rangle = \langle p_1,\ldots,p_r, \xi+p_i,$ $\gamma_1(t),\ldots,\gamma_{k-r-1}(t)\rangle$. H yields a homotopy between $\alpha_{r,i}$ and $\beta_{r,i}$ for $i>1$. (2): Define an embedding $I_r\colon S^n \to R^{n+1}-Q_{k-1}$ whose image is given by the following picture:

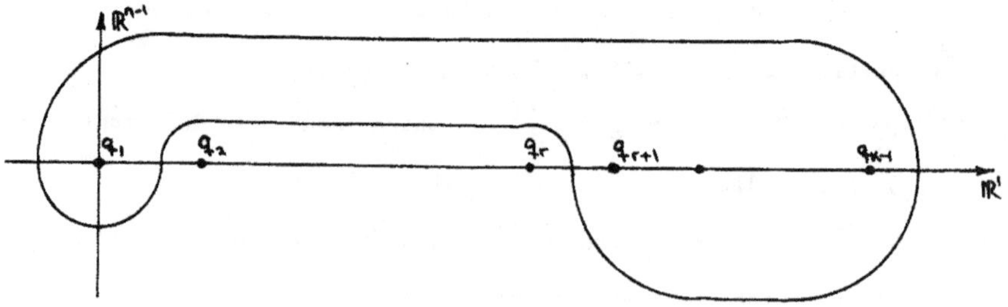

Figure 3. Embedding of S^n in $R^{n+1}-Q_{k-1}$

Let ν_i: $S^n \to R^{n+1}-Q_{k-1}$ be given by $\nu_i<\xi> = \xi+q_i$.

Subdividing the (n+1)-disc enclosed by the image of I_r, we have

$$(I_r)_* = (\nu_1)_* + \sum_{j=r+1}^{k-1} (\nu_j)_*.$$

Define G: $R^{n+1}-Q_{k-1} \to F(R^{n+1},k)$ by

$G<x> = <q_1,\ldots,q_r, x, q_{r+1},\ldots,q_{k-1}>$. Notice that

$$G\cdot I_r<\xi> = <q_1,\ldots,q_r, I_r\xi, q_{r+1},\ldots,q_{k-1}>$$

and $\beta_{r,1}<\xi>= <q_1,\ldots,q_r, \xi+q_1, y_1,\ldots,y_{k-r-1}>$. Using the paths of part (1) and "stretching" the unit sphere centered at the origin, we obtain a homotopy between $\beta_{r,1}$ and $G\cdot I_r$. But $G\cdot\nu_1 = \alpha_{k,1}$. If $j\geq r+1$, define a homotopy H: $S^n \times I \to F(R^{n+1},k)$ by

$H<\xi,t> = <q_1,\ldots,q_r, q_j+(1-t)\xi, q_{r+1},\ldots,q_j-t\xi,q_{j+1},\ldots,q_{k-1}>$. Then clearly $(G\cdot\nu_j)_* = (-1)^{n+1}(\alpha_{j,r+1})_*$. Hence

$$(\beta_{r,1})_* = (G\cdot I_r)_* = G_*[(\nu_1)_* + \sum_{j=r+1}^{k-1} (\nu_i)_*] = (\alpha_{r,1})_* + \sum_{j=r+1}^{k-1} (-1)^{n+1}(\alpha_{j,r+1})_*,$$

and the lemma is proved.

We can now easily determine the action of τ_r on $\{\alpha_{i,j} | k-1\geq i\geq j\geq 1\}$.

<u>Proposition 7.2.</u> For $F(R^{n+1},k)$,

(1) $\tau_r\alpha_{r-1,j} = \alpha_{r,j}$, hence $\tau_r\alpha_{r,j} = \alpha_{r-1,j}$ if $j < r$,

(2) $\tau_r\alpha_{i,r+1} = \alpha_{i,r}$, hence $\tau_r\alpha_{i,r} = \alpha_{i,r+1}$ if $i > r$,

(3) $\tau_r\alpha_{r,r} = (-1)^{n+1}\alpha_{r,r}$, and

(4) $\tau_r\alpha_{ij} = \alpha_{ij}$ otherwise.

<u>Proof:</u>

(1): $\tau_r\alpha_{r-1,j}<\xi>= \tau_r<q_1,\ldots,q_j,\ldots,q_{r-1}, \xi+q_j,q_r,\ldots>$

$= <q_1,\ldots,q_j,\ldots,q_{r-1},q_r, \xi+q_j,\ldots>$

$$= \alpha_{r,j}<\xi>.$$

(2): $\tau_r\alpha_{i,r+1}<\xi> = \tau_r<q_1,\ldots,q_r,\ q_{r+1},\ldots,q_i,\ \xi+q_{r+1},q_{i+1},\ldots,q_{k-1}>$

$$= <q_1,\ldots,q_{r+1},\ q_r,\ldots,q_i,\ \xi+q_{r+1},\ q_{i+1},\ldots,q_{k-1}>.$$

So $\tau_r\alpha_{i,r+1}$ is homotopic to $\alpha_{i,r}$.

(3): $\tau_r\alpha_{r,r}<\xi> = \tau_r<q_1,\ldots,q_r,\ \xi+q_r,\ q_{r+1},\ldots,q_{k-1}>$

$$= <q_1,\ldots,\xi + q_r,\ q_r,\ q_{r+1},\ldots> .$$

Define a homotopy $H: I \times S^n \to F(R^{n+1},k)$ by

$H<t,\xi> = <q_1,\ldots,\ q_{r-1},\ q_r + (1-t)\xi,\ \ q_r-t\xi,\ <q_{r+1},\ldots>$. Hence $\tau_r(\alpha_{r,r})_*$ equals $(-1)^{n+1}(\alpha_{r,r})_*$.

(4): $\tau_r\alpha_{ij}<\xi> = \tau_r<q_1,\ldots,q_j,\ldots,q_i,\ \xi+q_j,\ q_{i+1},\ldots>$. So $\tau_r\alpha_{ij}$ is homotopic to α_{ij} if $j \neq r,\ r+1$ and $i \neq r,\ r-1$.

A technical result useful for later calculations is

<u>Lemma 7.3.</u> Let $\rho \in \Sigma_k$ be such that $\rho(2i+1) = m$, $\rho(2i+2) = \ell$, where $m \neq 2i+1,\ 2i+2$. Then, for $F(\mathbb{R}^{n+1},k)$,

$$\rho\alpha_{2i+1,\ 2i+1} = \begin{cases} (-1)^{n+1}\alpha_{m-1,\ell} & \text{if } m > \ell \\ \alpha_{\ell-1,m} & \text{if } \ell > m \end{cases}$$

<u>Proof:</u> Immediate from the proofs in proposition 7.2 and the definition of α_{ij}.

Let α^*_{ij} be the dual of α_{ij} with induced left Σ_k-action given by $(\tau\alpha^*)(\beta) = \alpha^*(\tau^{-1}\beta)$ for $\tau \in \Sigma_k$. By inspection, we have

<u>Corollary 7.4.</u> For $F(R^{n+1},k)$,

(1) $\tau_r\alpha^*_{r-1,j} = \alpha^*_{r,j}$, hence $\tau_r\alpha^*_{r,j} = \alpha^*_{r-1,j}$ if $j < r$,

(2) $\tau_r\alpha^*_{i,r+1} = \alpha^*_{i,r}$, hence $\tau_r\alpha^*_{i,r} = \alpha^*_{i,r+1}$ if $i > r$,

(3) $\tau_r\alpha^*_{rr} = (-1)^{n+1}\alpha^*_{rr}$ and

(4) $\tau_r\alpha^*_{ij} = \alpha^*_{ij}$ otherwise.

An extremely useful result for later calculations is the action of $\sigma = \tau_1 \circ \ldots \circ \tau_{k-1}$, a permutation of order k, on the indecomposable elements α^*_{ij}.

<u>Corollary 7.5</u>. For $F(R^{n+1},k)$,

(1) $\sigma\alpha^*_{ij} = \alpha^*_{i+1,j+1}$ if $i<k-1$, and

(2) $\sigma\alpha^*_{k-1,j} = (-1)^{n+1}\alpha^*_{j,1}$.

<u>Proof</u>:

(1): let $k-1>i\geq j\geq 1$; then

$$\begin{aligned}\tau_1 \circ \ldots \circ \tau_{k-1}\alpha^*_{ij} &= \tau_1 \circ \ldots \circ \tau_{i+1}\alpha^*_{ij}\\ &= \tau_1 \circ \ldots \circ \tau_i\, \alpha^*_{i+1,j}\\ &= \tau_1 \circ \ldots \circ \tau_j\, \alpha^*_{i+1,j}\\ &= \tau_1 \circ \ldots \circ \tau_{j-1}\alpha^*_{i+1,j+1}\\ &= \alpha^*_{i+1,j+1}\end{aligned}$$

(2): Let $k-1 = i\geq j\geq 1$, then

$$\begin{aligned}\tau_1 \circ \ldots \circ \tau_{k-1}\alpha^*_{k-1,j} &= \tau_1 \circ \ldots \circ \tau_{k-2}\alpha^*_{k-2,j}\\ &= \tau_1 \circ \ldots \circ \tau_{j+1}\alpha^*_{j+1,j}\\ &= \tau_1 \circ \ldots \circ \tau_j\alpha^*_{jj}\\ &= (-1)^{n+1}\tau_1 \circ \ldots \circ \tau_{j-1}\alpha^*_{j,j}\end{aligned}$$

$$= (-1)^{n+1} \tau_1 \circ \cdots \circ \tau_{j-1} \alpha^*_{j,j-1}$$

$$= (-1)^{n+1} \alpha^*_{j,1}.$$

Finally, we obtain the information about products required to carry out the details of computation in the next section.

Lemma 7.6. The following set is a Z-basis for $H^{\ell n}F(R^{n+1},k)$, where $1 \leq \ell \leq k-1$:

$$\{\alpha^*_{i_1,j_1} \cdots \alpha^*_{i_\ell,j_\ell} \mid 1 \leq i_1 < \dots < i_\ell \leq k-1\}.$$

Proof: By a slight modification of the proof of Lemma 6.2, a basis for $E^2_{r,s}$, $r+s = \ell n$, is given by

$$\{\beta_{i_1,j_1} \otimes \dots \otimes \beta_{i_\ell,j_\ell} \mid 1 \leq i_1 < \dots < i_\ell \leq k-1\}.$$

By Lemma 7.1, $\beta_{i,j} = \alpha_{i,j}$ for $j > 1$ and

$$\beta_{i,1} = \alpha_{i,1} + \sum_{j=i+1}^{k-1} (-1)^{n+1} \alpha_{j,i+1}.$$

Consequently $\{\alpha^*_{i_1,j_1} \otimes \dots \otimes \alpha^*_{i_\ell,j_\ell} \mid 1 \leq i_1 < \dots < i_\ell \leq k-1\}$ is a basis for $E_2^{r,s}$, $r + s = \ell n$. Since the Serre spectral sequence is a spectral sequence of algebras, the result follows.

Lemma 7.7. For $F(R^{n+1},k)$

(1) $(\alpha^*_{ij})^2 = 0$, and

(2) $\alpha^*_{ij}\alpha^*_{ik} = -\alpha^*_{k-1,j}(\alpha^*_{ij} - \alpha^*_{ik})$ for $j < k$.

Proof: By equivariance, it suffices to check that (1) $(\alpha^*_{11})^2 = 0$, and (2) $\alpha^*_{21}\alpha^*_{22} = -\alpha^*_{11}(\alpha^*_{21} - \alpha^*_{22})$.

For (1), note that the following diagram is commutative

(1):

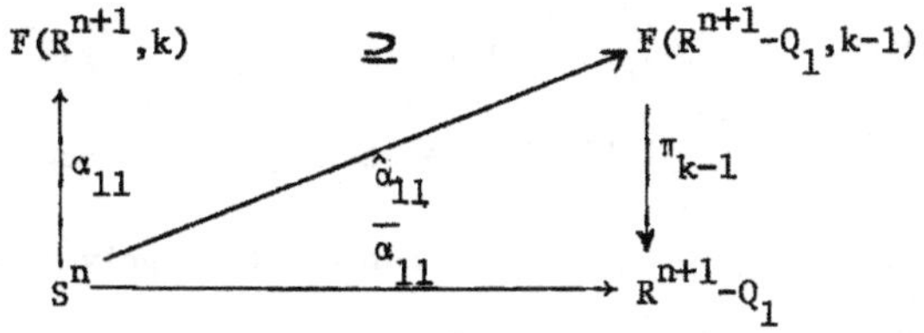

where $\hat{\alpha}_{11}<\xi> = <\xi+q_1,q_2,\ldots,q_{k-1}>$ and $\bar{\alpha}_{11}<\xi> = \xi+q_1$. Let ι^* denote the fundamental class of $\mathbb{R}^{n+1}-Q_1$ in cohomology. Further, let $\tilde{\alpha}_{ij}\colon S^n \to F(\mathbb{R}^{n+1}-Q_1,k_1)$ be given by the formula $\tilde{\alpha}_{ij}(\xi) = <q_2,\ldots,q_j,\ldots,q_i,\xi+q_j,q_{i+1},\ldots,q_{k-1}>$. Evidently,

$$\pi^*_{k-1}(\iota^*)(\tilde{\alpha}_{ij}(\iota)) = \iota^*(\pi_{k-1*}\tilde{\alpha}_{ij}(\iota)) = \begin{cases} 1 & \text{if } i = 1 \\ 0 & \text{if } i \text{ or } j \neq 1. \end{cases}$$

Hence $\pi^*_{k-1}(\iota^*) = \alpha^*_{11}$; consequently, $(\alpha^*_{11})^2$ is zero.

For (2), consider the map $\pi_{k-2}\colon F(R^{n+1}-Q_2,k-2) \to R^{n+1}-Q_2$. Let $\gamma^*_{2i} = \pi^*_{k-2}(\hat{\gamma}^*_{2i})$ where $\hat{\gamma}^*_{2i} \in H^*(\mathbb{R}^{n+1}-Q_2)$ is such that $(\hat{\gamma}^*_{2i})^2 = 0$ and $\hat{\gamma}^*_{21}\hat{\gamma}^*_{22} = 0$. Under the inclusion $F(R^{n+1}-Q_2,k-2) \subseteq F(R^{n+1},k)$, α^*_{2i} restricts to γ^*_{2i} in $H^*F(R^{n+1}-Q_2,k-2)$ in view of the commutative diagram

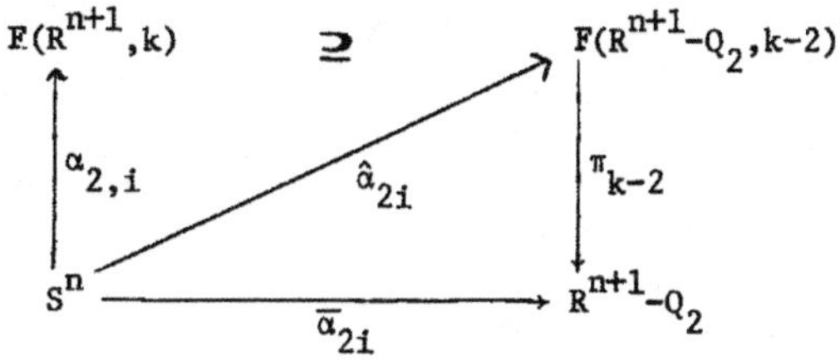

where $\hat{\alpha}_{2,i}<\xi> = <\xi+q_i,q_3,\ldots,q_{k-1}>$ and

$$\bar{\alpha}_{2,i}<\xi> \; = \; \xi+q_i.$$

Consequently, $\alpha_{21}^*\alpha_{22}^*$ restricts to zero in $H^*F(R^{n+1}-Q_2,k-2)$ and must lie in the principal ideal generated by α_{11}^*. So $\alpha_{21}^*\,\alpha_{22}^* = \sum_{i>1} X_{ij}\alpha_{11}^*\alpha_{ij}^*$ for some constants X_{ij} by Lemma 7.6. Applying τ_1 to both sides and quoting Lemma 7.5, we find

$$\alpha_{21}^*\alpha_{22}^* = \sum_{i>1} X_{i1}\,[\alpha_{11}^*\alpha_{i,1}^* - \alpha_{11}^*\alpha_{i,2}^*]$$

because $X_{i1} = -X_{i2}$ and $X_{ij} = 0$ if $j > 2$.

Applying τ_2 to both sides and again quoting Lemma 7.5 and 7.6, we find $X_{ij} = 0$ if $i > 2$ and $X_{21} = -1$. So $\alpha_{21}^*\,\alpha_{22}^* = -\alpha_{11}^*[\alpha_{21}^* - \alpha_{22}^*]$ and the result follows.

Proposition 7.8. The Steenrod operations are trivial on $H^*(F(\mathbb{R}^{n+1},j);\mathbb{Z}_p)$.

Proof: By the proof of Lemma 7.7(1), $\pi_{k-1}^*(\iota) = \alpha_{11}^*$. Hence the Steenrod operations are trivial on α_{11}^*. By equivariance and the internal Cartan formula, the Steenrod operations are trivial on any monomial in the α_{ij}^*.

8. The spectral sequence

To calculate $H^*(\mathrm{Hom}_{\Sigma_p}(C_*F(R^{n+1},p);\ Z_p(q)))$, we use the spectral sequence for a covering [7; p.335]: let G be a group, M a G-module, and X a space on which G acts properly. Then there is a spectral sequence such that $E_2^{**} = H^*(G;H^*(X;M))$ and $\{E_r\}$ converges to $H^*(\mathrm{Hom}_G(C_*X;M))$. Furthermore, if $M \otimes M' \to M''$ is a pairing of G-modules, with $M \otimes M'$ given diagonal G-operators, there is a cup product pairing $E_r \otimes E_r' \to E''$ of the associated spectral sequences. In our calculation, $G = \Sigma_p$, $X = F(R^{n+1},p)$, and $M = Z_p(q)$ as defined in section 5. The Σ_p-module structure of $H^*F(R^{n+1},p)$ has been identified in section 7.

Instead of attempting to evaluate E_2^{**} directly, where $\{E_r\}$ is the spectral sequence such that $E_2^{**} = H^*(\Sigma_p;\ H^*(F(R^{n+1},p);\ Z_p(q)))$ and E_r converges to $H(\mathrm{Hom}_{\Sigma_p}(C_*F(R^{n+1},p);\ Z_p(q)))$, we study $E_2'^{**}$, where E_2' is the spectral sequence obtained by replacing Σ_p by π_p, the cyclic group of order p. Then the restriction $i(\pi_p:\Sigma_p)$: $\pi_p \to \Sigma_p$ induces a morphism of spectral sequences, which, by the following theorem, is a monomorphism on the E_2-level.

<u>Theorem 8.1</u> [7; p.259] Let A be a G-module and p a prime. Let $\hat{H}(F,A,p)$ denote the p-primary component of $\hat{H}(G,A)$ and let π_p be a p-Sylow subgroup of G; then

$$i^*(\pi_p:\Sigma_p)\colon\ \hat{H}(G;A,p) \to \hat{H}(\pi_p,A) \text{ is a monomorphism.}$$

Since Tate cohomology agrees with ordinary cohomology in positive degrees, Theorem 8.1 obviously applies to the map $E_2^{r,*} \longrightarrow E_2^{\prime r,*}$ for $r > 0$. The case $r = 0$ is obvious.

We can now immediately identify almost all of E_2 and E_2' by the Vanishing Theorem:

Theorem 8.2. [Vanishing Theorem] In the spectral sequences $\{E_r\}$ and $\{E_r'\}$ for $F(R^{n+1},p)$, $E_2^{\prime s,t} = E_2^{s,t} = 0$ for $s > 0$ and $0 < t < n(p-1)$.

The proof of this theorem is held in abeyance until section 10.

Since $H^qF(R^{n+1},p) = 0$ for $q > n(p-1)$, we obviously have $E_2^{\prime s,t} = E_2^{s,t} = 0$ if $s > 0$ and $t > n(p-1)$. Combining this fact with the Vanishing Theorem, we see that the only possible non-trivial differentials in $\{E_r\}$ are

(1) $d_r\colon\ E_r^{0,r-1} \to E_r^{r,0}$ for $r \leq n(p-1) + 1$, and

(2) $d_{n(p-1)+1}\colon\ E_{n(p-1)+1}^{s,n(p-1)} \longrightarrow E_{n(p-1)+1}^{s+n(p-1)+1,0}$.

Comparing (1) and (2), we see that

(3) $E_2^{s,t} = E_{n(p-1)+1}^{s,t}$ unless $s = 0$, $t = 0$, and $s < n(p-1)+1$;

(4) $E_{n(p-1)+2}^{s,t} = E_\infty^{s,t}$ for all s and t.

But since $F(R^{n+1},p)$ and $\dfrac{F(R^{n+1},p)}{\Sigma p}$ are $p(n+1)$-dimensional manifolds, no classes of total degree greater than $p(n+1)$ can survive to E_∞. Hence for $s + n(p-1) > p(n+1)$ the differentials, (2), must be vector space isomorphisms. Of course, these formulas and remarks are also valid with $\{E_r\}$ replaced by $\{E_r'\}$.

Recalling that $H^*(\pi_p;\, Z_p) = E[u] \otimes P[\beta u]$, where u is a class of degree 1, and recalling Proposition 5.1, we see from (4) that

(5) $E_2^{\prime s,n(p-1)} = Z_p$ for $s + n(p-1) > p(n+1)$

(6) $E_2^{s,n(p-1)} = Z_p$ for $s + n(p-1) > p(n+1)$, and

$$s + n(p-1) + 1 = \begin{cases} 2j(p-1)-\varepsilon & q \text{ even}, \ \varepsilon = 0,1 \\ (2j+1)(p-1)-\varepsilon & q \text{ odd}, \ \varepsilon = 0,1. \end{cases}$$

To determine the classes for $s + (p-1) \leq p(n+1)$, define the p-period of a group G to be q if $\hat{H}^i(G,A)$ and $\hat{H}^{i+q}(G,A)$ have isomorphic p-primary components for all i and for all A. It is well known that π_p has period 2 and that the periodicity isomorphism is given by cup product with βu; Σ_p has p-period $2(p-1)$ by Swan's theorem:

<u>Theorem 8.3.</u> [23] Suppose p is odd and the p-Sylow subgroup of π is cyclic. Let π_p be a p-Sylow subgroup and let Φ_p be the group of automorphisms of π_p induced by inner automorphisms of π. Then the p-period of π is twice the order of Φ_p.

When specialized to our cyclic group π_p and to Σ_p, Theorem 8.3 can be expressed in the following explicit form:

<u>Theorem 8.4.</u> For any π_p-module M, let βu: $H^s(\pi_p; M) \to H^{s+2}(\pi_p;M)$ be given by cup product with $\beta u \in H^2(\pi_p; Z_p)$. Then βu is an isomorphism for all $s > 0$. For any Σ_p-module N, let βv: $H^s(\Sigma_p; N) \to H^{s+2(p-1)}(\Sigma_p;N)$ be given by the cup product with $\beta v \in H^{2(p-1)}(\Sigma_p;Z_p)$. Then βv is an isomorphism for all $s > 0$.

In short, formulas (5) and (6) remain valid for all $s > 0$.

Let $\alpha_I^* = \alpha_{11}^* \alpha_{33}^* \cdots \alpha_{2i+1,2i+1}^* \cdots \alpha_{p-2,p-2}^*$ and H denote the subgroup of elements of Σ_p which fix, up to sign, the class α_I^*. Let $\mu_1 H, \ldots, \mu_r H$, $r = [\Sigma_p : H]$, be the left cosets of H in Σ_p. With this notation, we state the following theorem which will be proven in section 9 and will complete the additive determination of E_2^{**}:

Theorem 8.5. In the spectral sequence $\{E_r\}$ for $F(R^{n+1}, p)$, $E_2^{0,*}$ is given as follows, as an algebra if q is even:

$$\text{if } p > 3,\ E_2^{0,*} = \begin{cases} E[\bar{\alpha}] & n+1 \text{ is even, } q \text{ is even} \\ Z_p & n+1 \text{ is odd, } q \text{ is even} \\ 0 & n+1 \text{ is even, } q \text{ is odd} \\ Z_p \cdot \bar{\lambda} & n+1 \text{ is odd, } q \text{ is odd,} \end{cases}$$

$$\text{if } p = 3,\ E_2^{0,*} = \begin{cases} \dfrac{E[\bar{\alpha}] \otimes \dfrac{P[\delta]}{(\delta^2)}}{(\bar{\alpha} \cdot \delta)} & n+1 \text{ is even, } q \text{ is even} \\ Z_3 & n+1 \text{ is odd, } q \text{ is even} \\ 0 & n+1 \text{ is even, } q \text{ is odd} \\ Z_3 \cdot \bar{\lambda} \oplus \mathbb{Z}_3 \cdot \delta & n+1 \text{ is odd, } q \text{ is odd,} \end{cases}$$

where $\bar{\alpha} = \sum_{p-1 \geq i > j \geq 1} \alpha_{ij}^* \in H^n F(\mathbb{R}^{n+1}, p)$

$$\bar{\lambda} = \sum_{i=1}^{r} (\text{sign } \mu_i) \mu_i(\alpha_I^*) \in H^{n(\frac{p-1}{2})} F(\mathbb{R}^{n+1}, p),$$

and if $p = 3$, $\delta = \alpha_{11}^* \alpha_{21}^* + \alpha_{11}^* \alpha_{22}^*$. Furthermore, if δ is considered as a fixed point of the π_3-module $H^{2n} F(\mathbb{R}^{n+1}, 3)$, then $(\beta u)^j \cdot \delta$ represents the non-zero class in $H^{2j}(\pi_3; H^{2n}(F(\mathbb{R}^{n+1}, 3); \mathbb{Z}_3))$ for $j > 0$.

Having determined E_2^{**} additively, we proceed to exhibit the differentials. We have $E_2^{*,0} = H^*(\Sigma_p; Z_p(q))$ and $E_2'^{*,0} = H^*(\pi_p; Z_p)$. If q is even, the classes $(\beta v)^s\beta^\varepsilon v$ form a Z_p-basis for $E_2^{*,0}$ in positive degrees; if q is odd, the classes $(\beta v)^s\beta^\varepsilon v'$ form such a basis $(\deg v' = p-2)$. The classes $(\beta u)^s\beta^\varepsilon u$ form a basis for $E_2'^{*,0}$. We first note that the only possible non-trivial differential on $\bar{\alpha}$ is d_{n+1}. But $d_{n+1}\alpha$ must be zero since $\beta u\cdot\bar{\alpha} = \beta v\cdot\bar{\alpha} = 0$ by the vanishing theorem. A similar argument shows that $d_i\bar{\lambda} = 0$ for all i. To consider the other possible non-trivial differentials, let G ambiguously denote π_p or Σ_p and let x_s denote the basis element for $E_2'^{s,n(p-1)}$ or $E_2^{s,n(p-1)}$ determined by (2) and periodicity. Explicitly if m is the p-period of G, then, in (2), set $s = im+j$ for $i \leq 0$, $1\leq j\leq m$. The differentials, (2), are isomorphisms for s sufficiently large. We read off the answer for $i = 0$ from the answer for i large:

For $G = \Sigma_p$,

$$d_{n(p-1)+1}x_s = \begin{cases} (\beta v)^i\beta^\varepsilon v & \text{for } q \text{ even and } s = (2i-n+2)(p-1)+\varepsilon-2 \\ (\beta v)^i\beta^\varepsilon v' & \text{for } q \text{ odd and } s = (2i-n+1)(p-1)+\varepsilon-2. \end{cases}$$

The case $s = 0$ requires some comment. By Theorem 8.5, when $p = 3$ and $n + q + 1$ is even, there is a possible non-trivial differential

$$d_{2n+1}\colon E_{2n+1}^{0,2n} \longrightarrow E_{2n+1}^{2n-1,0} .$$

The periodicity theorem as stated cannot be directly applied to $E_2^{0,2n}$. So we consider $E_2^{'0,2n}$. Here $d_{2n+1}((\beta u)^j \cdot \delta) = (\beta u)^{j+n} \cdot u$ by (2), hence $d_{2n+1}\delta = (\beta u)^n \cdot u$. By naturality, we see that, in this exceptional case, we have

$$d_{2n+1}\delta = \begin{cases} (\beta v)^i \cdot v & \text{if } n+1 \text{ even, } q \text{ even, } i = \frac{n-1}{2}, \\ (\beta v)^i \cdot v' & \text{is } n+1 \text{ odd, } q \text{ odd, } i = \frac{n}{2}. \end{cases}$$

Recall that according to our notation, x_i denotes the class in bidegree $(i, n(p-1))$. For $p = 3$, δ is x_0, and for primes larger than 3, we have that x_0 is zero.

Clearly this information determines E_∞, but for the sake of completeness, we determine E_2 as a $H^*(\Sigma_p;Z_p)$-module. For q even, $d_{n(p-1)+1}v \cdot x_s = -v \cdot d_{n(p-1)+1}X_s$. Clearly we have

$$(7) \qquad v \cdot x_s = \begin{cases} -x_{s+2(p-1)-1} & \text{if } s = (2j-n)(p-1)-1, \\ 0 & \text{if } s = (2j-n)(p-1)-2. \end{cases}$$

For q odd, the obvious modification is

$$(8) \qquad v \cdot x_s = \begin{cases} -x_{s+2(p-1)-1} & \text{if } s = (2j-n+1)(p-1)-1 \\ 0 & \text{if } s = (2j-n+1)(p-1)-2. \end{cases}$$

These results are summarized in the following theorems:

<u>Theorem 8.7</u>. Consider $E_2^{**} = H^*(\Sigma_p; H^*(F(\mathbb{R}^{n+1},p);\mathbb{Z}_p(q)))$. If q is even,

$$E_2^{**} = A_{n+1} \pi B_{n+1} \quad \text{as a connected algebra where}$$

$$A_{n+1} = \begin{cases} E[\alpha] & \text{if } n+1 \text{ is even} \\ Z_p & \text{if } n+1 \text{ is odd,} \end{cases}$$

and

$$B_{n+1} = \frac{E[v, x_{s-1}, x_s] \otimes P[\beta v]}{I};$$

here I is the ideal generated by the set

$$\{x_{s-1} \cdot x_s, \ v \cdot x_{s-1}, \ v \cdot x_s + (\beta v) \cdot x_{s-1}\}$$

$$\text{where } s = \begin{cases} (p-1)-1 & \text{if } n+1 \text{ is even} \\ 2(p-1)-1 & \text{if } n+1 \text{ is odd.} \end{cases}$$

If q is odd,

$$E_2^{**} = M_{n+1} \oplus B_{n+1} \quad \text{as an } H^*(\Sigma_p;Z_p)\text{-module, where}$$

$$M_n = \begin{cases} 0 & \text{if } n+1 \text{ is even} \\ Z_p \cdot \bar{\lambda} & \text{if } n+1 \text{ is odd,} \end{cases}$$

and B_{n+1} is generated by v', $\beta v'$, x_{s-1} and x_s with relations $v \cdot v' = 0$, $v \cdot x_{s-1} = 0$, and $(\beta v) \cdot x_{s-1} + v \cdot x_s = 0$.

$$\text{where } s = \begin{cases} 2(p-1)-1 & \text{if } n+1 \text{ is even} \\ (p-1)-1 & \text{if } n+1 \text{ is odd.} \end{cases}$$

Let x_{s-1} and x_s denote the classes in $'E_2^{**}$ of bidegree $(s,n(q-1))$ analogous to the class $\overline{x}_{s-1}$ and $\overline{x}_s$ in E_2^{**} which are required by periodicity. By the previous methods, we have

Theorem 8.8. Consider $'E_2^{**} = H^*(\pi_p; H^*(F(\mathbb{R}^{n+1},p); \mathbb{Z}_p))$.

$'E_2^{**} = 'A_{n+1}\pi\ 'B_{n+1}$ as a connected algebra, where A_{n+1} is a subalgebra of classes which restrict to fixed points in $H^*(F(\mathbb{R}^{n+1},p); \mathbb{Z}_p)$ under the action of π_p and where

$$'B_{n+1} = \frac{E[u,\overline{x}_0,\overline{x}_1] \otimes P[\beta u]}{I} \quad ;$$

here I is the two sided ideal generated by the set $\{\overline{x}_0 \cdot \overline{x}_1, u \cdot \overline{x}_0, u \cdot \overline{x}_1 + (\beta u) \cdot \overline{x}_0\}$.

Remark: We are deliberately incomplete in our description of $'A_{n+1}$ because $\overline{x}_0$ certainly restricts to a fixed point in $H^*(F(\mathbb{R}^{n+1},p); \mathbb{Z}_p)$, and $\overline{x}_0$ is a fixed point which does not persist to E_∞. (See the previous calculation of $d_{2n+1}: E_{2n+1}^{0,2n} \longrightarrow E_{2n+1}^{2n-1,0}$ for example.)

Theorem 8.9. The differentials in the spectral sequence $\{E_r\}$ are given by

(1) $d_j\overline{\alpha} = 0$ for all j,

(2) $d_j\overline{\lambda} = 0$ for all j,

(3) $d_j x_s = 0$ for $j \leq n(p-1)$, and

(4) $$d_{n(p-1)+1} x_s = \begin{cases} (\beta v)^k \beta^\varepsilon v & \text{if } q \text{ is even and} \\ & s = (2k-n+2)(p-1)+\varepsilon-2 \\ (\beta v)^k \beta^\varepsilon v' & \text{if } q \text{ is odd and} \\ & s = (2k-n+1)(p-1)+\varepsilon-2 \end{cases}$$

Theorem 8.10. The differentials in the spectral sequence $\{'E_r\}$ are given by

(1) $d_j \gamma = 0$ for all j and $\gamma \in {'A}_{n+1}$ provided γ has no summands of $\bar{x}_0$,

(2) $d_j \bar{x}_s = 0$ for $j \leq n(p-1)$, and

(3) $d_{n(p-1)+1} \bar{x}_s = (\beta u)^k \beta^\varepsilon u$ for $s = 2k + \varepsilon - n(p-1) + 1$.

Remark: The additive results stated in Theorems 5.2, 5.3 and 5.4 are immediate from the form of E_∞ implied by Theorems 8.7 through 8.10.

9. $E_2^{0,*}$

We identify $E_2^{0,*} = H^0(\Sigma_p; H^*(F(R^{n+1},p); Z_p(q)))$ as those classes in the Σ_p-module $H^*(F(R^{n+1},p); Z_p(q))$ which are fixed under the action of Σ_p. We must prove Theorem 8.5.

To study $H^0(\Sigma_p; H^*(F(R^{n+1},p); Z_p(q))$, we decompose $H^*F(R^{n+1},p)$ into a direct sum of $\mathbb{Z}_p$-modules and consider the fixed points which have summands in each submodule. This method is carred out for $p > 3$, but requires modification for the case $p = 3$.

Let I be a sequence of integers, $I = (i_1,j_1,\ldots,i_k,j_k)$. I is allowable if $1 \leq i_1 < i_2 < \ldots < i_k \leq p-1$ and $1 \leq j_r \leq i_r$, $1 \leq r \leq k$. α_I^* denotes the class $\alpha^*_{i_1,j_1} \ldots \alpha^*_{i_k,j_k}$. We define the length of I by $\ell(I) = k$ and, by convention, $\alpha_I^* = 1$ if $\ell(I) = 0$.

Define F to be the graded Z_p-module whose generators are α_I^*, I allowable, where

(1) $I = (i_1,j_1,\ldots,i_k,j_k)$

and for each $m \leq k$,

(2) $j_m \neq i_{m-x} + 1$ for all x, $1 \leq x \leq m-1$ and

(3) $j_m \neq j_{m-x}$ for all x, $1 \leq x \leq m-1$.

Let T be the graded Z_p-module whose generators are α_I^*, I allowable, where

(1) $I = (i_1,j_1,\ldots,i_k,j_k)$ and

(2) $j_m = i_{m-x} + 1$ or $j_m = j_{m-x}$ for some m and x.

<u>Proof of Theorem 8.5</u>: Clearly $H^*F(R^{n+1},p) = F \oplus T$ as a Z_p-module. Since a check of the four obvious cases reveals that a transposition, up to sign,

permutes the monomials of F, F is a Σ_p-submodule of $H^*F(\mathbb{R}^{n+1},p)$.

The calculation is now divided into two sections:

(1) we show that no monomial in T can be summand of a fixed point under the Σ_p-action (with either twisted or untwisted action on Z_p).

(2) We calculate the fixed points contained in the submodule F.

(1): We show first that no allowable monomial of the form $\alpha_I^*\alpha_{j,j}^* \alpha_{j+1,j}^*$ or $\alpha_I^*\alpha_{jj}^* \alpha_{j+1,j+1}$ can be a non-zero summand of a fixed point for $\ell(I)\geq 0$. Using this information, we show that no monomial satisfying the axioms for T can be non-trivial summand of a fixed point.

Let $V_{j,I}$ denote the Z_p-vector space spanned by $\alpha_I^*\alpha_{jj}^*\alpha_{j+1,j}^*$ and $\alpha_I^*\alpha_{jj}^*\alpha_{j+1,j+1}$ for fixed I and j. (These monomials are assumed allowable.) Then $\tau_{j+1}(V_{j,I}) \subseteq V_{j,I}$. Let $\overline{V}_{j,I'}$ be the Z_p-space spanned by

$$\alpha_{I'}^*(\alpha_{j-1,r}^*)^\delta\alpha_{j,r}^*\alpha_{j+1,j+\varepsilon}^*, \quad \varepsilon = 0,\ 1 \text{ and } \delta = 0,1;$$

$$\alpha_{I'}^*(\alpha_{j-1,r}^*)^\delta\alpha_{jj}^*\alpha_{j+1,j+\varepsilon}^*, \quad \varepsilon = 0,\ 1 \text{ and } \delta = 0,1$$

for fixed I' and r. Again, $\tau_j(\overline{V}_{j,I'})\subseteq \overline{V}_{j,I'}$.

Now suppose that $\alpha_I^*\alpha_{jj}\alpha_{j+1,j}$ is a summand of a fixed point. Then by application of τ_{j+1} to $V_{j,I}$, we must have that

$$C\alpha_I^*\alpha_{jj}^*\alpha_{j+1,j}^* + D\alpha_I^*\alpha_{jj}^*\alpha_{j+1,j+1}^*$$

is also a summand of the same fixed point where $C = (-1)^n\,[C + D]$. Application of τ_j to $\overline{V}_{j,I}$ forces $(-1)^{n+1}C = D$. So $C = (-1)^n\,[C + (-1)^{n+1}C]$; thus if $p \neq 3$, $C = 0$.

Let $\alpha_I^* = \alpha_{I'}^* \alpha_{i_r,j_r}^* \ldots \alpha_{i_s,j_s}^* \ldots \alpha_{i_k,j_k}^*$ be a summand of a fixed point, where $j_s = i_r + 1$ or $j_s = j_r$ (that is, an instance of axiom 2 for T.) We may ssume that $j_{r+\ell} \neq i_{r+\ell}$ for $\ell > 0$. Let τ denote τ_{i_s+1} and let $V_{\alpha_I^*}$ and $V_{\tau\alpha_I^*}$ be the one-dimensional subspaces spanned by α_I^* and $\tau\alpha_I^*$ respectively. Since $\tau^2 = 1$, we may decompose $H^*F(\mathbb{R}^{n+1},p)$ as a direct sum in two ways:

$$H^*F(\mathbb{R}^{n+1},p) = V_{\alpha_I^*} \oplus W_{\alpha_I^*} = V_{\tau\alpha_I^*} \oplus W_{\tau\alpha_I^*}$$

where

$$\tau(V_{\alpha_I^*}) \subseteq V_{\tau\alpha_I^*}$$

and

$$\tau(V_{\tau\alpha_I^*}) \subseteq V_{\alpha_I^*} .$$

We suppose that $j_r = j_s$ and remark that the case $j_s = i_r+1$ is checked in essentially the same manner. Iterating the above procedure for the permutation $(\tau_{p-2} \circ \ldots \circ \tau_{i_r+1}) \circ (\tau_{p-1} \circ \ldots \circ \tau_{i_s+1})$ applied to α_I^*, we see that a monomial of the form $\alpha_J^* \alpha_{p-2,j_r}^* \alpha_{p-1,j_s}^*$ is a non-zero summand of a fixed point. Applying the permutation $\tau_{p-3} \circ \ldots \circ \tau_{i_s}$ to $\alpha_J^* \alpha_{p-2,j_r}^* \alpha_{p-1,j_s}^*$ and quoting the argument for τ_{i_s} above, we see that a monomial of the form $\alpha_{J'}^* \alpha_{p-2,p-2}^* \alpha_{p-1,p-2}^*$ must be a non-zero summand of a fixed point. Therefore, by the previous remarks, an element fixed by Σ_p can have no non-zero monomial summands satisfying the axioms for T. The modifications necessary for the case of twisted Z_p-coefficients are obvious.

(2): Application of appropriate permutations indicates that any allowable monomial, α_I^*, of F, $\ell(I) = k+1$, can be permuted, up to sign, to

$$\alpha_{11}^* \alpha_{33}^* \cdots \alpha_{2k+1,2k+1}^*, \quad 2k+1 \leq p-2.$$

Hence the subspace of elements of degree $n(k+1)$ in F is generated as a Σ_p-module by $\alpha_{11}^*\alpha_{33}^* \cdots \alpha_{2k+1,\ 2k+1}^*$.

We first calculate the fixed point set in F for $n+1$ even. Suppose that q is even. Here, Σ_p just permutes the indecomposables α_{ij}^* and clearly

$$\sum_{p-1 \geq i \geq j \geq 1} \alpha_{ij}^*$$

is a fixed point. We claim that there are no non-zero fixed points in F concentrated in degree $n(k+1)$ if $k>0$. Suppose there is such a fixed point, γ. Then by the above paragraph, γ is in the Σ_p-module generated by

$$\alpha_I^* = \alpha_{11}^*\alpha_{33}^* \cdots \alpha_{2i+1,\ 2i+1}^* \cdots \alpha_{2k+1,\ 2k+1}^*.$$

By applying τ_2 to γ, we see that $\alpha_{21}^*\alpha_{32}^*\alpha_{55}^* \cdots \alpha_{2k+1,\ 2k+1}^*$ must be a summand of γ. We apply $\tau_1 o \tau_3$ to see that γ must have

$$\alpha_{32}^*\alpha_{21}^*\alpha_{55}^* \cdots \alpha_{2k+1,2k+1}^* = -\alpha_{21}^*\alpha_{32}^*\alpha_{55}^* \cdots \alpha_{2k+1,2k+1}^*$$

as a summand. There are clearly no non-zero fixed points which have this property. We now suppose that q is odd. Here, we see that $\tau_1(\alpha_{11}^*\alpha_{33}^* \cdots \alpha_{2k+1,2k+1}^*) = -\alpha_{11}^*\alpha_{33}^* \cdots \alpha_{2k+1,2k+1}^*$ for $k \geq 0$.

Consequently

$$H^{o}(\Sigma_p; H^*(F(R^{n+1},p); Z_p(2q+1))) = 0$$

if $n + 1$ is even.

We next study the case $n + 1$ is odd. For the moment we assume that q is even. Since $\tau_1\alpha^*_{11} = (-1)^{n+1}\alpha^*_{11}$, we have

$$\tau_1\alpha^*_I = -\alpha^*_I \quad \text{for} \quad \alpha^*_I = \alpha^*_{11}\alpha^*_{33}\cdots\alpha^*_{2k+1,2k+1}.$$

Consequently there are no fixed points in F for $n + 1$ and q even. We now consider the case when q is odd. Again, any fixed point must be in the Σ_p-module generated by α^*_I. If $2k + 1 < p - 2$, then $\tau_{p-1}\cdot\alpha^*_I = -\alpha^*_I$. Hence there are no non-zero fixed points in F concentrated in degree jn, $1 \leq j < \frac{p-1}{2}$. However, we claim that

$$\alpha^*_I = \alpha^*_{11}\cdots\alpha^*_{2i+1,2i+1}\cdots\alpha^*_{p-2,p-2}$$

does in fact generate a Σ_p-fixed point for $n + 1$ and q both odd. By definition of the α^*_{ij} we see that if $\rho \in \Sigma_p$ and ρ fixes, up to sign, each element of the set

$$\{\alpha^*_{11}, \alpha^*_{33},\ldots,\alpha^*_{2i+1,2i+1},\ldots, \alpha^*_{p-2,p-2}\},$$

then ρ is either the identity of Σ_p or a product of the transpositions τ_{2i+1}, $0 \leq i \leq \frac{p-3}{2}$.

Now, let H denote the subgroup of Σ_p which is generated by the elements that fix α^*_I up to sign. We claim that H is generated by the set

$$\{\eta \in \Sigma_p | \eta = \tau_{2j+1} \quad \text{or} \quad \eta = \tau_{2j+2}\,\tau_{2j+3}\,\tau_{2j+1}\,\tau_{2j+2} \quad 0 \leq j < \tfrac{p-3}{2}\}.$$

This fact is essentially immediate from the previous observation. The peculiar permutation $\tau_{2j+2}\tau_{2j+3}\tau_{2j+1}\tau_{2j+2}$ just interchanges

$$\alpha_{2j+1,2j+1} \text{ and } \alpha_{2j+3,2j+3}.$$

If $g \in H$, we can clearly choose a product of the permutations, η, which interchanges precisely the same classes which g interchanges. Denote this product by h. Then gh^{-1} must fix, up to sign, each indecomposable $\alpha^*_{2i+1,2i+1}$. Hence, by the previous observation, gh^{-1} must be the identity or a multiple of τ_{2j+1}, $0 \leq j \leq \frac{p-3}{2}$. We use these generators for H to finish calculating the fixed points for $n+1$ and q odd. Here

$$\tau_i o \alpha^*_I = (-1)\tau_i(\alpha^*_I)$$

where

$$\tau_i(\alpha^*_I)$$

is determined by the Σ_p-action defined on $H^*F(R^{n+1},p)$. Since

$$\tau_i o \alpha^*_{ii} = (-1)(-1)^{n+1}\alpha^*_{ii} = \alpha^*_{ii}$$

for $n+1$ and q odd, it follows that $g \cdot \alpha^*_I = \alpha^*_I$ for all $g \in H$. Let

$$\mu_1 H, \ldots, \mu_r H, \; r = [\Sigma_p : H],$$

be the left cosets of H in Σ_p and let

$$\overline{\lambda} = \sum_{i=1}^{r} (\text{sign } \mu_i)\mu_i(\alpha^*_I) \in H^{n(\frac{p-1}{2})}F(\mathbb{R}^{n+1},p).$$

If we show $\overline{\lambda}$ is independent of the choice of coset representatives $\{\mu_i | i = 1,\ldots,r\}$, then it will be obvious that $\tau \cdot \overline{\lambda} = \overline{\lambda}$ for all $\tau \in \Sigma_p$. So we suppose that $\mu_j \mu_j^{-1} = g \in H$. Since $g \cdot \alpha_I^* = \alpha_I^*$ by the above remarks, it follows immediately that

$$(\text{sign } \mu_j)\mu_j(\alpha_I^*) = (\text{sign } \overline{\mu}_j)\overline{\mu}_j(\alpha_I^*).$$

For the case $p = 3$, let $\delta = x\alpha_{11}^*\alpha_{21}^* + y\alpha_{11}^*\alpha_{22}^*$ be a Σ_3-fixed point. Then

$$\delta = \tau_1\delta = (-1)^q[(-1)^{n+1} x\, \alpha_{11}^*\alpha_{22}^* + (-1)^{n+1} y\alpha_{11}^*\alpha_{21}^*]$$

forces $y = (-1)^{q+n+1}x$ and

$$\delta = \tau_2\delta = x\tau_2[\alpha_{11}^*\alpha_{21}^* + (-1)^{q+n+1}\alpha_{11}^*\alpha_{22}^*] = (-1)^q x[\alpha_{21}^*\alpha_{11}^* + (-1)^{q+2n+2}\alpha_{21}^*\alpha_{22}^*].$$

Since

$$\alpha_{21}^*\alpha_{22}^* = -\alpha_{11}^*[\alpha_{21}^* - \alpha_{22}^*], \text{ by Lemma 7.7,}$$

we have that δ is fixed if and only if $n + q + 1$ is even. To check the fixed points concentrated in degree n, we notice that $\overline{\alpha} = \alpha_{11}^* + \alpha_{21}^* + \alpha_{22}^*$ is fixed if and only if q and $n + 1$ are both even, and $\overline{\lambda} = \alpha_{11} - \alpha_{21} + \alpha_{22}$ is fixed if and only if q and $n + 1$ are both odd.

To check that $(\beta u)^j \cdot \delta$ represents the non-zero class in

$$H^{2j}(\pi_3; H^{2n}(F(R^{n+1},3); Z_3(q))$$

we note that

$$(1+\sigma+\sigma^2)\alpha_{11}^*\alpha_{21}^* = (1+\sigma+\sigma^2)\alpha_{11}^*\alpha_{22}^* = 0.$$

Since $(\beta u)^j \cdot \delta$ must be a non-zero, non-cobounded cocyle, we have the conclusion.

10. Vanishing of E_2^{**}

We must prove Theorem 8.1, which states that $E_2^{\prime s,t} = 0$ if $s > 0$ and $t \neq 0$, $n(p-1)$.

Proof: All monomials, α_I^*, are assumed to be "allowable" in the sense of section 9. In addition, we define an admissible sequence, I, to have the properties

(1) $I = (i_1, j_1, \ldots, i_k, j_k)$, I allowable and

(2) for every m, $1 \leq m \leq k$,

$j_m = i_{m-x} + 1$ for some x, $1 \leq x \leq m - 1$.

or

$j_m = j_1$.

Define the height of I, h(I), to be j_1. By convention, let

$$I + y = (i_1+y, j_1+y, \ldots, i_k+y, j_k+y)$$

if $i_k + y \leq p - 1$.

Lemma 10.1. Every allowable monomial in H^*F is, up to sign, a product of admissible monomials.

Clearly the "factorization" of Lemma 10.1 is not unique; we can certainly write the monomial $\alpha_{11}^*\alpha_{21}^*$, for instance, in two ways as products of admissibles. The following definition strengthens the notion of an admissible monomial to the point where we can compute. Let $\alpha_{I_1}^*, \ldots, \alpha_{I_r}^*$ be admissible monomials such that

(1) $\alpha^*_{I_1} \dots \alpha^*_{I_r} = \pm\alpha^*_I$ for I allowable,

(2) $1 \leq h(I_1) < h(I_2) < \dots < h(I_r) \leq p-1$, and

(3) α^*_I cannot be written as a product of fewer admissible monomials.

Then we say that $\alpha^*_{I_1} \dots \alpha^*_{I_r}$ is a maximal admissible decomposition for α^*_I. The following lemma is obvious:

Lemma 10.2. Every allowable monomial has, up to sign, a unique maximal admissible decomposition.

The key to our proof of the vanishing theorem is that the notion of a maximal admissible decomposition enables us to partition the set of allowable monomials into very nice equivalence classes, which, when "enlarged" to Z_p-modules are stable under σ of Corollary 7.4. This stability condition gives a simple method of calculating the kernel of $(\sigma-1)^*$ where $(\sigma-1)^*$ is the "even dimensional" differential for the minimal resolution of π_p. So the next stage is to define the appropriate equivalence relation:

Let J^1 and J^2 be allowable. We say $\alpha^*_{J^1} \sim \alpha^*_{J^2}$ if

(1) $\alpha^*_{J^r}$ has a maximal admissible decomposition

$$\alpha^*_{I^r_1} \dots \alpha^*_{I^r_k}$$

(2) $I^r_t = (i^r_1, j^r_1, \dots, i^r_{n_\ell}, j^r_{n_\ell})$, $1 \leq t \leq k$.

(3) $i^1_\times = i^2_\times$ for all ℓ and $1 \leq \times \leq n_\ell$.

(4) $h(I^1_\ell) = h(I^2_\ell)$ for all ℓ.

The following lemma is obvious.

Lemma 10.3. "∿" is an equivalence relation.

Let $S_{\sigma^m I}$, $0 \le m < p$, be the Z_p-module spanned by those α_J^* such that α_J^* is equivalent (in the sense of ∿) to some non-trivial summand of $\sigma^m(\alpha_I^*)$.

Proposition 10.4. Write $S_I = S_{\sigma^0(I)}$. Then for $F(R^{n+1},p)$

(1) $\sigma^m(S_I) \subseteq S_{\sigma^m I}$, and

(2) $S_I \cap S_{\sigma^m I} = (0)$ if $\ell(I) < p-1$ and $1 \le m \le p-1$.

Proof.

(1): We may write α_I^* uniquely, up to sign, as a maximal decomposition $\alpha_{I_1}^* \cdots \alpha_{I_k}^*$. By Corollary 7.4,

$$\sigma\alpha_{ij}^* = \alpha_{i+1,j+1}^* \quad \text{if} \quad i < p-1$$

and

$$\sigma\alpha_{p-1,j}^* = (-1)^{n+1}\alpha_{j,1}^*$$ and so it is enough to check

that $\sigma(S_I) \subseteq S_{\sigma I}$ where $\alpha_{p-1,x}^*$ occurs in some admissible $\alpha_{I_\ell}^*$. But then

$$\alpha_I^* = \pm\alpha_{I_1}^* \cdots \alpha_{I_\ell}^* \cdots \alpha_{I_k}^* \quad \text{and}$$

$$\sigma\alpha_I^* = \pm\alpha_{I_1+1}^* \cdots \alpha_{I_{\ell-1}+1}^*\alpha_{I_{\ell+1}+1}^* \cdots \alpha_{I_k+1}^*\sigma\alpha_{I_\ell}^* .$$

Furthermore,

$$\alpha_{I_\ell}^* = \alpha_{i_1,j_1}^* \cdots \alpha_{i_q,j_q}^* \alpha_{p-1,x}^*$$

and so

$$\sigma\alpha^*_{I_\ell} = (-1)^{n+1}\alpha_{i_1+1,j_1+1}\cdots\alpha_{i_q+1,j_q+1}\,\alpha_{x,1}$$

where $x = j_1$ or i_t+1, by the definition of an admissible monomial. Applying the formula

$$\alpha^*_{ij}\alpha^*_{ik} = -\alpha^*_{k-1,j}\,[\alpha^*_{ij}-\alpha^*_{ik}],\ j < k,$$

(see Lemma 7.7), we have

$$\sigma\alpha^*_I = \Sigma \pm \alpha^*_{j_1,1}\alpha^*_{i_1+1,x_1}\,\alpha^*_{i_2+1,\ x_2}\cdots\alpha^*_{i_q+1,\ x_q}$$

for $x_i = 1,\ j_1+1$ or i_r+2 for $r < q$. Fix some monomial appearing as a summand of $\sigma\alpha^*_{I_\ell}$, say $\alpha^*_{J_\ell}$, $J_\ell = (j_1,1,\ i_1+1,\ x_1,\ldots,\ i_q+1,x_q)$. Note that all such summands of $\sigma\alpha^*_{I_\ell}$ are equivalent, under $\sim$, so it does not matter which we choose. Then by the definition of a maximal admissible decomposition,

$$\alpha^*_{J'_\ell}\ \alpha^*_{I_1+1}\cdots\alpha^*_{I_k+1}$$

is a maximal admissible decomposition for any allowable monomial summand of $\sigma\alpha^*_I$ for $J'_\ell \sim J_\ell$. Hence $\sigma\ S_I \subseteq S_{\sigma I}$.

(2): We check the case when α^*_I is a product of one admissible monomial. The case that α^*_I has a maximal admissible decomposition with more than one admissible monomial occurring is essentially the same.

It is enough to check that α^*_I is not a summand of $S_{\sigma^m I}$ for $1\leq m\leq p-1$ and $I = (i_1,1,i_2,j_2,\ldots,i_k,j_k)$, $\ell(I)<p-1$, since, for some t, we have

$$\sigma^t \alpha_I^* = \Sigma \pm \alpha_{I'}^*$$

where

$$I' = (i_1', 1, i_2', j_2', \ldots, i_k', j_k').$$

The idea of the following proof is that if $\ell(I)<p-1$, then the string of indecomposables, α_I^*, has too many "gaps" to appear as a summand of $S_\sigma m_I$ for $1\leq m\leq p-1$.

Let α_I^* be admissible, $\ell(I)<p-1$, $I = (i_1,1,i_2,j_2,\ldots,i_k,j_k)$ and suppose that for some m, $1\leq m<p-1$, we have that α_I^* is, in fact, a summand of $S_{\sigma m_I}$. By induction, the only time a "1" can occur in the j_1-coordinate of a summand for $S_{\sigma m_I}$ is when $m = p-i_{k-r}$ for $0\leq r\leq k-1$; also, if $m = p-i_{k-r}$, then I must have the form

$$I = (i_1',1,i_2',j_2',\ldots,i_r',j_r',i_{r+1}',j_{r+1}',\ldots,i_k',j_k')$$

where

$$i_\lambda' = i_{k-r+\lambda}-i_{k-r} \quad \text{for} \quad 1\leq\lambda\leq r,$$

$$i_{r+1}' = p-i_{k-r}, \quad \text{and}$$

$$i_{r+1+\lambda}' = i_\gamma+p-i_{k-r} \quad \text{for} \quad 1\leq\gamma\leq k-r-1.$$

Since

$$I = (i_1,1,i_2,j_2,\ldots,i_k,j_k),$$

we have

$$i_\lambda = i_{k-r+\lambda}-i_{k-r} \quad \text{for} \quad 1\leq\lambda\leq r,$$

$$i_{r+1} = p-i_{k-r}, \quad \text{and}$$

$$i_{r+1+\gamma} = i_\gamma+p-i_{k-r} \quad \text{for} \quad 1\leq\gamma\leq k-r-1.$$

Rearranging, we find

$$i_{k-r} = i_{k-r+\lambda} - i_\lambda \quad \text{for} \quad 1 \le \lambda \le r,$$

$$i_{k-r} = p - i_{r+1}, \text{ and}$$

$$i_{k-r} = i_\gamma + p - i_{r+1+\gamma} \quad \text{for } 1 \le \gamma \le k-r-1.$$

Summing, we have

$$k(i_{k-r}) = \left(\sum_{s=1}^{k} i_s\right) - i_{k-r} - \left(\sum_{s=1}^{k} i_s\right) + (k-r)p$$

Hence $(k+1)(i_{k-r}) = (k-r)p$. Since $k+1, i_{k-r} \neq 0$, and $k \le p-1, i_{k-r} \le p-1$, we must have $k+1=p$. But then $\ell(I) = k = p-1$. This is a contradiction because we assumed that $\ell(I) < p-1$. Hence α_I^* cannot lie in $S_\sigma m_I$ for $1 \le m \le p-1$ and the proposition is proved.

Remark: By Proposition 10.4,

$$S_I + S_{\sigma I} + \ldots + S_{\sigma^{p-1} I} = S_I \oplus S_{\sigma I} \oplus \ldots \oplus S_{\sigma^{p-1} I} \quad \text{as a}$$

Z_p-module. Since "$\sim$" is an equivalence relation on maximal admissible decompositions, we may decompose $H^{kn}(F(\mathbb{R}^{n+1},p);\mathbb{Z}_p)$ into $V_1 \oplus \cdots \oplus V_p$ as Z_p-module for $1 \le k \le p-2$ such that $\sigma(V_i) \subseteq V_{i+1}$ if $i<p$ and $\sigma(V_p) \subseteq V_1$.

Recall that the minimal resolution of Z_p considered as a trivial π_p-module, where π_p is generated by σ, has the form

$$\cdots \to Z_p\pi \to Z_p\pi \xrightarrow{\sigma-1} Z_p\pi \xrightarrow{N} Z_p\pi \xrightarrow{\sigma-1} Z_p\pi \xrightarrow{\varepsilon} \mathbb{Z}_p \to 0, \; N = 1+\sigma+\ldots+\sigma^{p-1}.$$

Let M denote $H^{\ell n}(F(\mathbb{R}^{n+1},p);\mathbb{Z}_p)$, $1 \le \ell \le p-2$. Then for the cochain complex

$$\cdots \leftarrow \mathrm{Hom}_{\pi_p}(Z_p\pi_p,M) \xleftarrow{(\sigma-1)^*} \mathrm{Hom}_{\pi_p}(Z_p\pi_p,M) \xleftarrow{N^*} \mathrm{Hom}_{\pi_p}(Z_p\pi_p,M) \xleftarrow{(\sigma-1)^*} \cdots$$

we use the decomposition of the previous paragraph to show that $\mathrm{Ker}(\sigma-1)^* = \mathrm{Im}N^*$. Since

$$\mathrm{Hom}_{\pi_p}(Z_p\pi_p, M) = \mathrm{Ker}(\sigma-1)^* \oplus \mathrm{Im}(\sigma-1)^* = \mathrm{Ker}\ N^* \oplus \mathrm{Im}N^*$$

as Z_p-modules, we have that $\mathrm{Ker}(\sigma-1)^* = \mathrm{Im}N^*$ implies $\mathrm{Ker}\ N^* = \mathrm{Im}(\sigma-1)^*$ by the obvious vector space dimension considerations. This information clearly implies Theorem 8.2.

To show $\mathrm{Ker}(\sigma-1)^* = \mathrm{Im}N^*$, we appeal to the following lemma:

Lemma 10.5. Let V be a finite dimensional vector space over a field. Suppose $\sigma\colon V \to V$ is a linear transformation such that

(1) $\sigma^p = 1$,

(2) $V = V_1 \oplus \cdots \oplus V_p$, and

(3) $\sigma(V_i) \subsetneq V_{i+1}$ if $i<p$, and $\sigma(V_p) \subseteq V_1$.

Then if $\sigma v = v$ for $v \in V$, we have $v = (1+\sigma+ \dots +\sigma^{p-1})v'$ for some $v' \in V$.

Proof: Let $v = \sum_1^p x_i v_i$ for $v_i \in V_i$. If $\sigma v = v$, then

$$v = (1+\sigma+ \dots +\sigma^{p-1})(x_1 v_1).$$

11. Steenrod operations and product structure

We show that there exists an element $\alpha \in H^n(B(R^{n+1},p);\mathbb{Z}_p)$ for $n+1$ even such that α restricts to $\bar{\alpha} \in H^n(F(\mathbb{R}^{n+1},p);\mathbb{Z}_p)$ specified in Theorem 8.5, and such that

$$H^*(B(R^{n+1},p);Z_p) = A_{n+1}\pi \mathrm{Im} f^*$$

as a connected Z_p-algebra, where

$$A_{n+1} = \begin{cases} E[\alpha] & \text{if } n+1 \text{ is even} \\ Z_p & \text{if } n+1 \text{ is odd;} \end{cases}$$

Moreover, the Steenrod operations are trivial on α.

Secondly, we show that there exists an element

$$\lambda \in H^{n(\frac{p-1}{2})}(B(\mathbb{R}^{n+1},p);\ \mathbb{Z}_p(2q+1) \qquad \text{for } n+1 \text{ odd}$$

such that λ restricts to the element $\bar{\lambda} \in H^{n(\frac{p-1}{2})}(F(\mathbb{R}^{n+1},p)\mathbb{Z}_p)$ specified in Theorem 8.5 and such that

$$H^*(F(R^{n+1},p);Z_p) = M_{n+1} \oplus \mathrm{Im} f^*$$

as a $H^*(\Sigma_p;\ Z_p)$-module, where

$$M_{n+1} = \begin{cases} 0 & \text{if } n+1 \text{ is even} \\ Z_p\cdot\lambda & \text{if } n+1 \text{ is odd.} \end{cases}$$

Define a map $S\colon\ R^{n+1} \to R^{n+1}$ by

$S\ (x_1,\ldots,x_{n+1}) = (x_1,\ -x_2,\ x_3,\ldots,x_{n+1})$. Since $B(-,p)$ is a functor defined on the category of toplogical spaces with morphisms 1-1 continuous maps, we have the obvious induced morphisms $B(S,p)\colon B(R^{n+1},p) \to B(R^{n+1},p)$. For convenience, we denote $B(S,p)$ by S. Let π_S denote the group of order 2 generated by S. There is an evident π_S-action induced on $B(R^\infty,p)$. Note that this action is certainly neither free nor trivial. However, we can calculate the action of π_S on $H^*(B(R^\infty,p);Z_p)$.

Proposition 11.1. For p odd, π_S acts trivially on $H^*(B(R^\infty,p);Z_p)$.

Proof: Let π_p act on $F(R^{n+2},p)$ by

$$\sigma\cdot\langle x_1,\dots,x_p\rangle = \langle x_p,x_1,\dots,x_{p-1}\rangle.$$

Obviously S commutes with the π_p-action. Now we consider the inclusion of R^n in R^{n+2} given by $x \to (0,0,x)$. Give $F(R^n,p)$ a trivial π_S-action. Then the induced inclusion $\frac{F(R^n,p)}{\pi_p} \subset \frac{F(R^{n+2},p)}{\pi_p}$ is π_S-equivariant. By our previous calculations, π_S must act trivially on $H^*(F\frac{(R^\infty,p)}{\pi_p};Z_p)$. Since the evident map $\frac{F(R^\infty,p)}{\pi_p} \to \frac{F(R^\infty,p)}{\Sigma_p}$ is π_S-equivariant, the result follows.

We use the above information to calculate the product structure and Steenrod operations in $H^*(B(R^{n+1},p);Z_p)$. To do these calculations efficiently, we need one fact about $F(R^{n+1},p)$.

Define $\gamma: S^n \to F(R^{n+1},p)$ by $\gamma\langle\xi\rangle = \langle 0,\xi, 2\xi,\ldots,(p-1)\xi\rangle$. Clearly γ is π_S-equivariant where the π_S-action on S^n is induced from that on R^{n+1}. Let ι denote the fundamental class of S^n.

<u>Lemma 11.4</u>. The dual of the class $\gamma_*(\iota)$ is the class

$$\sum_{p-1\geq i>j\geq 1} \alpha^*_{ij} \in H^n\,(F(R^{n+1},p);Z_p) \quad \text{for} \quad n+1 \quad \text{even.}$$

<u>Proof</u>: Define $\hat{\gamma}: S^n \to F(R^{n+1}-0,p-1)$ and $\bar{\gamma}: S^n \to \mathbb{R}^{n+1}-0$ by

$$\hat{\gamma}\langle\xi\rangle = \langle\xi,2\xi,\ldots,(p-1)\xi\rangle$$

and

$$\bar{\gamma}\langle\xi\rangle = \xi.$$

The following diagram commutes on the nose:

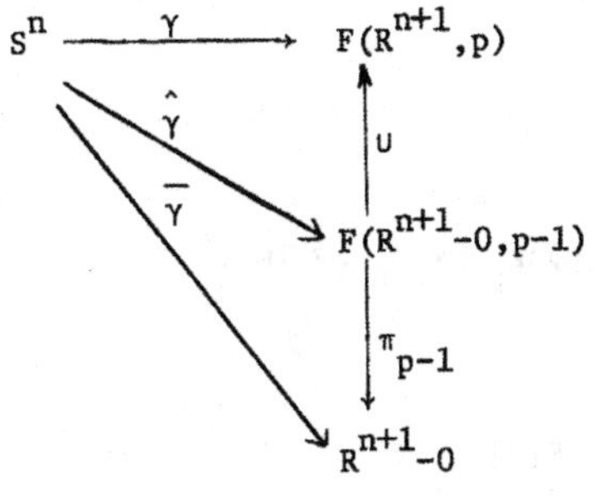

But clearly $(\pi_{p-1})_*[\hat{\gamma}_*-(\alpha_{11})_*] = 0$. Since $\gamma_* = \hat{\gamma}_*$ (recall that $F(\mathbb{R}^{n+1}-0,p-1) \subseteq F(\mathbb{R}^n,p)$ is a homotopy equivalence), we have that $\gamma_*(\iota) = \alpha_{11}+x$ where x is a linear combination of primitives in which α_{11} does not occur. If τ_r is a transposition, then $\tau_r \circ \gamma$ is clearly homotopic to γ if $n+1$ is even. Consequently Σ_p fixes $\alpha_*(\iota)$. But since Σ_p just permutes the primitives α_{ij}, it follows directly that $\gamma_*(\iota) = \sum_{p-1\geq i\geq j\geq 1} \alpha_{ij}$ and thus that $(\gamma_*(\iota))^* = \sum_{p-1\geq i\geq j\geq 1} \alpha_{ij}^*$ in the dual basis.

Recall that $\sum_{p-1\geq i\geq j\geq 1} \alpha_{ij}^*$ is denoted by $\bar{\alpha}$.

<u>Theorem 11.6</u>. For p odd, and $n+1$ even, there exists a class

$$\alpha \in H^n(B(\mathbb{R}^{n+1},p);\mathbb{Z}_p)$$

uniquely specified by the following two conditions:

(1) α restricts to $\bar{\alpha} \in H^n(F(\mathbb{R}^{n+1},p);\mathbb{Z}_p)$, and

(2) $S\alpha = -\alpha$.

The proof of Theorem 11.6 is held in abeyance until after the statement of Theorem 11.7.

<u>Theorem 11.7</u>. For p an odd prime,

$$H^*(B(R^{n+1},p);Z_p) = A_{n+1} \pi \operatorname{Im} f^*$$

as an algebra where

$$A_{n+1} = \begin{cases} E[\alpha] & \text{if } n+1 \text{ is even} \\ Z_p & \text{if } n+1 \text{ is odd}. \end{cases}$$

α is the class specified in Theorem 11.6, and the Steenrod operations are trivial on α.

<u>Proof of Theorem 11.6</u>: By Theorems 8.7 and 8.8, there is a class $\alpha' \in H^n(B(\mathbb{R}^{n+1},p);\mathbb{Z}_p)$ which restricts to $\bar{\alpha} \in H^*(F(R^{n+1},p);\mathbb{Z}_p)$. Since the maps γ (of Lemma 11.4) and $F(R^{n+1},p) \to B(R^{n+1},p)$ are π_S-equivariant, and $S\iota = -\iota$ for ι the fundamental class of S^n, we have $S\alpha' = -\alpha' + v_B$ for $v_B \in \operatorname{Im} f^*$. Let $\alpha = \alpha' - \frac{v_B}{2}$. By Proposition 11.1, S fixes v_B. Hence $S\alpha = -\alpha$.

To check uniqueness, we suppose that there exists another class $\alpha_1 \in H^n(B(\mathbb{R}^{n+1},p);\mathbb{Z}_p)$ such that α_1 restricts to $\bar{\alpha} \in H^n(F(\mathbb{R}^{n+1},p);\mathbb{Z}_p)$ and $S\alpha_1 = -\alpha_1$. Then $\alpha_1 - \alpha = v_C$ for $v_C \in \operatorname{Im} f^*$. By applying S to both sides of this equation, we see that $v_C = 0$.

<u>Proof of Theorem 11.7:</u> Define α by Theorem 11.6. We first prove the indicated product structure. By the form of E_∞ required by Theorems 8.7 and 8.9, it is clearly enough to show $\alpha \cdot v_B = 0$ for $v_B \in \operatorname{Im} f^*$.

Since v_B restricts to zero in $H^*(F(R^{n+1},p);Z_p)$, we have $\alpha \cdot v_B = v_C$ for $v_C \in \operatorname{Im} f^*$. We apply S to both sides of this equation, and conclude that $-v_C = v_C$, hence $v_C = 0$.

To show that the Steenrod operations are trivial on α, we note that $P^i\alpha = v_B$ for $v_B \in \operatorname{Im} f^*$. We apply S to both sides of this equation and again conclude that $v_B = 0$.

To calculate the module structure of

$$H^*(B(R^{n+1},p);\ Z_p(2q+1))$$

over $H^*(\Sigma_p;\ Z_p)$ for $n + 1$ odd, we note that by degree considerations,

$$H^*(B(R^{n+1},p);Z_p(2q+1)) = Z_p\cdot\lambda \oplus \operatorname{Imf}^* \text{ as an } H^*(\Sigma_p;\mathbb{Z}_p)\text{-module}$$

$$\text{if } mn \neq (2t+1)(p-1), \text{ where } m = \frac{p-1}{2}$$

and λ is the class, necessarily unique, which restricts to $\bar{\lambda}$ of Theorem 8.5. If $nm = (2t+1)(p-1)$, we exploit the map

$$i^*: H^*(\operatorname{Hom}_{\Sigma_p}(C_*F;Z_p(q)) \to H^*(\operatorname{Hom}_{\pi_p}(C_*F;Z_p)), \; F = F(\mathbb{R}^{n+1},p),$$

which is a monomorphism by previous remarks. In this case let $\lambda' \in H^{nm}(B(\mathbb{R}^{n+1},p);\mathbb{Z}_p(2q+1))$ be an element which restricts to $\bar{\lambda}$. If $\beta u\cdot i^*(\lambda') \neq 0$, then clearly $\beta u\cdot i^*(\lambda') = \beta u\cdot u_B$ for some u_B in the image of

$$H^*(\frac{F(R^\infty,p)}{\pi_p};Z_p) \to H^*(\operatorname{Hom}_{\pi_p}(C_*F;Z_p)).$$

Obviously, we may choose a class

$$\lambda \in H^{nm}(B(\mathbb{R}^{n+1},p);\mathbb{Z}_p(2q+1)),$$

such that $i^*\lambda = i^*\lambda'-u_B$. Clearly (1) λ restricts to $\bar{\lambda}$, and (2) $\beta u\cdot i^*(\lambda) = 0$. We claim that these two conditions uniquely determine λ. For suppose that λ_1 restricts to $\bar{\lambda}$ and $\beta u\cdot i^*(\lambda_1) = 0$; then $\lambda-\lambda_1 = u_C \in \operatorname{Imf}^*$ and $\beta u\cdot i^*(\lambda-\lambda_1) = 0$. However, if $u_C \neq 0$, then since the degree of $\beta u\cdot i^*(u_C)$ is $2 + nm \leq n(p-1)$, $\beta u\cdot i^*(u_i) \neq 0$. Consequently, $u_C = 0$ and the uniqueness property is proved. Furthermore, u must annihilate $i^*\lambda$ for if $u\cdot i^*\lambda = u_D \neq 0$, then, again by degree considerations, we have $\beta u\cdot u_D \neq 0$. This is a contradiction and so $u_D = 0$. Since i^* is a monomorphism and the restriction

$$i^*(\pi_p;\Sigma_p):\ H^*(\Sigma_p;Z_p) \to H^*(\pi_p;Z_p)$$

is a map of rings, we have that all elements of positive degree in $H^*(\Sigma_p;Z_p)$ annihilate λ. In summary, we have

Theorem 11.8. For q and $n+1$ odd, there exists an element

$$\lambda \in H^{nm}(B(\mathbb{R}^{n+1},p);\ \mathbb{Z}_p(q))$$

uniquely specified by the following two conditions:

(1) λ restricts to $\bar{\lambda} \in H^{nm}(F(\mathbb{R}^{n+1},p);\ \mathbb{Z}_p)$ and

(2) $\beta u \cdot i^*(\lambda) = 0$.

Furthermore, $H^*(B(R^{n+1},p);\ Z_p(q)) = M_{n+1} \oplus \mathrm{Im} f^*$ as an $H^*(\Sigma_p;Z_p)$-module, where

$$M_{n+1} = \begin{cases} 0 & \text{if } n+1 \text{ is even} \\ Z_p \cdot \lambda & \text{if } n+1 \text{ is odd.} \end{cases}$$

We close this section with a proof of the product structure described in Theorem 5.4.

By abuse of notation, we let α denote $i^*(\pi_p;\Sigma_p)(\alpha)$ and λ denote $i^*(\pi_p;\Sigma_p)(\lambda)$. To show that $\alpha \cdot \mathrm{Im} f^* = \lambda \cdot \mathrm{Im} f^* = 0$ in $H^*(\frac{F(\mathbb{R}^{n+1},p)}{\pi_p};\mathbb{Z}_p)$, it suffices to show that $\alpha \cdot u = \alpha \cdot \beta u = 0$ for $n+1$ even and $\lambda \cdot u = \lambda \cdot \beta u = 0$ for $n+1$ odd, where u is the one dimensional class in the image of $H^*(B\pi_p;\ \mathbb{Z}_p)$. Since u and βu go to zero under the map $H^*(\frac{F(\mathbb{R}^{n+1},p)}{\pi_p};\mathbb{Z}_p) \to H^*(F(\mathbb{R}^{n+1},p);\mathbb{Z}_p)$ and α is an odd dimensional class, it follows immediately that $\alpha \cdot u = k(\beta u)^j$ for $j = \frac{n+1}{2}$

and some k. Since $\beta\alpha = 0$, we have $\beta i^*(\pi_p;\Sigma_p)(\alpha) = 0$. This information together with the equation $\alpha\cdot u = k(\beta u)^j$ implies that $\alpha\cdot\beta u = 0$. To show that $k = 0$, we first observe that $i^*(\pi_p;\Sigma_p)(v) = u(\beta u)^{p-2}$. Hence

$$0 = i^*(\pi_p;\Sigma_p)(\alpha\cdot v) = \alpha\cdot u\cdot(\beta u)^{p-2}, \text{ and}$$

$$0 = k(\beta u)^{j+p-2}$$

If $n > 2$, then $\frac{n+1}{2} + p - 2 = j + p - 2 \leq n(\frac{p-1}{2})$ and consequently $(\beta u)^{j+p-2}$ is non-zero in $H^*(\frac{F(\mathbb{R}^{n+1},p)}{\pi_p};\mathbb{Z}_p)$ and $k = 0$. The case $n = 1$ is easily disposed of by use of the map S; the details are left to the reader.

Since $\lambda \in H^{nm}(B(\mathbb{R}^{n+1},p);\mathbb{Z}_p(2q+1))$, $n + 1$ odd, $p > 2$, and the Steenrod operations are trivial in $H^*F(\mathbb{R}^{n+1},p)$ [Prop. 7.8] it follows that $u\cdot\lambda = ru(\beta u)^{\ell}$ and $\beta\lambda = tu(\beta u)^{\ell}$, $\ell > 0$. But $0 = \beta^2\lambda = \beta(tu(\beta u)) = t(\beta u)^{\ell+1}$. Hence $t = 0$ and $\beta\lambda = 0$. The conclusion that $r = 0$ is in Theorem 11.8. Thus $u \cdot \lambda = 0$, and $\beta u \cdot \lambda = 0$ follows.

12. Auxiliary calculations

We present some auxiliary calculations to the previous 7 sections. These calculations provide a natural setting in which to proceed to the derivation of many of the formulas of section 1. Our main geometric lemma [12.1] allows us to calculate the map

$$\gamma_* : H_*\mathcal{C}_{n+1}(k) \otimes H_*\mathcal{C}_{n+1}(i_1) \otimes \cdots \otimes H_*\mathcal{C}_{n+1}(i_k) \to H_*\mathcal{C}_{n+1}(j),$$
$$j = i_1 + \cdots + i_k,$$

on primitives. Dualization of this information yields a proof of Lemma 12.4 which we translate into conceptually useful results in terms of Browder operations and Pontrjagin products [Theorem 12.3]. The final result of this section is the calcuation of $H_*\left(\frac{\mathcal{C}_{n+1}(p+1)}{\pi_p}\right)$ and the map

$\gamma_*: H_*(\mathcal{C}_{n+1}(2) \times \frac{\mathcal{C}_{n+1}(p)}{\pi_p} \times \mathcal{C}_{n+1}(1)) \to H_*\frac{\mathcal{C}_{n+1}(p+1)}{\pi_p}$. This information allows us to determine the formulas for $\lambda_n(\beta^\varepsilon Q^s x,y)$ and $\lambda_n(\zeta_n x,y)$ in the next section.

To begin, we let $\alpha_{r,s}$ denote the element in $H_n\mathcal{C}_{n+1}(\ell)$ given by the map $S^n \xrightarrow{\alpha_{r,s}} F(\mathbb{R}^{n+1},\ell) \xrightarrow{f_\ell} \mathcal{C}_{n+1}(\ell)$. ($f_\ell$ is the equivariant embedding of $F(\mathbb{R}^{n+1},\ell)$ in $\mathcal{C}_{n+1}(\ell)$ defined in [G;4.8].) We now define a map

$$\phi_t: \mathcal{C}_{n+1}(i_t) \to \mathcal{C}_{n+1}(k) \times \mathcal{C}_{n+1}(i_1) \times \cdots \times \mathcal{C}_{n+1}(i_k),\ 1 \le t \le k$$

by fixing points $c_k \in \mathcal{C}_{n+1}(k)$, $c_{i_m} \in \mathcal{C}_{n+1}(i_m)$, $m \neq t$ and setting

$\phi_t(x) = (c_k, c_1, \ldots, c_{i_t-1}, x, c_{i_t+1}, \ldots, c_k)$. We define

$$\psi\colon \zeta_{n+1}(k) \to \zeta_{n+1}(k) \times \zeta_{n+1}(i_1) \times \ldots \times \zeta_{n+1}(i_k)$$

similarly. Denote $\phi_{t*}(\alpha_{r,s})$ by α_{r,s,i_t} and $\psi_*(\alpha_{r,s})$ by $\alpha_{r,s,k}$.

<u>Lemma 12.1</u>. (1) $\gamma_*(\alpha_{r,s,i_t}) = \alpha_{r+x,s+x}$ where $x = \sum_{j=1}^{t-1} i_j$ (where, by convention, we set $x = 0$ if $t = 1$.)

(2) $\gamma_*(\alpha_{r,s,k}) = \sum_{\ell=0}^{i_{r+1}-1} \sum_{m=1}^{i_s} \alpha_{i_1+\cdots+i_r+\ell,\, i_1+\cdots+i_{s-1}+m}$.

<u>Proof:</u> (1): Consider the composite

$$S^n \xrightarrow{\alpha_{r,s}} F(\mathbb{R}^{n+1}, i_t) \xrightarrow{f_{i_t}} \zeta_{n+1}(i_t) \xrightarrow{\phi_t} \zeta_{n+1}(k) \times \zeta_{n+1}(i_1) \times \ldots$$
$$\times \zeta_{n+1}(i_k) \xrightarrow{\gamma} \zeta_{n+1}(j) \xrightarrow{g_j} F(\mathbb{R}^{n+1}, j)$$

where g_j is the equivariant retraction of $\zeta_{n+1}(j)$ onto $F(\mathbb{R}^{n+1}, j)$ [G;4.8]. A picture of the composite is instructive:

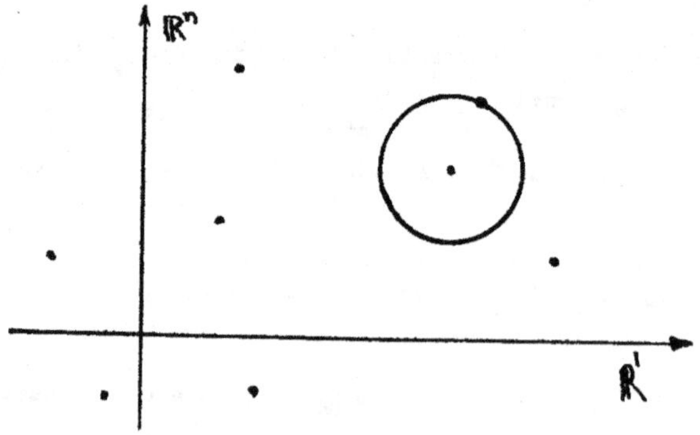

Figure 5.

By definition of the class $\alpha_{r,s} \in H_*F(\mathbb{R}^{n+1},j)$, we visibly have the formula

$$(g_j o \gamma o\ \phi_t \circ f_{i_t} o \alpha_{r,s})_*(\iota) = \alpha_{r+x,s+x}$$

where $x = i_1+\ldots+i_{t-1}$ and ι is a fixed fundamental class of S^n. Since g_j is an equivariant homotopy equivalence, (1) is verified.

(2): As above, we consider the composite

$$S^n \xrightarrow{\alpha_{r,s}} F(\mathbb{R}^{n+1},k) \xrightarrow{f_k} \mathcal{C}_{n+1}(k) \xrightarrow{\psi} \mathcal{C}_{n+1}(k) \times \mathcal{C}_{n+1}(i_1) \times \ldots$$
$$\times \mathcal{C}_{n+1}(i_k) \xrightarrow{\gamma} \mathcal{C}_{n+1}(j) \xrightarrow{g_j} F(\mathbb{R}^{n+1},j).$$

We again appeal to a picture:

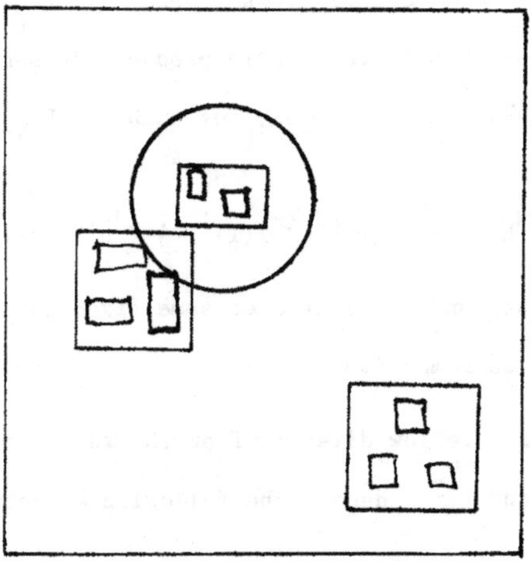

Figure 6.

Visibly, $(g_j o \gamma o \psi o\ f_k o \alpha_{r,s})_*(\iota) = F_0 + \ldots + F_{i_{r+1}-1}$ where

$$F_v = \sum_{m=0}^{i_s} \alpha_{i_1+\cdots+i_r+v,\ i_1+\cdots+i_{s-1}+m}\ , \quad v = 0,\ldots, i_{r+1}-1.$$

We are done.

Remark 12.2. The homology of $\mathcal{C}_{n+1}(i_t)$, $1 \le t \le k$, embeds very nicely in $H_*\mathcal{C}_{n+1}(j)$ via translations of the classes $\alpha_{r,s} \in H_*\mathcal{C}_{n+1}(i_t)$. By Lemma 12.1, the classes in $\gamma_* H_*\mathcal{C}_{n+1}(k)$ represent an algebraic "amalgamation" of the pieces $H_*\mathcal{C}_{n+1}(i_t)$ which corresponds to the geometric amalgamation given by γ. It is because of this amalgamation that the Browder operations behave so well.

Theorem 12.3. Let $\eta \in H_{sn}\mathcal{C}_{n+1}(k)$. Then $\theta_{k*}(\eta \otimes x_1 \ldots \otimes x_k)$ is an operation in k variables which is natural with respect to maps of $\mathcal{C}_{n+1}$-spaces and is given by a sum of classes, each of which is given by s Browder operations and $k-s-1$ Pontrjagin products in some order on the variables $x_1,\ldots,x_k$. In particular, for each $\nu \in H_{(k-1)n}\mathcal{C}_{n+1}(k)$,

$$\theta_{k*}(\nu \otimes x_1 \otimes \ldots \otimes x_k) = \sum c_{\nu,\sigma}\, ad_n(x_{\sigma(1)}) \ldots ad_n(x_{\sigma(k-1)})(x_{\sigma(k)})$$

where $c_{\nu,\sigma}$ are constants and σ runs over some fixed subset of Σ_k. (Compare Theorem 12.3 to Lemma 4.3.)

Theorem 12.3 will follow directly from the following lemma.

Lemma 12.4. Let γ^1 and γ^2 denote the following structure maps of operads:

$$\gamma^1 \colon \mathcal{C}_{n+1}(2) \times \mathcal{C}_{n+1}(k-1) \times \mathcal{C}_{n+1}(1) \longrightarrow \mathcal{C}_{n+1}(k), \quad \text{and}$$

$$\gamma^2 \colon \mathcal{C}_{n+1}(2) \times \mathcal{C}_{n+1}(k-2) \times \mathcal{C}_{n+1}(2) \longrightarrow \mathcal{C}_{n+1}(k).$$

If $\alpha_I \in H_{sn}\mathcal{C}_{n+1}(k)$, $1 \le s \le k-1$, is the dual basis element to the admissible monomial α_I^*, then $\alpha_I = \sum \sigma_J \alpha_J$ for some collection of elements $\sigma_J \in \text{Image } \gamma_*^i$, $i = 1, 2$, and $\sigma_J \in \Sigma_k$.

We prove 12.4 by use of Lemma 12.1 and a sequence of algebraic lemmas, after which we prove 12.3. We must first recover information concerning the action of Σ_k on the dual basis elements α_I dual to the admissible monomials α_I^*. Given an admissible sequence $I = (i_1, j_1, \ldots, i_m, j_m)$, we may read off the action on the classes α_I^* by Lemma 7.5 and the product structure specified in Lemma 7.7. Dualization via the Kronecker pairing $\langle \tau\alpha_I, x^* \rangle = \langle \alpha_I, \tau^{-1} x^* \rangle$ yields the desired result.

<u>Lemma 12.5</u>. Let τ denote the transposition τ_{i_ℓ}.

$$\tau\alpha_I = \begin{cases} (-1)^n \alpha_J, \text{ where } J = (i_1, j_1, \ldots, i_{\ell-1}, j_\ell, i_\ell, j_{\ell-1}, i_{\ell+1}, \tau j_{\ell+1}, \ldots, i_m, \tau_j m) \\ \quad \text{if } j_\ell \neq j_{\ell-1}, i_{\ell-1}+1 = i_\ell \text{ and } j_\ell < i_\ell. \\ (-1)^n(\alpha_K + \alpha_L) \text{ where } K = (i_1, \ldots, i_{\ell-1}, j_{\ell-1}, i_\ell, j_{\ell-1}, i_{\ell+1}, \tau j_{\ell+1}, \ldots, i_m, \tau i_m) \\ \quad \text{and } L = (i_1, \ldots, i_{\ell-1}, j_{\ell-1}, i_\ell, i_\ell, i_{\ell+1}, \tau j_{\ell+1}, \ldots, i_m, \tau j_m) \\ \quad \text{if } i_{\ell-1}+1 = i_\ell, \text{ and } j_{\ell-1} = j_\ell, \\ (-1)^{n+1}\alpha_M \text{ where } M = (i_1, \ldots, i_{\ell-1}, j_{\ell-1}, i_\ell, i_\ell, i_{\ell+1}, \tau j_{\ell+1}, \ldots, i_m, \tau j_m) \\ \quad \text{if } i_\ell = j_\ell \text{ and } i_{\ell-1} + 1 \leq i_\ell, \\ \alpha_N \text{ where } N = (i_1, \ldots, i_{\ell-1}, j_{\ell-1}, i_{\ell-1}, j_\ell, i_{\ell+1}, \tau j_{\ell+1}, \ldots, i_m, \tau j_m) \\ \quad \text{otherwise} \end{cases}$$

<u>Lemma 12.6</u>. Let $\beta = \sum_{x=1}^{k-1} \alpha_{I_x}$ where $I_x = (1, 1, 2, j_2, \ldots, k-2, j_{k-1}, k-1, x)$

for fixed j_i, $2 \leq i \leq k-2$. Fix integers r and i such that $1 \leq k-r \leq i-1$. Let $\lambda = \tau_{k-r} \circ \tau_{k-r+1} \circ \dots \circ \tau_{k-1}$ and $\lambda_i = \tau_{k-r} \circ \dots \circ \tau_{i-1}$. Then $\lambda\beta = (-1)^{nr} \sum_{x=1}^{k-r-1} \alpha_{J_x}$ where

$J_x = (1,1,2,j_2,\dots,k-r-2,j_{k-r-2},x,x-r,j_{k-r-1},k-r+1,\lambda_{k+r+1}(j_{k-r}),\dots,k-1,\lambda_{k-1}(j_{k-}$

The proof of 12.6 is immediate from 12.5 and induction on r.

Proof of 12.4: We break up our proof into two cases. First suppose that $s < k-1$. By 12.1, $\gamma^1_*(e_0 \otimes \alpha_{ij} \otimes e_0) = \alpha_{ij}$, $\gamma^2_*(e_0 \otimes \alpha_{ij} \otimes e_0) = \alpha_{ij}$, and $\gamma^2_*(e_0 \otimes e_0 \otimes \alpha_{11}) = \alpha_{k-1,k-1}$. By the obvious double dualization argument, $\alpha_I \in \operatorname{Im} \gamma^1_*$ provided $I = (i_1,j_1,\dots,i_s,j_s)$ for $i_s < k-1$. Again by double dualization, it follows that $\gamma^2_*(e_0 \otimes \alpha_I \otimes \alpha_{11}) = \alpha_J$ for $J = (I,k-1,k-1)$. Let $V_{I,r}$ denote the $\mathbb{Z}_p$-subspace spanned by elements α_K, $K = (I,k-1,r)$, K admissible. By Lemma 12.5, $\eta(V_{I,k-1}) \subset V_{I,j}$ for $\eta = \tau_j \circ \dots \circ \tau_{k-2}$. Since η is an isomorphism of vector spaces, the lemma is proved provided $1 \leq s < k-1$.

We proceed to the case $s = k-1$. By double dualization arguments, we see that $\gamma^2_*(\alpha_{11} \otimes \alpha_I \otimes \alpha_{11}) = \pm\alpha_J$ for appropriate I where J has the form $(1,1,2,j_2,\dots,k-2,j_{k-2},k-1,k-1)$. The result follows from previous remarks.

Proof of 12.3: We recall that the diagram below commutes [G]:

$$\begin{array}{ccc}
\mathcal{C}_{n+1}(2) \times \mathcal{C}_{n+1}(k-j) \times \mathcal{C}_{n+1}(j) \times X^k \times \mathcal{C}_{n+1}(k) \times X^k & \searrow & \\
\downarrow & & X \\
\mathcal{C}_{n+1}(2) \times \mathcal{C}_{n+1}(k-j) \times X^{k-j} \times \mathcal{C}_{n+1}(j) \times X^j \rightarrow \mathcal{C}_{n+1}(2) \times X^2 & \nearrow &
\end{array}$$

By the definition of the operations λ_n, Lemma 12.4, and the obvious induction, Theorem 12.3 is demonstrated.

We now calculate $H^*\left(\frac{\mathcal{C}_{n+1}(p+1)}{\pi_p}; \mathbb{Z}_p\right)$. Give $\mathcal{C}_{n+1}(p+1)$ the Σ_p-action defined by the inclusion $\Sigma_p = \Sigma_p \times \{1\} \leq \Sigma_{p+1}$ and the evident action of Σ_{p+1} on $\mathcal{C}_{n+1}(p+1)$. Let $\sigma = \tau_1 \circ \ldots \circ \tau_{p-1}$. By Proposition 7.2

$$(\#) \qquad \sigma\alpha_{ij} = \begin{cases} \alpha_{i+1,j+1} & \text{if } i < p-1 \\ (-1)^{n+1}\alpha_{j,1} & \text{if } i = p-1 \\ \alpha_{p,j+1} & \text{if } j < i = p \\ \alpha_{p,1} & \text{if } j = i = p. \end{cases}$$

By dualizing, we observe that $H^*\mathcal{C}_{n+1}(p) = A \oplus B$ as a $\mathbb{Z}_p\pi_p$-module where π_p is generated by σ, A has an additive basis given by

$$\{\alpha_J^* \mid J \text{ admissible, } J = (i_1,j_1,\ldots,i_k,j_k), i_k < p\}$$

and B has an additive basis given by

$$\{\alpha_I^* \mid I \text{ admissible, } I = (i_1,j_1,\ldots,i_{k_j}j_k), i_k = p\}.$$

It is trivial to see that the cyclic group π_p generated by σ acts freely on B. Hence $H^*(\pi_p;B) = B^{\pi_p}$, the classes in B fixed under the action of π_p. $H^*(\pi_p;A)$ has been calculated in section 8. Using the spectral sequence for a covering, the fact that $\frac{F(\mathbb{R}^{n+1},p)}{\pi_p}$ is a p(n+1)-manifold, and the requisite periodicity of the differentials in the spectral sequence (see section 8 for details), we trivially have

Lemma 12.7. If $n \geq 1$,

$$H^*\left(\frac{\mathcal{C}_{n+1}(p+1)}{\pi_p};\mathbb{Z}_p\right) = \mathrm{Imf}^*\pi\ C, \text{ additively,}$$

where Imf* is a subalgebra over the Steenrod algebra and is given by the image of the classifying map $f^* : H^*(B\pi_p;\mathbb{Z}_p) \to H^*\left(\frac{\mathcal{C}_{n+1}(p+1)}{\pi_p};\mathbb{Z}_p\right)$ specified in section 5, Kerf* is the ideal of $H^*(B\pi_p;\mathbb{Z}_p)$ which consists of all elements of degree greater than n(p-1), and C is a subalgebra of classes in $H^*\mathcal{C}_{n+1}(p)$ fixed under π_p. Furthermore, an additive basis may be chosen which extends the standard basis for Imf* and is such that $\beta x = 0$ for x a basis element which is not in Imf*.

As in section 5, we are deliberately incomplete in our description of C.

Note that the second statement of Lemma 12.7 follows directly from the action of the Bockstein on $H^*(B\pi_p;\mathbb{Z}_p)$ and the fact that the Steenrod operations are trivial in $H^*\mathcal{C}_{n+1}(p+1)$ [Prop. 7.8]. We remark that with some added work, the precise algebra extension over the Steenrod algebra can be calculated, but this extra information is irrelevant to our work.

By the equivariance conditions [G;1.1] satisfied by γ, we see that γ induces a map of quotient spaces

$$\gamma : \mathcal{C}_{n+1}(2) \times \frac{\mathcal{C}_{n+1}(p)}{\pi_p} \times \mathcal{C}_{n+1}(1) \to \frac{\mathcal{C}_{n+1}(p+1)}{\pi_p} .$$

We obtain information about γ_* here.

Lemma 12.8. (1) $\gamma_*(\alpha_{11} \otimes e_i \otimes e_0) = 0$ if $i < n(p-1)$, and $i = k(p-1) - \varepsilon$, $\varepsilon = 0,1$. (2) $\gamma_*(\alpha_{11} \otimes e_{n(p-1)} \otimes e_0)$ is in the image of the map

$$H_{np}\mathcal{C}_{n+1}(p+1) \to H_{np}\frac{\mathcal{C}_{n+1}(p+1)}{\pi_p} .$$

Proof. We break up the proof of (1) into three cases. If $n = 1$, it is enough to show that $\gamma_*(\alpha_{11} \otimes e_i \otimes e_0)$ when $i = 0$ or, if $p > 2$, when $i = p - 2$. If $x \in H^{p-1}(\frac{\mathcal{C}_2(p+1)}{\pi_p}; \mathbb{Z}_p)$, $p > 2$, then $\beta x = 0$ by Lemma 12.7. Since $\beta(\alpha_{11} \otimes e_{p-2} \otimes e_0)^* \neq 0$ for p odd where $(\alpha_{11} \otimes e_{p-2} \otimes e_0)^*$ is the class in $H^{p-1}(\mathcal{C}_2(2) \times \frac{\mathcal{C}_2(p)}{\pi_p})$ dual to $\alpha_{11} \otimes e_{p-2} \otimes e_0$, we have $\gamma_*(\alpha_{11} \otimes e_{p-2} \otimes e_0) = 0$. Let $i = 0$; we calculate $\gamma_*(\alpha_{11} \otimes e_0 \otimes e_0)$ by the commutative diagram

$$\begin{array}{ccc}
\mathcal{C}_{n+1}(2) \times \mathcal{C}_{n+1}(p) \times \mathcal{C}_{n+1}(1) & \xrightarrow{\gamma} & \mathcal{C}_{n+1}(p+1) \\
\downarrow{\pi'} & & \downarrow{\pi} \\
\mathcal{C}_{n+1}(2) \times \frac{\mathcal{C}_{n+1}(p)}{\pi_p} \times \mathcal{C}_{n+1}(1) & \xrightarrow{\gamma} & \frac{\mathcal{C}_{n+1}(p+1)}{\pi_p}
\end{array} .$$

$\gamma_*(\alpha_{11} \otimes e_0 \otimes e_0) = \gamma_*\pi'_*(\alpha_{11} \otimes e_0 \otimes e_0) = \pi_*\gamma_*(\alpha_{11} \otimes e_0 \otimes e_0) = \pi_*(\alpha_{p,1}+\ldots+\alpha_{p,p})$,

by Lemma 12.1. Recall [table (#) or Lemma 7.6] that

$$\sigma\alpha_{p,i} = \begin{cases} \alpha_{p,i+1} & \text{if } i < p \\ \alpha_{p,1} & \text{if } i = p. \end{cases}$$

It follows easily that $\pi_*\alpha_{p,i} = \pi_*\alpha_{p,1}$, $i \geq 1$ [This fact is checked by recalling the $E^2_{0,*}$-term of the spectral sequence for a covering and the definition of $H_o(\pi_p;M)$.] Consequently $\gamma_*(\alpha_{11} \otimes e_0 \otimes e_0) = 0$.

If $n > 2$, suppose that $\gamma^*(x) = c(\alpha_{11} \otimes e_i \otimes e_0)^* +$ other terms for $x \in H^*\frac{\zeta_{n+1}(p+1)}{\pi_p}$. By Lemma 12.7 and the algebra structure of Theorem 5.4, there exist $\varepsilon = 0,1$ and $s \geq 0$ such that $(e_1^*)^\varepsilon \cdot (\beta(e_1^*))^s \cdot x = 0$, $(e_1^*)^\varepsilon(\beta(e_1^*))^s \cdot (\alpha_{11} \otimes e_i \otimes e_0)^* \neq 0$, and $n(p-1) < \varepsilon + 2s + n + i < np$ for $i < n(p-1)$. It follows that $c = 0$, and consequently $\gamma_*(\alpha_{11} \otimes e_i \otimes e_0) = 0$ for $i < n(p-1)$.

If $n = 2$, it is easy to see that $\gamma_*(\alpha_{11} \otimes e_i \otimes e_0) = 0$ when $i = 0$, $p - 2$ or $2p-3$ by similar arguments to those used in case $n = 1$. If $i = p - 1$, we use arguments similar to those used in cases $n > 2$.

(2): By lemma 12.7, the only classes in $H_{np}\frac{\zeta_{n+1}(p+1)}{\pi_p}$ are in the image of the map $H_{np}\zeta_{n+1}(p+1) \to H_{np}\frac{\zeta_{n+1}(p+1)}{\pi_p}$.

Remarks 12.9. Observe that we may use Σ_p instead of π_p in our arguments (with troublesome modifications necessary in the case of twisted coefficients). By degree considerations, it is immediate that $\gamma_*(\alpha_{11} \otimes e_i \otimes e_0) = 0$ if $n + i \not\equiv 0(n)$ or $(p-1-\varepsilon)$, $\varepsilon = 0,1$. Hence our calculations are at least plausible.

13. Geometry of Browder operations

Let $g\colon \Sigma_2 \to \Sigma_p$ be a non-trivial homomorphism such that $g(\text{generator})$ interchanges j and k. Further suppose that $h\colon J^{n+1}\Sigma_p \to \mathcal{C}_{n+1}(p)$ is Σ_p-equivariant where $J^{n+1}G$ denotes Milnor's $(n+1)$-st join of the group G. Transparently the composite $J^{n+1}\Sigma_2 \xrightarrow{J^{n+1}g} J^{n+1}\Sigma_p \xrightarrow{h} \mathcal{C}_{n+1}(p) \xrightarrow{\pi} \mathcal{C}_{n+1}(2)$ is Σ_2-equivariant where $\pi\langle c_1,\ldots,c_p\rangle = \langle c_j, c_k\rangle$. It is trivial to show, by use of the spectral sequence for a covering, that if X and Y are $\mathbf{Z}_p$-homology spheres equipped with free π_p-actions and $f\colon X \to Y$ is π_p-equivariant, then $f_* \neq 0$. Since $J^{n+1}\Sigma_2$ and $\mathcal{C}_{n+1}(2)$ are homology spheres (see section 5), $\pi_* \circ g_* \circ f_*$ is non-zero. Consequently h cannot be equivariantly extended to a map from $J^{n+2}\Sigma_p$ into $\mathcal{C}_{n+1}(p)$. This observation indicates that the operations described in Theorems 1.1 and 1.3 cannot be defined in the entire range by the method of Dyer and Lashof (for odd primes). In fact, since the following diagram commutes, (where σ_i is defined in [G; 2.3])

$$\begin{array}{ccc}
\mathcal{C}_{n+1}(p) \times X^2 \times *^{p-2} & \xrightarrow{\ \theta_p\ } & X \\
\Big\downarrow {\scriptstyle \sigma_3 \circ \cdots \circ \sigma_p \times 1} & \nearrow_{\theta_2} & \\
\mathcal{C}_{n+1}(2) \times X^2 & &
\end{array}$$

we observe that it is precisely the presence of Browder operations which prevents all of the $\beta^{\varepsilon}Q^s$ and $\beta^{\varepsilon}\xi_n$ to be defined by use of the join.

We investigate further the properties of Browder operations using methods naturally suggested by the structure map of the little cubes operad.

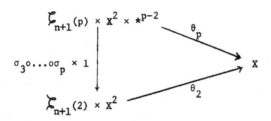

If we wish to calculate a particular formula, we need only substitute appropriate numbers in the composite map $\theta_{k*} \circ (\theta_{i_1 *} \otimes \dots \otimes \theta_{i_k *}) \circ (1 \otimes t_* \otimes 1)$ for the diagram

$$\begin{array}{ccccc}
\mathcal{C}_{n+1}(k)\times\mathcal{C}_{n+1}(i_1)\times\dots\times\mathcal{C}_{n+1}(i_k)\times X^j & \xrightarrow{\gamma\times 1} & \mathcal{C}_{n+1}(j)\times X^j & & \\
\downarrow{\scriptstyle 1\times t\times 1} & & & \searrow{\scriptstyle \theta_j} & \\
& & & & X \\
& & & \nearrow{\scriptstyle \theta_k} & \\
\mathcal{C}_{n+1}(k)\times\mathcal{C}_{n+1}(i_1)\times X^{i_1}\times\dots\times\mathcal{C}_{n+1}(i_k)\; X^{i_k} & \xrightarrow{1\times\theta_{i_1}\times\dots\times\theta_{i_k}} & \mathcal{C}_{n+1}(k)\times X^k & &
\end{array}$$

DIAGRAM 13.1

We use commutativity of this diagram [G; 1.4] and our homological calculations [§5-12] to achieve the desired results. It should be observed that the structure map, γ, of the little cubes operad carries all the information, quite elegantly and beautifully, sufficient for a complete theory of homology operations on $\mathcal{C}_{n+1}$-spaces. Observe first of all, that the properties in Theorem 1.2(1)-(6) except the internal Cartan formula have already been demonstrated in [A; §6]. (Recall that 1.2(6) is commutation with homology suspension).

We already know the map γ_* on the primitives [12.1]. We use this information to precisely identify, in terms of Browder operations and Pontrjagin products, the operations determined by the classes α and λ in the cohomology of braid spaces. The method of proof here is representative of the spirit of our proofs throughout this section. Let

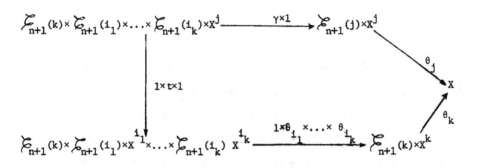

α_* and λ_* denote the basis elements which are dual to the basis elements α and λ specified in Theorems 5.2 and 5.3.

Corollary 13.2.

(1) $\theta_*(\alpha_* \otimes x^p) = -\lambda_n(x,x) * x^{p-2}$ if degree (x) is even and n is odd.

(2) $\theta_*(\lambda_* \otimes x^p) = (-1)^{\frac{p-1}{2}} (\lambda_n(x,x))^{\frac{p-1}{2}} * x$ if degree (x) is odd and n is even.

Remark 13.3. Visibly, the operations in Corollary 13.2 (1) and (2) are non-trivial; up to constant multiples, these are the only operations in one variable other than the $\beta^{\varepsilon} Q^s$ and $\beta^{\varepsilon} \xi_n$ which can occur in the p-th filtration of $H_* C_{n+1} X$[see section 4]. It is amusing to observe that the somewhat artificial looking classes α and λ are, for the above reasons, precisely the classes in the cohomology of braid spaces which cannot be in the image of $H^*(B\Sigma_p; \mathbb{Z}_p(q))$.

Corollary 13.2 will follow directly from the following corollary of the geometric calculations in 12.1.

Corollary 13.4.

(1) Consider γ_*: $H_* \mathcal{C}_{n+1}(2) \otimes H_* \mathcal{C}_{n+1}(k-1) \otimes H_* \mathcal{C}_{n+1}(1) \to H_* \mathcal{C}_{n+1}(k)$.

$\gamma_*(e_0 \otimes \alpha_{11} \otimes e_0) = \alpha_{11}$ where e_0 is the evident zero dimensional class.

(2) Consider γ_*: $H_* \mathcal{C}_{n+1}(k) \otimes H_* \mathcal{C}_{n+1}(2)^j \otimes H_* \mathcal{C}_{n+1}(k-2j) \to H_* \mathcal{C}_{n+1}(k)$.

$\gamma_*(e_0 \otimes (\alpha_{11})^j \otimes e_0) = \alpha_I$, $I = (1,1,3,3,5,5,\ldots 2j-1,2j-1)$ where α_I is the basis element in $H_* \mathcal{C}_{n+1}(k)$ dual to the admissible monomial α_I^* in $H^* \mathcal{C}_{n+1}(k)$.

Proof of 13.4: (1) is immediate from Lemma 12.1 for statement (2), we note that $\gamma_*(e_0 \otimes e_0^s \otimes \alpha_{11} \otimes e_0^{j-s-1}) = \alpha_{2s-1,2s-1}$ by Lemma 12.1. Dualizing, we observe that $\gamma(\alpha_J^*) = \varepsilon(e_0 \otimes \alpha_{11} \otimes \cdots \otimes \alpha_{11} \otimes e_0)^* +$ other terms, for α_J^* an admissible monomial where $\varepsilon = 0$ if $\alpha_J^* \neq \alpha_{11}^* \alpha_{33}^* \cdots \alpha_{2j-1,2j-1}^*$ and $\varepsilon = 1$ if $\alpha_J^* = \alpha_{11}^* \alpha_{33}^* \cdots \alpha_{2j-1,2j-1}^*$.

Proof of 13.2: Recall that the class $\alpha \in H^n(B(\mathbb{R}^{n+1},p);\mathbb{Z}_p)$ restricts to $\sum_{p-1 \geq i \geq j \geq 1} \alpha_{ij}^*$ in $H^*\mathcal{C}_{n+1}(p)$ and the class $\lambda \in H^{nm}(B(\mathbb{R}^{n+1},p);\mathbb{Z}_p(2q+1))$ restricts to $\sum g(\alpha_{11}^* \alpha_{33}^* \cdots \alpha_{p-2,p-2}^*)$, $m = \frac{p-1}{2}$, and g runs over a complete set of distinct coset representatives for H in Σ_p. [See section 8 for details.] Consequently, the dual class α_* is the image of α_{11} under the map $H_n\mathcal{C}_{n+1}(p) \to H_n B(\mathbb{R}^{n+1},p)$; similarly the dual class λ_* is the image of $\alpha_I \in H_{nm}\mathcal{C}_{n+1}(p)$ where $I = (1,1,3,3,\ldots,2j+1,2j+1,\ldots,\frac{p-1}{2},\frac{p-1}{2})$.

By 13.4(1), and commutativity of diagram 13.1, we have that

$$\text{(i)}\quad \theta_{p*}(\alpha_* \otimes x^p) = \theta_{p*}\circ\gamma_*(e_0 \otimes \alpha_{11} \otimes e_0 \otimes x^p)$$
$$= \theta_{2*}(1 \otimes \theta_{2*} \otimes \theta_{1*}^{p-2})(1 \otimes t_* \otimes 1)(e_0 \otimes \alpha_{11} \otimes e_0 \otimes x^p).$$

By the definition of λ_n, we have the formula

$$\text{(ii)}\quad \theta_{2*}(1 \otimes \theta_{2*} \otimes \theta_{1*}^{p-2})(1 \otimes t_* \otimes 1)(e_0 \otimes \alpha_{11} \otimes e_0 \otimes x^p) = (-1)^{n|x|+1}\theta_{2*}(e_0 \otimes \lambda_n(x,x) \otimes x^{p-2})$$
$$= (-1)^{n|x|+1}\lambda_n(x,x) * x^{p-2}$$

together, (i) and (ii) yield 13.2(1). Similarly, by 13.4(2) and commutativity of 13.1 we have that

$$\text{(iii)}\quad \theta_{p*}(\lambda_* \otimes x^p) = \theta_{p*}\,\gamma_*(e_0 \otimes \alpha_{11}^m \otimes e_0 \otimes x^p)$$

$$= \theta_{\frac{p+1}{2}*}(1 \otimes \theta_{2*}^m \otimes \theta_{1*})(1 \otimes t_* \otimes 1)(e_0 \otimes \alpha_{11}^m \otimes e_0 \otimes x^p).$$

By definition of λ_n we have the formula

(iv)

$$\theta_{\frac{p+1}{2}*}(1\otimes\theta_{2*}^{m}\otimes\theta_{1*})(1\otimes t_*\otimes 1)(e_0\otimes\alpha_{11}^{m}\otimes e_0\otimes x^p) = (-1)^{(n|x|+1)m}\theta_{\frac{p+1}{2}*}(e_0\otimes\lambda_n(x,x)^m\otimes x).$$

together, (iii) and (iv) yield 13.2(2).

<u>Proof of Theorem 1.2(7)</u>, the Jacobi identity:

Specializing diagram 13.1 to

$$\begin{array}{ccccc}
H_*\mathcal{C}_{n+1}(2)\otimes H_*\mathcal{C}_{n+1}(2)\otimes H_*\mathcal{C}_{n+1}(1)\otimes(H_*X)^3 & \xrightarrow{\gamma_*\otimes 1} & H_*\mathcal{C}_{n+1}(3)\otimes(H_*X)^3 & \xrightarrow{\theta_{3*}} & \\
\Big\downarrow 1\otimes t_*\otimes 1 & & & & H_*X \\
H_*\mathcal{C}_{n+1}(2)\otimes H_*\mathcal{C}_{n+1}(2)\otimes(H_*X)^2\otimes H_*\mathcal{C}_{n+1}(1)\otimes H_*X & \xrightarrow{1\otimes\theta_{2*}\otimes\theta_{1*}} & H_*\mathcal{C}_{n+1}(2)\otimes(H_*X)^2 & \xrightarrow{\theta_{2*}} &
\end{array}$$

we observe that

(i) $\theta_{2*}(1\otimes\theta_{2*}\otimes\theta_{1*})(1\otimes t_*\otimes 1)(\alpha_{11}\otimes\alpha_{11}\otimes e_0\otimes x\otimes y\otimes z) = (-1)^{n|y|+n}\lambda_n[\lambda_n(x,y),z].$

To take advantage of commutativity, we calculate $\gamma_*(\alpha_{11}\otimes\alpha_{11}\otimes e_0)$. By Lemma 12.1, $\gamma_*(\alpha_{11}\otimes e_0\otimes e_0) = \alpha_{21}+\alpha_{22}$ and $\gamma_*(e_0\otimes\alpha_{11}\otimes e_0) = \alpha_{11}$. Dualizing this information, we use the cup product structure to retrieve the formula

(ii) $(-1)^n\gamma^*(\alpha_{11}^*\alpha_{21}^*) = (\alpha_{11}\otimes e_0\otimes e_0)^*\cdot(e_0\otimes\alpha_{11}\otimes e_0)^* = (\alpha_{11}\otimes\alpha_{11}\otimes e_0)^*.$

Direct dualization of formula (ii) yields the desired result:

(iii) $\gamma_*(\alpha_{11}\otimes\alpha_{11}\otimes e_0) = (-1)^n[(\alpha_{11}^*\alpha_{21}^*)_* + (\alpha_{11}^*\alpha_{22}^*)_*]$

where $(\alpha_{11}^{*}\alpha_{2i}^{*})_{*}$ is the element dual to $\alpha_{11}^{*}\alpha_{2i}^{*}$. Let α_I denote $(\alpha_{11}^{*}\alpha_{21}^{*})_{*}$ and α_J denote $(\alpha_{11}^{*}\alpha_{22}^{*})_{*}$. We combine formulas (i) - (iii) and observe that

(iv) $\theta_{3*}[(\alpha_I + \alpha_J) \otimes x \otimes y \otimes z] = (-1)^{n|y|}\lambda_n[\lambda_n(x,y),z]$.

We let $\sigma = \tau_1\tau_2 \in \Sigma_3$ and recall that the action of σ on the dual basis is given by the Kronecker pairing

$<\sigma\alpha_K, x^{*}> = <\alpha_K, \sigma^{-1}x^{*}>$ for α_K arbitrary:

$$<\alpha_I, \sigma^{-1}x^{*}> = \begin{cases} 0 & \text{if } x^{*} = \alpha_{11}^{*}\alpha_{21}^{*} \\ 1 & \text{if } x^{*} = \alpha_{11}^{*}\alpha_{22}^{*} \end{cases}, \text{ and}$$

$$<\alpha_J, \sigma^{-1}x^{*}> = \begin{cases} -1 & \text{if } x^{*} = \alpha_{11}^{*}\alpha_{21}^{*} \\ -1 & \text{if } x^{*} = \alpha_{11}^{*}\alpha_{21}^{*}. \end{cases}$$

Combining this information with commutativity of 13.1 and the requisite equivariance, we have

(v) $\theta_{3*}(\alpha_I \otimes x \otimes y \otimes z) = \theta_{3*}[\sigma^2\alpha_I \otimes \sigma^{-2}(x \otimes y \otimes z)]$

$= (-1)^{1+|x|(|y|+|z|)}\theta_{3*}[(\alpha_I+\alpha_J) \otimes y \otimes z \otimes x]$ and

(vi) $\theta_{3*}(\alpha_J \otimes x \otimes y \otimes z) = \theta_{3*}[\sigma\alpha_J \otimes \sigma^{-1}(x \otimes y \otimes z)]$

$= (-1)^{1+|z|(|x|+|y|)}\theta_{3*}[(\alpha_I+\alpha_J) \otimes z \otimes x \otimes y]$.

Combining formulas (iv)-(vi) together with the formula $\lambda_n(x,y) = (-1)^{|x||y|+1+n(|x|+|y|+1)}\lambda_n(y,x)$, we get the Jacobi identity.

<u>Proof of Theorem 1.2(5).</u> The internal Cartan formula for λ_n:

We specialize diagram 13.1 to

$$\begin{array}{ccccc}
H_*\mathcal{C}_{n+1}(2)\otimes(H_*\mathcal{C}_{n+1}(2))^2\otimes(H_*X)^4 & \xrightarrow{\gamma_*\otimes 1} & H_*\mathcal{C}_{n+1}(4)\otimes(H_*X)^4 & \xrightarrow{\theta_{4*}} & H_*X \\
\downarrow{\scriptstyle 1\otimes t_*\otimes 1} & & & & \uparrow{\scriptstyle \theta_{2*}} \\
H_*\mathcal{C}_{n+1}(2)\otimes H_*\mathcal{C}_{n+1}(2)\otimes(H_*X)^2\otimes H_*\mathcal{C}_{n+1}(2)\otimes(H_*X)^2 & & \xrightarrow{1\otimes\theta_{2*}\otimes\theta_{2*}} & & H_*\mathcal{C}_{n+1}(2)\otimes(H_*X)^2
\end{array}$$

By Lemma 12.1, $\gamma_*(e_0\otimes\alpha_{11}\otimes e_0) = \alpha_{11}$ and consequently $\theta_{4*}(\alpha_{11}\otimes x\otimes y\otimes u\otimes v) = (-1)^{n|x|+1}\lambda_n(x,y)uv$. We also observe that $\gamma_*(\alpha_{11}\otimes e_0\otimes e_0) = \alpha_{21}+\alpha_{22}+\alpha_{31}+\alpha_{32}$ by Lemma 13.1 and

(i) $\theta_{4*}\gamma_*(\alpha_{11}\otimes e_0\otimes e_0\otimes x\otimes y\otimes u\otimes v) = (-1)^{n(|x|+|y|)+1}\lambda_n(xy,uv)$.

We recall the Σ_4-action on the classes α_{ij} [Prop. 7.2] and use commutativity of the diagram

$$\begin{array}{ccc}
\mathcal{C}_{n+1}(4)\times X^4 & \xrightarrow{\theta_4} & \\
\downarrow{\scriptstyle \rho\times\rho^{-1}} & & X, \quad \rho\in\Sigma_4 \\
\mathcal{C}_{n+1}(4)\times X^4 & \xrightarrow{\theta_4} &
\end{array}$$

to calculate

(ii) $\theta_{4*}(\alpha_{21}\otimes x\otimes y\otimes u\otimes v) = (-1)^{|y||u|}\theta_{4*}(\alpha_{11}\otimes x\otimes u\otimes y\otimes v)$

$= (-1)^{|y||u|+n|x|+1}\lambda_n(x,u)yv,$

(iii) $\theta_{4*}(\alpha_{22}\otimes x\otimes y\otimes u\otimes v) = (-1)^{|x||y|}\theta_{4*}(\alpha_{21}\otimes y\otimes x\otimes y\otimes v)$

$= (-1)^{|x||y|+|x||u|+n|y|+1}\lambda_n(y,u)xv$

(iv) $\theta_{4*}(\alpha_{31}\otimes x\otimes y\otimes u\otimes v) = (-1)^{|u||v|}\theta_{4*}(\alpha_{21}\otimes x\otimes y\otimes v\otimes u)$

$= (-1)^{|u||v|+|y||v|+n|x|+1}\lambda_n(x,v)yu,$ and

(v) $\theta_{4*}(\alpha_{32}\otimes x\otimes y\otimes u\otimes v) = (-1)^{|u||v|}\theta_{4*}(\alpha_{22}\otimes x\otimes y\otimes v\otimes u)$

$= (-1)^{|u||v|+|x||y|+|x||v|+n|y|+1}\lambda_n(y,v)xu.$

Combining formulas (i)-(v), we have

$$\lambda_n(xy,uv) = (-1)^{|y|(n+|u|)}\lambda_n(x,u)yv + (-1)^{|x|(n+|y|+|u|)}\lambda_n(y,u)xv$$
$$+ (-1)^{|y|(n+|v|)+|u||v|}\lambda_n(x,v)yu + (-1)^{|x|(n+|v|+|y|)+|u||v|}\lambda_n(y,v)xu.$$

The formula in 1.2(5) follows.

We note that the internal Cartan formula does not follow from the external Cartan formula because the multiplication $X\times X\to X$ is not a morphism of $\mathcal{C}_{n+1}$-spaces.

Proof of Theorem 1.2(7), the Nishida relation for λ_n:

Since λ_n is defined in terms of $\theta_2:\mathcal{C}_{n+1}(2)\times X^2\to X$, the Steenrod operations on $\lambda_n(x,y)$ are completely determined by the external (dual) Cartan formula:

$$P^r_*\lambda_n(x,y) = (-1)^{n|x|+1}P^r_*\theta_{2*}(\iota\otimes x\otimes y) = (-1)^{n|x|+1}\sum_{i\geq 0}\theta_{2*}(\iota\otimes P^i_*x\otimes P^{r-i}_*y).$$

Proof of Theorem 1.2(8), the relations $\lambda_n(Q^s x,y) = 0 = \lambda_n(\zeta_n z,y)$:

We specialize diagram 13.1 to

$$\begin{array}{ccc}
H_*[C_*\mathcal{C}_{n+1}(2)\otimes C_*\mathcal{C}_{n+1}(p)\otimes_{\pi_p\times 1} C_*\mathcal{C}_{n+1}(1)\otimes(H_*X)^{p+1}] & \xrightarrow{\gamma_*\otimes 1} & H_*[C_*\mathcal{C}_{n+1}(p+1)\otimes_{\pi_p}(H_*X)^{p+1}] \\
\Big\downarrow 1\otimes t_*\otimes 1 & & \searrow \theta_{p+1*} \quad H_*X \\
H_*[C_*\mathcal{C}_{n+1}(2)\otimes C_*\mathcal{C}_{n+1}(p)\otimes_{\pi_p}(H_*X)^p\otimes C_*\mathcal{C}_{n+1}(1)\otimes H_*X] & \xrightarrow{1\otimes\theta_{p*}\otimes\theta_{1*}} & H_*[C_*\mathcal{C}_{n+1}(2)\otimes H_*X)^2] \nearrow \theta_{2*}
\end{array}$$

By Lemma 12.8, $\gamma_*(e_0\otimes e_i\otimes e_0\otimes x^p\otimes y) = 0$ if $0 \leq i < n(p-1)$. By commutativity of 13.1, we have $\lambda_n(Q^s x,y) = 0 = \lambda_n(\zeta_n z,y)$ for x, y, $z \in H_*X$.

<u>Remark 13.5</u>: We note that the internal Cartan formula implies that $0 = \lambda_n(x^p,y) = \lambda_n(Q^s x,y)$ for $n > 0$ and degree $(x) = 2s$ [degree $(x) = s$] as required for consistency.

We present, finally, an algebraic proof of the commutativity of the diagram in section 1. Recall that the Whitehead product $[\iota_{p+n+1},\iota_{q+n+1}]$ in $\pi_{p+q+2n+1} S^{p+n+1}\vee S^{q+n+1}$ may be described as the generator of the kernel of

$$\pi_{p+q+2n+1}(S^{p+n+1}\vee S^{q+n+1}) \to \pi_{p+q+2n+1}(S^{p+n+1}\times S^{q+n+1})$$

where ι_k denotes the fundamental class of S^k [13]. It is easy to see that $\pi_*(S^{p+n+1}\vee S^{q+n+1}) \to \pi_*(S^{p+n+1}\times S^{q+n+1})$ is an epimorphism and that $[\iota_{p+n+1},\iota_{q+n+1}]$ is given in the obvious way by the short exact sequence

$$0 \to \pi_r(S^s\times S^t, S^s\vee S^t) \to \pi_{r-1}(S^s\vee S^t) \to \pi_{r-1}(S^s\times S^t) \to 0$$

where $r = p+q+2n+2$, $s = p+n+1$ and $t = q+n+1$. We can also regard $[\iota_{p+n+1}, \iota_{q+n+1}]$ as given by the kernel of

$$\pi_{p+q+n}\,\Omega^{n+1}(S^s \vee S^t) \to \pi_{p+q+n}\Omega^{n+1}(S^s \times S^t) \to 0$$

which is $\pi_{p+q+n+1}(\Omega^{n+1}(S^s \times S^t), \Omega^{n+1}(S^s \vee S^t)) = \mathbf{Z}$. Clearly $H_i(S^s \times S^t, S^s \vee S^t) = \begin{cases} 0 & \text{if } i < p+q+2n+2 \\ \mathbf{Z} & \text{if } i = p+q+2n+2 \end{cases}$. We observe that

$\phi_*(\sigma_*)^n[\iota_{p+n+1}, \iota_{q+n+1}] = \phi_*\partial(1)$ [where $\sigma_*:\pi_*X \xrightarrow{\cong} \pi_{*-1}\Omega X$] since $\phi_*: \pi_{p+q+n+1}(\Omega^{n+1}(S^s \times S^t), \Omega^{n+1}(S^s \vee S^t)) \to H_{p+q+n+1}(\Omega^{n+1}(S^s \times S^t), \Omega^{n+1}(S^s \vee S^t)$ is an isomorphism, $\pi_{p+q+n+1}(\Omega^{n+1}(S^s{\times}S^t), \Omega^{n+1}(S^s{\vee}S^t)) \cong \mathbf{Z}$, and the following diagram commutes:

$$\begin{array}{ccccccccc} 0 \to & \mathbf{Z} & \xrightarrow{\partial} & \pi_{p+q+n}\Omega^{n+1}(S^s \vee S^t) & \to & \pi_{p+q+n}\Omega^{n+1}(S^s \times S^t) & \to 0 \\ & \downarrow{\phi_*} & & \downarrow{\phi_*} & & \downarrow{\phi_*} & \\ \cdots \to & \mathbf{Z} & \to & H_{p+q+n}\Omega^{n+1}(S^s \vee S^t) & \to & H_{p+q+n}\Omega^{n+1}(S^s \times S^t) & \to \cdots \end{array}$$

To calculate $\phi_*\partial(1)$, it suffices to calculate the kernel of the map

$$f:\ H_{p+q+n}\Omega^{n+1}(S^s \vee S^t) \to H_{p+q+n}\Omega^{n+1}(S^s \times S^t).$$

Clearly $\Omega^{n+1}(S^s \vee S^t) = \Omega^{n+1}\Sigma^{n+1}(S^p \vee S^q)$ and $\Omega^{n+1}(S^s \times S^t) = \Omega^{n+1}\Sigma^{n+1}S^p \times \Omega^{n+1}\Sigma^{n+1}S^q$. Under the inclusion of $(n+1)$-fold loop spaces $i:\ \Omega^{n+1}(S^s \vee S^t) \to \Omega^{n+1}(S^s \times S^t)$, it is clear that $i_*\iota_p = \iota_p$ and $i_*\iota_q = \iota_q$. By our calculations mod p, Ker f_* is generated by $\lambda_n(\iota_p, \iota_q)$, It is easy to see that Ker f_*, rationally, is generated by $\lambda_n(\iota_p, \iota_q)$. It is clear that $\lambda_n(\iota_p, \iota_q)$ must generate Ker f_* integrally. Hence $\phi_*\partial(1) = \pm\lambda_n(\iota_p,\iota_q)$.

To check the correct sign, we recall that Samelson [21] has shown that

$$\phi_*\sigma_*[\iota_{p+1},\iota_{q+1}] = \iota_p * \iota_q - (-1)^{pq}\iota_q * \iota_p = \lambda_0(\iota_p,\iota_q).$$

Since the Hurewicz map commutes with σ_*, it must follow that $\phi_*\partial(1) = \lambda_n(\iota_p,\iota_q)$.

The diagram relating the Whitehead product and λ_n in section 1 follows directly, by naturality.

Remark. Our arguments for an (n+1)-fold loop space should be compared to Samelson's [21] for a first loop space. Of course we are using Samelson's sign convention for the Whitehead product here.

14. An algebraic lemma and a sketch of methods

Before proceeding to details, we sketch the methods used in the following three sections. Since the diagram

$$\begin{array}{ccc} \mathcal{C}_n(j) \times X^j & \xrightarrow{\theta_{n,j}} & \\ \Big\downarrow{\scriptstyle \sigma_n \times 1^j} & & X \\ \mathcal{C}_{n+1}(j) \times X^j & \xrightarrow[\theta_{n+1,j}]{} & \end{array}$$

equivariantly homotopy commutes [G; 1.4], all properties of operations derived from the $\mathcal{C}_n$-action must hold a fortiori for the evident induced operations associated to the $\mathcal{C}_{n+1}$-action. However, since X is a $\mathcal{C}_{n+1}$-space, all terms involving λ_{n-1} vanish; new operations are born from the $\mathcal{C}_{n+1}$-action which are not present from the $\mathcal{C}_n$-action, namely, λ_n, ξ_n, and, if $p > 2$, ζ_n. Clearly the lion's share of our work consists of analyzing the properties of these new operations.

Most of the properties of the λ_n have already been determined. The properties of $\xi_n X$ and $\zeta_n X$ follow, up to error terms involving the λ_n, from the stable results for $Q^{\frac{n+q}{2}} x[Q^{n+q}x]$ and $\beta Q^{\frac{n+q}{2}} x$, $x \in H_q X$. We then apply several ad hoc tricks to calculate the error terms precisely.

To determine our formulas, it suffices, by Lemmas 3.4 and 3.5, to check them for $H_* C_{n+1} X$ and $H_* \Omega^{n+1} \Sigma^{n+1} X$. By Theorems 3.1 and 3.2, $\alpha_{n+1*}\colon H_* C_{n+1} X \to H_* \Omega^{n+1} \Sigma^{n+1} X$ is a monomorphism of allowable $AR_n \Lambda_n$-Hopf algebras. Hence we need only verify our formulas for

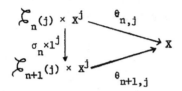

$H_*\Omega^{n+1}\Sigma^{n+1}X$. Here we show that the unstable error terms lie in a submodule, M_X, of GW_nH_*X if $n > 0$ and of W_0H_*X otherwise. We construct a simple "external" operation which detects the elements of M_X.

We begin with a definition. Let M_X be the $\mathbf{Z}_p$-subspace of GW_nH_*X, $n > 0$ or of W_0H_*X spanned by elements of the form

$$A * \lambda_I^m$$

where (1) λ_I is a basic λ_n-product for some fixed I, (2) $p\nmid m$ and (3) A has no additional factors λ_I. We consider the maps

$$GW_n(j_*)\colon\ GW_nH_*X \longrightarrow GW_nH_*(X \vee S^\ell) \quad \text{or}$$

$$W_0(j_*)\colon\ W_0H_*X \longrightarrow W_0H_*(X \vee S^\ell)$$

induced by the inclusion $j\colon X \to X \vee S^\ell$, $\ell > 0$. Clearly $GW_n(j_*)$ and $W_0(j_*)$ are monomorphisms. By abuse of notation, we identify M_X with $GW_n(j_*)(M_X)$ or $W_0(j_*)(M_X)$. Let ι denote the image of the fundamental class of S^ℓ via the standard inclusion

$$H_*S^\ell \to GW_nH_*(X \vee S^\ell).$$

Our main algebraic result is

<u>Lemma 14.1</u>. The homomorphism defined by

$$\lambda_n(-,\iota)|_{M_X}\colon\ M_X \longrightarrow GW_nH_*(X \vee S^\ell) \quad \text{if } n > 0 \text{ or}$$

$$\lambda_0(-,\iota)|_{M_X}\colon\ M_X \longrightarrow W_0H_*(X \vee S^\ell)$$

is a monomorphism.

<u>Proof</u>: We first consider the case $n > 0$. Let $\{A_1 * \lambda_{I_1}^{m_1}, \ldots, A_k * \lambda_{I_k}^{m_k}\}$ be an arbitrary finite set of linearly independent elements in M_X. It suffices to show that $\lambda_n(-, \iota)$ is a monomorphism when restricted to the subspace generated by these elements. Fix one of the $\lambda_{I_s} = \lambda$. Then by definition of M_X, we may write each term $A_\ell * \lambda_{I_\ell}^{m_\ell}$, $\ell = 1, \ldots, k$ as $\hat{A}_\ell * \lambda^{n_\ell}$ where $\hat{A}_\ell$ has no factors of λ, $n_\ell \geq 0$. Clearly $\{\hat{A}_1 * \lambda^{n_1}, \ldots, \hat{A}_k * \lambda^{n_k}\}$ is a set of linearly independent vectors which span the same subspace as $\{A_1 * \lambda_{I_1}^{m_1}, \ldots, A_k * \lambda_{I_k}^{m_k}\}$.

Suppose $\lambda_n(V, \iota) = 0$ where $V = \sum_{i=1}^{k} a_i \hat{A}_i * \lambda^{n_i}$. By the internal Cartan formula for λ_n [Theorem 1.2], we have

$$\lambda_n(v, \iota) \equiv \sum_{i=1}^{k} n_i a_i \hat{A}_i * \lambda^{n_i - 1} * \lambda n(\lambda, \iota) \quad \text{modulo}$$

terms which have no factor of the form $\lambda_n(\lambda, \iota)$. By definition of $GW_n H_* X$, it is clear that $n_j a_j = 0$, $j = 1, \ldots, k$. But for our fixed choice of $\lambda = \lambda_{I_s}$, we have $n_s = m_s$ and $p \nmid m_s$. Hence $a_s \equiv 0 \bmod p$. It follows easily that $a_j = 0$, $j = 1, \ldots, k$ by a similar argument.

The case $n = 0$ is trivial.

15. The formula $\lambda_n(x,\xi_n y) = ad_n^p(y)(x)$ and commutation with homology suspension

We present an amusing proof of the formulas $\sigma_*\xi_n = \xi_{n-1}\sigma_*$, and $\sigma_*\zeta_n = -\zeta_{n-1}\sigma_*$ if $n > 1$. Our proof does not require construction of chain level operations and the requisite explicit equivariant chain approximation for $C_*\mathscr{E}_{n+1}(p)$. The ingredients are that (1) $\sigma_*\xi_n \equiv \xi_{n-1}\sigma_*$ modulo error terms generated by Browder operations, (2) $\sigma_*\zeta_n \equiv -\zeta_{n-1}\sigma_*$ as in (1) if $n > 1$, (3) the errors are approximated by an application of [G; Theorem 6.1], and (4) the formula $\lambda_n(x,\xi_n y) = ad_n^p(y)(x)$ detects the possible error terms. Of course, this method requires a derivation of the formula for $\lambda_n(x,\xi_n y)$ which is logically independent of the fact that the top operation commutes with suspension on the nose.

Recall that the adjoint of the identity map on $\Omega^{n+1}X$ yields $\phi_{n+1}\colon \Sigma^{n+1}\Omega^{n+1}X \to X$ and a map of fibrations:

$$\begin{array}{ccc}
\Omega^{n+1}\Sigma^{n+1}\Omega^{n+1}X & \xrightarrow{\Omega^{n+1}\phi_{n+1}(1)} & \Omega^{n+1}X \\
\downarrow\cap & & \downarrow\cap \\
P\Omega^{n}\Sigma^{n+1}\Omega^{n+1}X & \xrightarrow{P\Omega^{n}\phi_{n+1}(1)} & P\Omega^{n}X \\
\downarrow P_n & & \downarrow P_n \\
\Omega^{n}\Sigma^{n+1}\Omega^{n+1}X & \xrightarrow{\Omega^{n}\phi_{n+1}(1)} & \Omega^{n}X
\end{array}$$

Since $\Omega^{n+1}\phi_{n+1}(1)_*$ is an epimorphism, it suffices to verify our results in the left-hand fibration.

Let I_{n-1} denote the ideal of $GW_{n-1}H_*(\Sigma\Omega^{n+1}X)$ generated by Browder operations of weight greater than one and the iterations of the operations $\beta^\varepsilon Q^s$ and $\beta^\varepsilon \xi_{n-1}$ on these Browder operations.

<u>Lemma 15.1.</u> (1) $\sigma_*\xi_n \equiv \xi_{n-1}\sigma_* \; (I_{n-1})$ if $n \geq 1$,

(2) $\sigma_*\zeta_n \equiv -\zeta_{n-1}\sigma_*(I_{n-1})$ if $n \geq 2$, and

(3) $\sigma_*\zeta_1 \equiv 0 \; (I_0)$.

<u>Proof</u>: Let $j_n(X) : \Omega^n\Sigma^n X \to QX$ be the standard inclusion of $\Omega^n\Sigma^n X$ in $QX = \varinjlim_n \Omega^n\Sigma^n X$. We now recall that the following commutative diagram yields a map of fibrations:

$$\begin{array}{ccc}
\Omega^{n+1}\Sigma^{n+1}\Omega^{n+1}X & \xrightarrow{\Omega(j_n(\Sigma\Omega^{n+1}X))} & \Omega Q\Sigma\Omega^{n+1}X = Q\Omega^{n+1}X \\
\downarrow \cap & & \downarrow \cap \\
P\Omega^n\Sigma^{n+1}\Omega^{n+1}X & \xrightarrow{P(j_n(\Sigma\Omega^{n+1}X))} & PQ\Sigma\Omega^{n+1}X \\
\downarrow P_n & & \downarrow P \\
\Omega^n\Sigma^{n+1}\Omega^{n+1}X & \xrightarrow{j_n(\Sigma\Omega^{n+1}X)} & Q\Sigma\Omega^{n+1}X
\end{array}$$

Since the operations commute with suspension in the right hand fibration [A; §3], we know that our formulas in 15.1 are correct modulo the kernel of $j_n(\Sigma\Omega^{n+1}X)_*$. But then our calculations of $H_*\Omega^{n+1}\Sigma^{n+1}X$ in section 3 are correct at least as algebras and visibly $I_{n-1} = \ker j_n(\Sigma\Omega^{n+1}X)_*$.

By 15.1, $\sigma_*\xi_n x = \xi_{n-1}\sigma_* x + \Delta$, $\sigma_*\zeta_n x = -\zeta_{n-1}\sigma_* x + \Gamma$ if $n > 1$, and $\sigma_*\zeta_1 x = \Phi$. We estimate Δ, Γ, and Φ; the crucial point being that these terms have no non-trivial summands of Dyer-Lashof operations (and in particular, no pth powers).

Lemma 15.2. For the fibration $\Omega^{n+1}\Sigma^{n+1}X \to P\Omega^n\Sigma^{n+1}X \to \Omega^n\Sigma^{n+1}X$, Δ, Γ, and Φ are given by $\sum \lambda_{I_1} * \dots * \lambda_{I_j}$, $w(\lambda_{I_k}) > 0$, $w(\lambda_{I_1}) + \dots + w(\lambda_{I_j}) \leq p$, λ_{I_k} is a λ_{n-1}-product on classes in $H_*\Sigma X$, and if $\lambda_{I_1} = \dots = \lambda_{I_j}$, then $j < p$.

Proof: By constructions 2.4 and 6.6. [G], the spaces $C_{n+1}X$, $E_{n+1}(TX,X)$, and $C_n\Sigma X$ are filtered. We observe that the inclusion $C_{n+1}X \longrightarrow E_{n+1}(TX,X)$ and the projection $\pi_{n+1} : E_{n+1}(TX,X) \to C_n\Sigma X$ restrict to maps of filtered spaces. Now consider the following diagram whose lower left hand rectangle commutes by the above observation,

$$
\begin{array}{ccccccc}
\mathcal{C}_{n+1}(p)\times X^{p} & & & \xrightarrow{\quad 1\times\eta^{p}_{n+1}\quad} & & & \mathcal{C}_{n+1}(p)\times(\Omega^{n+1}\Sigma^{n+1}X)^{p} \\
\downarrow & & & & & & \downarrow{\scriptstyle \theta_{n+1}} \\
F_{p}C_{n+1}X & \xrightarrow{\subset} & C_{n+1}X & \xrightarrow{C_{n+1}\eta_{n+1}} & C_{n+1}\Omega^{n+1}\Sigma^{n+1}X & \xrightarrow{\theta_{n+1}} & \Omega^{n+1}\Sigma^{n+1}X \\
\downarrow{\scriptstyle i} & & \downarrow{\scriptstyle \cap} & & \downarrow{\scriptstyle \cap} & & \downarrow{\scriptstyle \cap} \\
F_{p}E_{n+1}(TX,X) & \longrightarrow & E_{n+1}(TX,X) & \xrightarrow{E_{n}\tilde{\eta}_{n}} & E_{n+1}(P\Omega^{n}\Sigma^{n+1}X,\Omega^{n+1}\Sigma^{n+1}X) & \xrightarrow{\tilde{\theta}_{n}} & P\Omega^{n}\Sigma^{n+1}X \\
\downarrow{\scriptstyle \pi_{n+1}} & & \downarrow{\scriptstyle \pi_{n+1}} & & \downarrow{\scriptstyle p_{n}} & & \downarrow{\scriptstyle p} \\
F_{p}C_{n}\Sigma X & \longrightarrow & C_{n}\Sigma X & \xrightarrow{C_{n}\eta_{n}} & C_{n}\Omega^{n}\Sigma^{n+1}X & \xrightarrow{\theta_{n}} & \Omega^{n}\Sigma^{n+1}X
\end{array}
$$

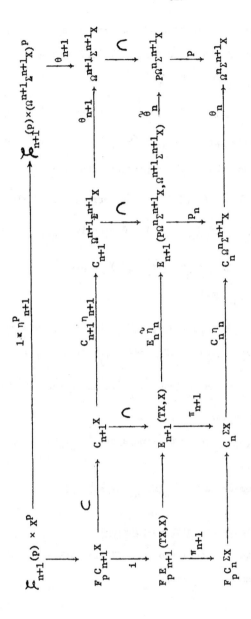

The top rectangle commutes by definition of the action θ_{n+1}; the rest of commutativity follows from [G; 6.9 and 6.11]. By definition $\xi_n(x) = (-1)^{\frac{n+q}{2}} \nu(q)\theta_*(e_{n(p-1)} \otimes x^p)[\theta_*(e_n \otimes x^2)]$ and $\zeta_n(x) = (-1)^{\frac{n+q}{2}} \nu(q)\theta_*(e_{n(p-1)-1} \otimes x^p)$ which, by commutativity of the diagram is just $(-1)^{\frac{n+q}{2}} \nu(q)\theta_{n+1*}(C_{n+1}\eta_{n+1})_*(e_{n(p-1)-\varepsilon} \otimes x^p)$, $\varepsilon = 0,1$. Let $D_\varepsilon \in C_*F_pC_{n+1}X$ be such that D_ε represents the cycle $e_{n(p-1)-\varepsilon} \otimes x^p$. Because $F_pE_{n+1}(TX,X)$, is contractible [G; 7.1], $i_*D_\varepsilon = \partial C_\varepsilon$ for some $C_\varepsilon \in C_*F_pE_{n+1}(TX,X)$. Obviously $\pi_{n+1*}C_\varepsilon$ is a cycle such that $(-1)^{\frac{n+q}{2}} \nu(q)\theta_{n*}(C_n\eta_n)_*(\pi_{n+1*}C_\varepsilon)$ represents $\sigma_*((-1)^{\frac{n+q}{2}} \nu(q)\theta_{n+1*}(e_{n(p-1)-\varepsilon} \otimes x^p))$.

By section 4, we see that $\theta_{n*}(C_n\eta_n)_*(H_*F_pC_n\Sigma X)$ is spanned by classes of the form $\beta^\xi Q^s y$, $\xi_{n-1}y$, $\zeta_{n-1}y$, and $\lambda_{I_1} * \ldots * \lambda_{I_j}$, $w(\lambda_{I_1}) + \ldots + w(\lambda_{I_j}) \leq p$, $y \in H_*\Sigma X$ and λ_{I_k} is an iterated Browder operation on classes from $H_*\Sigma X$.

Since $\sigma_*\xi_n x$ and $\sigma_*\zeta_n x$ are in the image of $H_*F_pC_n\Sigma X$, the lemma follows [see [A; §3]].

We assume for the moment that $\lambda_n(x,\xi_n y) = ad_n^p(y)(x)$.

Proof of Theorem 1.4. (Commutation with suspension):

Let j denote the standard inclusion of ΣX in $\Sigma(X \vee S^\ell)$, ℓ fixed. Obviously j induces a map of fibrations

$$\begin{array}{ccc}
\Omega^{n+1}\Sigma^{n+1}X & \xrightarrow{\Omega^{n+1}\Sigma^{n}(j)} & \Omega^{n+1}\Sigma^{n+1}(X \vee S^{\ell}) \\
\downarrow \cap & & \downarrow \cap \\
P\Omega^{n}\Sigma^{n+1}X & \xrightarrow{P\Omega^{n}\Sigma^{n}(j)} & P\Omega^{n}\Sigma^{n+1}(X \vee S^{\ell}) \\
\downarrow p_n & & \downarrow p_n \\
\Omega^{n}\Sigma^{n+1}X & \xrightarrow{\Omega^{n}\Sigma^{n}\ (j)} & \Omega^{n}\Sigma^{n+1}(X \vee S^{\ell}).
\end{array}$$

Let ι denote the image of the fundamental class of S^{ℓ} in $H_*\Omega^{n+1}\Sigma^{n+1}(X \vee S^{\ell})$.

With these preliminaries, we prove Theorem 1.4 by induction on n. If $n = 0$, there is nothing to prove. Hence we begin with the case $n = 1$. Then

(i) $\sigma_*\lambda_1(\iota,\xi_1 x) = \lambda_0(\sigma_*\iota,\sigma_*\xi_1 x)$,

(ii) $\lambda_0(\sigma_*\iota,\sigma_*\xi_1 x) = \lambda_0(\sigma_*\iota,\xi_0\sigma_* x) + \lambda_0(\sigma_*\iota,\Delta)$, and

(iii) $\sigma_*\lambda_1(\iota,\xi_1 x) = \sigma_* ad_1^p(x)(\iota) = ad_0^p(\sigma_* x)(\sigma_*\iota) = \lambda_0(\sigma_*\iota,\xi_0\sigma_* x)$.

Together, (i)-(iii) yield $\lambda_0(\sigma_*\iota,\Delta) = 0$. By the definition of M_X [see section 15] and Lemma 15.2, we have $\Delta \in M_X$. By Lemma 14.1 and the fact that $\lambda_0(\sigma_*\iota,\Delta) = 0$, we have $\Delta = 0$. Since $\lambda_1(\iota,\zeta_1 x) = 0$ by Theorem 1.3 (the proof being in section 13), it follows from 14.1 and 15.2 that $\sigma_*\zeta_1 x = 0$.

To check the assertions (1)-(3) of Theorem 1.4, we observe that

(iv) $\sigma_*\lambda_n(\iota,\xi_n x) = \lambda_{n-1}(\sigma_*\iota,\sigma_*\xi_n x)$,

(v) $\lambda_{n-1}(\sigma_*\iota,\sigma_*\xi_n x) = \lambda_{n-1}(\sigma_*\iota,\xi_{n-1}\sigma_* x) + \lambda_{n-1}(\sigma_*\iota,\Delta)$,

(vi) $\sigma_*\lambda_n(\iota,\xi_n x) = \sigma_* ad_n^p(x)(\iota) = ad_{n-1}^p(\sigma_* x)(\sigma_*\iota) = \lambda_{n-1}(\sigma_*\iota,\xi_{n-1}\sigma_* x)$

(vii) $0 = \sigma_*\lambda_n(\iota,\zeta_n x) = \lambda_{n-1}(\sigma_*\iota,\sigma_*\zeta_n x)$, and

(viii) $\lambda_{n-1}(\sigma_*\iota,\sigma_*\zeta_n x) = \lambda_{n-1}(\sigma_*\iota,\Gamma)$.

Together, (iv)-(vi) and (vii)-(viii) yield $\lambda_{n-1}(\sigma_*\iota,\Delta) = \lambda_{n-1}(\sigma_*\iota,\Gamma) = 0$. By definition of M_X, Lemmas 14.1 and 15.2, we have $\Delta = \Gamma = 0$.

Now let y be a class in $H_*\Omega^2 X$ represented by the cycle b and let a be a chain in $C_* P\Omega X$ whose boundary is $i_*(b)$. ($i\colon \Omega^2 X \to P\Omega X$) Let C_1 be the chain previously constructed whose boundary is the cycle $i_*\theta_*(e_{p-2} \otimes b^p)$. It is not hard to see that our previous construction together with the results of [A; p 171] imply that

$C_1 = \theta_*(ke_0 \otimes a^{p-1} \otimes b)$ + terms of lower filtration where $k \in \mathbf{Z}_p$, e_0 is a zero dimensional chain in $C_*\mathcal{C}_2(p)$, and a and b are as given above. By the hypotheses in [A; 3.4] we have that C_1 represents $k\{p_{2*}i_*a^{p-1}\} \otimes y$ in E^2 of the Serre spectral sequence for the path fibration. Theorem 1.4(4) follows directly.

Finally, we derive the formula $\lambda_n(y,\xi_n x) = ad_n^p(x)(y)$ using only the approximate information of Lemma 15.1.

<u>Theorem 1.3(4)</u>. $\lambda_n(y,\xi_n x) = ad_n^p(x)(y)$.

<u>Proof:</u> To take advantage of calculations in section 12, we calculate $\lambda_n(\xi_n x,y)$ and use the formula $\lambda_n(x,y) = (-1)^{|x||y|+1+n(|x|+|y|+1)}\lambda_n(y,x)$.

The definitions of λ_n and ξ_n give that

$$\lambda_n(\xi_n x,y) = (-1)^{n|\xi_n x|+1+\frac{n+|x|}{2}}\nu(|x|)\theta_{2*}(\theta_{p*}(e_{n(p-1)} \otimes x^p) \otimes y).$$

$$[\text{or } \theta_{2*}(\theta_{2*}(e_n \otimes x^2) \otimes e_0 \otimes y) \text{ if } p=2]$$

Hence we calculate $\lambda_n(\xi_n x,y)$ via the commutative diagram

$$\begin{array}{ccccc}
\mathcal{C}_{n+1}(2) \times \mathcal{C}_{n+1}(p) \times \mathcal{C}_{n+1}(1) \times X^{p+1} & \xrightarrow{\gamma\times 1} & \mathcal{C}_{n+1}(p+1) \times X^{p+1} & \xrightarrow{\theta_{p+1}} & \\
\big\downarrow 1 \times t \times 1 & & & & X . \\
\mathcal{C}_{n+1}(2) \times \mathcal{C}_{n+1}(p) \times X^p \times \mathcal{C}_{n+1}(1) \times X & \xrightarrow{1\times\theta_p\times\theta_1} & \mathcal{C}_{n+1}(2) \times X^2 & \xrightarrow{\theta_2} &
\end{array}$$

By Lemma 12.8(2) we have that $\gamma_*(\iota \otimes e_{n(p-1)} \otimes e_0)$ is in the image of the map $H_{np}\mathcal{C}_{n+1}(p+1) \to H_{np}\dfrac{\mathcal{C}_{n+1}(p+1)}{\pi_p}$. By Theorem 12.3 and the fact that the diagram

$$\begin{array}{ccc}
\mathcal{C}_{n+1}(p+1) \times X^{p+1} & \xrightarrow{\theta_{p+1}} & \\
\big\downarrow & & X \\
\mathcal{C}_{n+1}(p+1) \times_{\pi_p} X^{p+1} & \xrightarrow{\theta_{p+1}} &
\end{array}$$

commutes, we have that $\theta_{p+1*}(\gamma_* \otimes 1)(\iota \otimes e_{n(p-1)} \otimes e_0 \otimes x^p \otimes y)$ is given by p-fold iterates of the λ_n on p occurrences of x and one occurrence of y. Since $\lambda_n(x,x) = 0$ if $\xi_n x$ is defined, it follows that

$$\theta_{p+1*}(\gamma_* \otimes 1)(\iota \otimes e_{n(p-1)} \otimes e_0 \otimes x^p \otimes y) = k' \mathrm{ad}_n^p(x)(y)$$

for some fixed constant k'. Hence $\lambda_n(y,\xi_n x) = k\ \mathrm{ad}_n^p(x)(y)$ for some fixed constant k.

To calculate k, we consider the path-space fibration $\Omega^{n+1}X \to P\Omega^n X \to \Omega^n X$ for appropriate X. Let $X = \Sigma^{n+1}S^m$ for $n+m$ even and m large. Let ι_m denote the image of the fundamental class of S^m in $H_*\Omega^{n+1}\Sigma^{n+1}S^m$. By Lemma 15.1(1), $\sigma_*\xi_n\iota_m = \xi_{n-1}\sigma_*\iota_m + \Delta$. Since the Browder operations in $H_*\Omega^n\Sigma^{n+1}S^m$ are all trivial when $n+m$ is even, it follows that $\sigma_*\xi_n\iota_m = \xi_{n-1}\sigma_*\iota_m$. Now let $X = \Sigma^{n+1}(S^r \vee S^m)$, $r > 0$. By naturality, we have the formula $\sigma_*\lambda_n(\iota_r,\xi_n\iota_m) = \lambda_{n-1}(\sigma_*\iota_r,\xi_{n-1}\sigma_*\iota_m)$. That $k = 1$ follows immediately by induction on n and the formula $\lambda_0(y,\xi_0 x) = \mathrm{ad}_0^p(x)(y)$ for a first loop space. [see Jacobson [14] for the calculations in the case $n = 0$.]

16. Additional properties of the $\beta^{\varepsilon}Q^s$, ξ_n, and ζ_n

In this section, we determine the remaining properties of Theorems 1.1 and 1.3 except for the Nishida relations. Properties (1)-(3) of Theorem 1.1 and (1) of Theorem 1.3 follow immediately from the definitions of the operations. Additional properties are proved in the following order: (1) deviation from linearity of ξ_n and the linearity of Q^s, (2) Cartan formulas, (3) Adem relations, and (4) commutation with conjugation.

We require an observation due to Steenrod.

Observation 16.1. Let H_*X have homogeneous basis $\{x_i\}$. Then

$H_*(C_*\mathcal{C}_{n+1}(p) \otimes_{\pi_p} (C_*X)^p) \cong H_*(C_*\mathcal{C}_{n+1}(p) \otimes_{\pi_p} (H_*X)^p)$ and

$H_*C_*\mathcal{C}_{n+1}(p) \otimes_{\pi_p} (H_*X)^p \cong H_*\underline{\mathcal{C}_{n+1}(p)} \otimes A \oplus H_*\mathcal{C}_{n+1}(p) \otimes B$, additively,
$\qquad\qquad\qquad\qquad\qquad\qquad\qquad\pi_p$

where A has basis $\{x \otimes \ldots \otimes x \mid x \in \{x_i\}\}$ and B has a basis

$\{x_{i_{\sigma(1)}} \otimes \ldots \otimes x_{i_{\sigma(p)}} \mid i_1 \leq i_2 \leq \ldots \leq i_p,\ i_1 < i_p,\ \sigma \in K\}$ and K

is a complete set of distinct left coset representatives for π_p in Σ_p. (See [8] or 4.2).

We next show

Proof of Theorem 1.3(5), and the linearity of $\beta^{\varepsilon}Q^s$, the formula for $\xi_n(x+y)$:

By Jacobson [12], we know that $\xi_0(x+y) = \xi_0 x + \xi_0 y + \sum d_0^i(y)(x)$. To calculate $\xi_n(x+y)$, we observe that

$e_i \otimes (x+y)^p = e_i \otimes x^p + e_i \otimes y^p + e_i \otimes NF(x,y)$ by 16.1 where

$N = 1 + \sigma + \ldots + \sigma^{p-1}$ and $F(x,y)$ is a function of x and y. Since e_i is a chain in $C_*\mathcal{C}_{n+1}(p)$ which projects to a cycle in $C_*\mathcal{C}_{n+1}(p)/\pi_p$, the transfer homomorphism shows that Ne_i is a cycle in $C_*\mathcal{C}_{n+1}(p)$. If $i = n(p-1)$, then $e_i \otimes (x+y)^p = e_i \otimes x^p + e_i \otimes y^p + \gamma \otimes F(x,y)$ where γ is a cycle in $C_*\mathcal{C}_{n+1}(p)$ of degree $n(p-1)$. By Theorem 12.3, $\theta_*(\{\gamma\} \otimes f(x,y)) = \sum c_{\gamma,\mu} ad_n(x_{\mu(1)}) \ldots ad_n(x_{\mu(p-1)})(x_{\mu(p)})$, $\mu \in \Sigma_p$ and $x_i = x$ or y. Since this class suspends non-trivially to $H_*\Omega^n\Sigma^{n+1}X$, the obvious induction argument yields the formula $\xi_n(x+y) = \xi_n(x) + \xi_n(y) + \sum_{i=1}^{p-1} d_n^i(x)(y)$.

The linearity of $\beta^\varepsilon Q^s$ follows directly since $\lambda_j(x,y) = 0$ if $j < n$.

Proof of Theorems 1.1(4) and 1.3(2); the Cartan formulas:

We first determine the external and diagonal Cartan formulas and then derive the internal Cartan formula which, like its analogue for $\lambda_n(xy,zw)$, has "extra" terms not predicted by the external formula. These extra terms arise because the multiplication in X is not a morphism of $\mathcal{C}_{n+1}$-spaces, but only of $\mathcal{C}_n$-spaces and, of course, these terms are unstable.

By Theorem 5.4, we read off the coproduct in $H_*(\mathcal{C}_{n+1}(p)/\pi_p; \mathbb{Z}_p)$ on the classes e_i. If $p = 2$, $\psi e_i = \sum_{j+k=i} e_j \otimes e_k$. If $p > 2$, then

$$\psi(e_{2i+1}) = \sum_{j+k=2i+1} e_j \otimes e_k + \sum x \otimes y \quad \text{and} \quad \psi(e_{2i}) = \sum_{j+k=i} e_{2j} \otimes e_{2k} + \sum x' \otimes y'$$

where x, y, x' and y' are never multiples of the classes α_* or λ_*, the classes in the image of $H_*(\mathcal{C}_{n+1}(p);\mathbb{Z}_p) \to H_*(\mathcal{C}_{n+1}(p)/\Sigma_p;\mathbb{Z}_p(q))$. If X and Y are $\mathcal{C}_{n+1}$-spaces, we calculate the external Cartan formula with the above information and the method of [A; 2.6]. That is, the map, $\theta_p:\mathcal{C}_{n+1}(p)\times_{\pi_p} X^p \to X$, factors through $\mathcal{C}_{n+1}(p)\times_{\Sigma_p} X^p$ and the diagram below commutes by definition of the action on $X\times Y$:

$$\begin{array}{ccc}
\mathcal{C}_{n+1}(p)\times(X\times Y)^p & \xrightarrow{\ \theta_p\ } & X\times Y \\
\Big\downarrow{\scriptstyle \psi\times 1} & & \Big\uparrow{\scriptstyle \theta_p\times\theta_p'} \\
\mathcal{C}_{n+1}(p)\times\mathcal{C}_{n+1}(p)\times(X\times Y)^p & \xrightarrow{\ 1\times u\ } & \mathcal{C}_{n+1}(p)\times X^p\times\mathcal{C}_{n+1}(p)\times Y^p
\end{array}$$

The external Cartan formula follows. Similarily, the diagonal Cartan formula is immediate.

We calculate $\xi_n(x*y)$ in $H_*\Omega^{n+1}\Sigma^{n+1}X$. By the calculation of $H_*\Omega^{n+1}\Sigma^{n+1}X$, it is immediate that
$$\xi_n(x*y) = \sum_{\substack{r+s=\frac{n+|x|+|y|}{2} \\ [r+s=n+|x|+|y|]}} Q^r x * Q^s y + X_{(x,y)}$$

where $X_{(x,y)}$ is a sum of unstable error terms. We recall that the diagram

$$\begin{array}{ccccc}
\mathcal{C}_{n+1}(p)\times\mathcal{C}_{n+1}(2)^p\times X^{2p} & \xrightarrow{\ \gamma\times 1\ } & \mathcal{C}_{n+1}(2p)\times X^{2p} & \xrightarrow{\ \theta_{n+1,2p}\ } & X \\
\Big\downarrow{\scriptstyle 1\times t\times 1} & & & & \Big\| \\
\mathcal{C}_{n+1}(p)\times(\mathcal{C}_{n+1}(2)\times X^2)^p & \xrightarrow{\ 1\times\theta^p_{n+1,2}\ } & \mathcal{C}_{n+1}(p)\times X^p & \xrightarrow{\ \theta_{n+1,p}\ } & X
\end{array}$$

commutes and that $\xi_n(x*y) = (-1)^{\frac{n+|x|+|y|}{2}} \nu(|x|+|y|)\, \theta_{p*}\gamma_*(e_{n(p-1)} \otimes e_0^p \otimes (x \otimes y)^p)$ $[\xi_n(x*y) = \theta_{2*}\gamma_*(e_n \otimes (x \otimes y)^2)]$. It follows that $X_{(x,y)}$ is a sum of unstable operations on p occurrences of x and p occurrences of y. By the definition of our operations, it follows that $X_{(x,y)}$ is a sum of elements in M_X (of 14.1) and possibly $k\beta^\varepsilon Q^s \lambda_n(x,y)$ $k \in \mathbb{Z}_p$. But $\beta^\varepsilon Q^s \lambda_n(x,y)$ has degree higher than $\xi_n(x*y)$. Hence $X_{(x,y)} \in M_X$. If $n > 0$, we write $X_{(x,y)} = \sum_{\substack{0 \le i,j \le p \\ 0 \le i+j \le p}} x^i y^j \Gamma_{ij}$.

We calculate some of the Γ_{ij} by the formulas

(i) $\lambda_n(z, \xi_n(x*y)) = \mathrm{ad}_n^p(x*y)(z)$,

(ii) $\lambda_n(z, \xi_n(x*y)) = \lambda_n\Big(z, \sum_{\substack{r+s = \frac{n+|x|+|y|}{2} \\ [r+s = n+|x|+|y|]}} Q^r x * Q^s y + \sum_{\substack{0 \le i,j \le p \\ 0 \le i+j \le p}} x^i y^j \Gamma_{ij}\Big)$,

(iii) the internal Cartan formula for λ_n, and

(iv) Lemma 14.1.

We list some values for Γ_{ij}. If $p = 2$, $\Gamma_{0,0} = \Gamma_{1,0} = \Gamma_{0,1} = 0$, and $\Gamma_{1,1} = \mathrm{ad}_n(x)(y)$; if $p > 2$, and $|x|$, $|y|$ are both even, then $\Gamma_{1,1} = [\mathrm{ad}_n(x)(y)]^{p-1}$, $\Gamma_{i,p-i} = d_n^i(x)(y)$, $i \ne 0,p$, where $d_n^i(x)(y)$ have already been defined.

We also have the following additional formulas

$$
\begin{aligned}
0 = \lambda_n(x*y) \quad \xi_n(x*y)) &= x^p y\, \mathrm{ad}_n^p(y)(x) + x y^p\, \mathrm{ad}_n^p(x)(y) \\
&+ \sum_{0 \le i,j < p} x^i y^{j+1}\, \mathrm{ad}_n(\Gamma_{ij})(x) + x^{i+1} y^j \mathrm{ad}_n(\Gamma_{ij})(y) \\
&+ \sum_{0 \le i,j < p} (i-j) x^i y^j \Gamma_{ij} \mathrm{ad}_n(y)(x)
\end{aligned}
$$

and the resulting simplification in GW_nH_*X,

$$ad_n(\Gamma_{i,j-1})(x) + ad_n(\Gamma_{i-1,j})(y) + (i-j)\Gamma_{ij}\ ad_n(y)(x) = 0.$$

This last formula allows an inductive calculation of the Γ_{ij} in terms of $\Gamma_{i,i}$. Other coefficients may be calculated for different values of i and j, and in case some of n, x, and y are odd. It does not appear fruitful to specify the Γ_{ij} more directly.

The case $n = 0$ is deliberately omitted; the reader may wish to observe the amusing complications here.

<u>Proof of Theorems 1.1(5) and 1.3(2),</u> the Adem relations:

We initially consider the following commutative diagram:

$$\begin{array}{ccc}
\dfrac{\mathcal{C}_{n+1}(p) \times \mathcal{C}_{n+1}(p)^p}{\pi_p \int \pi_p} & \xrightarrow{\quad\gamma\quad} & \dfrac{\mathcal{C}_{n+1}(p^2)}{\Sigma_{p^2}} \\
\Big\downarrow{\scriptstyle \sigma \times \sigma^p} & & \Big\downarrow{\scriptstyle \sigma} \\
\dfrac{\mathcal{C}_{\infty}(p) \times \mathcal{C}_{\infty}(p)^p}{\pi_p \int \pi_p} & \xrightarrow{\quad\gamma\quad} & \dfrac{\mathcal{C}_{\infty}(p^2)}{\Sigma_{p^2}}
\end{array}$$

where σ is the equivariant inclusion of $\mathcal{C}_{n+1}(p)$ in $\mathcal{C}_{\infty}(p)$.

Assume, for the moment, that $p > 2$. Let

$$\begin{aligned}\Gamma = &\sum_k (-1)^k \nu(s)(k,[s/2]-pk) e_{r+(2pk-s)(p-1)} \otimes e^p_{s-2k(p-1)} \\ &-\delta(r)\delta(s-1)\sum_k (-1)^k \nu(s-1)(k,[\tfrac{s-1}{2}]-pk) e_{r+p+(2pk-s)(p-1)} \otimes e^p_{s-2k(p-1)-1}\end{aligned}$$

and

$$\Delta = (-1)^{rs+mq}(\sum_j (-1)^j\nu(r)(j,[\tfrac{r}{2}]-pj)e_{s+(2pj-r)(p-1)} \otimes e^p_{r-2j(p-1)}$$

$$-\delta(s)\delta(r-1)\sum_j (-1)^j\nu(r-1)(j,[\tfrac{r-1}{2}]-pj)e_{s+p+(2pj-r)(p-1)} \otimes e^p_{r-2j(p-1)-1})$$

for r, s, q fixed integers and where our notation is that of [A; p.176]. (Recall that $m = \frac{p-1}{2}$.) Then by the stable results [A; 4.4 and 4.6], we have $\gamma_*(\sigma_* \otimes \sigma^p_*)(\Gamma-\Delta) = 0$. Let $x \in H_qX$. Then $\theta_{p^2*}\gamma_*((\Gamma-\Delta)\otimes x^{p^2}) = \sum X_\alpha$ where

$$X_\alpha = C_\alpha(Q^{I_1}x)^{n_1}*\dots*(Q^{I_k}x)^{n_k}*(Q^{J_1}\lambda_n(x,x))^{m_1}*\dots*(Q^{J_\ell}\lambda_n(x,x))^{m_\ell}*(\lambda_n(x,x))^r*x^s$$

for some choice of I_t, J_t and some $m_i > 0$, $s > 0$ or $r > 0$, $C_\alpha \in \mathbf{Z}_p$, $p \nmid r$ or $p \nmid s$ with $\ell(I_t) = (J_t) = 1$, and $[(n_1+\dots+n_k+2m_1+\dots+2m_\ell)p+2r+s] = p^2$.

To calculate ξ_nQ^sx, we observe that if $p > 2$, $n + |x|$ is even, and hence $\lambda_n(x,x) = 0$. Hence we may assume that $X_\alpha = 0$. The calculation of ξ_nQ^sx follows directly from the proof of [A; Theorem 4.7].

To calculate $\xi_n\beta Q^sx$, we first observe that $\lambda_n(\iota,\xi_n\beta Q^sx) = ad^p_n(\beta Q^sx)(\iota) = 0$. Hence no terms, X_α, can be in M_X. It follows that we may assume that $r = s = 0$. Assume that x is primitive. It is an easy exercise in the definition of GW_nH_*X to verify that $X_\alpha = 0$. We proceed as before.

The case $p = 2$ follows from the above remarks. Here we let

$$\Gamma = \sum_k (k,s-2k)\, e_{r+2k-s} \otimes e^2_{s-k} \quad \text{and}$$

$$\Delta = \sum_j (j,r-2j)e_{s+2j-r} \otimes e^2_{r-j}.$$

Since $\lambda_n(x,x) = 0 \mod 2$, we again appeal to [A; 4.7].

<u>Proof of Proposition 1.5, commutation with conjugation</u>:

By [G; 5.8] and section 14, it suffices to show that $\chi\xi_n = \xi_n\chi$ and $\chi\lambda_n(y,z) = -\lambda_n(\chi y, \chi z)$. We require some preliminary information.

Define $\bar{c}\colon I^{n+1} \to I^{n+1}$ by the formula $c(t_1,\dots,t_{n+1}) = (1-t_1,t_2,\dots,t_{n+1})$. Note that $\bar{c}$ is not a "little cube". Further define $\bar{\chi}\colon \mathcal{C}_{n+1}(j) \to \mathcal{C}_{n+1}(j)$ by setting $\bar{\chi}\langle c_1,\dots,c_j\rangle = \langle \bar{c}^{-1}\circ c_1\circ\bar{c},\dots,\bar{c}^{-1}\circ c_j\circ\bar{c}\rangle$. It is trivial to verify that $\bar{\chi}\langle c_1,\dots,c_j\rangle$ is in fact in $\mathcal{C}_{n+1}(j)$. [See G; p. 30].
Let c denote the standard inverse in $\Omega^{n+1}X$.

<u>Lemma 16.2.</u> The following diagram equivariantly commutes:

$$\begin{array}{ccc}
\mathcal{C}_{n+1}(j) \times (\Omega^{n+1}X)^j & \xrightarrow{\theta_{n+1,j}} & \Omega^{n+1}X \\
\Big\downarrow{\scriptstyle \bar{\chi}\times c^j} & & \Big\downarrow{\scriptstyle c} \\
\mathcal{C}_{n+1}(j) \times (\Omega^{n+1}X)^j & \xrightarrow{\theta_{n+1,j}} & \Omega^{n+1}X
\end{array}$$

<u>Proof</u>: Let $(\langle c_1,\dots,c_j\rangle, y_1,\dots,y_j) \in \mathcal{C}_{n+1}(j) \times (\Omega^{n+1}X)^j$. Then

$$c \circ \theta_{n+1}(\langle c_1,\dots,c_j\rangle, y_1,\dots,y_j)(v) = \begin{cases} y_r(u) & \text{if } c_r(u) = cv \\ * & \text{if } cv \notin \operatorname{Im} c_j, \end{cases}$$ and

$$\begin{aligned}
\theta_{n+1}\circ(\bar{\chi}\times c^j)(\langle c_1,\dots,c_j\rangle, y_1,\dots,y_j)(v) &= \theta_{n+1}\circ(\langle \bar{c}^{-1}\circ c_1\circ\bar{c},\dots,\bar{c}^{-1}\circ c_j\circ\bar{c}\rangle, cy_1,\dots,cy_j)(v) \\
&= \begin{cases} cy_r(u) & \text{if } \bar{c}^{-1}\circ c_j\circ\bar{c}(u) = v \\ * & \text{if } v \notin \operatorname{Im} \bar{c}^{-1}\circ c_j\circ\bar{c}. \end{cases} \\
&= \begin{cases} y_r(u) & \text{if } \bar{c}^{-1}\circ c_j(u) = v \\ * & \text{if } v \notin \operatorname{Im} \bar{c}^{-1} c_j. \end{cases}
\end{aligned}$$

Hence the diagram commutes. Equivariance is evident.

Proof of Proposition 1.5:

By the defining fromula $(\eta\varepsilon = \phi(1\otimes\chi)\psi)$ for the conjugation in a Hopf algebra, it is easy to calculate that (1) $\chi\xi_0 y = \chi(y^p) = \xi_0(\chi y)$, and (2) $\chi\lambda_0(y,z) = \chi[y*z-(-1)^{|y||z|}z*y] = -\lambda_0(\chi y,\chi z)$. To calculate $\chi\xi_n y$ and $\chi\lambda_n(y,z)$, $n > 0$, we observe that

(3) $\chi\theta_{p*}(e_{n(p-1)} \otimes y^p) = \theta_{p*}(\overline{\chi}_* e_{n(p-1)} \otimes (\chi y)^p)$ and

(4) $\chi\theta_{p*}(\iota\otimes y \otimes z) = \theta_{p*}(\overline{\chi}_*\iota \otimes \chi y \otimes \chi z)$ by 16.2. Since $\overline{\chi}$ is an order 2 equivariant automorphism of $\mathcal{C}_{n+1}(j)$, it follows that

$$\overline{\chi}_* e_{n(p-1)} = k\, e_{n(p-1)} \quad \text{and} \quad \overline{\chi}_*\iota = \ell\iota \quad \text{where } k^2 = \ell^2 = 1.$$

Combining the formulas $\sigma_*\chi\xi_n y = \chi\xi_{n-1}\sigma_* y$ and $\sigma_*\chi\lambda_n(y,z) = \chi\lambda_{n-1}(\sigma_* y,\sigma_* z)$ with (1) and (2) above and an evident induction, we observe that $k = 1$ and $\ell = -1$. The result follows.

17. The Nishida relations

We prove Theorems 1.1(7) and 1.3(3) by calculating the A-action on the operation ξ_n and, if $p > 2$, on the operation ζ_n by induction on n. Evidently, this information suffices to calculate inductively the A-action on $\beta^{\varepsilon}Q^s$.

If $n = 0$, the class $e_0 \otimes x^p \in H_*\mathcal{C}_1(p) \otimes_{\pi_p} (H_*X)^p$ is in the image of the map $H_*\mathcal{C}_1(p) \otimes (H_*X)^p \to H_*\mathcal{C}_1(p) \otimes_{\pi_p} (H_*X)^p$. We calculate the A-action on $e_0 \otimes x^p$ by using the dual of the external Cartan formula and naturality: $P^r_*(e_0 \otimes x^p) = \sum_{i_1+\ldots+i_p=r} e_0 \otimes P^{i_1}_* x \otimes \ldots \otimes P^{i_p}_* x$.
Evidently $P^r_*(e_0 \otimes x^p) = e_0 \otimes (P^{[r/p]}_* x)^p + \sum e_0 \otimes P^{i_{\sigma(1)}}_* x \otimes \ldots \otimes P^{i_{\sigma(p)}}_* x$ where $P^{[r/p]}_* x = 0$ if $p \nmid r$, $P^{[r/p]}_* x = P^{r/p}_* x$ if $p|r$, and the sum runs over sequences $(i_1,\ldots,i_p)$ such that $i_1 + \ldots + i_p = r$, $i_1 = \ldots = i_{n_1}$, $i_{n_1+1} = \ldots = i_{n_2}, \ldots, i_{n_{k-1}+1} = \ldots = i_{n_k} = i_p$, $i_{n_1} < \ldots < i_{n_k}$, $k > 1$, and σ runs over a complete set of distinct left coset representative for $\Sigma_{n_1} \times \Sigma_{n_2-n_1} \times \ldots \times \Sigma_{n_k-n_{k-1}}$ in Σ_p. Consider $\theta_1 : \mathcal{C}_1(p) \times X^p \to X$. Since $\theta_{1*}P^r_*(e_0 \otimes x^p) = P^r_*\theta_{1*}(e_0 \otimes x^p)$ and $\theta_{1*}(e_0 \otimes x_1 \otimes \ldots \otimes x_p) = x_1 * \ldots * x_p$, we observe that $P^r_*\xi_0 x = \xi_0 P^{[r/p]}_* x + \sum (P^{i_{\sigma(1)}}_* x) * \ldots * (P^{i_{\sigma(p)}}_* x)$ where the sum runs over sequences $(i_1,\ldots,i_p)$ described above. By Lemma 17.1 which is stated and proved directly after this proof , this sum is given by

(p-1)-fold commutators as

$$\sum \frac{1}{n_1} \quad ad_0(P_*^{i_{\sigma(1)}}x) \ \ldots \ ad_0(P_*^{i_{\sigma(p-1)}}x)(P_*^{i_1}x)$$

where the sum runs over sequences $(i_1,\ldots,i_p)$ described in Theorem 1.3(5).

To calculate $P_*^r \xi_n x$, $n > 0$, observe that it suffices to do the calculations in $H_*\Omega^{n+1}\Sigma^{n+1}X$. For convenience, we let

$$S_x = \sum_i (-1)^{r+i}(r-pi, \tfrac{n+q}{2}(p-1)-pr+pi)Q^{\frac{n+q}{2}-r+i} P_*^i x$$

$$[\sum_i (r-pi, (n+q)(p-1)-pr+pi)Q^{n+q-r+i}P_*^i x]$$

By our calculations of $H_*\Omega^{n+1}\Sigma^{n+1}X$ and the Nishida relations for the stable case [A; §10], it follows that $P_*^r \xi_n x - S_x \in \operatorname{Ker} j_{n+1}(X)_*$ where $j_{n+1}(X)$ denotes the inclusion of $\Omega^{n+1}\Sigma^{n+1}X$ in QX. We will show that $P_*^r \xi_n x - S_x = \Gamma$ where Γ is given by the sum of (p-1)-fold iterated Browder operations specified in Theorem 1.3(5).

To obtain an inital estimate of $P_*^r \xi_n x - S_x$, we use the following commutative diagram:

$$\begin{array}{ccc}
\mathcal{C}_{n+1}(p) \times X^p & \xrightarrow{1 \times \eta_{n+1}^p} & \mathcal{C}_{n+1}(p) \times (\Omega^{n+1}\Sigma^{n+1}X)^p \\
\downarrow & & \downarrow \theta_{n+1,p} \\
F_p C_{n+1}X \xrightarrow{\subseteq} C_{n+1}X & \xrightarrow{C_{n+1}\eta_{n+1}} C_{n+1}\Omega^{n+1}\Sigma^{n+1}X & \xrightarrow{\theta_{n+1}} \Omega^{n+1}\Sigma^{n+1}X
\end{array}$$

By the defintion of $\xi_n x$ and naturality of P^r_*

$P^r_*\xi_n x - S_x \in \theta_{n+1*} \circ (C_{n+1}\eta_{n+1})_* \circ h_*(H_*(\mathcal{C}_{n+1(p)} \times_{\Sigma_p} X^p))$. Observe that no Dyer-Lashof operations may occur as summands of $P^r_*\xi_n x - S_x$. Consequently

$$P^r_*\xi_n x - S_x = \sum \lambda_{I_1} * \cdots * \lambda_{I_j}$$

where $w(\lambda_{I_1}) + \cdots + w(\lambda_{I_j}) \leq p$ and if $\lambda_{I_1} = \cdots = \lambda_{I_j}$ then $j < p$.

We recall the definition of the submodule M_X of $GW_n H_*(X \vee S^\ell)$ [section 14]. Clearly $P^r_*\xi_n X - S_x \in M_X$. We claim that $P^r_*\xi_n x - S_x$ has no decomposable summands. Let $P^r_*\xi_n x - S_x = \Gamma$. Collecting this information, we have the formulas

(i) $P^r_*\lambda_n(\iota, \xi_n x) = \lambda_n(\iota, P^r_*\xi_n x) = \lambda_n(\iota, S_x + \Gamma)$,

(ii) $\lambda_n(\iota, S_x) = 0$ if $p \nmid r$ (Observe that no top operations occur in S_x.),

(iii) $\lambda_n(\iota, S_x) = ad^p_n(P^i_* x)(\iota)$ if $r = ip$, and

(iv) $P^r_*\lambda_n(\iota, \xi_n x) = P^r_*\, ad^p_n(x)(\iota) = \sum_{i_1 + \cdots + i_{p-1} = r} ad_n(P^{i_1}_* x) \cdots ad_n(P^{i_{p-1}}_* x)(\iota)$.

Hence $\lambda_n(\iota, \Gamma)$ has no decomposable summands. A glance at the proof of Lemma 14.1, reveals that if $\Gamma \neq 0$, $n > 0$, and Γ has decomposable summands (with respect to the Pontrjagin product), then $\lambda_n(\iota, \Gamma)$ has non-trivial decomposable summands. Consequently Γ is a sum of iterated Browder operations which must suspend nontrivially to $H_*\Omega^n\Sigma^{n+1}X$. The formula for $P^r_*\xi_n x$ follows from the formula for $P^r_*\xi_0 x$, induction on n, and the formula $\sigma_* P^r_*\xi_n x = P^r_*\xi_{n-1}\sigma_* x$.

To calculate the A-action on $\zeta_n x$, we use the above technique: any unstable error term, Γ, must lie in the image of the map

$$H_*(\mathcal{C}_{n+1}(p) \times_{\Sigma_p} X^p) \xrightarrow{(1\times\eta^p_{n+1})_*} H_*(\mathcal{C}_{n+1}(p) \times_{\Sigma_p} (\Omega^{n+1}\Sigma^{n+1}X)^p) \xrightarrow{\theta_{n+1,p*}} H_*\Omega^{n+1}\Sigma^{n+1}X.$$

It follows that $\Gamma \in M_X$. Since no top operations (ξ_n) can occur in the stable summand in the Nishida relation for $P^r_*\zeta_n x$, it follows that $\lambda_n(\iota,\Gamma) = 0$. (Recall here the equations $\lambda_n(x,\zeta_n y) = 0 = \lambda_n(x,\beta^{\varepsilon}Q^s z)$). An application of Lemma 14.1 indicates that there are no unstable error terms.

We next show that the definitions of $\zeta_n X$ given in Theorem 1.3 and in section 5.7 are consistent. Let $\xi_n x$ and $\zeta_n x$ be defined as in section 5.7. It suffices to show that $\beta(\xi_n x) = \zeta_n(x) + \mathrm{ad}^p_n(x)(\beta x)$ where β is the mod p homology Bockstein. Again by the above technique, we have that $\beta(\xi_n x) = \zeta_n x + \Delta$ where $\Delta \in \theta_{n+1*} \circ (1 \times \eta^p_{n+1})_*(H_*\mathcal{C}_{n+1}(p) \times_{\Sigma_p} X^p)$. Clearly $\Delta \in M_X$.

Combining the formulas

(i) $\beta\lambda_n(\iota,\xi_n x) = \pm\lambda_n(\iota,\beta\xi_n x) = \pm\lambda_n(\iota,\Delta)$ and

(ii) $\beta\lambda_n(\iota,\xi_n x) = \beta\mathrm{ad}^p_n(x)(\iota) = \sum \pm\, \mathrm{ad}^i_n(x)\, \mathrm{ad}_n(\beta x)\mathrm{ad}^{p-i-1}_n(x)(\iota)$,

with the proof of Lemma 14.1, we see that Δ has no decomposable summands if $n \geq 1$. Hence Δ is a sum of iterated Browder operations which must suspend non-trivially to $H_*\Omega^n\Sigma^{n+1}X$. Obviously,

$\beta\xi_0 x = \beta(x^p) = \sum_{p-1\geq i\geq 0} x^i * \beta x * x^{p-i-1} = ad_0^{p-1}(x)(\beta x)$ (see Jacobson [12]).

Since $\sigma_*\beta\xi_n x = -\beta\xi_{n-1}\sigma_* x$ and $\sigma_*\beta = -\beta\sigma_*$, the result follows by induction on n and the result for $n = 0$.

Finally, we must demonstrate the identity for a restricted Lie algebra required for the calculation of $P_*^r\xi_0 x$. Consider the tensor algebra, TV, of the graded $\mathbf{Z}_p$ vector space space V, where V is generated by variables $x_1,\ldots,x_k$, all of even degree. Fix non-negative integers $n_1,\ldots,n_k$ such that $n_1 +\ldots+ n_k = p$. We consider the polynomial $P(x_1,\ldots,x_k) = \sum y_1 * \ldots * y_p$ where the sum is taken over all monomials $y_1 * \ldots * y_p$ with n_i factors of x_i. We express $P(x_1,\ldots,x_k)$ in terms of commutators.

<u>Lemma 17.1</u>. $P(x_1,\ldots,x_k) = \frac{1}{n_1}\sum ad(y_{1_{\sigma(p-1)}})\ldots ad(y_{i_{\sigma(1)}})(x_1)$ where summation is over sequences $(y_1,\ldots,y_{p-1})$ such that

$y_1 = \ldots = y_{n_1-1} = x_1$, $y_{n_1} = \ldots = y_{n_2+n_1} = x_2,\ldots,$

$y_{n_{k-1}+\ldots+n_1+k-2} = \ldots = y_{n_k+n_{k-1}+\ldots+n_1+k-2} = x_k$, and σ runs over a complete set of distinct left coset representatives for $\Sigma_{n_1-1}\times\Sigma_{n_2}\times\ldots\times\Sigma_{n_k}$ in Σ_{p-1}.

<u>Proof</u>: Let $z = x_2 + \ldots + x_k$. Observe that $P(x_1,\ldots,x_k)$ is a summand of $(x_1+z)^p = x_1^p + z^p + \sum d_0^i(x_1)(z)$. From the definition of the d_0^i, we observe that $P(x_1,\ldots,x_k)$ is a summand of $d_0^{n_1}(x_1)(z)$. Expanding

$d_0^i(x_i)(z)$ using bilinearity of $\lambda_0(-,-)$, we have

$$n_1 d^{n_1}(x_1)(z) = \sum ad_0^{j_1}(x_1)ad_0^{k_1}(z) \;\ldots\; ad_0^{j_r}(x_1)ad_0^{k_r}(z)(x_1)$$

$$= \sum ad_0^{j_1}(x_1)ad_0(y_{1,k_1})\ldots ad_0(y_{k_1,k_1})ad_0^{j_r}(x_1)\ldots ad_0(y_{k_r,k_r})(x_1).$$

for $y_{i,j} \in \{x_2,\ldots,x_k\}$. By inspection

$$P(x_1,\ldots,x_k) = \frac{1}{n_1}\sum ad_0(y_{i_{\sigma(p-1)}})\, ad_0(y_{i_{\sigma(p-2)}})\ldots ad_0(y_{i_{\sigma(1)}})(x_1)$$

where $y_1 = \ldots = y_{n_1-1} = x_1$, $y_{n_1} = \ldots = y_{n_2+n_1} = x_2,\ldots,$

$y_{n_{k-1}+\ldots+n_1+k-2} = \ldots = y_{n_k+n_{k-1}+\ldots+n_1+k-2} = x_k$ and σ runs over

a complete set of distinct left coset representatives for

$\Sigma_{n_1-1} \times \Sigma_{n_2} \times \ldots \times \Sigma_{n_k}$ in Σ_{p-1}.

Appendix: Homology of the classical braid groups

Our description of $H_*C_2S^0$ yields a calculation of the homology of the braid groups, B_r, on r strings. Here we describe these results with coefficients in $\mathbb{Z}_p$, $\mathbb{Z}$, and $\mathbb{Q}$ (all with trivial action). When homology is taken with $\mathbb{Z}_p$-coefficients, the action of the Steenrod algebra, A, is also completely described. In case coefficients are taken in $\mathbb{Z}_2$, the additive results here have been described by Fuks [29].

Theorem A.1. (a) Let $p = 2$. Then $H_*(B_r;\mathbb{Z}_2)$ is isomorphic as a module over A to the algebra over A

$$\frac{P[\xi_j]}{I}$$

where (i) $|\xi_j| = 2^j-1$, and

(ii) I is the two sided ideal generated by

$$(\xi_{j_1})^{u_1} \cdots (\xi_{j_t})^{k_t}$$

where $\sum_{i=1}^{t} k_i 2^{j_i} > r$.

Furthermore, the A action is completely described by requiring that P^r_* act trivially if $r > 1$ and that $P^1_*(\xi_{j+1}) = (\xi_j)^2$ if $j \geq 1$ and $P^1_* \xi_1 = 0$;

(b) Let $p > 2$. Then $H_*(B_r;\mathbb{Z}_p)$ is isomorphic as a module over A to the algebra over A

$$\frac{E[\lambda] \otimes E[\xi_j] \otimes P[\beta\xi_j]}{I}$$

where (i) $|\lambda| = 1$

(ii) $|\beta^{\varepsilon}\xi_j| = 2p^j - 1 - \varepsilon$, and

(iii) I is the two-sided ideal generated by

$$(\lambda)^{\ell} \cdot (\beta^{\varepsilon_1}\xi_{j_1})^{k_1} \cdots (\beta^{\varepsilon_t}\xi_{j_t})^{k_t}$$

where
$$2(\ell + \sum_{i=1}^{t} k_i p^{j_i}) > r.$$

Furthermore the A-action is described by requiring that P_*^r act trivially and that $\beta\lambda = 0$ and $\beta(\xi_j) = \beta\xi_j$.

Remark A.2. The classes ξ_j in case $p = 2$ correspond to the elements $\underbrace{\xi_1 \cdots \xi_1}_{j}([1])$ in the homology of C_2S^0; the classes $\beta^{\varepsilon}\xi_j$ in case $p > 2$ correspond to the elements $\beta^{\varepsilon}\underbrace{\xi_1 \cdots \xi_1}_{j}\lambda_1([1],[1])$ while λ corresponds to $\lambda_1([1],[1])$ in the homology of C_2S^0.

We may read off $H_*(B_r;\mathbb{Z})$ and $(H_*(B_r;\mathbb{Q})$ from the action of the Bocksteins and the results in sections 3 and 4.

Corollary A.3. If $r \geq 2$, $H_*(B_r;\mathbb{Q}) = H_*(S^1;\mathbb{Q})$ and $H_1(B_r;\mathbb{Z}) = \mathbb{Z}$.

To compute $H_*(B_r;\mathbb{Z})$ we have

Corollary A.4. The p-torsion in $H_*(B_r;\mathbb{Z})$ is all of order p. In particular the p-torsion subgroup of $H_*(B_r;\mathbb{Z})$ in degrees greater than one is additively isomorphic to the following:

(i) If $p = 2$, the free strictly commutative algebra ξ_1 and $(\xi_j)^2$, $j > 1$, subject to the conditions of Theorem A.1, and

(ii) If $p > 2$, the free commutative algebra on λ and the $\beta\xi_j$ subject to the conditions of Theorem A.1.

These corollaries follow immediately from Theorems 3.12 and A.1. To prove A.1, it suffices only to recall the results in section 4 and that

$$0 \longrightarrow \tilde{H}_*F_jC_{n+1}X \longrightarrow \tilde{H}_*F_{j+1}C_{n+1}X \longrightarrow \tilde{H}_*E^0_{j+1}C_{n+1}X \longrightarrow 0$$

is short exact. Here we set $n + 1 = 2$ and $X = S^0$. The computation of the Steenrod operations is also immediate from the A-action specified by Theorems 1.2, 1.3, and Lemma 3.10. .

Bibliography

1. S. Araki and T. Kudo, Topology of H_n-spaces and H-squaring operations. Mem. Fac. Sci. Kyûsyû Univ. Ser. A, 1956, 85-120.

2. V. I. Arnold, The cohomology ring of the colored braid group. Math. Zametki, 2(1969) , 227-231.

3. V. I. Arnold, On a class of algebraic functions and cohomologies of fiber tails. Uspeki Mat. Nauk S.S.S.R., 23(1968), 247-248.

4. E. Artin, Theory of braids. Annals of Math.,48(1947), 101-126.

5. R. Bott and H. Samelson, On the Pontrjagin product in spaces of paths. Comment. Math. Helv., 27(1953), 320-337.

6. W. Browder, Homology operations and loop spaces. Ill. J. Math., 4(1960). 347-357.

7. H. Cartan and S. Eilenberg, Homological Algebra. Princeton: Princeton University Press, 1956.

8. E. Dyer and R. K. Lashof, Homology of iterated loop spaces. Amer. J. Math., 1962, 35-88.

9. E. Fadell and L. Neuwirth, Configuration spaces. Math. Scand.,10(1962), 119-126.

10. R. Fox and L. Neuwirth, The braid groups. Math. Scand., 10(1962), 119-126.

11. M. Hall, A basis for free Lie rings and higher commutators in free groups. Proc. Am. Soc.,1(1950), 575-581.

12. P. Hilton, On the homotopy groups of a union of spheres. Comment. Math. Helv., 29(1955), 59-92.

13. P. Hilton, Homotopy Theory and Duality. Gordon and Breach, 1965.

14. N. Jacobson, Lie Algebras. Interscience Tracts in Pure and Applied Math., no. 10, Interscience, New York, 1962.

15. S. MacLane, Homology. Academic Press, 1963

16. (A) J. P. May, A general algebraic approach to Steenrod operations. The Steenrod Algebra and its Applications, Lecture Notes in Math., Vol. 168, Springer Verlag, 1970, 153-231.

17. (G) J. P. May, The Geometry of Iterated Loop Spaces, Lecture Notes in Math., vol. 271, Springer Verlag, 1972.

18. (G') J. P. May, E_∞ ring spaces and permutative categories.

19. (I) J. P. May, The homology of E_∞ spaces.

20. R. J. Milgram, Iterated loop spaces, Ann. of Math., 84(1966), 386-403.

21. H. Samelson, A connection between the Whitehead and the Pontrjagin product. Am. J. Math. 75(1953), 744-752.

22. G. Segal, Configuration spaces and iterated loop spaces. Invent. Math., 21(1973), 213-221.

23. R. G. Swan, The p-period of a finite group. Ill. J. Math. 4(1960), 341-346.

24. W. Browder, Torsion in H-spaces. Ann. of Math. 74(1961), 24-51.

25. V. Snaith, A stable decomposition for $\Omega^n\Sigma^n X$. J. London Math. Soc., 8(1974).

26. F. Cohen, Cohomology of braid spaces. B.A.M.S. 79(1973), 761-764.

27. F. Cohen, Homology of $\Omega^{n+1}\Sigma^{n+1}X$ and $C_{n+1}X$, $n > 0$. B.A.M.S., 79(1973), 1236-1241.

28. F. Cohen, Splitting certain suspensions via self-maps. To appear in Ill. J. Math..

29. D. B. Fuks, Cohomologies of the braid groups mod 2. Functional Analysis and its Applications, 4 (1970), 143-151.

30. F. Cohen and L. Taylor, The cohomology of configuration spaces and of braid spaces on manifolds. (In preparation)

THE HOMOLOGY OF SF(n+1)

Fred Cohen

This paper contains a computation of the Hopf algebra structure of $H_*(SF(n+1);Z_p)$ where $SF(n+1)$ is the space of based degree one self-maps of S^n. The point of this computation is that it provides the essential first step necessary to obtain information about the homology of certain other monoids, such as G(n+2), $\widetilde{PL}$(n+2), Top(n+2) and of their classifying spaces.

Much information is already known in this direction due to May II, Milgram [2], and Tsuchiya [5]. However their methods fail to give the requisite information in case $p > 2$ and n is odd. Consequently much of the work in this paper is devoted to this case.

Section 1 contains the basic results concerning the composition pairing in homology together with the characterization of the Pontrjagin ring $H_*(SF(n+1);Z_p)$ for all n and p.

The geometric diagrams required for our computations are described in section 2. These diagrams are described in terms of the little cubes operads [G] and suffice to give complete formulas for the composition pairing in the homology of finite loop spaces.

The homological corollaries of section 2 are described, and are for the most part proven, in section 3; the formulas for $x \circ Q^s y$ and $x \circ \xi_n y$ are more delicate and are proven in section 4.

Section 5 contains a catalogue of special formulas for the homology of SF(n+1) along with the application of these formulas to the study of the associated graded algebra for $H_* SF(n+1)$.

In section 6, we show that $H_*\Omega^{n+1}S^{n+1}$ is not universal for Dyer-Lashof operations defined via the composition pairing if $n < \infty$. This is not merely an interesting exercise, which contrasts with the case $n = \infty$, but provides the key to the proof that the Pontrjagin ring $H_*SF(n+1)$ is commutative.

To carry out the technical details of the proof of commutativity and other statements, it is first necessary to describe a certain sub-algebra of $H_*\Omega^{n+1}S^{n+1}$ together with some of its properties. This is done in section 7; as a first application of these properties we give the proof of the expansion of $(Q^I[m]*[1-ap^j]) \circ Q^J\lambda_n([1],[1])$ stated in section 5.

Since SF(n+1) is not homotopy commutative and we don't know how to embed $H_*SF(n+1)$ as a sub-algebra of an algebra which we know a priori is commutative (if n is odd), we must resort to computing commutators in $H_*SF(n+1)$. This step is carried out in section 8 using the results of the previous six sections together with III 1.1-1.5.

I wish to thank Peter May and Kathleen Whalen for their constant encouragement during the preparation of this paper.

Finally, I owe Neurosurgeon Jim Beggs an ineffable sense of gratitude; without his skill and compassion, this paper would probably not have appeared.

The author was partially supported by NSF grant MPS 72-05055 .

CONTENTS

§1. THE HOMOLOGY OF SF(n+1), n > 0

As usual, SF(n+1) is the space of based degree one self-maps of S^{n+1}; SF(n+1) is an associative H-space with identity, where multiplication is given by composition of maps. We give a complete description of the Hopf algebra structure of $H_* SF(n+1)$, where all homology is taken with $\mathbb{Z}_p$-coefficients for p an odd prime. Recall from III that $\Omega^{n+1}_\phi S^{n+1}$ denotes the component of the base point in $\Omega^{n+1}S^{n+1}$.

Theorem 1.1. The Pontrjagin ring $H_* SF(n+1)$ is isomorphic as an algebra to $H_*\Omega^{n+1}_\phi S^{n+1}$ for all $n > 0$ and all odd primes.

Remark 1.2. The coproduct on the algebra generators for $H_* SF(n+1)$ is determined from the list of generators given in Lemma 1.7 here and the diagonal Cartan formulas given in III.1.1, 1.2, and 1.3.

We observe that the algebra isomorphism in Theorem 1.1 cannot be realized by an H-map if $n < \infty$ because SF(n+1) is not homotopy commutative [4], while $\Omega^{n+1}_\phi S^{n+1}$ is evidently homotopy commutative.

Remark 1.3. The structure of the Pontrjagin ring is studied to determine the unstable analogues of the stable spherical characteristic classes, II and [2, 5]. Furthermore there are well-known maps (where any successive two form a fibration)

$$SF(n+1) \to SG(n+2) \to S^{n+1} \to BSF(n+1) \to BSG(n+2) .$$

Hence the cohomology of BSG(n+2) follows from that of BSF(n+1). Consequently, we do not require an explicit computation of the Pontrjagin ring $H_* SG(n+2)$. However the passage from $H_* SF(n+1)$ to $H_* BSF(n+1)$ is not yet understood and will not be discussed here.

The crux of all these problems lies in showing

Theorem 1.4. The Pontrjagin ring $H_*SF(n+1)$ is commutative for all n and all primes p.

Remark 1.5. In case $p = 2$, Theorem 1.4 was first proven by Milgram [2] and follows directly from the facts that the natural map

$$i_{n+1*} : H_*SF(n+1) \to H_*SF$$

is an algebra monomorphism and that SF is an infinite loop space (and obviously homotopy commutative). In case $p > 2$, i_{n+1*} is again a monomorphism provided n is even; if n is odd, the kernel of i_{n+1*} consists of the ideal generated by the Browder operation, $\lambda_n([1],[1])*[-1]$ and certain sequences of Dyer-Lashof operations applied to the Browder operation, $Q^I\lambda_n([1],[1])*[1-2p^{\ell(I)}]$, III. §3. Consequently, the structure of $H_*SF(n+1)$ follows directly from the work of May II and Tsuchiya [5] provided n is even.

Much of the work of this paper is directed toward the case in which $p > 2$ and n is odd. However the results to be proven on the composition pairing in sections 2 through 4 and 6 apply to the homology of any $(n+1)$-fold loop space and any prime p. Homological modifications required for the case $p = 2$ are stated in brackets in these sections. When specialized to $H_*SF(n+1)$ with n odd and $p > 2$, they yield the formulas which are the heart of the calculation of the Pontrjagin algebra.

Theorem 1.1 results from a statement most conveniently given in a slightly more general setting. Consider the space of all based self-maps of S^{n+1}, denoted by $\widetilde{F}(n+1)$. Clearly $\widetilde{F}(n+1)$ is an associative H-space when given the multiplication defined by composition of maps, and, as a space, $\widetilde{F}(n+1)$ is $\Omega^{n+1}S^{n+1}$. Let $\widetilde{F}_i(n+1)$ denote the component of $\widetilde{F}(n+1)$ consisting of those maps of degree i. Then $SF(n+1) = \widetilde{F}_1(n+1)$ and $F(n+1) = \widetilde{F}_1(n+1) \cup \widetilde{F}_{-1}(n+1)$, and the inclusions of these monoids into $\widetilde{F}(n+1)$ are homomorphisms. We let $\circ$ denote the composition product in homology.

The homology of $\Omega^{n+1}S^{n+1}$ has been studied in III, where it is shown that there is a non-zero homology class given by $\lambda_n([1],[1])$ (if n is odd and $p > 2$), which we abbreviate in this paper by λ_n. Furthermore there are additional operations, Q^s and ξ_n, generalizing those of Dyer and Lashof, defined in the homology of $\Omega^{n+1}\Sigma^{n+1}X$. As in III, we denote iterations of these operations by Q^I and carry over to this paper the associated notations concerning sequences of operations and degrees of elements. Notice that we write $Q^{\frac{n+|x|}{2}}x$ $[Q^{n+|x|}x]$ for $\xi_n x$.

We define a weight function on $H_*\widetilde{F}(n+1)$ by the formulas

(i) $w(Q^I[1]*[m]) = p^{\ell(I)}$ if $Q^I[1]$ is defined,

(ii) $w(\lambda_n*[m]) = 2$,

(iii) $w(Q^I\lambda_n*[m]) = 2p^{\ell(I)}$ if $Q^I\lambda_n$ is defined,

(iv) $w(x*y) = w(x) + w(y)$, and

(v) $w(x+y) = \text{minimum}\{w(x), w(y)\}$.

We filter $H_*\widetilde{F}_i(n+1)$ by defining $F_jH_*\widetilde{F}_i(n+1)$ to be the vector space spanned by those elements of weight at least j and prove

Theorem 1.6. Composition in $H_*\widetilde{F}(n+1)$ is filtration preserving and, modulo higher filtration, is given by the formula

$$(x*[1])\circ(y*[1]) = x*y*[1]$$

for $x, y \in H_*\widetilde{F}_0(n+1)$.

Proof of Theorem 1.1: Define a morphism of algebras

$$\alpha : H_*\Omega_\phi^{n+1}S^{n+1} \to H_*SF(n+1)$$

by $\alpha(x) = x*[1]$ on elements of the generating set specified by III. §3 and by the requirement that α be a morphism of (commutative!) algebras. This makes sense because $H_*SF(n+1)$ is commutative by Theorem 1.4 and because $H_*\Omega^{n+1}_\phi S^{n+1}$ is free commutative. The previous theorem implies that α is filtration preserving and induces an isomorphism on E^0, the associated graded algebra. Thus Theorem 1.1 will be proven once Theorems 1.4 and 1.6 are proven.

The structure of the associated graded algebra is proven without use of the commutativity of $H_*SF(n+1)$ and is already sufficient to prove the following lemma, which provides a list of the algebra generators for $H_*SF(n+1)$.

Lemma 1.7. Assume that $p > 2$ and n is odd. Then the following set generates $H_*SF(n+1)$ as an algebra (under the composition pairing)

(i) $\lambda_n * [-1]$,

(ii) $Q^I\lambda_n * [1-2p^{\ell(I)}]$, and

(iii) $Q^J[1]*[1-p^{\ell(J)}]$

where I and J are specified by III. §3.

Proof of Lemma 1.7: Let $x, y \in H_*\widetilde{F}_0(n+1)$. Then by Theorem 1.6, we have the formula $(x*[1]) \circ (y*[1]) = x*y*[1]$ modulo terms of higher filtration. Using this formula together with the additive structure of $H_*\widetilde{F}_1(n+1)$ given in III. §3, we see that the projections of the elements specified in Lemma 1.7 generate the associated graded algebra for $H_*SF(n+1)$. Lemma 1.7 now follows directly from the following lemma which we state without proof.

Lemma 1.8. Let A be a positively graded connected filtered algebra. Let $\{x_\alpha | \alpha \in I\} \subseteq A$ be such that the projections of the x_α form a collection of algebra generators for the associated graded algebra. Then the x_α generate A as an algebra.

§2. THE COMPOSITION PAIRING AND THE LITTLE CUBES

It is convenient to work in a more general setting than that obtained by restricting attention to SF(n+1). As preliminaries to this section (and the others), we assume that all spaces in sight are compactly generated Hausdorff with non-degenerate base points. Hence the results of I, III, and [G] apply and we take them for granted.

We consider the (right) composition of $\widetilde{F}(n+1)$ on $\Omega^{n+1}X$,

$$c_{n+1} : \Omega^{n+1}X \times \widetilde{F}(n+1) \to \Omega^{n+1}X$$

where c_{n+1} is given on points by $c_{n+1}(f, g) = f \circ g$. We record in this section the requisite commutative diagrams which relate the composition pairing, c_{n+1}, to the action of the little cubes, θ_{n+1}. The geometric setting provided by the little cubes is especially convenient for several reasons. The relevant diagrams equivariantly commute on the nose (not just up to equivariant homotopies) and consequently our proofs are simple and easy to visualize. Most importantly, complete results on the composition pairing for finite loop spaces are obtained.

Recall that $\Omega^{n+1}X$ is identified as the space of continuous based maps from S^{n+1} to X, $S^{n+1} = I^{n+1}/\partial I^{n+1}$. It is convenient to recall here several of the maps defined in [G]. First we consider the map

$$\alpha_{n+1} : C_{n+1}X \to \Omega^{n+1}\Sigma^{n+1}X .$$

By definition, α_{n+1} is the composite

$$C_{n+1}X \xrightarrow{C_{n+1}\eta} C_{n+1}\Omega^{n+1}\Sigma^{n+1}X \xrightarrow{\theta_{n+1}} \Omega^{n+1}\Sigma^{n+1}X$$

where η is the evident inclusion of X into $\Omega^{n+1}\Sigma^{n+1}X$. Furthermore, the map θ_{n+1} is induced by

$$\theta_{n+1} : \mathcal{C}_{n+1}(j) \times (\Omega^{n+1}X)^j \to \Omega^{n+1}X$$

where

$$\theta_{n+1}(<c_1,\ldots,c_j>, f_1,\ldots,f_j)(u) = \begin{cases} f_j(v) & \text{if } c_j(u) = v, \\ * & \text{otherwise.} \end{cases}$$

Since $C_{n+1}S^0 = \coprod_{j\geq 0} \frac{\mathcal{C}_{n+1}(j)}{\Sigma_j}$, the map

$$a_{n+1} : C_{n+1}S^0 \to \Omega^{n+1}S^{n+1} = \widetilde{F}(n+1)$$

yields a particularly simple description:

$$(a_{n+1}<c_1,\ldots,c_j>)(u) = \begin{cases} v & \text{if } c_j(v) = u, \\ * & \text{otherwise.} \end{cases}$$

The following picture provides a visualization of this map:

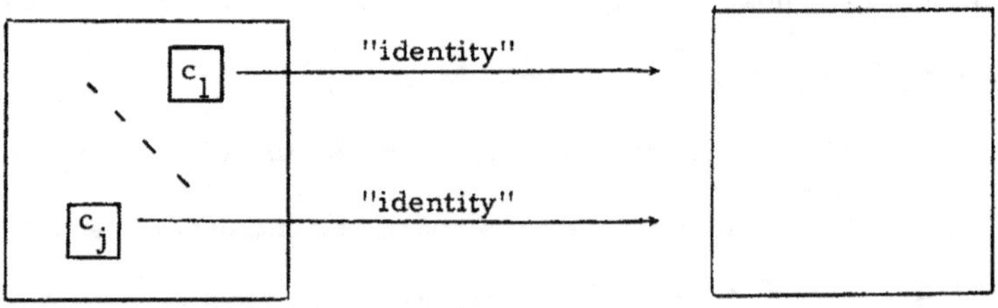

Combining these descriptions with the definition of the composition pairing yields

Lemma 2.1. The following Σ_j-equivariant diagram commutes

$$\begin{array}{ccc}
\Omega^{n+1}X \times \mathcal{C}_{n+1}(j) \times (\widetilde{F}(n+1))^j & \xrightarrow{1\times\theta_{n+1}} & \Omega^{n+1}X \times \widetilde{F}(n+1) \\
\downarrow \Delta\times 1\times 1 & & \downarrow c_{n+1} \\
(\Omega^{n+1}X)^j \times \mathcal{C}_{n+1}(j) \times (\widetilde{F}(n+1))^j & & \Omega^{n+1}X \\
\downarrow \text{Shuff} & & \uparrow \theta_{n+1} \\
\mathcal{C}_{n+1}(j) \times (\Omega^{n+1}X \times \widetilde{F}(n+1))^j & \xrightarrow{1\times c^j_{n+1}} & \mathcal{C}_{n+1}(j) \times (\Omega^{n+1}X)^j
\end{array}$$

where $\Delta(x) = \overbrace{(x, \ldots, x)}^{j\text{-times}}$.

This diagram allows us, in principle, to compute the composition pairing locally; that is, one operation in $H_*\widetilde{F}(n+1)$ at a time. Evidently, it would be convenient to have an analogous diagram which would facilitate the computation (in homology) of the composite

$$\mathcal{C}_{n+1}(j) \times (\Omega^{n+1}X)^j \times \widetilde{F}(n+1) \xrightarrow{\theta_{n+1} \times 1} \Omega^{n+1}X \times \widetilde{F}(n+1) \xrightarrow{c_{n+1}} \Omega^{n+1}X$$

in terms of θ_{n+1}, c_{n+1}, and diagonal maps. It seems unlikely that an easily visualized diagram of this ilk exists. However, we do have analogues in case $\widetilde{F}(n+1)$ is replaced by $C_{n+1}S^0$ and $\Omega^{n+1}X$ by $C_{n+1}X$.

We recall the map

$$c_{n+1} : C_{n+1}X \times C_{n+1}S^0 \to C_{n+1}X$$

defined in [G] and the following proposition which is proved there.

Proposition 2.2. The following diagram commutes:

$$\begin{array}{ccc} C_{n+1}X \times C_{n+1}S^0 & \xrightarrow{c_{n+1}} & C_{n+1}X \\ \Big\downarrow{\scriptstyle \alpha_{n+1} \times \alpha_{n+1}} & & \Big\downarrow{\scriptstyle \alpha_{n+1}} \\ \Omega^{n+1}\Sigma^{n+1}X \times \widetilde{F}(n+1) & \xrightarrow{c_{n+1}} & \Omega^{n+1}\Sigma^{n+1}X \end{array}$$

Corollary 2.3. $C_{n+1}S^0$ is an associative monoid with multiplication given by c_{n+1}. The map α_{n+1} is a homomorphism of monoids.

The following few observations indicate that $C_{n+1}S^0$ has even more structure.

Set

$$F_r C_{n+1} S^0 = \coprod_{j \geq 0} \frac{\mathcal{C}_{n+1}(r^j)}{\Sigma_{r^j}} \quad \text{if } r > 1, \text{ and}$$

$$F_1 C_{n+1} S^0 = F_0 C_{n+1} S^0 = C_{n+1} S^0 .$$

Then we have the following obvious corollary to Proposition 2.2, which will be useful in the study of $H_*BSF(n+1)$.

Corollary 2.4. $F_r C_{n+1} S^0$ is an associative H-space with identity and the inclusions

$$F_{r^k} C_{n+1} S^0 \to F_r C_{n+1} S^0 \to C_{n+1} S^0$$

are all homomorphisms.

Proof: It is easy to check that $F_r C_{n+1} S^0$ is closed under the pairing defined by c_{n+1}.

Proposition 2.5. The following $\Sigma_j \times \Sigma_k$ – equivariant diagram commutes, where i is the natural map of $\mathcal{C}_{n+1}(k)$ into $C_{n+1}S^0$:

$$\begin{array}{ccc}
\mathcal{C}_{n+1}(j) \times (C_{n+1}X)^j \times \mathcal{C}_{n+1}(k) & \xrightarrow{\theta_{n+1} \times i} & C_{n+1}X \times C_{n+1}S^0 \\
\downarrow{\scriptstyle \Delta \times \Delta^j \times 1} & & \downarrow{\scriptstyle c_{n+1}} \\
\mathcal{C}_{n+1}(j)^k \times (C_{n+1}X)^{jk} \times \mathcal{C}_{n+1}(k) & & C_{n+1}X \\
\downarrow{\scriptstyle \text{Shuff}} & & \uparrow{\scriptstyle \theta_{n+1}} \\
\mathcal{C}_{n+1}(k) \times \mathcal{C}_{n+1}(j)^k \times (C_{n+1}X)^{jk} & \xrightarrow{\gamma \times 1} & \mathcal{C}_{n+1}(jk) \times (C_{n+1}X)^{jk}
\end{array}$$

Proof: This follows from an obvious check of definitions.

Remark 2.6. Together with an obvious modification of Lemma 2.1, Proposition 2.5 shows that a two sided distributivity law is satisfied by the products $*$ and c_{n+1*} in $C_{n+1}S^0$. This contrasts with the case of $\widetilde{F}(n+1)$.

§3. FORMULAS FOR THE COMPOSITION PAIRING

We consider the composition pairing, c_{n+1}, which was defined in section 2. This section is a catalogue of homological information concerning c_{n+1*} which is required in the following sections.

Remark 3.1. If $n = \infty$, the composition pairing has been studied by Madsen [1], May II, Milgram [2], and Tsuchiya [5]. Their results give insufficient information in case $n < \infty$ for our purposes. In particular, their methods give no information at all about the classes $Q^I \lambda_n$.

Theorem 3.2. c_{n+1*} gives $H_*\Omega^{n+1}X$ the structure of Hopf algebra over the Hopf algebra $H_*\widetilde{F}(n+1)$. Furthermore for $x \in H_*\Omega^{n+1}X$ and $y, z \in H_*\widetilde{F}(n+1)$ the following formulas hold:

(i) $x \circ (y*z) = \Sigma(-1)^{|x''||y|}(x'\circ y)*(x''\circ z)$ where $\psi x = \Sigma x' \otimes x''$,

(ii) $x \circ Q^s y = \sum_i Q^{s+i}(P^i_* x\circ y)$ if $Q^s y$ is defined,

(iii) $x \circ \xi_n y = \sum_i Q^{\frac{n+|y|}{2}+i}(P^i_* x\circ y) + \Delta$ if $\xi_n y$ is defined,

where Δ is given by at least two fold iterations of the operation $\lambda_n(-,-)$ on elements of $H_*\Omega^{n+1}X$ [$x \circ \xi_n y = \Sigma Q^{n+|y|+i}(P^i_* x\circ y) + \Delta$ where Δ is given in terms of the (non-iterated) operation $\lambda_n(-,-)$ on elements of $H_*\Omega^{n+1}X$], and

(iv) $x \circ \lambda_n = \Sigma(-1)^{|x''|}\lambda_n(x',x'')$ where $\psi x = \Sigma x' \otimes x''$.

Remark 3.3. The error Δ in 3.2(iii) can be determined precisely with some additional work; the result stated here has the advantage that it is both sufficient for our purposes and follows directly from the methods in III.

Using the results of Theorem 3.2, we prove the following two results in section 6.

Theorem 3.4. Let y be a spherical homology class in $H_*\Omega^{n+1}\Sigma^{n+1}X$ such that $|y| \geq n$ if n is odd or $|y| > n$ if n is even [$|y|>n$]. Let $z \in H_*\widetilde{F}(n+1)$ be such that $|z| > 0$ and let $Q^I y$ be defined. Then

$$y \circ z = 0 \qquad \text{and} \qquad Q^I y \circ z = 0.$$

Proposition 3.5. Let $Q^I\lambda_n$, $Q^J\lambda_n$, and $Q^K[1]$ be defined in $H_*\tilde{F}(n+1)$ and let $m = 1 - ap^s$. Then

(i) $([m] * Q^I\lambda_n) \circ Q^K[1] = 0$, and

(ii) $([m] * Q^I\lambda_n) \circ Q^J\lambda_n = 0$.

Remark 3.6. This last theorem and proposition indicate an interesting contrast between the composition pairings in homology for finite and infinite loop spaces. This contrast provides the key to Theorem 1.4 and is discussed more thoroughly in section 6.

We require some additional information for which we recall that ε denotes the counit for $H_*\tilde{F}(n+1)$. The following proposition is an evident modification of the analogous result in I. §1, and the details of proof are left to the reader.

Proposition 3.7. Let $x \otimes y \in H_*\Omega^{n+1}X \otimes H_*\tilde{F}(n+1)$. Then:

(i) $\phi \circ y = \varepsilon(y)\phi$ where ϕ is the class of the base point in $H_*\Omega^{n+1}X$,

(ii) $P^k_*(x \circ y) = \Sigma P^i_* x \otimes P^{k-i}_* y$, and

(iii) $\beta(x \circ y) = \beta x \circ y + (-1)^{|x|} x \circ \beta y$

Remark 3.8. Some of the formulas in Theorem 3.2 are similar to those given in the stable case in I. §1 and II. §2. Observe, however, that the formulas there are transposed from ours: we compute $x \circ (y * z)$ and $x \circ Q^s y$ rather than $(y * z) \circ x$ and $Q^s y \circ x$. Stably there is no real difference; unstably the distinction is vital since only one distributive law holds geometrically.

Before proceeding to the proofs of the results in this section, we require an observation concerning the Nishida relations for $P^r_* Q^s$ and $P^r_* \xi_n$. Observe that the following lemma is trivially true if $n = \infty$, and is false in case $n < \infty$, without the additional hypothesis that $\beta^\varepsilon P^r_* y = 0$, $r > 0$.

<u>Lemma</u> 3.9. Let $y \in H_* \Omega^{n+1} X$ be such that $\beta^\varepsilon P^r_* y = 0$ for all $r > 0$ and assume that $Q^I y$ is defined. Then $P^s_* Q^I y = \Sigma c_{I'} Q^{I'} y$ for some $c_{I'} \in Z_p$ and I' such that $\ell(I) = \ell(I')$.

<u>Proof</u>: Write $Q^I y = \beta^{\varepsilon_k} Q^{s_k} \ldots \beta^{\varepsilon_1} Q^{s_1} y$. If none of the Q^{s_j} is the " top operation", ξ_n, then the lemma follows directly from III.1.1. If some of the Q^{s_j} are equal to ξ_n, there are unstable errors given in terms of $\lambda_n(-,-)$ for the Nishida relations. In this case, we present an inductive proof of 3.9.

If $\ell(I) = 1$, the formula follows directly because the unstable errors are given in terms of the Steenrod operations on y. We assume the result for those I of length k and show that the result is true for those I of length $k+1$. Here the result is easily checked using the formulas $\lambda_n(x, Q^s z) = 0 = \lambda_n(x, \zeta_n z)$ and $\lambda_n(x, \xi_n z) = \mathrm{ad}^p_n(z)(x)$ of III.1.2 and 1.3.

We now derive the homological properties of the composition pairing implicit in section 2. The formulas concerning $x \circ \xi_n y$ and $x \circ Q^s y$ require some special attention and their proofs are postponed until section 4.

<u>Proof of Theorem</u> 3.2(i), $x \circ (y * z) = \Sigma (-1)^{|x''||y|} (x' \circ y) * (x'' \circ z)$:

We specialize the diagram of Lemma 2.1 to

$$\begin{array}{ccc}
\Omega^{n+1}X \times \mathcal{C}_{n+1}(2) \times \widetilde{F}(n+1)^2 & \xrightarrow{1 \times \theta_{n+1}} & \Omega^{n+1}X \times \widetilde{F}(n+1) \\
\downarrow{\scriptstyle \Delta \times 1 \times 1} & & \downarrow{\scriptstyle c_{n+1}} \\
(\Omega^{n+1}X)^2 \times \mathcal{C}_{n+1}(2) \times \widetilde{F}(n+1)^2 & & \Omega^{n+1}X \\
\downarrow{\scriptstyle \mathrm{Shuff}} & & \uparrow{\scriptstyle \theta_{n+1}} \\
\mathcal{C}_{n+1}(2) \times (\Omega^{n+1}X \times \widetilde{F}(n+1))^2 & \xrightarrow{1 \times c^2_{n+1}} & \mathcal{C}_{n+1}(2) \times (\Omega^{n+1}X)^2
\end{array}$$

Evidently, we have the equation

(i) $$x \circ (y*z) = c_{n+1*}(x \otimes \theta_{n+1*}(e_0 \otimes y \otimes z))$$

where the right hand side may be computed by the above commutative diagram to obtain

(ii) $$\theta_{n+1*}(1 \otimes c^2_{n+1*})(\mathrm{Shuff})(\psi \otimes 1 \otimes 1)(x \otimes e_0 \otimes y \otimes z) = c_{n+1*}(x \otimes \theta_{n+1*}(e_0 \otimes y \otimes z))$$

Combining (i) and (ii) together yields the desired result $x \circ (y* z) = \Sigma(-1)^{|x''||y|}(x' \circ y)*(x'' \circ z)$.

Proof of Theorem 3.2(iv), $x \circ \lambda_n = \Sigma(-1)^{|x''|}\lambda_n(x', x'')$:

For this computation, we appeal to the commutative diagram used in the preceding proof and replace e_0 by ι, the fundamental class of $\mathcal{C}_{n+1}(2)$ described in III. §5. Here we have $\lambda_n = \lambda_n([1],[1])$ and by the definition of $\lambda_n(-,-)$, the following formula holds;

(i) $$x \circ \lambda_n = (-1)^n c_{n+1*}(x \otimes \theta_{n+1*}(\iota \otimes [1] \otimes [1])).$$

Taking advantage of commutativity of the above diagram, we have the additional formula

(ii) $$(-1)^n c_{n+1*}(x \otimes \theta_{n+1*}(\iota \otimes [1] \otimes [1])) =$$
$$(-1)^n \theta_{n+1*}(\iota \otimes c^2_{n+1*})(\mathrm{Shuff})(\psi \otimes 1 \otimes 1)(x \otimes \iota \otimes [1]^2).$$

Evidently

(iii) $$(-1)^n \theta_{n+1*}(\iota \otimes c^2_{n+1*})(\mathrm{Shuff})(\psi \otimes 1 \otimes 1)(x \otimes \iota \otimes [1]^2) = (-1)^{n+|x|}\Sigma\theta_{n+1*}(\iota \otimes x' \otimes x'')$$

where $\psi x = \Sigma x' \otimes x''$. Since $\lambda_n(x,y) = (-1)^{n|x|+1}\theta_{n+1*}(\iota \otimes x \otimes y)$, we may combine formulas (i), (ii), and (iii) to obtain the result.

§4. THE FORMULAS FOR $x \circ \xi_n y$ AND $x \circ Q^s y$:

Consider the following commutative diagram where j_{n+1} is the natural inclusion:

$$\begin{array}{ccc}
\Omega^n \xi^n X \times \widetilde{F}(n) & \xrightarrow{\quad c_n \quad} & \Omega^n \Sigma^n X \\
\Big\downarrow {\scriptstyle j_{n+1} \times j_{n+1}} & & \Big\downarrow {\scriptstyle j_{n+1}} \\
\Omega^n \Sigma^n X \times \widetilde{F}(n) & \xrightarrow{\quad c_{n+1} \quad} & \Omega^{n+1} \Sigma^{n+1} X .
\end{array}$$

By remarks similar to those of III. §14, and the above commutative diagram, the formulas for $x \circ Q^s y$ follow from those for $x \circ \xi_{n-i}$, $i > 0$. To compute $x \circ \xi_n y$, we appeal to Lemma 2.1 and a homological computation depending on the methods in III. We remark here that the results we obtain for $x \circ \xi_n y$ are only approximate; we have more accurate results, but the ones here are both sufficient for our purposes and are much easier to prove.

The requisite homological data concerns the map

$$d_{n+1} : X \times \mathcal{C}_{n+1}(p) \times_{\pi_p} Y^p \rightarrow \mathcal{C}_{n+1}(p) \times_{\pi_p} (X \times Y)^p$$

where π_p is the cyclic group of order p acting in the natural diagonal fashion and d_{n+1} is given on points by the formula

$$d_{n+1}(x, c, y_1, \ldots, y_p) = (c, x \times y_1, \ldots, x \times y_p).$$

We begin by stating

<u>Lemma</u> 4.1. Let $x \otimes y \in H_s X \otimes H_t Y$ and $x \otimes e_r \otimes y^p \in H_{r+s+pt}(X \times \mathcal{C}_\infty(p) \times_{\pi_p} Y^p)$.

Then

(i) if $p = 2$, $d_{\infty *}(x \otimes e_r \otimes y^p) = \sum_k e_{r+2k-t} \otimes (P^k_* x \otimes y)^2$,

(ii) if $p > 2$, $d_{\infty *}(x \otimes e_r \otimes y^p) = \dfrac{\nu(s+t)}{\nu(t)} \Sigma(-1)^{k+sr} e_{r+(2pk-s)(p-1)} \otimes (P^k_* x \otimes y)^p$

$$- \delta(r) \frac{\nu(s+t-1)}{\nu(t-1)} \Sigma(-1)^{k+sr} e_{r+p+(2pk-s)(p-1)} \otimes (P^k_* \beta x \otimes y)^p,$$

where $\delta(-)$ and $\nu(-)$ are as given in [A; §9].

Proof: This result follows directly from the commutative diagram

$$\begin{array}{ccc} X \times \mathcal{C}_\infty(p) \times_{\pi_p} Y & \xrightarrow{\text{Shuff}} & \mathcal{C}_\infty(p) \times_{\pi_p} (X \times Y) \\ \downarrow 1 \times 1 \times \Delta & & \downarrow 1 \times \Delta \\ X \times \mathcal{C}_\infty(p) \times_{\pi_p} Y^p & \xrightarrow{d_\infty} & \mathcal{C}_\infty(p) \times_{\pi_p} (X \times Y)^p \end{array}$$

where π_p acts trivially on the right hand factors of the spaces on the top line and by induction on the degree of x using the methods in [A] or [3].

Recalling the description of the homology of $\mathcal{C}_{n+1}(p) \times_{\pi_p} Y^p$ given in III. §16 and combining this information with Lemma 4.1, we obviously have

Lemma 4.2. Let $x \otimes y \in H_s X \otimes H_t Y$ and $x \otimes e_r \otimes y^p \in H_{r+s+pt}(X \times \mathcal{C}_{n+1}(p) \times_{\pi_p} Y^p)$.

Then

(i) if $p = 2$, $d_{n+1*}(x \otimes e_r \otimes y^2) = \sum_k e_{r+2k-t} \otimes (P^k_* x \otimes y)^2 + \Gamma$, and

(ii) if $p > 2$, $$d_{n+1*}(x \otimes e_r \otimes y^p) = \frac{\nu(s+t)}{\nu(t)} \sum_k (-1)^{k+rs} e_{r+(2pk-s)(p-1)} \otimes (P^k_* x \otimes y)^p$$
$$- \delta(r) \frac{\nu(s+t-1)}{\nu(t-1)} \sum_k (-1)^{k+rs} e_{r+p+(2pk-s)(p-1)} \otimes (P^k_* \beta x \otimes y)^p$$
$$+ \Gamma$$

where Γ is in the image of the natural map induced by the covering projection

$$H_*(\mathcal{C}_{n+1}(p) \times (X \times Y)^p) \to H_*(\mathcal{C}_{n+1}(p) \times_{\pi_p} (X \times Y)^p).$$

To compute Γ, we recall that we may replace $\mathcal{C}_{n+1}(p)$ by $F(\mathbb{R}^{n+1}, p)$ in Lemma 4.2 [G, §4]. In III. §11, there is a π_S-action (generated by an element S of order 2) defined on $F(\mathbb{R}^{n+1}, p)$ which commutes with the natural Σ_p-action on $F(\mathbb{R}^{n+1}, p)$. From the definition of the map S, and the replacement of $\mathcal{C}_{n+1}(p)$ by $F(\mathbb{R}^{n+1}, p)$ together with the evident modification of the map d_{n+1}, we have the following lemma.

Lemma 4.3. Let π_S denote the group which acts by S as in III. §11. Then the map

$$d_{n+1} : X \times F(\mathbb{R}^{n+1}, p) \times_{\pi_p} Y^p \to F(\mathbb{R}^{n+1}, p) \times_{\pi_p} (X \times Y)^p$$

is π_S-equivariant.

Proof: Immediate from the definition of S and d_{n+1}.

Lemma 4.4. Let Γ be as in Lemma 4.2 and assume that $p > 2$. Then Γ is in the image of the natural map

$$\sum_{j \geq 0} H_{2j\,n} \mathcal{C}_{n+1}(p) \otimes H_*(X \times Y)^p \to H_*(\mathcal{C}_{n+1}(p) \times_{\pi_p} (X \times Y)^p) .$$

Proof: We write S for S_* throughout the remainder of this section.

As remarked above, we replace $\mathcal{C}_{n+1}(p)$ by $F(\mathbb{R}^{n+1}, p)$. Consider the element $x \otimes e_r \otimes y^p \in H_*(X \times F(\mathbb{R}^{n+1}, p) \times_{\pi_p} Y^p)$. By III.11.1, $x \otimes e_r \otimes y^p$ is fixed by π_S (acting on homology). Similarly, the elements $e_i \otimes (x' \otimes y)^p \in H_*(F(\mathbb{R}^{n+1}, p) \times_{\pi_p} (X \times Y)^p$ and $x' \in H_* X$ and $i \geq 0$, must be fixed by π_S. By Lemma 4.4, d_{n+1} is π_S-equivariant and consequently Γ must be fixed by the π_S-action.

We shall show in Lemma 4.5 that those elements concentrated in $H_{2j\,n} F(\mathbb{R}^{n+1}, p)$, $j = 0, 1, \ldots$, are precisely those elements fixed by π_S. Furthermore, we show that if $z \in H_{(2j+1)n} F(\mathbb{R}^{n+1}, p)$, then $S(z) = -z$. Since Γ must be fixed by the π_S action on homology we see that the form of Γ required by Lemma 4.2 forces Γ to be as asserted in Lemma 4.4.

Lemma 4.5. Let $z \in H_{kn}(F(\mathbb{R}^{n+1}, p); \mathbb{Z})$. Then

$$S(z) = \begin{cases} z & \text{if } k \text{ is even and} \\ -z & \text{if } k \text{ is odd.} \end{cases}$$

Proof: Let α_{ij} be as in III. §6. Observe that the map S defined in III. §11 restricts to an automorphism of degree -1 on the S^n standardly embedded in $\mathbb{R}^{n+1}$. It is obvious that the map

$$\alpha_{ij} : S^n \to F(\mathbb{R}^{n+1}, p)$$

is π_S-equivariant when S^n is given the previous π_S-action. Hence $S(\alpha_{ij}) = -\alpha_{ij}$.

We obtain a similar formula upon dualization: $S(\alpha_{ij}^*) = -\alpha_{ij}^*$. The lemma follows directly from the structure of the cohomology algebra $H^*F(\mathbb{R}^{n+1}, p)$ given in III. §6, and dualization arguments.

Proof of Theorem 3.2(ii) and (iii):

As remarked at the beginning of this section, it suffices to compute $x \circ \xi_n y$. Here we specialize the commutative diagram given by Lemma 2.1 to

$$\begin{array}{ccc}
\Omega^{n+1}X \times \mathcal{C}_{n+1}(p) \times \widetilde{F}(n+1)^p & \xrightarrow{1 \times \theta_{n+1}} & \Omega^{n+1}X \times \widetilde{F}(n+1) \\
\big\downarrow {\scriptstyle \Delta \times 1 \times 1} & & \big\downarrow {\scriptstyle c_{n+1}} \\
(\Omega^{n+1}X)^p \times \mathcal{C}_{n+1}(p) \times \widetilde{F}(n+1)^p & & \Omega^{n+1}X \\
\big\downarrow {\scriptstyle \text{Shuff}} & & \big\uparrow {\scriptstyle \theta_{n+1}} \\
\mathcal{C}_{n+1}(p) \times (\Omega^{n+1}X \times \widetilde{F}(n+1))^p & \xrightarrow{1 \times c^p_{n+1}} & \mathcal{C}_{n+1}(p) \times (\Omega^{n+1}X)^p .
\end{array}$$

The formula for $x \circ \xi_n y$, except for the errors involving λ_n, now follows directly from Lemma 4.2 and the definition of the operations $Q^s z$. We leave this part as an exercise for the reader.

We compute the unstable error terms involving the λ_n if $p > 2$. Observe, by Lemmas 4.2 and 4.4, that these terms are all in the image of

$$\theta_{n+1*} : H_{2r(n)}\mathcal{C}_{n+1}(p) \otimes H_*(\Omega^{n+1}X)^p \to H_*\Omega^{n+1}X.$$

By III.12.1, the elements in the image here are given by 2r-fold iterates of the λ_n. The result follows.

If $p = 2$, the result follows by similar (but easier) considerations.

§5. SPECIAL FORMULAS IN $H_*\widetilde{F}(n+1)$ AND THE PROOF OF THEOREM 1.6.

We specialize the results of section 3 to the case $X = S^{n+1}$ and obtain certain corollaries; the proof of Theorem 1.6, which depends heavily on these corollaries, is also given in this section.

Theorem 5.1. For $x, y, z \in H_*\widetilde{F}(n+1)$ the following formulas hold:

(i) $x \circ (y * z) = \Sigma(-1)^{|x''||y|}(x' \circ y)*(x'' \circ z)$ where $\psi x = \Sigma x' \otimes x''$,

(ii) $x \circ Q^s y = \Sigma Q^{s+i}(P_*^i x \circ y)$ if $Q^s y$ is defined,

(iii) $x \circ \xi_n y = \Sigma Q^{\frac{n+|y|}{2}+i}(P_*^i x \circ y)$ if $\xi_n y$ is defined,

$[= \Sigma Q^{n+|y|+i}(P_*^i x \circ y)]$

(iv) $x \circ \lambda_n = \Sigma(-1)^{|x''|}\lambda_n(x', x'')$ where $\psi x = \Sigma x' \otimes x''$.

Proof: Formulas (i), (ii), and (iv) follow directly by specializing Theorem 3.2 to the case $X = S^{n+1}$. Formula (iii) also follows from Theorem 3.2 together with the observation that two-fold iterations of Browder operations are zero in $H_*\widetilde{F}(n+1)$ in case $p > 2$. If $p = 2$, the result is obvious since $\lambda_n = 0$ here.

We state the following lemma without proof since the results are evident specializations of those given in III.1.1, 1.2, and 1.3 together with the observation that all 2-fold iterations of the operation $\lambda_n(-,-)$ in $H_*\widetilde{F}(n+1)$ are zero.

Lemma 5.2. Let x and y in $H_*\widetilde{F}(n+1)$ be such that $\xi_n x$ and $\xi_n y$ or, for (iii), $\xi_n(x*y)$ are defined. Then

(i) $\xi_n(x+y) = \xi_n x + \xi_n y$,

(ii) $\lambda_n(x, \xi_n y) = 0$,

(iii) $\xi_n(x*y) = \sum_{i+j=\frac{n+|x|+|y|}{2}} Q^i x * Q^j y$

$[= \sum_{i+j=n+|x|+|y|} Q^i x * Q^j y]$, and

(iv) the unstable errors in the Nishida relations for $\beta^\varepsilon P_*^r \xi_n x$ involving $\lambda_n(-,-)$

re zero.

Recall from I. 3. 2, the conventions on sums $I + J$ of sequences of the ame length and observe that the diagonal Cartan formulas in III.1.1 and 1.3, imply hat

$$\psi Q^I x = \sum_{I'+I''=I} \sigma(Q^{I'}x', Q^{I''}x'')Q^{I'}x' \otimes Q^{I''}x''$$

where $\psi x = \Sigma x' \otimes x''$ and $\sigma(Q^{I'}x', Q^{I''}x'') = \pm 1$. We require the notation $\sigma(-,-)$ or future referencing of signs.

In order to obtain more precise formulas, we record the following trivial specialization of the internal Cartan formulas given in III.1.1 and 1.3, and Lemma 5.2 to $H_*\widetilde{F}(n+1)$. (Note that this formula is generally false in $H_*\Omega^{n+1}\Sigma^{n+1}X$ f $n < \infty$ and X is not a homology sphere).

Lemma 5.3. Let $x \in H_*\widetilde{F}(n+1)$ be such that $\psi x = x \otimes [m] + [m] \otimes x$. Then

$$\sum_{I'+I''=I} (-1)^{|x||I''|} Q^{I'}x * Q^{I''}[r] = Q^I(x * [r])$$

or any $r \in \mathbb{Z}$.

We will complete this section by obtaining more precise formulas in $H_*\widetilde{F}(n+1)$ by coupling Theorem 5.1 with the results given in III, and using these formulas to prove Theorem 1.6.

Theorem 5.4. Let $x \in H_*\widetilde{F}(n+1)$ and write the m-fold iterated coproduct on x as $\psi^m x = \Sigma x^{(1)} \otimes \dots \otimes x^{(m)}$. Then

(i) if $m > 0$, $x \circ [m] = \Sigma x^{(1)} * \dots * x^{(m)}$,

(ii) $x \circ [m] = \chi(x) \circ [-m]$ where χ is the conjugation,

(iii) if $m \geq 0$, $Q^I x \circ [m] = Q^I(x \circ [m])$ provided $Q^I x$ is defined,

(iv) $(\lambda_n * [k]) \circ [m] = m\ \lambda_n * [(k+2)(m-1)+k]$ for all m and $k \in \mathbb{Z}$,

(v) $(Q^I x * [1-ap^r]) \circ \lambda_n = \Sigma Q^I(x^{(1)} * x^{(2)}) * \lambda_n * [-2ap^r]$ provided $Q^I x$ is defined,

(vi) $(Q^J\lambda_n * [1-ap^r]) \circ [1-bp^s] = Q^J(\lambda_n * [-2bp^s]) * [(1-ap^r)(1-bp^s)]$ provided $Q^J\lambda_n$ is defined,

(vii) $([m]*\lambda_n) \circ Q^I x = 0$ for any $m \in \mathbb{Z}$ provided $Q^I x$ is defined,

(viii) $([m]*Q^J\lambda_n) \circ Q^I x = 0$ for any $m \in \mathbb{Z}$ provided $Q^I x$ and $Q^I\lambda_n$ are defined, and

(ix) $\left(Q^I[m]*[1-ap^r]\right) \circ Q^J\lambda_n = \sum_{J'+J''=J} \sigma(Q^{J'}[2], Q^{J''}\lambda_n)\left\{Q^I[2m] \circ Q^{J'}[1]\right\} * Q^{J''}(\lambda_n * [-2ap$

for any $m \in \mathbb{Z}$ provided $Q^I[m]$ and $Q^J\lambda_n$ are defined.

<u>Proof of Theorem</u> 5.4:

5.4(i) <u>and</u> (ii), <u>the formula for</u> $x \circ [m]$:

If $m > 0$, the formula $x \circ [m] = \Sigma x^{(1)} * \ldots * x^{(m)}$ follows immediately from Theorem 5.1(i) and induction on m. If $m < 0$, we apply the formula $y \circ [-1] = \chi(y)$.

5.4(iii), <u>the formula for</u> $Q^I x \circ [m]$:

Notice that the result here is generally false if $\widetilde{F}(n+1)$ is replaced by an arbitrary $(n+1)$-fold loop space. However, the result obtained here by specializing to $\widetilde{F}(n+1)$ follows by an evident induction on m together with the diagonal and internal Cartan formulas of III.1.1 and 1.3, and Lemma 5.3.

5.4(iv), <u>the formula for</u> $(\lambda_n * [k]) \circ [m]$:

If $m > 0$, the formula

(i) $(\lambda_n * [k]) \circ [m] = \sum_i [i(2+k)] * \lambda_n * [(m-i-1)(2+k)]$

follows by inspection of the coproduct for λ_n given in III.1.2. Hence we have

(ii) $(\lambda_n * [k]) \circ [m] = \quad m\ \lambda_n * [(2+k)(m-1)+k]$.

In case $m < 0$, then $(\lambda_n * [k]) \circ [m] = \chi(\lambda_n * [k]) \circ [-m]$ by Lemma 5.4(ii). By III.1.5, we have $\chi(\lambda_n) = -\lambda_n([-1],[-1])$. It is easy to compute $\lambda_n([-1],[-1])$

y the internal Cartan formula for $\lambda_n(-,-)$ given in III.1.2. We obtain the
ormula $\chi(\lambda_n) = (-1)\lambda_n*[-4]$. Consequently we have

ii) $(\lambda_n*[k]) \circ [m] = (-1)(\lambda_n*[-4-k]) \circ [-m] = \quad m\ \lambda_n*[(2+k)(m-1)+k]$, and

e are done.

.4(v), <u>the formula for</u> $(Q^I x*[1-ap^r]) \circ \lambda_n$:

Assume that $Q^I x$ is defined. We derive this formula by induction on $\ell(I)$.
irst assume that $\ell(I) = 1$ and consider $Q^I x = Q^s x$. Then by Theorem 5.1(iv)
which gives the formula for $x \circ \lambda_n$) we have

.) $(Q^s x*[1-ap^r]) \circ \lambda_n = \Sigma(-1)^{|x''|}\lambda_n(Q^i x'*[1-ap^r], Q^{s-i}x''*[1-ap^r])$.

xpanding the right side of this equation by the internal Cartan formula for
$_n(-,-)$ given in III.1.2, observing that $[p^r]$ is a p-th power, and that $\lambda_n(Q^i x',-)$
$0 = \lambda_n(-, Q^{s-i}x'')$ by III.1.2 and 1.3; and using Lemma 5.2(ii), we obtain the
ormula

i) $\lambda_n(Q^i x'*[1-ap^r], Q^{s-i}x''*[1-ap^r]) = (-1)^{|Q^i x'|}\lambda_n*Q^i x'*Q^{s-i}x''*[-2ap^r]$.

ombining (i) and (ii) together yields

ii) $(Q^s x*[1-ap^r]) \circ \lambda_n = \Sigma(-1)^{|x''|+|x'|}\lambda_n*Q^i x'*Q^{s-i}x''*[-2ap^r]$.

ombining (iii) with the internal Cartan formulas for Q^s and ξ_n, we have

iv) $(Q^s x*[1-ap^r] \circ \lambda_n = \Sigma Q^s(x'*x'')*\lambda_n*[-2ap^r]$

The details in case $Q^I x = \beta Q^s x$ are similar and the case $\ell(I) > 1$
ollows by an obvious induction.

.4(vi), <u>the formula for</u> $(Q^I\lambda_n*[1-ap^r]) \circ [1-bp^s]$:

The proof of this formula follows immediately from Theorem 5.4(i)-(iv).

.4(vii)-(viii) <u>the formulas</u> $([m]*\lambda_n) \circ Q^I x = 0$ <u>and</u> $([m]*Q^J\lambda_n) \circ Q^I x = 0$:

By III.§1, λ_n and $[m]*\lambda_n$ are spherical homology classes of degree n.
ince n is odd, the result follows from Theorem 3.4 and Proposition 3.5.

Remark 5.5. The proof of Theorem 5.4(ix) is more delicate and depends on some additional machinery; consequently, we have postponed this proof until section 7.

Throughout the remainder of this section, we assume that n and p are both odd. We show

Theorem 1.6. Composition in $H_*\widetilde{F}(n+1)$ is filtration preserving and modulo filtration is given by the formula

$$(x*[1]) \circ (y*[1]) = x*y*[1]$$

for $x, y \in H_*\widetilde{F}_0(n+1)$.

Proof: Let $x = Q^{I_1}[1]*\ldots*Q^{I_r}[1]*Q^{J_1}\lambda_n*\ldots*Q^{J_s}\lambda_n*\lambda_n^{\varepsilon}*[a]$ where $\varepsilon = 0,1$, λ_n^0 is defined to be ϕ, $a \in \mathbb{Z}$, and $I_i = (\varepsilon_{n_i}, s_{k_i}, \ldots, \varepsilon_{1_i}, s_{1_i})$ for $s_{1_i} > 0$.

By the definition of the weight function, ω, defined on $H_*\widetilde{F}(n+1)$ (given in section 1) and the internal Cartan formula, it is obvious that

(i) $\omega(Q^t x) \geq p\,\omega(x)$.

We next compute the weight of $x \circ \lambda_n$:

First observe that if $\varepsilon = 1$, then $x \circ \lambda_n = 0$. So we assume that $\varepsilon = 0$. By an inspection of the diagonal Cartan formulas in III.1.1, 1.2, and 1.3, the formula for $x \circ \lambda_n$ given in Theorem 5.1(iv), the internal Cartan formula for $\lambda_n(-,-)$ given in III.1.2, and the definition of ω, we see that

(ii) $\omega(x \circ \lambda_n) \geq 2(\omega(x)+1)$.

We observe that lower bounds for $\omega(x \circ Q^L[1])$ and $\omega(x \circ Q^K\lambda_n)$ for any sequences L and K may be computed from (i) and (ii) above. Furthermore since $\omega(x) = \omega(x \circ [-1])$, it is obvious that

(iii) $\omega(x) = \omega(x \circ [k])$ for $k \in \mathbb{Z}$.

Let $y = Q^{L_1}[1]*\dots*Q^{L_t}[1]*Q^{K_1}\lambda_n*\dots*Q^{K_u}\lambda_n*\lambda_n^{\delta}*[\gamma]$, where $\delta = 0,1$ and $\gamma \in \mathbf{Z}$. Recall that the m-fold iterated coproduct for x is given as $\psi^m x = \Sigma x^{(1)} \otimes \dots \otimes x^{(m)}$. We expand $x \circ y$ by Theorem 5.1 and compute the weight of $x \circ y$ via formulas (i)-(iii) above:

$$\text{(iv)}\quad w(x\circ y) \geq \text{minimum}\left\{\sum_i p^{\ell(L_i)}(w(x^{(i)})\ ) + \sum_j 2p^{\ell(K_j)}(w(x^{(t+j)})+1) + 2\delta(w(x^{(t+u+\delta)})+1) + w(x^{(t+u+\delta+1)})\right\}$$

where the minimum is taken over all terms in the $(t+u+\delta+1)$-fold coproduct for x. Consequently, we see that

$$\text{(v)}\quad w(x\circ y) \geq w(x) + w(y)$$

and the first part of the theorem is demonstrated.

To finish the proof of 1.6, we let α and γ be such that $x, y \in H_*\widetilde{F}_0(n+1)$ and compute $(x*[1])\circ(y*[1])$ using the formula in (iv) above. By Theorem 5.1(i) we have the formula

$$\text{(vi)}\quad (x*[1])\circ(y*[1]) = \Sigma \pm \{(x^{(1)}*[1])\circ y\} * x^{(2)}*[1].$$

By formula (iv) above, it is apparent that the summand of least weight in the right hand side of (vi) is given by $\pm\{[1]\circ y\}*x*[1]$. Checking the sign dictated by Theorem 5.1(i) for this last summand, we see that, modulo terms of higher filtration, we have

$$\text{(vii)}\quad (x*[1])\circ(y*[1]) = x*y*[1].$$

§6. THE NON-UNIVERSALITY OF THE COMPOSITION PAIRING, $n < \infty$

An alternative method for the definition of natural homology operations defined on the homology of an iterated loop space is provided by the composition pairing. In particular, each element x in $H_*\tilde{F}(n+1)$ can be used to define a homology operation on any element $y \in H_*\Omega^{n+1}X$, namely the operation given by $y \circ x$. In case $n = \infty$, all Dyer-Lashof operations on infinite loop spaces can in fact be defined in terms of the operations given by the composition pairing. In this sense, $H_*\tilde{F}$ is universal for all homology operations defined for the homology of any infinite loop space. It is natural to expect that similar results should hold in case $n < \infty$. The fact that $H_*\tilde{F}(n+1)$ fails to be universal for homology operations, $n < \infty$, occupies much of this section and is in direct contrast to the stable case, $H_*\tilde{F}$. This last fact is crucial in the proof of Theorem 1.4 (commutativity of $H_* SF(n+1)$).

Throughout this section we assume that $n < \infty$. In addition, all proofs are carried out for odd primes. There are analogous results in case $p = 2$; the details of proof are obvious modifications of those already presented and are left to the reader. Our main result here is

Theorem 3.4. Let y be a spherical homology class in $H_*\Omega^{n+1}\Sigma^{n+1}X$ such that $|y| \geq n$ if n is odd or $|y| > n$ if n is even $[|y|>n]$. Let $z \in H_*\tilde{F}(n+1)$ be such that $|z| > 0$ and let $Q^I y$ be defined. Then

$$y \circ z = 0 \quad \text{and} \quad Q^I y \circ z = 0.$$

Remark 6.1. One is tempted to construct a slick (but fallacious) proof by writing

$$\sum_i Q^{s+i}(P^i_* x) = \sum_i Q^{s+i}(P^i_* x \circ [1]) = x \circ Q^s[1].$$

This formula is evidently correct in $H_*\widetilde{F}$ (and explains why $H_*\widetilde{F}$ is universal for homology operations). Furthermore, Theorem 3.2 guarantees that this formula is correct provided $Q^s[1]$ is defined in $H_*\widetilde{F}(n+1)$. Since the left hand expression makes sense for large s and large $|x|$, we see from III.1.1 and 1.3, that $Q^s[1]$ may not be defined and that this formula is wildly false in $H_*\widetilde{F}(n+1)$. Indeed, the fact that the Pontrjagin ring is commutative follows (eventually) from the fact that the above equation is generally false.

We present a simple, but interesting, test case of this type of phenomena before proceeding to the technical details of this section in

Remark 6.2. By Steer's results [4], we know that the Samelson product in $\pi_* SF(2n)$ defined for the adjoint of the element $\alpha_1 \in \pi_{2n+2p-3}S^{2n}$ and the Whitehead product $[\iota, \iota] \in \pi_{4n-1}S^{2n}$ is non-zero (where ι is the fundamental class of S^{2n}). Since the adjoints of the elements α_1 and $[\iota, \iota]$ have non-zero image under the Hurewicz homomorphism for $SF(2n)$, this suggests that the first interesting place to check commutativity of the Pontrjagin ring $H_* SF(2n)$ is on the elements $\lambda_n * [-1]$ and $Q^s[1] * [1-p]$.

By the formulas given in III.1.1, 1.2, and 1.3, and Theorem 5.1 it is easy to check that $(Q^s[1]*[1-p]) \circ (\lambda_n * [-1]) = Q^s[1]*\lambda_n*[-1-p]$ and $(\lambda_n*[-1]) \circ (Q^s[1]*[1-p]) = Q^s[1]*\lambda_n*[-1-p] + \{(\lambda_n*[-1]) \circ Q^s[1]\}*[1-p]$. Consequently, for commutativity to be satisfied in this case, it is both necessary and sufficient that $(\lambda_n*[-1]) \circ Q^s[1] = 0$. It is worthwhile to point out that the vanishing of $(\lambda_n*[-1]) \circ Q^s[1]$ is particularly easy to check: By the definition of $Q^s x$, $0 \leq 2s - |x| \leq n$; hence, if $Q^s[1]$ is defined and non-zero, then $0 \leq s \leq n/2$. Furthermore, $(\lambda_n*[-1]) \circ Q^s[1] = Q^s(\lambda_n*[-1])$ (by Theorem 5.1). If $Q^s(\lambda_n*[-1])$ is defined and non-zero then $0 \leq 2s - n \leq n$. Note that $2s - n \neq 0$ because n

is odd. The restrictions on s are obviously inconsistent and hence $Q^s(\lambda_n * [-1]) = 0$. Note also that this result follows from Theorem 3.4 because $\lambda_n * [-1]$ is a spherical homology class.

Another example which may be easily checked without recourse to lengthy computations is

Observation 6.3. The composition pairing

$$c_{2*} : H_*\Omega^2\Sigma^2 S^k \otimes H_*^+\widetilde{F}(2) \to H_*\Omega^2\Sigma^2 S^k$$

is zero when p and k are odd, where $H_*^+\widetilde{F}(2)$ is the subspace of positive degree elements in $H_*\widetilde{F}(2)$.

Proof: By III. §3, $H_*\widetilde{F}(2)$ is defined in terms of products of elements given by translates of the elements λ_1 and $\beta^\varepsilon \xi_1 \ldots \xi_1 \lambda_1$. If k is odd, then $H_*\Omega^2\Sigma^2 S^k$ has only trivial Browder operations by III. §3, and the result follows directly from Theorem 3.2.

The following lemma, which keeps track of the domain of definition of the Q^I, is useful.

Lemma 6.4. Write $I = (\varepsilon_k, s_k, \ldots, \varepsilon_1, s_1)$, $\varepsilon_i = 0, 1$, and assume that $Q^I x$ is defined in $H_*\Omega^{n+1}X$. Then

$$\frac{|x|}{2} p^{j-1} \leq s_j \leq \frac{n+|x|}{2} p^{j-1} .$$

Furthermore if $|x|$ is odd, then

$$\frac{|x|}{2} p^{j-1} < s_j .$$

Proof: We check the case where $|x|$ is odd and show that

$$\frac{|x|}{2} p^{j-1} < s_j \leq \frac{n+|x|}{2} p^{j-1}$$

y induction on k. The other case is left to the reader.

If $k = 1$, the result follows directly from the definition of the operations Q^s and ξ_n in III. § 1, and the fact that $|x|$ is odd.

We assume the result for k and check it for $k+1$: By definition of $Q^{s_{k+1}}\beta^{\varepsilon_k}Q^{s_k}\ldots Q^{s_1}x$, we have the inequality

(i) $$0 \leq 2s_{k+1} - |\beta^{\varepsilon_k}Q^{s_k}\ldots Q^{s_1}x| \leq n .$$

But $|\beta^{\varepsilon_k}Q^{s_k}\ldots Q^{s_1}x| = 2\sum_{i=1}^{k} s_i(p-1) - b + |x|$ where b is the number of non-trivial Bocksteins. Applying the induction hypothesis to (i), we have the additional inequality

(ii) $$(2k-1) + 2\sum_{j=1}^{k} \frac{|x|}{2} p^{j-1}(p-1) - b + |x| < 2s_{k+1} \leq n + |x| + \sum_{j=1}^{k} \frac{n+|x|}{2} p^{j-1} .$$

Clearly $k - b \geq 0$; (ii) reduces to

(iii) $$|x|p^k < 2s_{k+1} \leq (n+|x|)p^k ,$$

and we are done.

We use the last lemma to prove the following result, which directly implies Theorem 3.4.

Lemma 6.5. Let y be a spherical homology class in $H_*\Omega^{n+1}\Sigma^{n+1}X$. Assume that either $|y| \geq n$ if n is odd or $|y| > n$ if n is even $[|y| > n]$. Further assume that $Q^I y$ is defined and $Q^s[1]$ and $Q^r\lambda_n$ are defined and non-zero. Then

(i) $y \circ Q^s[1] = 0$,

(ii) $y \circ \lambda_n = 0$,

(iii) $y \circ Q^r\lambda_n = 0$

(iv) $Q^I y \circ Q^s[1] = 0$

(v) $Q^I y \circ \lambda_n = 0$

(vi) $Q^I y \circ Q^r \lambda_n = 0$

Using similar methods as those occurring in the proof of the previous lemma, we obtain

Lemma 6.6. Let ι be the fundamental class of S^k and let $k > n$. If $Q^{I_i}\iota$ is any monomial defined in $H_*\Omega^{n+1}\Sigma^{n+1}S^k$, $i = 1, \ldots, m$, then

$$(Q^{I_1}\iota * \ldots * Q^{I_m}\iota) \circ Q^s[1] = 0$$

provided $Q^s[1]$ is defined.

Remark 6.7. Since we do not have a left distributive law for the composition pairing associated to finite loop spaces, Lemma 6.6 does not follow from Lemma 6.5 as one would hope. The composition products here must be computed "bare hands".

Proof of Theorem 3.4:

We shall show that if y is a spherical homology class in $H_*\Omega^{n+1}\Sigma^{n+1}X$, $|y| \geq n$ if n is odd, and $z \in H_*\widetilde{F}(n+1)$, $|z| > 0$, then $Q^I y \circ z = 0$. The other cases are similar (and easier) and are left to the reader.

Let $f : S^k \to \Omega^{n+1}\Sigma^{n+1}X$ be such that $f_*(\iota) = y$ where ι is the fundamental class of S^k. Evidently, it suffices to show that $Q^I\iota \circ z$ is zero in $H_*\Omega^{n+1}\Sigma^{n+1}S^k$. Our first step is to show that $Q^I\iota \circ Q^J[1] = 0$, $Q^I\iota \circ \lambda_n = 0$ and $Q^I\iota \circ Q^{J'}\lambda_n = 0$ for any $Q^J[1]$ and $Q^{J'}\lambda_n$ which are defined. The second step is to show that $Q^I\iota \circ z = 0$. This follows directly from the first step and the distributivity law given in Theorem 3.2(i).

To show that $Q^I\iota \circ Q^J[1] = 0$, $Q^I\iota \circ \lambda_n = 0$, and $Q^I\iota \circ Q^{J'}\lambda_n = 0$, we first observe that the result is correct by Lemma 6.5(iv)-(vi) if $\ell(J) = \ell(J') = 1$. Assume that the result is true for all J and J' of length k; we shall prove the result for those J and J' of length $k+1$.

We expand the elements $Q^I\iota \circ Q^J[1]$ and $Q^I\iota \circ Q^{J'}\lambda_n$ by Theorem 3.2(ii)-
ii) and observe that since Δ is zero in $H_*\Omega^{n+1}\Sigma^{n+1}S^k$, our result follows
mmediately via the inductive hypothesis together with Lemma 3.9.

Proof of Lemma 6.5: We prove the lemma for the case where $|y| \geq n$ and n
s odd; the other case is similar and is left to the reader.

Since y is spherical, let $f : S^k \to \Omega^{n+1}\Sigma^{n+1}X$ be such that $f_*(\iota) = y$ where
is the fundamental class of S^k. As in the proof of 3.4, it suffices to prove
.5 for the case $\Omega^{n+1}\Sigma^{n+1}S^k$ and where y is replaced by ι.

In case (i), observe that $0 \leq 2s \leq n$. By Theorem 3.2 it follows
$\circ Q^s[1] = Q^s\iota$. If $Q^s\iota$ is non-zero, then $0 < 2s - k \leq n$ because k is odd and
y assumption $n \leq k$. Hence we have the inequality $\frac{n}{2} \leq \frac{k}{2} < s < \frac{n}{2}$ which contradicts
he non-vanishing of $\iota \circ Q^s[1]$.

In case (ii), we have the formula $y \circ \lambda_n = \lambda_n(y, \phi) \pm \lambda_n(\phi, y)$ by Theorem 3.2.
But by III.1.2, this sum is zero. Since $\iota \circ Q^r\lambda_n = Q^r(\iota \circ \lambda_n) = 0$, the result
ollows.

Case (iii) is an evident corollary of case (ii).

In case (iv), we assume that $Q^I\iota \circ Q^s[1] \neq 0$. We have the formula

i) $Q^I\iota \circ Q^s[1] = \Sigma\, Q^{s+r}P^r_* Q^I\iota$,

by Theorem 3.2(ii)-(iii) and the fact that $\Delta = 0$ here. Set $I = (\varepsilon_m, s_m, \ldots, \varepsilon_1, s_1)$.
Then by Lemma 3.9 and III.1.1 and 1.3, we have

ii) $P^r_* Q^I\iota = \Sigma\, c_{I'}\beta^{\delta_m} Q^{s_m - \ell_m} \ldots \beta^{\delta_2} Q^{s_2-\ell_2} \beta^{\delta_1} Q^{s_1-\ell_1}\iota$,

$c_{I'} \in Z_p$ where $\Sigma \ell_j = r$. Combining (i) and (ii) together with Lemma 6.4, we
obtain the inequalities

(iii) $\frac{k}{2}p^{j-1} < s_j - \ell_j \leq \frac{n+k}{2} p^{j-1}$ and

(iv) $\frac{k}{2}p^m < s + r \leq \frac{n+k}{2} p^m$.

Furthermore, by applying Lemma 6.4 to $Q^I\iota$, we have the additional inequality

(v) $\frac{k}{2}p^{j-1} < s_j \leq \frac{n+k}{2} p^{j-1}$.

(iii) and (iv) together yield

(vi) $\ell_j < s_j - \frac{k}{2}p^{j-1} \leq \frac{n}{2}p^{j-1}$.

Combining (iv) and (vi) together, we have

(vii) $\frac{k}{2}p^m - \sum_{j=1}^{m} \frac{k}{2}p^{j-1} < \frac{k}{2}p^m - \sum_{j=1}^{m} \ell_j = \frac{k}{2}p^m - r < s.$

Hence $\frac{k}{2}\{p^m - \frac{p^m-1}{p-1}\} < s$. But by definition of $Q^s[1]$, we see that $0 \leq s \leq n/2$.

Since $n \leq k$, we have the additional inequality

(viii) $\frac{n}{2}\{p^m - \frac{p^m-1}{p-1}\} < \frac{n}{2}$ which is of course a contradiction to the assumption that $Q^I\iota \circ Q^s[1] \neq 0$.

In case (v), we see that $Q^I\iota \circ Q^s\lambda_n = \Sigma Q^{s+k}(P^k_* Q^I\iota \circ \lambda_n)$. Clearly $(P^r_* Q^I\iota) \circ \lambda_n = P^r_*(Q^I\iota \circ \lambda_n)$ by III.1.2. Since ι is primitive of positive degree, we see that $Q^I\iota$ is also primitive. Evidently $Q^I\iota \circ \lambda_n = \lambda_n(Q^I\iota, \phi) \pm \lambda_n(\phi, Q^I\iota)$ by Theorem 3.2. This element is zero by III.1.2.

Case (vi) follows immediately from case (v) together with Theorem 3.2(i)-(iii) and the observation that $\Delta = 0$ in $H_*\Omega^{n+1}\Sigma^{n+1}S^k$.

Proof of Lemma 6.6: The proof is very similar to that used in case (iv) above. The details are purely mechanical and are left to the reader.

Proof of Proposition 3.5: Let $m = 1 - ap^s$. In order to show that $([m]*Q^I\lambda_n) \circ Q^K x = 0$, $x = [1]$ or λ_n, we claim that it suffices to prove the result for the case in which $\ell(K) = 1$. This claim follows from an obvious application of Theorem 5.1 together with an induction on the length of K. (See the proof of Theorem 3.4.) Furthermore, the identical argument used in the proof of Lemma 6.5(iv) can be used to prove the result in case $x = [1]$. Consequently we shall only include the requisite modifications for the case $x = \lambda_n$.

Assume that $([m]*Q^I\lambda_n) \circ Q^r\lambda_n \neq 0$ and consider the expansion

(i) $$([1-ap^s]*Q^I\lambda_n) \circ Q^r\lambda_n = 2\Sigma\, Q^{r+t}P_*^t(Q^I(\lambda_n*[2])*\lambda_n*[-2ap^s])$$

which follows by an application of Theorems 5.1(ii-iii) and 5.4(v). Let $\ell(I) = m$. By applying the internal Cartan formula together with Lemma 6.4, we see that

(ii) $$\frac{n}{2}(p^m+1) < r + t \leq np^m .$$

By arguments almost identical to those used in the proof of Lemma 6.5(iv) together with formula (ii) above, we see that

(iii) $$\frac{n}{2}(p^m+1) - \sum_{j=1}^{m} \frac{n}{2}p^{j-1} < r \leq n.$$

This is an obvious contradiction and we are done.

§7. THE ALGEBRA $\mathscr{S}_n$ AND THE FORMULA FOR $(Q^I[m]*[1-ap^r])\circ Q^J\lambda_n$

A certain sub-algebra, $\mathscr{S}_n$, of $H_*\widetilde{F}(n+1)$ occurs ubiquitously in our remaining work on $H_*SF(n+1)$. We define $\mathscr{S}_n$ in this section and observe some of its properties. Our first application is the derivation of the formula given in Theorem 5.4(ix) for expanding $(Q^I[m]*[1-ap^r])\circ Q^J\lambda_n$. Throughout the remaining sections, we write [?] for [m] whenever [m] is determined by the context.

Definition 7.1. $\mathscr{S}_n$ is the subspace of $H_*\widetilde{F}(n+1)$ spanned by all monomials in the $*$-product of $Q^I[1]$, I as defined in III. §3, and $[m]$, $m \in \mathbb{Z}$.

We note first that the formulas in Theorem 5.1 and Lemma 5.2 demonstrate that $\mathscr{S}_n$ is closed under the composition product. Since $\mathscr{S}_n$ maps monomorphically into $H_*\widetilde{F}$ via the natural map

$$\mathscr{S}_n \xrightarrow{\text{inclusion}} H_*\widetilde{F}(n+1) \to H_*\widetilde{F},$$

we have

Lemma 7.2. With the composition product, $\mathscr{S}_n$ is a commutative subalgebra of $H_*\widetilde{F}(n+1)$ and the natural map

$$\mathscr{S}_n \to H_*\widetilde{F}(n+1) \to H_*\widetilde{F}$$

is a monomorphism.

Proof of Theorem 5.4(ix); the formula

$$(Q^I[m]*[1-ap^r])\circ Q^J\lambda_n = \sum_{J'+J''=J} \sigma(Q^{J'}[2],Q^{J''}\lambda_n)\left\{Q^I[2m]\circ Q^{J'}[1]\right\}*Q^{J''}(\lambda_n*[-2ap^r]):$$

We assume that $Q^I[m]$ and $Q^J\lambda_n$ are defined; our proof follows by induction on the length of J.

First assume that $J = (s)$, $Q^J = Q^s$. Then

(i) $(Q^I[m]*[1-ap^r])\circ Q^s\lambda_n = \Sigma\, Q^{s+r}(P^r_*Q^I[m]*[1-ap^r])\circ\lambda_n$

r Theorem 5.1. Expanding the right side of (i) using the fact that the Steenrod erations act trivially on λ_n, and quoting Theorem 5.4(v) we have

i) $$(P^r_* Q^I[m]*[1-ap^r]) \circ \lambda_n = P^r_*((Q^I[m]*[1-ap^r]) \circ \lambda_n) = P^r_* Q^I[2m]*\lambda_n*[-2ap^r] .$$

ombining (ii) together with the internal Cartan formula for Q^s given in emma 5.2 and III.1.1, 1.3, we have

ii) $$(Q^I[m]*[1-ap^r]) \circ Q^s\lambda_n = \sum_{i\geq 0} \{Q^{s+r-i}P^r_* Q^I[2m]\} * Q^i(\lambda_n * [-2ap^r])$$

We write $Q^I[2m]$ as $\beta^{\varepsilon_k}Q^{s_k}\ldots\beta^{\varepsilon_1}Q^{s_1}[2m]$ and quote Lemma 3.9 to see at

v) $$Q^{s+r-i}P^r_* Q^I[2m] = Q^{s+r-i}(\Sigma c_{I'}\beta^{\delta_k}Q^{s_k-t_k+t_{k-1}}\ldots\beta^{\delta_2}Q^{s_2-t_2+t_1}\beta^{\delta_1}Q^{s_1-t_1}[2m])$$

here $t_k = r$ and $c_{I'} \in \mathbb{Z}_p$. By inspecting the coefficients appearing in the ishida relations, we see that

r) $$0 \leq s_j(p-1) - pt_j + pt_{j+1} - \delta_j, \qquad j = 1,\ldots,k .$$

umming over j, we have the additional inequality

ri) $$0 \leq \sum_{j\geq 1} \{s_j(p-1) - pt_j + pt_{j+1} - \delta_j\} .$$

etting b equal the number of non-zero Bocksteins in I and checking the efinition of $Q^{s+r-i}P^r_* Q^I[2m]$, we find the additional inequality

rii) $$0 \leq 2(s+r-i) - |P^r_* Q^I[2m]| .$$

Consequently, we have

riii) $$2\,\Sigma s_i(p-1) - 2r(p-1) - b - 2r \leq 2(s-i)$$

rhich evidently yields

(ix) $$2(\Sigma s_i(p-1)-rp-b) \leq 2\ \ s_i(p-1) - 2rp - b \leq 2(s-i)\ .$$

Since the left hand is just twice the sum expressed in (vi), we also have

(x) $$0 \leq s - i.$$

Consequently, by the internal Cartan formulas given in Lemma 5.2 and in III.1.1 and 1.3, (iii) reduces to

(xi) $$(Q^I[m]*[1-ap^r]) \circ Q^s\lambda_n = \sum_{\substack{j\geq 0\\ r\geq 0}} \{Q^{j+r}P^r_*Q^I[2m]\}*Q^{s-j}(\lambda_n*[-2ap^r])$$

(The crucial point above is the restriction on the indices of summation.) Since $\sum_{r\geq 0} Q^{j+r}P^r_*Q^I[2m] \in \mathcal{S}_n$, we may use Lemma 7.1 to rewrite this sum for fixed j as

(xii) $$Q^I[2m] \circ Q^j[1]\ .$$

Hence (xi) reduces to

(xiii) $$(Q^I[2m]*[1-ap^r]) \circ Q^s\lambda_n = \sum_j\{Q^I[2m] \circ Q^j[1]\}* Q^{s-j}(\lambda_n*[-2ap^r])\ .$$

The case where $Q^I\lambda_n = \beta Q^s\lambda_n$ is checked similarly.

Now we assume the result for all J of length ℓ and prove the result for those J of length $\ell+1$. For this step, we assume that $J = (1, s, J_1)$ and leave the simpler case where $J = (0, s, J_1)$ to the reader. First we expand $(Q^I[m]*[1-ap^r]) \circ \beta\ Q^sQ^{J_1}\lambda_n$ by Theorem 5.1 and Lemma 3.7 to obtain

(xiv) $$(Q^I[m]*[1-ap^r]) \circ \beta Q^sQ^{J_1}\lambda_n = (-1)^{|Q^I[m]|}\Sigma\beta Q^{s+r}((P^r_*Q^I[m]*[1-ap^r])\circ Q^{J_1}\lambda_n$$
$$+ (-1)^{|Q^I[m]|+1}\Sigma Q^{s+r}((P^r_*\beta Q^I[m]*[1-ap^r])\circ Q^{J_1}\lambda$$

We expand the right side of this equation by the Nishida relations and the requirements of Lemma 3.9 to obtain

$$\text{(xv)}\qquad (P^r_* Q^I[m]*[1-ap^r]) \circ Q^{J_1}\lambda_n = \Sigma c_{I_1}(Q^{I_1}[m]*[1-ap^r]) \circ Q^{J_1}\lambda_n, \quad \text{and}$$

$$\text{(xvi)}\qquad (P^r_* \beta Q^I[m]*[1-ap^r]) \circ Q^{J_1}\lambda_n = \Sigma d_{I_2}(Q^{I_2}[m]*[1-ap^r]) \circ Q^{J_1}\lambda_n$$

where c_{I_1}, $d_{I_2} \in \mathbf{Z}_p$. We now apply the inductive hypothesis to compute each of the sums in (xv) and (xvi):

$$\text{(xvii)}\quad \Sigma c_{I_1}(Q^I[m]*[1-ap^r]) \circ Q^{J_1}\lambda_n = \Sigma c_{I_1}\sigma(Q^{J'_1}[2], Q^{J''_1}\lambda_n)\{Q^{I_1}[2m] \circ Q^{J'_1}[1]\} * Q^{J''_1}(\lambda_n * [-2ap^r]),$$

and

$$\text{(xviii)}\quad \Sigma d_{I_2}(Q^{I_2}[m]*[1-ap^r]) \circ Q^{J_1}\lambda_n = \Sigma d_{I_2}\sigma(Q^{J'_1}[2], Q^{J''_1}\lambda_n)\{Q^{I_2}[2m] \circ Q^{J'_1}[1]\} * Q^{J''_1}(\lambda_n * [-2ap^r]).$$

But for fixed $Q^{J'_1}[1]$ and $Q^{J''_1}\lambda_n$, we apply the formula of Lemma 3.9 to see that the sums in (xvii) and (xviii) are respectively equal to

$$\text{(xix)}\qquad \Sigma\sigma(Q^{J'_1}[2], Q^{J''_1}\lambda_n)\{P^r_* Q^I[2m] \circ Q^{J'_1}[1]\} * Q^{J''_1}(\lambda_n * [-2ap^r]) \qquad \text{and}$$

$$\text{(xx)}\qquad \Sigma\sigma(Q^{J'_1}[2], Q^{J''_1}\lambda_n)\{P^r_* \beta Q^I[2m] \circ Q^{J'_1}[1]\} * Q^{J''_1}(\lambda_n * [-2ap^r]).$$

Combining the results in (xix) and (xx) together we obtain

$$\text{(xxi)}\qquad (Q^I[m]*[1-ap^r]) \circ \beta Q^s Q^{J_1}\lambda_n =$$

$$(-1)^{|Q^I[m]|}\Sigma\sigma(Q^{J'_1}[2], Q^{J''_1}\lambda_n)\beta Q^{s+r}\{(P^r_* Q^I[2m] \circ Q^{J'_1}[1]) * Q^{J''_1}(\lambda_n * [-2ap^r])\}$$

$$+ (-1)^{|Q^I[m]|+1}\Sigma\sigma(Q^{J'_1}[2], Q^{J''_1}\lambda_n)Q^{s+r}\{P^r_* \beta Q^I[2m] \circ Q^{J'_1}[1]) * Q^{J''_1}(\lambda_n * [-2ap^r])\}$$

We use an argument similar to that in the initial step of the induction together with the action of the Bockstein given by Lemma 5.2 and Proposition 3.7 to show that

(xxii) $$\sum_{r\geq 0} \sigma(Q^{J_1'}[2], Q^{J_1''}\lambda_n)\beta Q^{s+r}\{(P_*^r Q^I[2m] \circ Q^{J_1'}[1]) * Q^{J_1''}(\lambda_n * [-2ap^r])\}$$

$$= \beta \sum_{\substack{r\geq 0\\ j\geq 0}} \sigma(Q^{J_1'}[2], Q^{J_1''}\lambda_n)\{Q^{j+r}(P_*^r Q^I[2m] \circ Q^{J_1'}[1])\} * \{Q^{s-j}Q^{J_1''}(\lambda_n * [-2ap^r])\}$$

and

(xxiii) $$\sum_{r\geq 0} \sigma(Q^{J_1'}[2], Q^{J_1''}\lambda_n)Q^{s+r}\{(P_*^r \beta Q^I[2m] \circ Q^{J_1'}[1]) * Q^{J_1''}(\lambda_n * [-2ap^r])\}$$

$$= \sum_{\substack{r\geq 0\\ j\geq 0}} \sigma(Q^{J_1'}[2], Q^{J_1''}\lambda_n)\{Q^{j+r}(P_*^r \beta Q^I[2m] \circ Q^{J_1'}[1])\} * Q^{s-j}Q^{J_1''}(\lambda_n * [-2ap^r]) .$$

Since the terms $Q^{j+r}(P_*^r \beta^\varepsilon Q^I[2m] \circ Q^{J_1'}[1])$ are all in $\mathcal{S}_n$, we may apply Lemma 7.2 to obtain

(xxiv) $$\sum_{r\geq 0} Q^{j+r}\{P_*^r \beta^\varepsilon Q^I[2m] \circ Q^{J_1'}[1]\} = \beta^\varepsilon Q^I[2m] \circ Q^j Q^{J_1'}[1] .$$

Combining (xxi) together with (xxii)-(xxiii) and the action of the Bockstein given in Lemma 5.2 and Proposition 3.7, we obtain

(xxv) $$(Q^I[m] * [1-ap^r]) \circ \beta Q^s Q^{J_1}\lambda_n =$$

$$\sum \sigma(Q^{J_1'}[2], Q^{J_1''}\lambda_n)\{Q^I[2m] \circ \beta Q^j Q^{J_1'}[1]\} * Q^{s-j}Q^{J_1''}(\lambda_n * [-2ap^r])$$

$$+ \sum (-1)^{|Q^{J'}[1]|} \sigma(Q^{J_1'}[2], Q^{J_1''}\lambda_n)\{Q^I[2m] \circ Q^j Q^{J_1'}[1]\} * \beta Q^{i-j}Q^{J_1''}(\lambda_n * [-2ap^r]) .$$

Visibly this formula is that given in Corollary 5.3(ix) and we are done

Remark 7.3. We were careful to keep track of the indices of summation here in order to make certain that all operations used here are defined in $H_*\widetilde{F}(n+1)$. Compare this remark to remark 6.1.

§8. COMMUTATIVITY OF THE PONTRJAGIN RING $H_*SF(n+1)$

Throughout this section, we assume that n and p are both odd. A rief outline of our method follows. Because we have no analogue of a "maximal rus" for $SF(n+1)$ and no algebra monomorphism of $H_*SF(n+1)$ into an algebra hich we know a priori is commutative, we resort to the inelegant method of tually computing commutators in $H_*SF(n+1)$. The work is simplified greatly finding algebra generators for $H_*SF(n+1)$ (given in Lemma 1.7) and then nowing that the algebra generators commute. (Indeed, our original proof of ommutativity involved the computation of an arbitrary commutator!)

roof of Theorem 1.4; the commutativity of $H_*SF(n+1)$:

A collection of algebra generators for $H_*SF(n+1)$ has been specified in emma 1.7. There are three types of generators listed. That they commute is necked by the six evident cases. All sequences of Dyer-Lashof operations re assumed to be defined in $H_*F(n+1)$; the indices of summation are evident nd are consequently deleted.

ase I: $Q^I[1]*[1-p^{\ell(I)}]$ and $\lambda_n*[-1]$.

Applying Theorem 5.1(i) together with the definition of $\sigma(-,-)$, we obtain

) $$(Q^I[1]*[1-p^{\ell(I)}])\circ(\lambda_n*[-1])$$

$$= \Sigma\,\sigma(Q^{I'}[1],Q^{I''}[1])(-1)^{|Q^{I''}[1]|}\{(Q^{I'}[1]*[1-p^{\ell(I)}])\circ\lambda_n\}*\{(Q^{I''}[1]*([1-p^{\ell(I)}])\circ[-1]\}.$$

xpanding $(Q^{I'}[1]*[1-p^{\ell(I)}])\circ\lambda_n$ by Theorem 5.4(v), we obtain

i) $$(Q^{I'}[1]*[1-p^{\ell(I)}])\circ\lambda_n = Q^{I'}[2]*\lambda_n*[-2p^{\ell(I)}]\,.$$

xpanding $(Q^{I''}[1]*[1-p^{\ell(I)}])\circ[-1]$ by Theorem 5.4(ii), we obtain

ii) $$(Q^{I''}[1]*[1-p^{\ell(I)}])\circ[-1] = Q^{I''}[-1]*[-1+p^{\ell(I)}]\,.$$

Combining (i), (ii), and (iii) together with the fact that $|\lambda_n|$ is odd we find that

(iv) $(Q^I[1]*[1-p^{\ell(I)}]) \circ (\lambda_n*[-1])$

$$= \Sigma\sigma(Q^{I'}[1], Q^{I''}[1])(-1)^{|Q^{I''}[1]|+|Q^{I'}[2]|}\lambda_n * Q^{I'}[2]*Q^{I''}[-1]*[-1-p^{\ell(I)}] .$$

An application of Lemma 5.3 (the internal Cartan formula) to the right hand side of this equation yields

(v) $$(Q^I[1]*[1-p^{\ell(I)}]) \circ (\lambda_n*[-1]) = (-1)^{|Q^I[1]|}\lambda_n * Q^I[1]*[-1-p^{\ell(I)}] .$$

We now compute $(\lambda_n*[-1]) \circ (Q^I[1]*[1-p^{\ell(I)})$. Since the coproduct for λ_n is given by $\psi\lambda_n = \lambda_n \otimes [2] + [2] \otimes \lambda_n$, we may apply Theorem 5.1(i) to see that

(vi) $(\lambda_n*[-1]) \circ (Q^I[1]*[1-p^{\ell(I)}])$

$$= \{(\lambda_n*[-1]) \circ Q^I[1]\}*[1-p^{\ell(I)}] + (-1)^{|Q^I[1]|}Q^I[1]*\{(\lambda_n*[-1])\circ[1-p^{\ell(I)}]\} .$$

By Theorem 5.4(vii), we have $(\lambda_n*[-1]) \circ Q^I[1] = 0$. We expand the second summand in this equation by Theorem 5.4(iv) to obtain

(vii) $$(\lambda_n*[-1]) \circ (Q^I[1]*[1-p^{\ell(I)}]) = (-1)^{|Q^I[1]|}Q^I[1]*\lambda_n*[-1-p^{\ell(I)}] .$$

Comparing (v) and (vii), we observe that $\lambda_n*[-1]$ and $Q^I[1]*[1-p^{\ell(I)}]$ commute.

<u>Case</u> II: $Q^I\lambda_n*[1-2p^{\ell(I)}]$ <u>and</u> $\lambda_n*[-1]$

We use the distributivity law in Theorem 5.1(i) together with the definition of $\sigma(-,-)$ to see that

i) $(Q^I\lambda_n * [1-2p^{\ell(I)}]) \circ (\lambda_n * [-1])$

$$= \Sigma\sigma(Q^{I'}[2], Q^{I''}\lambda_n)(-1)^{|Q^{I''}\lambda_n|}\{(Q^{I'}[2]*[1-2p^{\ell(I)}])\circ\lambda_n\}*\{(Q^{I''}\lambda_n*[1-2p^{\ell(I)}])\circ[-1]\}$$

$$+ \Sigma\sigma(Q^{I'}\lambda_n, Q^{I''}[2])(-1)^{|Q^{I''}[2]|}\{(Q^{I'}\lambda_n*[1-2p^{\ell(I)}])\circ\lambda_n\}*\{(Q^{I''}[2]*[1-2p^{\ell(I)}])\circ[-1]\} .$$

Expanding $(Q^{I'}[2]*[1-2p^{\ell(I)}]) \circ \lambda_n$ by Theorem 5.4(v), we find that

ii) $$(Q^{I'}[2]*[1-2p^{\ell(I)}]) \circ \lambda_n = Q^{I'}[4]*\lambda_n*[-4p^{\ell(I)}] .$$

A similar expansion of $(Q^{I''}\lambda_n*[1-2p^{\ell(I)}]) \circ [-1]$ by Theorem 5.4(ii) and (iv) yields

iii) $$(Q^{I''}\lambda_n*[1-2p^{\ell(I)}]) \circ [-1] = (-1)Q^{I''}(\lambda_n*[-4])*[-1+2p^{\ell(I)}] .$$

We also expand the terms $(Q^{I'}\lambda_n*[1-2p^{\ell(I)}]) \circ \lambda_n$ and $(Q^{I''}[2]*[1-2p^{\ell(I)}]) \circ [-1]$ by Theorem 5.4(ii) and (v):

iv) $$(Q^{I'}\lambda_n*[1-2p^{\ell(I)}]) \circ \lambda_n = 2Q^{I'}(\lambda_n*[2])*\lambda_n*[-4p^{\ell(I)}],$$ and

v) $$(Q^{I''}[2]*[1-2p^{\ell(I)}]) \circ [-1] = Q^{I''}[-2]*[-1+2p^{\ell(I)}] .$$

In the following, we combine formulas (i)-(v) together.

vi) $(Q^I\lambda_n * [1-2p^{\ell(I)}]) \circ (\lambda_n * [-1])$

$$= \Sigma\sigma(Q^{I'}[2], Q^{I''}\lambda_n)(-1)^{|Q^{I''}\lambda_n|+1} Q^{I'}[4]*\lambda_n*Q^{I''}(\lambda_n*[-4])*[-1-2p^{\ell(I)}]$$

$$+ \Sigma 2\sigma(Q^{I'}\lambda_n, Q^{I''}[2])(-1)^{|Q^{I''}[2]|} Q^{I'}(\lambda_n*[2])*\lambda_n*Q^{I''}[-2]*[-1-2p^{\ell(I)}] .$$

A check of signs together with Lemma 5.3 (the internal Cartan formula) and the formula in (vi) above yields

vii) $$(Q^I\lambda_n*[1-2p^I]) \circ (\lambda_n*[-1]) = Q^I\lambda_n*\lambda_n*[-1-2p^{\ell(I)}] .$$

We now compute $(\lambda_n * [-1]) \circ (Q^I\lambda_n * [1-2p^{\ell(I)}]$ as in the above case by applying Theorem 5.1(i), and Theorem 5.4(iv) and (vii) to obtain

(viii) $$(\lambda_n * [-1]) \circ (Q^I\lambda_n * [1-2p^{\ell(I)}]) = (-1)^{|Q^I\lambda_n|} Q^I\lambda_n * \lambda_n * [-1-2p^{\ell(I)}] .$$

Commutativity of the elements in this case is checked by a comparison of formulas (vii) and (viii).

Case III: $(\lambda_n * [-1])^2$

It is easy to check that $(\lambda_n * [-1]) \circ (\lambda_n * [-1]) = 0$ by Theorem 5.1, Theorem 5.4, and III.1.2(6).

Case IV: $Q^I[1]*[1-p^{\ell(I)}]$ and $Q^J\lambda_n * [1-2p^{\ell(J)}]$

We compute $(Q^I[1]*[1-p^{\ell(I)}]) \circ (Q^J\lambda_n * [1-2p^{\ell(J)}])$ by Theorem 5.1(i).

(i) $$(Q^I[1]*[1-p^{\ell(I)}]) \circ (Q^J\lambda_n * [1-p^{\ell(J)}])$$
$$= \Sigma\sigma(Q^{I'}[1], Q^{I''}[1])(-1)^{|Q^{I''}[1]||Q^J\lambda_n|}\{(Q^{I'}[1]*[1-p^{\ell(I)}])\circ Q^J\lambda_n\}$$
$$* \{(Q^{I''}[1]*[1-p^{\ell(I)}]) \circ [1-2p^{\ell(J)}]\} .$$

Applying Theorem 5.4(ix), we find that

(ii) $$(Q^{I'}[1]*[1-p^{\ell(I)}]) \circ Q^J\lambda_n = \Sigma\sigma(Q^{J'}[2], Q^{J''}\lambda_n)\{Q^{I'}[2]\circ Q^{J'}[1]\} * Q^{J''}(\lambda_n * [-2ap^{\ell(I)}]) .$$

Combining (i) and (ii) together with the expansion of $(Q^{I''}[1]*[1-p^{\ell(I)}])\circ[1-2p^{\ell(J)}]$ implied by Theorem 5.4 we obtain

(iii) $$(Q^I[1]*[1-p^{\ell(I)}]) \circ (Q^J\lambda_n * [1-2p^{\ell(J)}])$$
$$= \Sigma\sigma(Q^{I'}[1], Q^{I''}[1])\sigma(Q^{J'}[2], Q^{J''}\lambda_n)(-1)^{|Q^{I''}[1]||Q^J\lambda_n|}$$
$$\{Q^{I'}[2]\circ Q^{J'}[1]\} * Q^{J''}(\lambda_n * [-2ap^{\ell(I)}]) * Q^{I''}[1-2p^{\ell(J)}] * [?] .$$

We commute $Q^{J''}(\lambda_n * [-2ap^{\ell(I)}])$ and $Q^{I''}[1-2p^{\ell(J)}]$ in the $*$-product to obtain the formula

v) $(Q^I[1]*[1-p^{\ell(I)}]) \circ (Q^J\lambda_n*[1-2p^{\ell(J)}])$

$$= \Sigma\sigma(Q^{I'}[1], Q^{I''}[1])\sigma(Q^{J'}[2], Q^{J''}\lambda_n)(-1)^{|Q^{I''}[1]||Q^{J'}[2]|}$$

$$\{Q^{I'}[2]\circ Q^{J'}[1]\}*Q^{I''}[1-2p^{\ell(J)}]*Q^{J''}(\lambda_n*[-2p^{\ell(I)}])*[?].$$

We compute $(Q^J\lambda_n*[1-2p^{\ell(J)}]) \circ (Q^I[1]*[1-p^{\ell(I)}])$ in a similar fashion:

) $(Q^J\lambda_n*[1-2p^{\ell(J)}]) \circ (Q^I[1]*[1-p^{\ell(J)}])$

$$= \Sigma\sigma(Q^{J'}[2], Q^{J''}\lambda_n)(-1)^{|Q^{J''}\lambda_n||Q^I[1]|}$$

$$\{(Q^{J'}[2]*[1-2p^{\ell(J)}])\circ Q^I[1]\}*\{(Q^{J''}\lambda_n*[1-2p^{\ell(J)}])\circ [1-p^{\ell(I)}]\}$$

$$+ \Sigma\sigma(Q^{J'}\lambda_n, Q^{J''}[2])(-1)^{|Q^{J''}[2]||Q^I[1]|}$$

$$\{(Q^{J'}\lambda_n*[1-2p^{\ell(J)}])\circ Q^I[1]\}*\{(Q^{J''}[2]*[1-2p^{\ell(J)}])\circ [1-p^{\ell(I)}]\} .$$

he second sum vanishes by Theorem 5.4(viii). Furthermore, the element $Q^{J'}[2]*[1-2p^{\ell(J)}]) \circ Q^I[1]$ is in $\mathcal{S}_n$ by Definition 7.1. By Lemma 7.2, we ay write

$$\text{vi)} \quad (Q^{J'}[2]*[1-2p^{\ell(J)}])\circ Q^I[1] = (-1)^{|Q^I[1]||Q^{J'}[2]|}Q^I[1]\circ(Q^{J'}[2]*[1-2p^{\ell(J)}])$$

hich may be expanded by Theorem 5.1(i) to obtain

$$\text{vii)} \quad (Q^{J'}[2]*[1-2p^{\ell(I)}])\circ Q^I[1] = \Sigma\sigma(Q^{I'}[1], Q^{I''}[1])(-1)^{|Q^{I'}[1]||Q^{J'}[2]|}$$

$$\{Q^{I'}[1]\circ Q^{J'}[2]\}*\{Q^{I''}[1-2p^{\ell(J)}]\} .$$

he element $(Q^{J''}\lambda_n*[1-2p^{\ell(J)}])\circ[1-p^{\ell(I)}]$ may be expanded by use of Theorem .4(vi) to obtain

$$\text{viii)} \quad (Q^{J''}\lambda_n*[1-2p^{\ell(J)}])\circ[1-p^{\ell(I)}] = Q^{J''}(\lambda_n*[-2p^{\ell(I)}])*[(1-2p^{\ell(J)})(1-p^{\ell(I)})] .$$

Using formulas (vii) and (viii) to substitute in formula (v), we find that

(ix) $(Q^J\lambda_n * [1-2p^{\ell(J)}]) \circ (Q^I[1] * [1-p^{\ell(I)}])$

$$= \Sigma\sigma(Q^{I'}[1], Q^{I''}[1])\sigma(Q^{J'}[2], Q^{J''}\lambda_n)(-1)^{(|Q^{J''}\lambda_n||Q^I[1]| + |Q^{J'}[2]||Q^{I'}[1]|)}$$

$$\{Q^{I'}[1] \circ Q^{J'}[2]\} * Q^{I''}[1-2p^{\ell(J)}] * Q^{J''}(\lambda_n * [-2p^{\ell(I)}]) * [?] .$$

Since $Q^{I'}[1] \circ Q^{J'}[2]$ is in $\mathcal{S}_n$, it is obvious that $Q^{I'}[1] \circ Q^{J'}[2] = Q^{I'}[1] \circ Q^{J'}[1] \circ [2] = Q^{I'}[2] \circ Q^{J'}[1]$. Substituting this result in (ix) together with the obvious commutation formulas for $*$-products and the fact that $|Q^{J''}\lambda_n| + |Q^{J'}[2]| = |Q^J\lambda_n|$, we have

(x) $(Q^J\lambda_n * [1-2p^{\ell(J)}]) \circ (Q^I[1] * [1-p^{\ell(I)}])$

$$= \Sigma\sigma(Q^{I'}[1], Q^{I''}[1])\sigma(Q^{J'}[2], Q^{J''}\lambda_n)(-1)^{(|Q^{J''}\lambda_n||Q^I[1]| + |Q^{J'}[2]||Q^{I'}[1]|)}$$

$$\{Q^{I'}[2] \circ Q^{J'}[1]\} * Q^{I''}[1-2p^{\ell(J)}] * Q^{J''}(\lambda_n * [-2p^{\ell(I)}]) * [?] .$$

Visibly, the formulas in (iv) and (x) agree modulo the appropriate commutation sign and we are done.

<u>Case</u> V: $Q^I\lambda_n * [1-2p^{\ell(I)}]$ <u>and</u> $Q^J\lambda_n * [1-2p^{\ell(J)}]$

We appeal to Theorem 5.1(i) again.

(i) $(Q^I\lambda_n * [1-2p^{\ell(I)}]) \circ (Q^J\lambda_n * [1-2p^{\ell(J)}])$

$$= \Sigma\sigma(Q^{I'}[2], Q^{I''}\lambda_n)(-1)^{|Q^{I''}\lambda_n||Q^J\lambda_n|}$$

$$\{(Q^{I'}[2] * [1-2p^{\ell(I)}]) \circ Q^J\lambda_n\} * \{(Q^{I''}\lambda_n * [1-2p^{\ell(I)}]) \circ [1-2p^{\ell(J)}]\}$$

$$+ \Sigma\sigma(Q^{I'}\lambda_n, Q^{I''}[2])(-1)^{|Q^{I''}[2]||Q^J\lambda_n|}$$

$$\{(Q^{I'}\lambda_n * [1-2p^{\ell(I)}]) \circ Q^J\lambda_n\} * \{(Q^{I''}[2] * [1-2p^{\ell(I)}]) \circ [1-2p^{\ell(J)}]\} .$$

he second sum vanishes by Theorem 5.4(ix) to get

i) $$(Q^I\lambda_n * [1-2p^{\ell(I)}]) \circ (Q^J\lambda_n * [1-2p^{\ell(J)}])$$

$$= \Sigma\sigma(Q^{I'}[2], Q^{I''}\lambda_n)(-1)^{|Q^{I''}[2]||Q^J\lambda_n|}$$

$$\{(Q^{I'}[2] * [1-2p^{\ell(I)}]) \circ Q^J\lambda_n\} * \{(Q^{I''}\lambda_n * [1-2p^{\ell(I)}]) \circ [1-2p^{\ell(J)}]\} .$$

/e expand $(Q^{I'}[2] * [1-2p^{\ell(I)}]) \circ Q^J\lambda_n$ and $(Q^{I''}\lambda_n * [1-2p^{\ell(I)}]) \circ [1-2p^{\ell(J)}]$ by heorem 5.4(ix) and (vi) to see that

ii) $$(Q^{I'}[2] * [1-2p^{\ell(I)}]) \circ Q^J\lambda_n = \Sigma\sigma(Q^{J'}[2], Q^{J''}\lambda_n)\{Q^{I'}[4] \circ Q^{J'}[1]\} * Q^{J''}(\lambda_n * [-4p^{\ell(I)}]),$$

nd

v) $$(Q^{I''}\lambda_n * [1-2p^{\ell(I)}]) \circ [1-2p^{\ell(J)}] = Q^{I''}(\lambda_n * [-4p^{\ell(J)}]) * [(1-2p^{\ell(I)})(1-2p^{\ell(J)})] .$$

ubstituting these last two results in (ii) above we see that

v) $$(Q^I\lambda_n * [1-2p^{\ell(I)}]) \circ (Q^J\lambda_n * [1-2p^{\ell(J)}])$$

$$= \Sigma\sigma(Q^{I'}[2], Q^{I''}\lambda_n)\sigma(Q^{J'}[2], Q^{J''}\lambda_n)(-1)^{|Q^{I''}[2]||Q^J\lambda_n|}$$

$$\{Q^{I'}[4] \circ Q^{J'}[1]\} * Q^{J''}(\lambda_n * [-4p^{\ell(I)}]) * Q^{I''}(\lambda_n * [-4p^{\ell(J)}]) * [?] .$$

)bserve that $Q^{I'}[4] \circ Q^{J'}[1]$ is in $\mathcal{S}_n$ and by Lemma 7.2, we have

vi) $$Q^{I'}[4] \circ Q^{J'}[1] = (-1)^{|Q^{I'}[4]||Q^{J'}[1]|} Q^{J'}[1] \circ Q^{I'}[4] .$$

Now by interchanging I and J in formulas (v) and (vi) and checking legrees, we see that the two generators in case V commute.

<u>Case</u> VI: $Q^I[1] * [1-p^{\ell(I)}]$ <u>and</u> $Q^J[1] * [1-p^{\ell(J)}]$

Since both of these elements lie in $\mathcal{S}_n$, they commute by Lemma 7.2.

Bibliography

1. I. Madsen. On the action of the Dyer-Lashof operations in $H_*(G)$. To appear in Pacific J. Math.

2. R. J. Milgram. The mod-2 spherical characteristic classes. Annals of Math. 92(1970), 238-261.

3. N. E. Steenrod and D. B. A. Epstein. Cohomology Operations. Princeton University Press, 1962.

4. B. Steer. Extensions of mappings into H-spaces. Proc. London Math. Soc. (3) 13(1963), 219-272.

5. A. Tsuchiya. Characteristic classes for spherical fiber spaces. Nagoya Math. J. 43(1971), 1-39.

STRONG HOMOTOPY ALGEBRAS OVER MONADS

Thomas J. Lada

Introduction: A topological space X will be called a loop space if there exists a space Y and a weak homotopy equivalence $X \to \Omega Y$; such a space Y is called a classifying space for X. Here the symbol ΩY denotes the set of continuous base-pointed functions from S^1, the 1-sphere, into X topologized with the compact open topology. There is a history of theorems that identify certain H-spaces as loop spaces. Milnor [9] showed that a topological group is a loop space. Sugawara [12], Dold and Lashoff [4] and Stasheff [11] extended this result to associative H-spaces and to strong homotopy associative (or A_∞) H-spaces. The fundamental point in each proof is the construction of a classifying space for the given space.

One is then confronted with the problem of whether a given space X is an n-fold loop space, i.e., whether there is a space Y and a weak homotopy equivalence $X \to \Omega^n Y$. In this case Y is called an nth classifying space for X. While it was essentially the strong homotopy associativity of the multiplication on X that enabled one to construct its classifying space in the 1-fold loop case, higher homotopy commutativity of the multiplication proved to be the key to n-fold loop spaces. As a special example, Dold and Thom [5] proved that a strictly associative, commutative H-space has the weak homotopy type of a product of Eilenberg-MacLane spaces. (It has been pointed out to the author that J. C. Moore also has an unpublished proof of this fact.) In general one must develop some method of keeping track of all the requisite higher associativity and commutativity homotopies on X.

Boardman and Vogt [2] showed that if a certain type of functor acted appropriately on a space X, they could then conclude that X was homotopy equivalent to an associative H-space Y and thus build BY and further, that another functor acted similarly on BY and they could then iterate their argument. Segal [10] was able to accomplish the same thing by using only one functor. He has his functor act not only on X but also on spaces of the homotopy type of X^n.

In category theory there is a concept of a functor being a monad or triple. Beck [1] had shown that if the monad $\Omega^n\Sigma^n$ acts on a space X in a certain manner, then an n-fold classifying space could be constructed. Although this theorem gives a procedure for identifying an iterated loop space, there are few spaces on which $\Omega^n\Sigma^n$ acts properly.

May [G] generalized this result to monads that look like $\Omega^n\Sigma^n$. He has two theorems along these lines. The first theorem makes precise the idea of "looks like" $\Omega^n\Sigma^n$. His second theorem tells how to construct the nth classifying space of X when one of these monads acts on X. One point that is missing in this theorem is homotopy invariance; if X is an n-fold loop space and Y is homotopy equivalent to X, this functor need not act on Y.

In this work we introduce the idea of a monad D acting on a space up to homotopy and study the theory of such spaces and maps between them. In this context May's recognition theorem is generalized up to homotopy. In addition a homotopy invariance theorem in the sense of Boardman and Vogt [3, p. 1] for this theory is proved.

Section 1 contains some motivation for and the definition of a strong homotopy D-space; it is this strong homotopy action of the monad that encodes the homotopies required for an n-fold loop space. The monad action in May's theorem is a special case of this strong homotopy action.

Given an s.h.D-space X, we construct a D-space UX in Section 2. UX contains X as a deformation retract. At this point May's recognition theorem can be generalized to an s.h.D-space X by applying his theorem to the D-space UX.

In Section 3 we introduce a conceptual definition of a strong homotopy D-map between s.h.D-spaces (called an SHD-map); such a map from X to Y will be essentially a D-map from UX to UY. SHD-maps form the collection of morphisms for a category whose objects are s.h.D-spaces. This section concludes with definitions of geometric strong homotopy D-maps from X to Y where one space is a D-space and the other is an s.h.D-space (these maps are called s.h.D-maps). These are the maps that frequently occur in nature; e.g., a homotopy equivalence between an arbitrary space and a D-space. Sections 4 and 5 provide some machinery required to link together our conceptual and geometric definitions of strong homotopy D-maps.

The following conceptual homotopy invariance theorem is proved in Section 6:

1) If Y is an s.h.D-space, $f: X \to Y$ a homotopy equivalence, then X is an s.h.D-space and f is an SHD-map.

2) If $f: X \to Y$ is an SHD-map between s.h.D-spaces and if $g \simeq f$, then g is an SHD-map and $g \simeq f$ as SHD-maps.

3) If $f: X \to Y$ is an SHD-map between s.h.D-spaces and is a homotopy equivalence with homotopy inverse g, then g is an SHD-map and $f \circ g \simeq 1$ and $g \circ f \simeq 1$ as SHD-maps.

This theorem is deduced from our geometric homotopy invariance theorem which consists of the above three statements restricted to D-spaces with SHD-maps replaced by s.h.D-maps. The proof of this geometric homotopy invariance theorem occupies Sections 7, 8 and 9.

In a concluding appendix, a geometric definition of a strong homotopy D-map between s.h.D-spaces is discussed. These details should convince the reader that the conceptual definition of such a map is both reasonable and desirable.

Throughout this work whenever mention is made of a category of topological spaces, it should be taken to mean the category of compactly generated weak Hausdorff spaces with non-degenerate base points. It will be denoted by the symbol T.

I am indebted to J. Stasheff who guided me in this work as part of my doctoral thesis for the University of Notre Dame and to D. Kraines who offered many valuable suggestions. I would like to thank J. P. May who influenced a good part of this work and is largely responsible for its present organization. It should be pointed out that Section 5 is joint work with P. Malraison.

I am also grateful to the Mathematics Departments of Temple University, Duke University, and North Carolina State University for financial support during the part of this work that was contained in my thesis (Sections 1, 2 and 7). The remainder was supported in part by a grant from the North Carolina State Engineering Foundation.

CONTENTS

1. Strong Homotopy Algebras over Monads.

We begin by recalling the definitions of a monad and an algebra over a monad.

Definition 1.1: A monad in any category C is a triple (D,μ,η) where $D: C \to C$ is a covariant functor, $\eta: 1 \to D$ and $\mu: DD \to D$ are natural transformations of functors such that if X is any object in C, then the following diagrams commute:

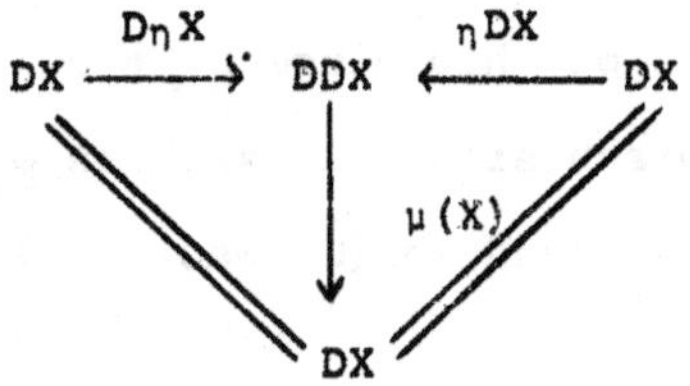

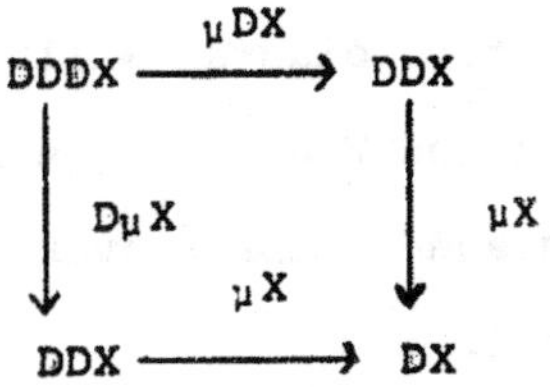

Now let X be a topological space and let (D,μ,η) be a monad in J, our category of topological spaces. Suppose also that $h_o: DX \to X$ is a continuous map in J.

Definition 1.2: The pair (X,h_o) is a D-space (or D-algebra) if the diagrams

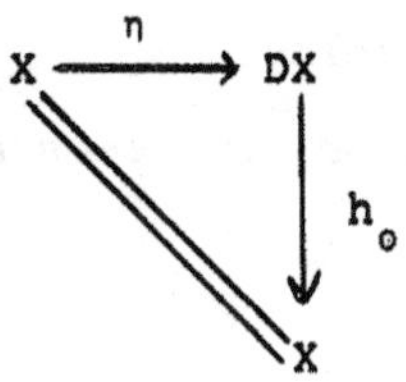

and

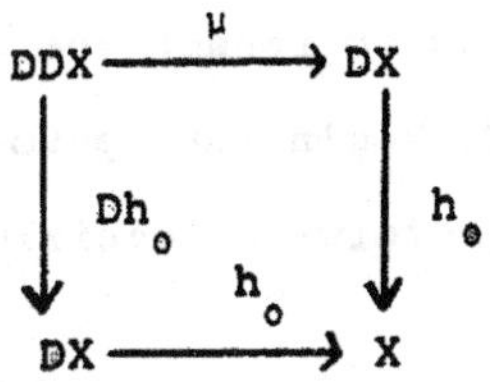

commute. If (X',g_o) is another D-space, then $f: X \to X'$ is a map of D-spaces if

$$\begin{array}{ccc} DX & \xrightarrow{Df} & DX' \\ {\scriptstyle h_o}\downarrow & & \downarrow{\scriptstyle g_o} \\ X & \xrightarrow{f} & X' \end{array}$$

commutes. It should be emphasized that f is required to preserve all of the D-structure of both X and X' in the sense that if we apply D to the diagram it will still commute; the same must be true if we apply higher iterates of D. This one commutative diagram will guarantee all of this since D is a functor and X and X' are D-spaces. Discussion and examples of monads and algebras over monads may be found in [G, Section 2] and in [1].

Since the definitions of monad and D-space involve commutative diagrams of maps between topological spaces, it appears that one might be able to generalize these definitions "up to homotopy". This generalization for a monad will not be pursued here; however, it will be shown that generalization of D-spaces does merit some attention. In all that follows, let us agree to write D^n for D iterated n times whenever n is so large as to render the previous notation unwieldy.

To begin our up-to-homotopy generalization, it seems a straightforward requirement that the diagram

$$\begin{array}{ccc} DDX & \xrightarrow{\mu(X)} & DX \\ {\scriptstyle Dh_o}\downarrow & & \downarrow{\scriptstyle h_o} \\ DX & \xrightarrow{h_o} & X \end{array}$$

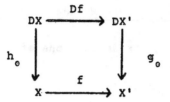

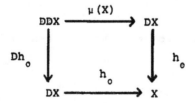

commute up to homotopy; i.e., that we have a homotopy $h_1: I \times D^2X \to X$ such that $h_1(0,y) = h_o \circ \mu(y)$ and $h_1(1,y) = h_o \circ Dh_o(y)$ where $y \in D^2X$. At first glance this appears to be a natural generalization of D-space, but we have not yet taken into account all of the D-structure on X. Implicit in the definition of D-space is the commutative diagram

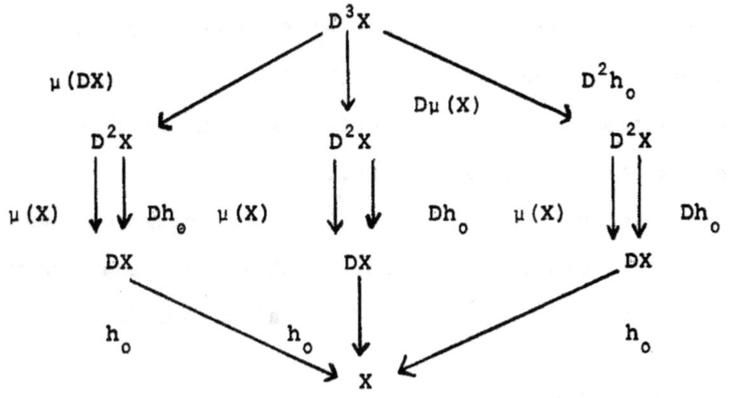

In other words, all possible ways of mapping $D^3X \to X$ via μ and h_o are equal. The homotopy problem now becomes more subtle. We have the six maps from $D^3X \to X$ given by

$$h_o \circ \mu \circ D\mu \qquad h_o \circ Dh_o \circ D^2h_o \qquad h_o \circ \mu \circ \mu D$$

$$h_o \circ Dh_o \circ D\mu \qquad h_o \circ \mu \circ D^2h_o \qquad h_o \circ Dh_o \circ \mu D$$

and the relations

$$h_o \circ \mu \circ D\mu \simeq h_o \circ Dh_o \circ D\mu \quad \text{via the homotopy} \quad h_1 \circ D\mu$$

$$h_o \circ \mu \circ \mu D \simeq h_o \circ Dh_o \circ \mu D \quad \text{via the homotopy} \quad h_1 \circ \mu D$$

$$h_o \circ Dh_o \circ D^2h_o \simeq h_o \circ \mu \circ D^2h_o \quad \text{via the homotopy} \quad h_1 \circ D^2h_o.$$

We also have that $h_0 \circ \mu \circ D\mu = h_0 \circ \mu \circ \mu D$ since (D, μ, η) is a monad and that $h_0 \circ Dh_0 \circ \mu D = h_0 \circ \mu \circ D^2 h_0$ since μ is a natural transformation of functors. In addition there is a homotopy between $h_0 \circ Dh_0 \circ D\mu$ and $h_0 \circ Dh_0 \circ D^2 h_0$ which is denoted by $h_0 \circ Dh_1$. The homotopy Dh_1 may be defined by $(Dh_1)(t) = D(h_1(t)): D^3 X \to DX$ for all $0 \leq t \leq 1$. It is a special property of D which enables us to piece together the $Dh_1(t)$'s to define the map $Dh_1: I \times D^3 X \to DX$. This property will be discussed later. The essential fact about Dh_1 is that it is a homotopy between $Dh_0 \circ D\mu$ and $Dh_0 \circ D^2 h_0$; in applications we may want to choose different homotopies between these two maps and denote them all by the symbol Dh_1.

Thus we have four copies of $I \times D^3 X$ and because of the above equalities we may join them together at their matching endpoints to obtain the space $\partial I^2 \times D^3 X$ where ∂I^2 is the boundary of I^2.

$$
\begin{array}{ccc}
\begin{array}{l} h_0 \circ \mu \circ D^2 h_0 \\ = h_0 \circ Dh_0 \circ \mu D \end{array} & h_1 \circ D^2 h_0 \qquad\qquad h_0 \circ Dh_0 \circ D^2 h_0 \\[1em]
h_1 \circ \mu D & \boxed{} \qquad\qquad h_0 \circ Dh_1 \\[1em]
\begin{array}{l} h_0 \circ \mu \circ \mu D \\ = h_0 \circ \mu \circ D\mu \end{array} & h_1 \circ D\mu \qquad\qquad h_0 \circ Dh_0 \circ D\mu
\end{array}
$$

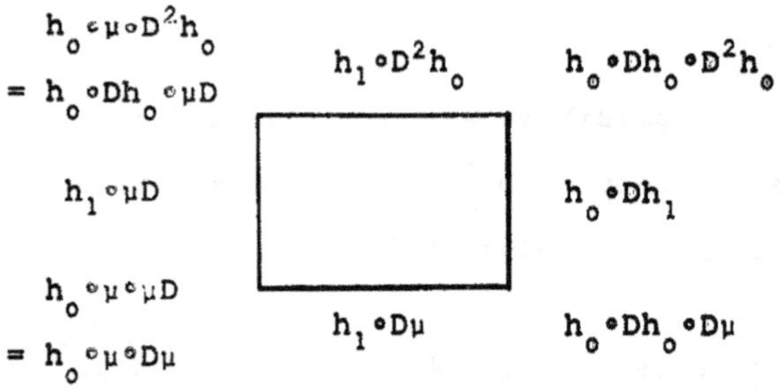

To have an appropriate generalization of D-space up to homotopy, we would like the above homotopies to be homotopic; i.e., we want to assume that the above 2-cube may be filled in by a map $h_2: I^2 \times D^3 X \to X$ such that

$$h_2(0,t,y) = h_1 \circ \mu D(t,y)$$

$$h_2(s,0,y) = h_1 \circ D\mu(s,y)$$

$$h_2(1,t,y) = h_o \circ Dh_1(t,y)$$

$$h_2(s,1,y) = h_1 \circ D^2 h_o(s,y).$$

It is apparent that for arbitrary $n > 0$, we would like to consider all of the homotopies between all of the maps $D^n X \to X$ and consider compatibility relations among them.

This discussion should motivate

Definition 1.3: Let $h_o: DX \to X$ be a map in $\mathcal{J}$ and (D,μ,η) be a monad in $\mathcal{J}$; then the pair $(X,\{h_q\})$ is a strong homotopy D-space (s.h.D-space or s.h.D-algebra) if the homotopies $h_q: I^q \times D^{q+1}X \to X$ satisfy the compatibility relations

$$h_q(t_1,\ldots,t_q,y) = h_{q-1}\circ(1 \times D^{j-1}\mu_{q-j})(t_1,\ldots,\hat{t}_j,\ldots t_q,y)$$

$$\text{if} \quad t_j = 0$$

and

$$h_q(t_1,\ldots,t_q,y) = h_{j-1}\circ(1 \times D^j h_{q-j})(t_1,\ldots,\hat{t}_j,\ldots,t_q,y)$$

$$\underline{\text{def}}\; h_{j-1}(t_1,\ldots,t_{j-1},D^j h_{q-j}(t_{j+1},\ldots,t_q,y))$$

$$\text{if} \quad t_j = 1.$$

Here, $j = 1,\ldots,q$, $q \geq 0$, $y \in D^{q+1}X$, and $\hat{t}_j$ means delete the coördinate t_j. We also require the commutative diagram

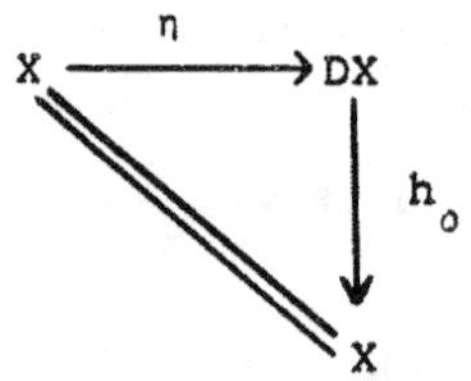

The symbol μ_{q-j} is used to denote the map $\mu(D^{q-j}X)\colon D^{q-j+2}X \to D^{q-j+1}$ in order to facilitate the notation. It is easy to see that a strict D-space may be regarded as an s.h.D-space by taking all of the higher homotopies to be constant. It is also not difficult to see by a counting argument, the properties of μ arising from the monad structure of D, and the fact that μ is a natural transformation of functors that all of the maps $D^{q+1}X \to X$ involving h_j's and μ's are taken into account by this definition.

To justify the homotopies $D^j h_{q-j}$, and for later use, we discuss the notion of continuity of a functor $F\colon \mathcal{T} \to \mathcal{T}$. This means that $F\colon \mathit{Mor}(X,Y) \to \mathit{Mor}(FX,FY)$ is a continuous map for all spaces X and Y, where $\mathit{Mor}(X,Y)$ is the function space of based maps $X \to Y$. Given a homotopy $h\colon I \times X \to Y$ and a continuous functor F, application of F to the collection of maps $h_t\colon X \to Y$, $0 \le t \le 1$, yields a family of maps $Fh_t\colon FX \to FY$ which "fit together" continuously to yield a map $I \times FX \to FY$. From an adjoint point of view we can think of our original homotopy h as a continuous map $I \to Y^X$. The continuity of F by definition means that $F\colon Y^X \to FY^{FX}$ is continuous. Thus the composite $F\circ h\colon I \to Y^X \to FY^{FX}$ is continuous. We will use the symbol Fh for the homotopy $I \times FX \to FY$ as well as for the usual map $F(I \times X) \to FY$ and hope that the context will be clear enough to avoid confusion. In practice the homotopy will factor through the usual map via a canonical map $I \times FX \to F(I \times X)$.

Thus, to validate the discussion above, we assume that D is a continuous functor. This holds, for example, if D is derived from an operad $\mathcal{D} = \{\mathcal{D}(j)\}$ [G, p. 1]. Here a canonical map $\delta: I \times DX \to D(I \times X)$ is induced by passage to quotients from the maps

$$\delta: I \times \coprod_j \mathcal{D}(j) \times X^j \to \coprod_j \mathcal{D}(j) \times (I \times X)^j$$

specified by

$$\delta(t,[d,x_1,\ldots,x_j]) = [d;(t,x_1),\ldots,(t,x_j)],$$

and $Dh: I \times DX \to DY$ is the composite of δ and $Dh: D(I \times X) \to DY$.

2. A Generalized Bar Construction.

A generalized bar construction for strong homotopy D-spaces is presented here and is used in the proof of a recognition theorem for these spaces. Let (D,μ,η) be a monad and (F,λ) a D-functor [G, p. 36]. Assume that D and F are continuous functors.

Before proceeding to our constructions, a few comments are needed concerning an appropriate category for which our main theorem is valid. A reasonable setting is the category of NDR pairs of the homotopy type of CW complexes. In this category a pair (Y,A) is said to be <u>retractile</u> if the homology exact sequence reduces to $0 \to H(A) \to H(Y) \to H(Y,A) \to 0$ [7]. It is not required that this sequence split. Retractile pairs have not only the homotopy extension property, but also in some sense a relative homotopy extension property. Stasheff [10, p. 291] has shown

<u>Proposition 2.1</u>: Let (X,m) be an H-space. If (Y,A) is a retractile pair, then given homotopic maps $f_0,f_1\colon Y \to X$ and a homotopy $g_t\colon A \to X$ such that $g_i = f_i|_A$ for $i = 0,1$, then g_t extends to a homotopy $f_t\colon Y \to X$.

We will make use of this proposition in Theorem 2.3.

Construction 9.6 [G, p. 88] may now be generalized for strong homotopy D-spaces to

<u>Construction 2.2</u>: Define a topological space that depends upon a monad (D,μ,η), a strong homotopy D-space $(X,\{h_q\})$, and a

D-functor (F,λ) by

$$\hat{B}(F,D,X) = \coprod_q I^q \times FD^qX/\sim$$

where the equivalence relation $\sim$ is defined by

$$(t_1,\ldots,t_q,x) \sim \begin{cases} (t_2,\ldots,t_q,\lambda(x)) \in I^{q-1} \times FD^{q-1}X & \text{if } t_1 = 0 \\ (t_1,\ldots,\hat{t}_j,\ldots,t_q,FD^{j-2}\mu_{q-j}(x)) \in I^{q-1} \times FD^{q-1}X & \text{if } t_j = 0 \\ (t_1,\ldots,t_{j-1},FD^{j-1}h_{q-j}(t_{j+1},\ldots,t_q,x)) \in I^{j-1} \times FD^{j-1}X & \text{if } t_j = 1 \end{cases}$$

where $x \in FD^qX$.

The primary example of such a space is given by taking the D-functor to be $(D^n,D^{n-1}\mu)$.

The key technical detail needed for our generalization of the recognition theorem and our homotopy invariance theorem is presented here as

<u>Theorem 2.3</u>: Let (DY,Y) be retractile, D come from an operad, and consider the D-functor (D,μ). Then $D\hat{B}(D,D,X) = \hat{B}(D^2,D,X)$.

<u>Proof</u>: We need the existence of a Σ_j-equivariant, 1-1, onto map

$$\mathcal{D}(j) \times (\coprod_q I^q \times D^{q+1}X/\sim)^j \to \coprod_n I^n \times (\mathcal{D}(j) \times (D^{n+1}X)^j)/\sim$$

for all j. May has proven such a theorem and has exhibited such a map for the strict D case [G, Theorem 12.2, p. 113, and also p. 126]. This map is essentially defined by using the concept of simplicial subdivision. The only difference between our required map and May's map is that his is defined on simplexes whereas ours must be defined on cubes. However, if one looks closely at the definition of $\tilde{B}(D,D,X)$ and thinks of cubes as "thickened" simplexes, the identifications in $\tilde{B}(D,D,X)$ collapse these extra faces. Although these faces are not collapsed to points in general, they have lost their parameters from the cube. We thus describe the required map as follows: to map

$$\mathcal{D}(j) \times I^{q_1} \times D^{q_1+1}X \times \dots \times I^{q_j} \times D^{q_j+1}X \to I^{\Sigma q_i} \times \mathcal{D}(j) \times (D^{\Sigma q_i+1}X)^j,$$

first calculate the map for simplexes. Here, it simplifies calculations if we define $\Delta n \subset R^n$ to be $\{(t_1,\dots,t_n) \in R^n$ such that $0 \le t_i \le 1$ and $t_1 \le \dots \le t_n\}$. It is necessary to subdivide the product $\Delta q_1 \times \dots \times \Delta q_j$ to define the map. Then "thicken" the appropriate faces of each Δq_i and obtain the cube $I^{\Sigma q_i}$ which contains $\Delta q_1 \times \dots \times \Delta q_j$. Now subdivide the cube $I^{\Sigma q_i}$ in exactly the same manner that $\Delta q_1 \times \dots \times \Delta q_j$ was subdivided and use exactly the same degeneracy maps on $D^{q_1+1}X \times \dots \times D^{q_j+1}X \to (D^{\Sigma q_i+1}X)^j$ that are used in the simplicial case. The equivalence relation in $\tilde{B}(D^2,D,X)$ will guarantee that our map is well-defined and continuous if we require our higher homotopies be relative homotopies with respect to the subspaces of D^qX given by the various η's: $D^{q-1}X \to D^qX$. Noting that each D^qX is an H-space and recalling the earlier comments about our category, we utilize Proposition 2.1 to guarantee that our higher homotopies behave properly on subspaces. ∎

<u>Corollary 2.4</u>: $\tilde{B}(D,D,X)$ is a D-space.

<u>Proof</u>: We take for the D-structure map on $\tilde{B}(D,D,X)$ the map $\tilde{B}(D^2,D,X) \to \tilde{B}(D,D,X)$ induced by the natural transformation $\mu: D^2 \to D$. We also denote this structure map by μ. ∎

It is perhaps instructive to examine a few examples of the procedure in 2.3:

<u>Example 1</u>: Let us map

$$\mathcal{D}(2) \times I \times D^2X \times I \times D^2X \to I^2 \times \mathcal{D}(2) \times D^3X \times D^3X$$

by

$$(d,s,x,t,y) \to \begin{cases} (s, \frac{t-s}{1-s}, d, D^2\eta(x), D\eta D(y)) & \text{if } s \le t \\ (t, \frac{s-t}{1-t}, d, D\eta D(x), D^2\eta(y)) & \text{if } s \ge t \end{cases}$$

It is clear that the coordinates of the cube in the range are just those of the thickened simplex; but for this thickening, the map is the same as in the strict D case. The image of the point $(d,1,x,1,y)$ is $(1, \frac{0}{0}, d, D^2\eta(x), D\eta D(y))$. Since this point is equivalent to $(d, Dh_1(\frac{0}{0}, D^2\eta(x)), Dh_1(\frac{0}{0}, D\eta D(y)))$, $\frac{0}{0}$ may be taken to be any $0 \le t \le 1$ if

$$Dh_1 | D^2\eta(D^2X) = Dh_1 | D\eta D(D^2X) = D^2h_o$$

for all t; this last equality always holds for $t = 0,1$ and the assumption that (DY,Y) is retractile allows us to alter Dh_1 for $0 < t < 1$ so that the equality holds for all t. This guarantees that the map is well-defined and continuous.

<u>Example 2</u>: Define the map

$$\mathcal{D}(2) \times I \times D^2X \times I^2 \times D^3X \to I^3 \times \mathcal{D}(2) \times D^4X \times D^4X$$

by

$$d,(s,x),(t_1,t_2,y) \to \begin{cases} (s,\frac{t_1-s}{1-s},t_2,d,D^3\eta\circ D^2\eta(x),D\eta D^2(y)) & \text{if } s \leq t_1 \\ (t_1,\frac{s-t_1}{1-t_1},\frac{t_2-s+t_2(1-s)}{1-s+t_2(1-s)},d,D^3\eta\circ D\eta D(x),D^2\eta D(y)) & \\ \qquad \text{if } t_1 \leq s \leq t_2 + t_1(1-t_2) & \\ (t_1,t_2,\frac{s-t_2-t_1(1-t_2)}{1-t_2-t_1(1-t_2)},d,D^2\eta D\circ D\eta D(x),D^3\eta(y)) & \\ \qquad \text{if } t_2 + t_1(1-t_2) \leq s \leq 1. & \end{cases}$$

In this case our subdivision of I^3 consists of the three regions that look like

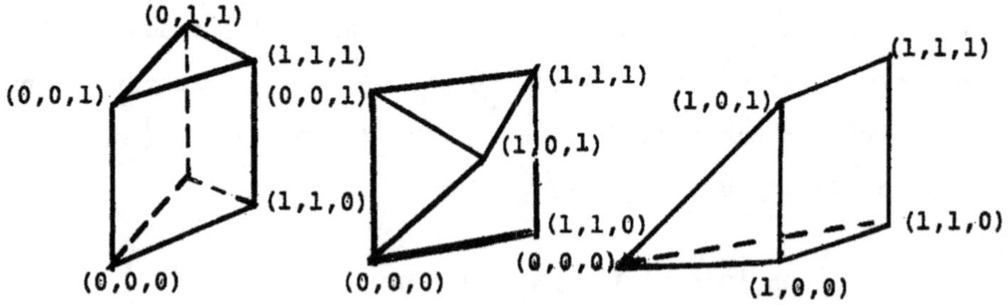

where the surface that the second and third regions have in common is defined by $s = t_2 + t_1(1 - t_2)$. This is again the generalization of the problem of subdividing $\Delta 1 \times \Delta 2$. Let us examine what occurs at the edge $(1,x)$, $(1,t_2,y)$; we will ignore the factor $\mathcal{D}(2)$ for the moment since it remains unchanged in these calculations. If we define $\frac{0}{0} = 0$ and $\frac{t_2-1}{0} = t_2$, we have

$$(1,x),(1,t_2,y) \to (1,\tfrac{0}{0},t_2,D^3\eta\circ D^2\eta(x),D\eta D^2(y))$$

$$\sim Dh_2(\tfrac{0}{0},t_2,D^3\eta\circ D^2\eta(x),D\eta D^2(y))$$

$$= Dh_1(t_2,Dh_o(x),y)$$

$$(1,x),(1,t_2,y) \to (1,\tfrac{0}{0},\tfrac{t_2-1}{0},D^3\eta\circ D\eta D(x),D^2\eta D(y))$$

$$\sim Dh_2(\tfrac{0}{0},\tfrac{t_2-1}{0},D^3\eta\circ D\eta D(x),D^2\eta D(y))$$

$$= Dh_1(t_2,Dh_o(x),y)$$

and

$$(1,x),(1,t_2,y) \to (1,t_2,\tfrac{0}{0},D^2\eta D\circ D\eta D(x),D^3\eta(y))$$

$$\sim Dh_2(t_2,\tfrac{0}{0},D^2\eta D\circ D\eta D(x),D^3\eta(y))$$

$$= Dh_1(t_2,Dh_o(x),y)$$

if

$$Dh_2(s,t,D^3\eta(y)) = Dh_1(s,y)$$

$$Dh_2(s,t,D\eta D^2(y)) = Dh_1(t,y)$$

and

$$Dh_2(0,t,D^2\eta D(y)) = Dh_1(t,y).$$

Again, this will be true under our restriction that Dh_2 be a homotopy relative to the subspace $D^3\eta(D^3X)$ union $D\eta D^2(D^3X)$; the assumption that (DX,X) be retractile guarantees this.

Notation: Throughout the remainder of this work, the strict D-space $\hat{B}(D,D,X)$ will be denoted by the symbol UX.

Proposition 2.5: Let $(X,\{h_q\})$ be a strong homotopy D-space. Then X is a deformation retract of UX.

Proof: Define a map $i: X \to UX$ by $i(x) = \eta(x) \in I^\circ \times DX \subset UX$. Now define a map $r: UX \to X$ by

$$r(t_1,\ldots,t_q,y) = h_q(t_1,\ldots,t_q,y).$$

To see that r is well-defined, suppose that $t_j = 0$. Then

$$\begin{aligned} r(t_1,\ldots,t_q,y) &= h_q(t_1,\ldots,t_j,\ldots,t_q,y) \\ &= h_{q-1}(t_1,\ldots,\hat{t}_j,\ldots,t_q,D^{j-1}\mu_{q-j}(y)) \end{aligned}$$

by the properties of h_q.

But

$$(t_1,\ldots,t_q,y) \sim (t_1,\ldots,\hat{t}_j,\ldots,t_q,D^{j-1}\mu_{q-j}(y))$$

nd

$$r(t_1,\dots,\hat{t}_j,\dots,t_q,D^{j-1}\mu_{q-j}(y)) = h_{q-1}(t_1,\dots,\hat{t}_j,\dots,t_q,D^{j-1}\mu_{q-j}(y))$$

gain. Now suppose that $t_j = 1$. Then

$$r(t_1,\dots,t_q,y) = h_q(t_1,\dots,t_q,y)$$

$$= h_{j-1}(t_1,\dots,t_{j-1},D^j h_{q-j}(t_{j+1},\dots,t_q,(y)).$$

On the other hand, since

$$(t_1,\dots,t_q,y) \sim (t_1,\dots,t_{j-1},D^j h_{q-j}(t_{j+1},\dots,t_q,y)) \in I^{j-1} \times D^j X,$$

$$r(t_1,\dots,t_{j-1},D^j h_{q-j}(t_{j+1},\dots,t_q,y))$$

$$= h_{j-1}(t_1,\dots,t_{j-1},D^j h_{q-j}(t_{j+1},\dots,t_q,y))$$

gain. Thus r is indeed well-defined. We also have that $r\circ i = id_X$ since

$$r\circ i(x) = r(\eta(x)) = h_0(\eta(x)) = x.$$

To show that $i\circ r$ is homotopic to the identity of UX, define a homotopy

$$F\colon I \times UX \to UX$$

by

$$F(s,t_1,\dots,t_q,y) = (s,t_1,\dots,t_q,\eta_{q+1}(y)) \in I^{q+1} \times D^{q+2}X$$

where $(s,t_1,\ldots,t_q,y) \in I \times I^q \times D^{q+1}X$ and $\eta_{q+1} = \eta D^{q+1}$. To see that F is well-defined, first let $t_j = 0$. Then

$$(s,t_1,\ldots,t_q,y) \sim (s,t_1,\ldots,\hat{t}_j,\ldots,t_q,D^{j-1}\mu_{q-j}(y))$$

and

$$*)\quad F(s,t_1,\ldots,t_q,y) = (s,t_1,\ldots,t_q,\eta_{q+1}(y))$$

$$\sim (s,t_1,\ldots,\hat{t}_j,\ldots,t_q,D^j\mu_{q-j+1}\circ\eta_{q+1}(y)).$$

Also,

$$**)\quad F(s,t_1,\ldots,\hat{t}_j,\ldots,t_q,D^{j-1}\mu_{q-j}(y))$$

$$= (s,t_1,\ldots,\hat{t}_j,\ldots,t_q,\eta_q D^{j-1}\mu_{q-j}(y)).$$

The equality of the two points on the right hand side of *) and **) follows from the equality $D^j\mu_{q-j+1}\circ\eta_{q+1} = \eta_q D^{j-1}\mu_{q-j}$ which is a consequence of the naturality of η. On the other hand, if $t_j = 1$,

$$(s,t_1,\ldots,t_q,y) \sim (s,t_1,\ldots,t_{j-1},D^j h_{q-j}(t_{j+1},\ldots,t_q,y))$$

and we have

$$*)\quad F(s,t_1,\ldots,t_q,y) = (s,t_1,\ldots,t_q,\eta_{q+1}(y))$$

$$\sim (s,t_1,\ldots,t_{j-1},D^{j+1}h_{q-j}(t_{j+1},\ldots,t_q,\eta_{q+1}(y))$$

and

$$**)\quad F(s,t_1,\ldots,t_{j-1},D^j h_{q-j}(t_{j+1},\ldots,t_q,y))$$

$$= (s,t_1,\ldots,t_{j-1},\eta_{j+1}\circ D^j h_{q-j}(t_{j+1},\ldots,t_q,y)).$$

Again, equality of the right hand sides of *) and **) follows from the naturality of η. Thus F is well-defined.

When $s = 0$, we have

$$F(0,t_1,\ldots,t_q,y) = (0,t_1,\ldots,t_q,\eta_{q+1}(y))$$

$$\sim (t_1,\ldots,t_q,\mu_{q+1}\eta_{q+1}(y))$$

$$= (t_1,\ldots,t_q,y).$$

Thus $F|\{0\} \times UX$ = identity on UX. When $s = 1$, we have

$$F(1,t_1,\ldots,t_q,y) = (1,t_1,\ldots,t_q,\eta_{q+1}(y))$$

$$\sim Dh_q(t_1,\ldots,t_q,\eta_{q+1}(y))$$

$$= \eta_0\circ h_q(t_1,\ldots,t_q,y)$$

$$= i\circ r(t_1,\ldots,t_q,y).$$

Thus $F|\{1\} \times UX = i\circ r$ and we are done. ∎

<u>Proposition 2.6</u>: Let (X,ξ) be a D-space. Then $r\colon (UX,\mu) \to (X,\xi)$ is a D-map.

<u>Proof</u>: $r\colon I^n \times D^{n+1}X \to X$ is the constant homotopy which we may take to be the map $\xi\circ D\xi\circ\ldots\circ D^n\xi$. To show that r is a D-map, it

suffices to show that the diagram

$$\begin{array}{ccc} I^n \times D^{n+2}X & \xrightarrow{\quad Dr \quad} & DX \\ {\scriptstyle 1 \times \mu_n}\downarrow & & \downarrow{\scriptstyle r} \\ I^n \times D^{n+1}X & \xrightarrow{\quad r \quad} & X \end{array}$$

commutes. This is clear since $r \circ Dr = \xi \circ D\xi \circ \ldots \circ D^{n+1}\xi$ $= \xi \circ D\xi \circ \ldots \circ D^n\xi \circ \mu_n = r \circ \mu_n$ since (X,ξ) is a D-space. ∎

<u>Remarks</u>: 1) We denote the category of D-spaces and D-maps by $D[T]$. We then have that U is a functor from $D[T]$ to itself. If $f: X \to Y$ is a D-map we define $U(f): UX \to UY$ by

$$U(f)(t_1,\ldots,t_q,z) = (t_1,\ldots,t_q,D^{q+1}f(z))$$

where $z \in D^{q+1}X$. That $U(f)$ is well-defined follows from the facts that f is a D-map and that μ is a natural transformation. The naturality of μ is also used to see that $U(f)$ is a D-map. That U respects composition and the identity map is obvious.
2) The retraction $r: UX \to X$ is a natural transformation $U \to 1$ in $D[T]$. This is easily verified by utilizing the fact that morphisms in $D[T]$ are D-maps.

The recognition principles for n-fold and infinite loop spaces developed in [G,G', and R] now generalize directly to s.h.D-spaces X over appropriate monads D. We need only apply the D-space recognition principles to $UX = \tilde{B}(D,D,X)$. When X is a D-space, $r: UX \to X$ is a D-map and a homotopy equivalence, hence the uniqueness results in [G] and [G'] for the de-loopings of

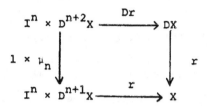

D-spaces apply to show that the de-loopings of X regarded as a D-space and of X regarded as an s.h.D-space are equivalent. In particular this remark applies to show that the de-loopings of UX and of UUX are equivalent for an s.h.D-space X. We also note that for the relevant monads, H_*DY is known as a functor of H_*Y and the pair (DY,Y) is always retractile.

3. Strong Homotopy Maps of Algebras over Monads

It is quite difficult to construct a category with s.h.D-spaces as objects. One would certainly expect a morphism in such a category to preserve the higher homotopy structures of both the domain and range spaces at least up to compatible higher homotopies. There are thus three types of homotopies to be remembered for such a map. Moreover, even after such maps are defined, composition of them is awkward since one must glue together the structure homotopies of each to define the higher homotopies of the composite; such a procedure is of course well-defined only up to homotopy. Associativity of such a composition presents even more difficulties.

We are able to bypass the above problems with the following definition.

Definition 3.1: Let X and Y be s.h.D-spaces. An SHD-map $X \to Y$ is a D-map $g: UX \to UY$. (Such a map may be thought of as the underlying map $\bar{g} = r \circ g \circ \eta: X \to Y$, where $r: UY \to Y$ is the retraction and $\eta: X \to UX$ is the inclusion as defined in Proposition 2.5 together with the additional information that g is a D-map.) We say that g represents $\bar{g}$ as an SHD-map.

Composition of SHD-maps is just composition of the representing D-maps and we have a category SHD[T] whose objects are s.h.D-spaces and morphisms are SHD-maps. We say that two SHD-maps are homotopic if the corresponding D-maps $UX \to UY$ are homotopic through D-maps. Thus we may define

$$[X,Y]_{SHD} = [UX,UY]_D$$

where [,] means homotopy classes of maps and the subscript refers to the appropriate category. It is obvious that with these definitions U is a fully faithful functor $hSHD[T] \to hD[T]$ where h preceding a symbol denoting a category means take as morphisms homotopy classes of the morphisms in that category. Hence $hSHD[T]$ is equivalent to the full subcategory of $hD[T]$ with objects all UX.

Of course, with this definition, appropriate de-loopability of SHD-maps between s.h.D-spaces is automatic from the naturality on D-maps of the constructions of [G] and [G']. It is also clear that the de-loopings of strict D-maps and of strict D-maps regarded via U as SHD-maps are consistent. Thus with Definition 3.1 the generalization of the recognition principles for iterated loop spaces is complete.

At this point the analysis of homotopy invariance in $SHD[T]$ begins. We are confronted with the problem of constructing SHD-maps from data that arise in nature, such as a homotopy equivalence $X \to Y$ when Y is an s.h.D-space. As Definition 3.1 is impractical for such a procedure, we complete this section with a direct homotopical definition of s.h.D-maps $X \to Y$ when either X or Y is a D-space and of a homotopy between two such maps when the range space is a strict D-space. In the next two sections we prove that certain maps in and out of the D-spaces UX are s.h.D-maps and analyze the relationship between the notions of s.h.D-maps and of SHD-maps. This analysis will allow us to reduce the proof of the homotopy invariance theorems on the SHD-map level to certain geometrical homotopy invariance theorems concerning s.h.D-maps between D-spaces. The latter results occupy the last three sections of this work. A geometrical alternative to Definition 3.1 is presented in the appendix.

Definition 3.2: Let $(X,\{\xi_n\})$ be an s.h.D-space and (Y,ϕ) a D-space. Then $f\colon X \to Y$ is said to be an s.h.D-map if there exists a collection of homotopies $\{f_n\}$ with

$$f_n\colon I^n \times D^n X \to Y$$

satisfying

$$f_n(t_1,\ldots,t_n,z) = \begin{cases} \phi\circ Df_{n-1}(t_2,\ldots,t_n,z) & \text{if } t_1 = 0 \\ f_{n-1}(t_1,\ldots,\hat{t}_j,\ldots,t_n,D^{j-2}\mu_{n-j}(z)) & \text{if } t_j = 0 \\ f\circ\xi_{n-1}(t_2,\ldots,t_n,z) & \text{if } t_1 = 1 \\ f_{j-1}(t_1,\ldots,t_{j-1},D^{j-1}\xi_{n-j}(t_{j+1},\ldots,t_n,z) & \text{if } t_j=1. \end{cases}$$

As it frequently occurs, if the domain space X happens to be a strict D-space, we use this as the definition of an s.h.D-map between two D-spaces where the ξ_n are constant homotopies.

The notion of a homotopy between such maps may also be defined.

Definition 3.3: Let $\{f_n\}$ and $\{g_n\}$ be s.h.D-maps from an s.h.D-space $(X,\{\xi_n\})$ to a D-space (Y,ϕ). Then a homotopy between $\{f_n\}$ and $\{g_n\}$ is a collection of maps

$$h_n\colon I^{n+1} \times D^n X \to Y$$

satisfying

$$h_n(t_1,\dots,t_{n+1},z) = \begin{cases} f_n(t_2,\dots,t_{n+1},z) & \text{if } t_1 = 0 \\ g_n(t_2,\dots,t_{n+1},z) & \text{if } t_1 = 1 \\ \phi\circ Dh_{n-1}(t_1,t_3,\dots,t_{n+1},z) & \text{if } t_2 = 0 \\ h_o\circ\xi_{n-1}(t_1,t_3,\dots,t_{n+1},z) & \text{if } t_2 = 1 \\ h_{n-1}(t_1,\dots,\hat{t}_j,\dots,t_{n+1},D^{j-2}\mu_{n-j}(z)) & \text{if } t_{j+1} = 0 \\ h_{j-1}(t_1,\dots,t_j,D^{j-1}\xi_{n-j}(t_{j+2},\dots,t_{n+1},z)) & \\ & \text{if } t_{j+1} = 1. \end{cases}$$

Note that $\{h_n|t_1 = c\}$ is an s.h.D-map from $(X,\{\xi_n\})$ to (Y,ϕ) in the sense of Definition 3.2 for each $0 \le c \le 1$. Moreover, with this definition, it is clear that homotopy is an equivalence relation between s.h.D-maps $X \to Y$. Again, if the domain space X is a strict D-space, we use the above as the definition of a homotopy between s.h.D-maps from one D-space to another by taking the structure homotopies ξ_n to be constant.

We shall also utilize one more type of s.h.D-map.

Definition 3.4: Let (X,ξ) be a D-space and $(Y,\{\phi_n\})$ an s.h.D-space. Then $f\colon X \to Y$ is said to be an s.h.D-map if there exists a collection of homotopies $\{f_n\}$ with

$$f_n\colon I^n \times D^n X \to Y$$

satisfying

$$f_n(t_1,\ldots,t_n,z) = \begin{cases} f_{n-1}(t_1,\ldots,t_{n-1},D^{n-1}\xi(z)) & \text{if } t_n = 0 \\ f_{n-1}(t_1,\ldots,\hat{t}_j,\ldots,t_n,D^{j-1}\mu_{n-j-1}(z)) & \text{if } t_j = 0 \\ \phi_{n-1}(t_1,\ldots,t_{n-1},D^n f(z)) & \text{if } t_n = 1 \\ \phi_{j-1}(t_1,\ldots,t_{j-1},D^j f_{n-j}(t_{j+1},\ldots,t_n,z)) & \text{if } t_j = 1. \end{cases}$$

We do not require the definition of a homotopy between such maps.

It is profitable to discuss the motivation for the above definitions to see exactly what these higher homotopies are doing. Let us examine Definition 3.4. Here, we have (X,ξ) a D-space, $(Y,\{\phi_n\})$ an s.h.D-space, and $f\colon X \to Y$ a map. For f to preserve the monad structure on the spaces, we require at a minimum that the diagram

$$\begin{array}{ccc} DX & \xrightarrow{Df} & DY \\ \downarrow{\scriptstyle \xi} & & \downarrow{\scriptstyle \phi} \\ X & \xrightarrow{f} & Y \end{array}$$

commute up to homotopy; i.e., we want a homotopy $f_1\colon I \times DX \to Y$ such that $f_1|0 = f\circ\xi$ and $f_1|1 = \phi\circ Df$. In addition we want to require that the diagram

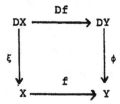

$$\begin{array}{ccc} I \times D^2X & \xrightarrow{\;1 \times D^2f\;} & I \times D^2Y \\ {\scriptstyle \xi_1}\downarrow & & \downarrow{\scriptstyle \phi_1} \\ X & \xrightarrow{\;f\;} & Y \end{array}$$

commute up to homotopy. This diagram provides us with a map

$$f_2 : I^2 \times D^2Y \to X$$

which says that the maps determined by the endpoints of ϕ_1 and ξ_1 are homotopic. For f_2 to be compatible with f_1 and f, we need to examine the diagrams

$$\begin{array}{ccc} D^2X & \xrightarrow{\;D^2f\;} & D^2Y \\ {\scriptstyle \mu}\downarrow & /\!/\!/ & \downarrow{\scriptstyle \mu} \\ DX & \xrightarrow{\;Df\;} & DY \\ {\scriptstyle \xi}\downarrow & \sim & \downarrow{\scriptstyle \phi} \\ X & \xrightarrow{\;f\;} & Y \end{array} \qquad \text{and} \qquad \begin{array}{ccc} D^2X & \xrightarrow{\;D^2f\;} & D^2Y \\ {\scriptstyle D\xi}\downarrow & \sim & \downarrow{\scriptstyle D\phi} \\ DX & \xrightarrow{\;Df\;} & DY \\ {\scriptstyle \xi}\downarrow & \sim & \downarrow{\scriptstyle \phi} \\ X & \xrightarrow{\;f\;} & Y \end{array}$$

They imply that f_2 is not naturally defined on I^2 but rather is defined on a pentagonal subset of the plane:

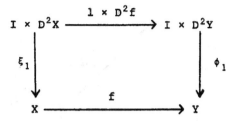

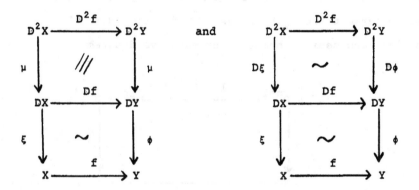

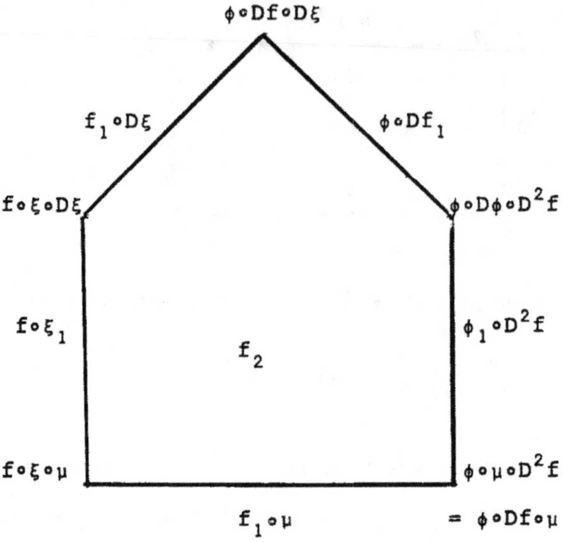

Since the edge $f \circ \xi_1$ is a constant map, we may collapse it to a point to parametrize f_2 by I^2 as in Definition 3.4. By examining each desired homotopy commutative diagram

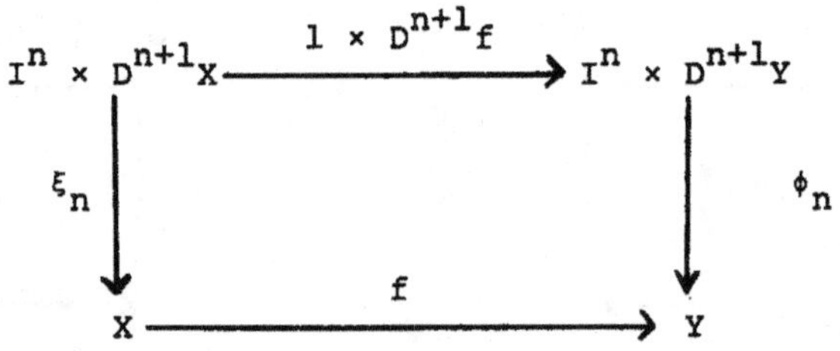

studying compatibility conditions, and collapsing constant faces, we arrive at Definition 3.4.

4. Examples of Strong Homotopy Maps

Various maps associated with Construction 2.2 are shown to be s.h.D-maps and a technical reparametrization lemma is proven.

Explicitly we have

Theorem 4.1: If $(X,\{\xi_n\})$ is an s.h.D-space, then the inclusion $\eta\colon X \to UX$ and the retraction $r\colon UX \to X$ are s.h.D-maps.

Proof: An s.h.D-structure is defined for the inclusion $\eta\colon X \to UX$ by

$$\eta_n\colon I^n \times D^nX \to UX$$

where

$$\eta_n(t_1,\ldots,t_n,z) = (t_1,\ldots,t_n,\eta D^n(z)).$$

Recall that ηD^n is the natural transformation $\eta\colon D^n \to DD^n$. We verify the conditions set out in Definition 3.2:

If $t_1 = 0$,

$$\begin{aligned} \eta_n(t_1,\ldots,t_n,z) &= (0,t_2,\ldots,t_n,\eta D^n(z)) \\ &\sim (t_2,\ldots,t_n,\mu D^{n-1}\circ\eta D^n(z)) \quad \text{by 2.2} \\ &= (t_2,\ldots,t_n,\mu D^{n-1}\circ D\eta D^{n-1}(z)) \quad \text{since D is a monad} \\ &= \mu D^{n-1}\circ D\eta_{n-1}(t_2,\ldots,t_n,z). \end{aligned}$$

If $t_j = 0$,

$$\eta_n(t_1,\ldots,t_n,z) = (t_1,\ldots,t_n,\eta D^n(z))$$
$$\sim (t_1,\ldots,\hat{t}_j,\ldots,t_n,D^{j-1}\mu D^{n-j}\circ\eta D^n(z))$$
$$= \eta_{n-1}\circ D^{j-2}\mu D^{n-j}(t_1,\ldots,\hat{t}_j,\ldots,t_n,z)$$

since η is natural.

If $t_1 = 1$,

$$\eta_n(t_1,\ldots,t_n,z) = (1,t_2,\ldots,t_n,\eta D^n(z))$$
$$\sim D\xi_{n-1}(t_2,\ldots,t_n,\eta D^n(z))$$
$$= \eta\circ\xi_{n-1}(t_2,\ldots,t_n,z) \qquad \text{since } \eta \text{ is natural.}$$

Finally, if $t_j = 1$,

$$\eta_n(t_1,\ldots,t_n,z) = (t_1,\ldots,t_n,\eta D^n(z))$$
$$\sim (t_1,\ldots,t_{j-1},D^j\xi_{n-j}(t_{j+1},\ldots,t_n,\eta D^n(z)))$$
$$= \eta D^{j-1}(t_1,\ldots,t_{j-1},D^{j-1}\xi_{n-j}(t_{j+1},\ldots,t_n,z))$$

again since η is natural.

To see that the retraction $r\colon UX \to X$ is an s.h.D-map we define

$$r_n\colon I^n \times D^nUX \to X$$

by

$$r_n(t_1,\ldots,t_n,z) = \xi_{q+n}(t_1,\ldots,t_n,s_1,\ldots,s_q,y)$$

where

$$z = (s_1,\ldots,s_q,y) \in I^q \times D^{q+n+1}X.$$

In this definition, we utilize the fact that $D^nUX = \mathcal{B}(D^{n+1},D,X)$. We now verify the compatibility requirements of Definition 3.4. First suppose that $t_n = 0$. Then

$$\begin{aligned} r_n(t_1,\ldots,t_{n-1},0,z) &= \xi_{q+n}(t_1,\ldots,t_{n-1},0,s_1,\ldots,s_q,y) \\ &= \xi_{q+n-1}(t_1,\ldots,t_{n-1},s_1,\ldots,s_q,D^{n-1}\mu_q(y)) \\ &= r_{n-1}(t_1,\ldots,t_{n-1},D^{n-1}\mu_q(z)). \end{aligned}$$

If $t_j = 0$, we have

$$\begin{aligned} r_n(t_1,\ldots,t_n,z) &= \xi_{q+n}(t_1,\ldots,t_n,s_1,\ldots,s_q,y) \\ &= \xi_{q+n-1}(t_1,\ldots,\hat{t}_j,\ldots,t_n,s_1,\ldots,s_q,D^{j-1}\mu_{q+n-j}(y)) \\ &= \xi_{q+n-1}(t_1,\ldots,\hat{t}_j,\ldots,t_n,s_1,\ldots,s_q,D^{j-1}\mu_{n-j-1}D^{q+1}(y)) \\ &= r_{n-1}(t_1,\ldots,\hat{t}_j,\ldots,t_n,D^{j-1}\mu_{n-j-1}(z)). \end{aligned}$$

On the other hand, if $t_n = 1$, we have

$$r_n(t_1,\ldots,t_n,z) = \xi_{q+n}(t_1,\ldots,1,s_1,\ldots,s_q,y)$$

$$= \xi_{n-1}(t_1,\ldots,t_{n-1},D^n\xi_q(s_1,\ldots,s_q,y))$$

$$= \xi_{n-1}(t_1,\ldots,t_{n-1},D^n r(z)).$$

Finally, if $t_j = 1$, we have

$$r_n(t_1,\ldots,t_n,z) = \xi_{q+n}(t_1,\ldots,t_n,s_1,\ldots,s_q,y)$$

$$= \xi_{j-1}(t_1,\ldots,t_{j-1},D^j\xi_{q-j+n}(t_{j+1},\ldots,t_n,s_1,\ldots,s_q,y))$$

$$= \xi_{j-1}(t_1,\ldots,t_{j-1},D^j r_{n-j}(t_{j+1},\ldots,t_n,z)).$$

One must also check that each r_n is well-defined; however, this is a straightforward consequence of the fact that D^nUX is constructed by means of the identities in the definition of an s.h.D-space and its proof is similar to that of Proposition 2.5. ∎

Another useful s.h.D-map is given in

<u>Theorem 4.2</u>: Let $\{f_n\}: (X,\{\xi_n\}) \to (Y,\{\phi_n\})$ be an s.h.D-map where either $(X,\{\xi_n\})$ or $(Y,\{\phi_n\})$ is a D-space (so that the ξ_n or ϕ_n are constant homotopies). Then the composition $\eta\circ f: X \to UY$ is an s.h.D-map.

<u>Proof</u>: By reparametrizing f_n, we define a map $f'_n: I^n \times D^nX \to Y$ such that

$$f'_n(t_1,\dots,t_n,z) = \begin{cases} f\circ\xi_{n-1}(t_2,\dots,t_n,z) & \text{if } t_1 = 1 \\ f_{j-1}(t_1,\dots,t_{j-1},D^{j-1}\xi_{n-j}(t_{j+1},\dots,t_n,z)) & \text{if } t_j = 1 \\ f_{n-1}(t_1,\dots,\hat{t}_j,\dots,t_n,D^{j-2}\mu_{n-j}(z)) & \text{if } t_j = 0 \\ \{\phi_{i-1}\circ D^i f'_{n-i}(2t_2,\dots,2t_i,2t_{i+1}-1,t_{i+2},\dots,t_n,z)\} & \\ \quad \text{where } t_1,\dots,t_i \le \frac{1}{2}\,,\ t_{i+1} \ge \frac{1}{2}\,,\ i = 1,\dots,n & \\ \quad \text{if } t_1 = 0. & \end{cases}$$

We do not give an explicit formula for f'_n on the interior of I^n since we have merely deformed the boundary formulas for f_n; the existence of f_n thus implies that of f'_n. Examination of the case $n = 2$ may clarify this. Here, depending on whether X or Y is a D-space, f_2 is given by

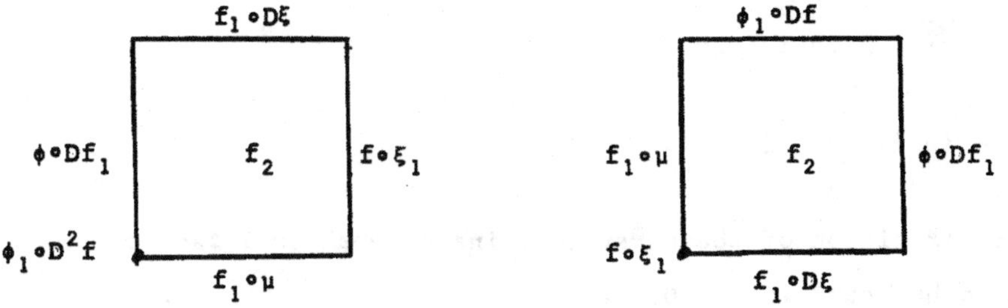

and f'_2 is given by

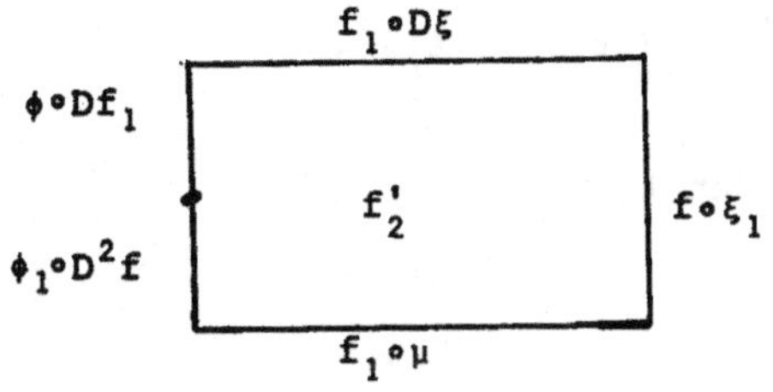

Now define a collection of homotopies

$$g_n: I^n \times D^n X \to UY$$

by

$$g_n(t_1,\ldots,t_n) = \{\eta_i \circ D^i f'_{n-i}(2t_1,\ldots,2t_i,2t_{i+1} - 1,t_{i+2},\ldots,t_n)\}$$

where $t_1,\ldots,t_i \leq \frac{1}{2}$, $t_{i+1} \geq \frac{1}{2}$, $i = 0,\ldots,n$. This amounts to taking the cube defined by f_n and glueing to its reparametrized $t_1 = 0$ face n other n-cubes. Note that we evaluate g_n by applying η_i to the first i coordinates and $D^i f'_{n-i}$ to the others. The coordinate $D^n X$ is omitted in these calculations.

To see that it makes sense to do this glueing, if $t_1 = \frac{1}{2}$ we have

$$g_n = \{\eta_i \circ D^i f'_{n-i}(1,2t_2,\ldots,2t_i,2t_{i+1} - 1,t_{i+2},\ldots,t_n)\}$$

with $i \geq 1$

$$= \{\eta \circ \phi_{i-1} \circ D^i f'_{n-i}(2t_2,\ldots,t_n)\}$$

by the definition of the first coordinate equal to 1 face of η_i. On the other hand if $i = 0$, we have

$$g_n = \{\eta \circ f_n'(0,t_2,\ldots,t_n)\}$$

$$= \{\eta \circ \phi_{i-1} \circ D^i f'_{n-i}(2t_2,\ldots,t_n)\}.$$

We thus have agreement when $t_1 = \frac{1}{2}$.

To see that it makes sense to glue together the n-cubes that are defined when $t_1 \leq \frac{1}{2}$, we have to consider the case when $t_j = \frac{1}{2}$ where $j \leq i + 1$. If $j < i + 1$, we have

$$g_n = \{\eta_i \circ D^i f'_{n-i}(2t_1, \ldots, 1, \ldots, 2t_{i+1} - 1, t_{i+2}, \ldots, t_n)\}$$

$$= \{\eta_{j-1} \circ D^{j-1} \phi_{i-j} \circ D^i f'_{n-i}(2t_1, \ldots, 2t_{j-1}, 2t_{j+1}, \ldots, 2t_i, 2t_{i+1} - 1, \ldots t_n)\}$$

by the definition of the s.h.D-relation on η. If $j = i + 1$,

$$g_n = \{\eta_i \circ D^i f'_{n-i}(2t_1, \ldots, 2t_i, 0, t_{i+2}, \ldots, t_n)\}$$

$$= \{\eta_i \circ D^i \phi_{k-1} \circ D^{i+k} f'_{n-i-k+1}\}$$

with $k = 0, \ldots, n - j - 1$. We re-index by letting $i = j + k + 1$ and get

$$g_n = \{\eta_{j-1} \circ D^{j-1} \phi_{i-j} \circ D^i f'_{n-i}\}$$

and thus have agreement.

It remains to show that $\{g_n\}$ as defined satisfies Definition 3.2. If $t_1 = 0$,

$$g_n(0, t_2, \ldots, t_n) = \{\eta_i \circ D^i f'_{n-i}(0, \ldots, 2t_i, 2t_{i+1} - 1, \ldots, t_n)\}$$

$$= \{\mu \circ D\eta_{i-1} \circ D^i f'_{n-i}(2t_2, \ldots, 2t_i, 2t_{i+1} - 1, \ldots, t_n)\}$$

$$= \{\mu \circ D(\eta_{i-1} \circ D^{i-1} f'_{n-i})(2t_2, \ldots, t_n)\}$$

$$= \{\mu \circ D(\eta_k \circ D^k f'_{n-1-k})\}$$

if we re-index by $k = i - 1$, $k = 0, \ldots, n-1$

$$= \mu \circ D g_{n-1}.$$

If $t_1 = 1$,

$$g_n(1,t_2,\ldots,t_n) = \eta\circ f_n'(1,t_2,\ldots,t_n)$$

$$= \eta\circ f\circ\xi_{n-1}$$

$$= g\circ\xi_{n-1}.$$

If $t_j = 0$, we have two cases:

<u>Case 1</u>: $j \leq i$. Then

$g_n(t_1,\ldots,0,\ldots,t_i,\ldots,t_n)$

$$= \{\eta_{i-1}\circ D^{j-2}\mu_{i-j}\circ D^i f'_{n-i}(\ldots 2\hat{t}_j\ldots)\}$$

$$= \{\eta_{i-1}\circ D^{i-1}f'_{n-i}\circ D^{j-2}\mu_{n-j}(\ldots 2\hat{t}_j\ldots)\}$$

by the naturality of μ

$$= \{\eta_k\circ D^k f'_{n-1-k}\circ D^{j-2}\mu_{n-j}\}$$

if we re-index via $k = i - 1$, $k = 0,\ldots,n - 1$

$$= g_{n-1}\circ D^{j-2}\mu_{n-j}.$$

<u>Case 2</u>: $j > i$. Let $j = i + k$ where $k = 1,\ldots,n - i$. Then

$$g_n = \{\eta_i\circ D^i(f'_{n-i-1}\circ D^{k-2}\mu_{n-i-k})\}$$ by the s.h.D-property of f

$$= \{\eta_i\circ D^i f'_{n-i-1}\circ D^{i+k-2}\mu_{n-i-k}\}$$

$$= \{\eta_i\circ D^i f'_{n-i-1}\circ D^{j-2}\mu_{n-j}\},\ i = 0,\ldots,n - 1$$

$$= g_{n-1}\circ D^{j-2}\mu_{n-j}.$$

Finally, if $t_j = 1$, we have that $j \geq i + 1$ and

$$g_n = \{\eta_i \circ D^i f'_{k-1} \circ D^{i+k-1} \xi_{n-i-k}\} \text{ where } j = i + k,\ k = 1,\ldots,n - i$$

$$= \{\eta_i \circ D^i f'_{k-1} \circ D^{j-1} \xi_{n-j}\}$$

$$= \{\eta_i \circ D^i f'_{j-i-1} \circ D^{j-1} \xi_{n-j}\},\ i = 0,\ldots,j - 1$$

$$= g_{j-1} \circ D^{j-1} \xi_{n-j}. \qquad \blacksquare$$

We complete this section with the following technical reparametrization lemma.

<u>Lemma 4.3</u>: Let $\{f_n\}: (X,\{\xi_n\}) \to (Y,\{\phi_n\})$ be an s.h.D-map where $(Y,\{\phi_n\})$ is a D-space. Then there is an s.h.D-map $\{\hat{f}_n\}: (X,\{\xi_n\}) \to (Y,\{\phi_n\})$ which is homotopic to $\{f_n\}$ via an s.h.D-homotopy and satisfies $\hat{f}_o = f_o$ and

$$\hat{f}_1 = \begin{cases} \phi \circ Df & \text{if } t \leq \frac{1}{2} \\ f_1(2t - 1) & \text{if } t \geq \frac{1}{2} \end{cases}$$

<u>Proof</u>: First construct $\{f'_n\}$ as specified in the beginning of the previous proof. Let us define

$$\hat{f}_n: I^n \times D^n X \to Y$$

by

$$\hat{f}_n(t_1,\ldots,t_n,z) = \begin{cases} f'_n(2t_1 - 1,t_2,\ldots,t_n) & \text{if } t_1 \geq \frac{1}{2} \\ f'_n(0,t_2,\ldots,t_n) & \text{if } t_1 \leq \frac{1}{2} \end{cases}$$

To see that $\{\hat{f}_n\}$ is an s.h.D-map, we first assume that $t_1 = 0$:

$$\hat{f}_n(0,t_2,\ldots,t_n) = f'_n(0,t_2,\ldots,t_n)$$

$$= \{\phi_{i-1}\circ D^i f'_{n-i}(2t_2,\ldots,2t_i,2t_{i+1} - 1,\ldots,t_n)\}$$

with $t_1,\ldots,t_i \leq \frac{1}{2}$, $t_{i+1} \geq \frac{1}{2}$, $i = 1,\ldots,n$;

but this last map is equal to

$$\phi\circ Df'_{n-1}(2t_2 - 1,t_3,\ldots,t_n) \quad \text{if } t_2 \geq \frac{1}{2}$$

and

$$\{\phi_{i-1}\circ D^i f'_{n-i}(2t_2,2t_3,\ldots,2t_i,2t_{i+1} - 1,\ldots,t_n)\}$$

$i = 2,\ldots,n$, if $t_2 \leq \frac{1}{2}$

$$= \{\phi_{i-1}\circ D^i f'_{n-i}(0,2t_3,\ldots,2t_i,2t_{i+1} - 1,\ldots,t_n)\}$$

since ϕ_1 is a constant homotopy

$$= \{\phi\circ D(\phi_{i-2}\circ D^{i-1} f'_{n-i})(\underline{\qquad})\}$$

$$= \{\phi\circ D(\phi_{j-1}\circ D^j f'_{n-j-1})(\underline{\qquad})\}$$

if we re-index via $j = i - 1$, $j = 1,\ldots,n - 1$.

Thus

$$\hat{f}_n(0,t_2,\ldots,t_n) = \phi \circ D\hat{f}_{n-1}.$$

Now let $t_j = 0$. If $t_1 \geq \frac{1}{2}$,

$$\hat{f}_n(t_1,\ldots,t_n) = f'_n(2t_1 - 1,\ldots,0,\ldots,t_n)$$

$$= f'_{n-1} \circ D^{j-2}\mu_{n-j}(2t_1 - 1,t_2,\ldots,\hat{t}_j,\ldots,t_n)$$

by the construction in 4.2.

If $t_1 \leq \frac{1}{2}$, we have two cases to consider:

<u>Case 1</u>: $j \leq i$. Here,

$$\hat{f}_n(t_1,\ldots,t_n) = f'_n(0,t_2,\ldots,t_n)$$

$$= \{\phi_{i-1} \circ D^i f'_{n-i}(2t_2,\ldots,0,\ldots,2t_i,2t_{i+1} - 1,\ldots,t_n)\}$$

$$= \{\phi_{i-2} \circ D^{j-2}\mu_{i-j} \circ D^i f'_{n-i}(2t_2,\ldots,2\hat{t}_j,\ldots,2t_i,\ldots)\}$$

by the definition of the s.h.D-structure on Y

$$= \{\phi_{i-2} \circ D^{i-1} f'_{n-i} \circ D^{j-2}\mu_{n-j}(\underline{\qquad})\}$$

by the naturality of μ

$$= \{\phi_{k-1} \circ D^k f_{n-k-1} \circ D^{j-2}\mu_{n-j}(\underline{\qquad})\}$$

by letting $k = i - 1$, $k = 1,\ldots,n - 1$

$$= f'_{n-1} \circ D^{j-2}\mu_{n-j}.$$

Note that as before we may take the first coordinate of f'_{n-1} to be 0 since ϕ_1 is a constant homotopy.

<u>Case 2</u>: $j > i$. Here

$$\hat{f}_n(t_1,\ldots,t_n) = f'_n(0,t_2,\ldots,0,\ldots,t_n)$$

$$= \{\phi_{i-1}\circ D^i f'_{n-i}(2t_2,\ldots,2t_i,2t_{i+1}-1,\ldots,0,\ldots,t_n)\}$$

which by letting $j = i + k$, $k = 1,\ldots,n - i$

$$= \{\phi_{i-1}\circ D^i(f'_{n-i-1}\circ D^{k-2}\mu_{n-i-k})(\underline{\qquad\hat{t}_j\qquad})\}$$

$$= \{\phi_{i-1}\circ D^i f'_{n-i-1}\circ D^{j-2}\mu_{n-j}(0,2t_3,\ldots,\hat{t}_j,\ldots)\}$$

$$= f'_{n-1}\circ D^{j-2}\mu_{n-j}.$$

We now suppose that $t_1 = 1$. Then

$$\hat{f}_n(t_1,\ldots,t_n) = f'_n(1,t_2,\ldots,t_n)$$

$$= f\circ\xi_{n-1}(t_2,\ldots,t_n).$$

Finally, if $t_j = 1$, we note that $j > i$ and

$$\hat{f}_n(t_1,\ldots,t_n) = f'_n(2t_1 - 1,t_2,\ldots,t_n) \qquad \text{if } t_1 \geq \tfrac{1}{2}$$

$$= f'_{j-1}(2t_1 - 1,t_2,\ldots,t_{j-1},D^{j-1}\xi_{n-j}(t_{j+1},\ldots,t_n))$$

and

$$= f'_n(0,t_2,\ldots,1,\ldots,t_n) \qquad \text{if } t_1 \leq \tfrac{1}{2}$$

$$= \{\phi_{i-1}\circ D^i f'_{n-i}(2t_2,\ldots,2t_i,2t_{i+1} - 1,\ldots,1,\ldots,t_n)\}$$

which by letting $j = i + k$, $k = 1,\ldots,n - i$

$$= \{\phi_{i-1}\circ D^i f'_{k-1}\circ D^{i+k-1}\xi_{n-i-k}(2t_2,\ldots,\hat{t}_k,\ldots,t_n)\}$$

$$= \{\phi_{i-1}\circ D^i f'_{j-i-1}\circ D^{j-1}\xi_{n-j}(0,2t_3,\ldots,\hat{t}_k,\ldots,t_n)\}$$

$$= \hat{f}_{j-1}\circ D^{j-1}\xi_{n-j}.$$

To complete the proof of Proposition 4.3 we construct the requisite s.h.D-homotopy. Define $h_n\colon I^{n+1} \times D^n X \to Y$ by

$$h_n(t_1,\ldots,t_{n+1}) = \begin{cases} \overline{f}_n\left(\dfrac{t_2 - \frac{1}{2}(1 - t_1)}{1 - \frac{1}{2}(1 - t_1)}, t_3,\ldots,t_{n+1}\right) & \text{if } t_2 \geq \frac{1}{2}(1 - t_1) \\ \phi\circ Dh_{n-1}(t_1,t_3,\ldots,t_{n+1}) & \text{if } t_2 \leq \frac{1}{2}(1 - t_1) \end{cases}$$

where $\overline{f}_n\left(\dfrac{t_2 - \frac{1}{2}(1 - t_1)}{1 - \frac{1}{2}(1 - t_1)}, t_3,\ldots,t_{n+1}\right)$ is the reparametrization in Theorem 4.2 using the constant $\frac{1}{2}(1 - t_1)$ in the place of the constant $\frac{1}{2}$. From Theorem 4.2 we have an explicit formula for the boundary of each $\overline{f}_n$ and we must assume that their interiors piece together in a continuous fashion as t_1 varies. To complete our inductive definition of h_n, we define

$$h_1(t_1,t_2) = \begin{cases} f_1\left(\dfrac{t_2 - \frac{1}{2}(1 - t_1)}{1 - \frac{1}{2}(1 - t_1)}\right) & \text{if } t_2 \geq \frac{1}{2}(1 - t_1) \\ \phi\circ Df & \text{if } t_2 \leq \frac{1}{2}(1 - t_1). \end{cases}$$

If $t_1 = 0$,

$$h_n = \begin{cases} \bar{f}_n(2t_2 - 1, t_3, \ldots, t_{n+1}) & \text{if } t_2 \geq \frac{1}{2} \\ \phi \circ Dh_{n-1}(0, t_3, \ldots, t_{n+1}) & \text{if } t_2 \leq \frac{1}{2} \end{cases}$$

$$= \begin{cases} f'_n(2t_2 - 1, t_3, \ldots, t_{n+1}) & \text{if } t_2 \geq \frac{1}{2} \\ f'_n(0, t_3, \ldots, t_{n+1}) & \text{if } t_2 \leq \frac{1}{2} \end{cases}$$

$$= \hat{f}_n.$$

If $t_1 = 1$,

$$h_n = f_n(t_2, \ldots, t_{n+1}).$$

Thus each h_n is the appropriate homotopy from $\hat{f}_n$ to f_n. Finally, an easy induction argument shows that $\{h_n\}$ is an s.h.D-homotopy. $\blacksquare$

5. Lifting s.h.D-Maps to D-Maps.

When $(X,\{\xi_n\})$ is an s.h.D-space and (Y,ϕ) is a strict D-space, we show that an s.h.D-map $f: X \to Y$ may be thought of as a strict D-map $UX \to Y$. This theorem will enable us to tie together the concepts of SHD-maps and s.h.D-maps. This section concludes with several examples of this theorem that are required in the proof of the conceptual homotopy invariance theorem.

Theorem 5.1: Let $(X,\{\xi_n\})$ be an s.h.D-space, (Y,ϕ) a strict D-space, and $\{f_n\}: X \to Y$ an s.h.D-map. Then there exists a unique D-map $\tilde{f}: UX \to Y$ such that $\tilde{f}\circ\eta = f$ as s.h.D-maps from X to Y. Moreover, if $f \simeq f'$ as s.h.D-maps from X to Y, then $\tilde{f} \simeq \tilde{f}'$ as D-maps.

Corollary 5.2: Let $(X,\{\xi_n\})$ be an s.h.D-space, (Y,ϕ) a strict D-space, and $\{f_n\}: X \to Y$ an s.h.D-map. Then there exists a unique D-map $Uf: UX \to UY$ such that the diagram

$$\begin{array}{ccc} UX & \xrightarrow{Uf} & UY \\ \uparrow{\scriptstyle\eta} & & \uparrow{\scriptstyle\eta} \\ X & \xrightarrow{f} & Y \end{array}$$

commutes as a diagram of s.h.D-maps.

Proof: By Theorem 4.2, $\eta\circ f: X \to UY$ is an s.h.D-map. Define $Uf = \widetilde{\eta\circ f}$ and apply Theorem 5.1. ∎

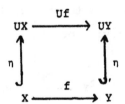

<u>Remarks</u>:

1) As the notation in Corollary 5.2 suggests, it can be shown that U is in fact a functor from, for example, the category of D-spaces and homotopy classes of s.h.D-maps to the category of D-spaces and homotopy classes of D-maps. As this fact is not required in this work, these details are omitted.

2) It will frequently be useful to know that the composition of an s.h.D-map with a strict D-map is an s.h.D-map. If X is an s.h.D-space, Y and Z strict D-spaces, $f: X \to Y$ an s.h.D-map, and $g: Y \to Z$ a D-map, we define an s.h.D-structure for the composition $g\circ f$ by $\{g\circ f_n\}$. The conditions set out in Definition 3.2 are easily verified.

<u>Proof of Theorem 5.1</u>: Define a map $\tilde{f}: UX \to Y$ by the collection $\{\phi\circ Df_n\}$ where $\phi\circ Df_n: I^n \times D^{n+1}X \to Y$. It is easy to see that this map respects the relation used in defining UX. To see that $\tilde{f}$ is in fact a D-map, we have to show that $\tilde{f}\circ\mu = \phi\circ D\tilde{f}$. This follows from the commutativity of

$$\begin{array}{ccccc}
I^n \times D^{n+2}X & \xrightarrow{D^2 f_n} & D^2Y & \xrightarrow{D\phi} & DY \\
\Big\downarrow {\scriptstyle 1\times\mu} & & \Big\downarrow {\scriptstyle \mu} & & \Big\downarrow {\scriptstyle \phi} \\
I^n \times D^{n+1}X & \xrightarrow{Df_n} & DY & \xrightarrow{\phi} & Y
\end{array}$$

for all n since (Y,ϕ) is a strict D-space, and μ is a natural transformation. Note that $\tilde{f}\circ\eta = \phi\circ Df\circ\eta = \phi\circ\eta\circ f = f$; the last equality follows from the definition of an algebra over a monad. Moreover, by Remark 2 above, $\tilde{f}\circ\eta$ has for its s.h.D-structure the collection $\{\phi\circ Df_n\circ\eta_n\}$. Again, $\tilde{f}\circ\eta_n = \phi\circ Df_n\circ\eta_n = \phi\circ\eta\circ f_n = f_n$. It is clear that $\tilde{f}$ is the unique map having this property.

To complete the proof it remains to show that this lifting of s.h.D-maps preserves homotopy. To this end we let $\{f_n\}$ and $\{g_n\}$ be s.h.D-maps from $(X,\{\xi_n\})$ to (Y,ϕ) and $\{h_n\}$ a homotopy between them. We claim that the $\{h_n\}$ induce a homotopy through D-maps between the D-maps $\{\phi\circ Df_n\}$ and $\{\phi\circ Dg_n\}$; define

$$\tilde{h}: I \times UX \to Y$$

by $\tilde{h}_n = \phi\circ Dh_n: I \times I^n \times D^{n+1}X \to Y$. Since $\{h_n|_{t_1=c}\}$ is an s.h.D-map from X to Y for each $c \in I$, $\{\phi\circ Dh_n|_{t_1=c}\}$ is a D-map from UX to Y. The continuity of D allows us to piece together these maps to obtain the desired homotopy. ■

Three useful examples of Theorem 5.1 are contained in

<u>Lemma 5.3</u>: a) Let $f: (X,\xi) \to (Y,\phi)$ be a D-map. Then $\tilde{f} = f\circ r_X$.

b) Let $(X,\{\xi_n\})$ be an s.h.D-space, (Y,ϕ) and (Z,ψ) D-spaces, $f: X \to Y$ an s.h.D-map, and $g: Y \to Z$ a D-map. Then $g\circ\tilde{f} = \widetilde{g\circ f}: UX \to Z$.

c) Let $(X,\{\xi_n\})$ be an s.h.D-space. Then the maps 1_{UX} and $Ur\circ U\eta$ are homotopic as D-maps from UX to UX.

<u>Proof</u>: a) Since X is a D-space and f is a D-map, we may take the collection of constant homotopies $\{f\circ\xi_n\}$ as the s.h.D-structure for f. Thus

$$\tilde{f} = \{\phi \circ Df \circ D\xi_n\}$$

$$= \{f \circ \xi \circ D\xi_n\} \quad \text{since } f \text{ is a D-map}$$

$$= \{f \circ \xi_n\} = f \circ r_X.$$

b) $g \circ \tilde{f}$ is given by the collection $\{g \circ \phi \circ Df_n\}$. On the other hand, $\widetilde{g \circ f}$ is given by the collection $\{\psi \circ D(g \circ f_n)\}$. But $\psi \circ Dg \circ Df_n = g \circ \phi \circ Df_n$ since g is a D-map.

c) We show that the s.h.D-maps from X to UX induced by the given D-maps are homotopic as s.h.D-maps. Theorem 5.1 will then apply to show that the D-maps themselves are homotopic in D[T]. Since $\eta \circ r_X \simeq 1_{UX}$, 6.1 (ii) and 5.1 yield $Ur = \widetilde{\eta \circ r} \simeq \tilde{1}_{UX}$ as D-maps. Thus $Ur \circ U\eta \simeq \tilde{1}_{UX} \circ U\eta$ as D-maps. Since $\tilde{1}_{UX} = r_{UX}$ by 5.3a, it suffices to show that $r_{UX} \circ U\eta \circ \eta \simeq 1_{UX} \circ \eta = \eta$ as s.h.D-maps from X to UX. Let $(t_1, \ldots, t_n, z) \in I^n \times D^n X$. Then

$$r_{UX} \circ U\eta \circ \eta_n(t_1, \ldots, t_n, z)$$

$$= \{\mu_i \circ \mu \circ D\eta_i \circ D^{i+1}\eta'_{n-i} \circ \eta D^n(2t_1, \ldots, 2t_i, 2t_{i+1} - 1, \ldots, t_n, z)\}$$

$$i = 0, \ldots, n$$

$$= \{\mu_i \circ D^{i+1}\eta'_{n-i} \circ \eta D^n(2t_1, \ldots, 2t_i, 2t_{i+1} - 1, \ldots, t_n, z)\}$$

$$= \{\mu_i \circ \eta_{n+1} \circ D^i \eta'_{n-i}(2t_{i+1} - 1, \ldots, t_n, z)\}$$

by the naturality of η and the definition of μ_i

$$= \{\mu_{i-1} \circ D^i \eta'_{n-i}(2t_{i+1} - 1, \ldots, t_n, z)\}$$

$$= \{\hat{\eta}_n\} \underset{s.h.D}{\simeq} \{\eta_n\} \quad \text{by Lemma 4.3.}$$

In this proof, we take $\mu_i = \mu_o \circ \ldots \circ \mu_{i-1} \circ \mu_i$ where $\mu_j \colon D^2D^n \to DD^n$ for each n. ∎

6. Homotopy Invariance Theorems

We begin with

Theorem 6.1 (Geometrical Homotopy Invariance Theorem):

(i) Let (Y,ϕ) be a D-space and $f: X \to Y$ a homotopy equivalence; then X is an s.h.D-space and f is an s.h.D-map.

(ii) Let $f: (X,\xi) \to (Y,\phi)$ be a D-map and suppose that g is homotopic to f; then g is an s.h.D-map and g is homotopic to f as an s.h.D-map.

(iii) Let $f: (X,\xi) \to (Y,\phi)$ be a D-map and a homotopy equivalence with homotopy inverse g; then g is an s.h.D-map and $f \circ g$ is homotopic to 1_Y as an s.h.D-map.

The proof of this theorem is deferred until Sections 7, 8 and 9. We may, however, use this theorem to deduce

Theorem 6.2 (Conceptual Homotopy Invariance Theorem)

(i) Let $(Y,\{\phi_n\})$ be an s.h.D-space and $f: X \to Y$ a homotopy equivalence; then X is an s.h.D-space and f is an SHD-map.

(ii) Let $f: (X,\{\xi_n\}) \to (Y,\{\phi_n\})$ be an SHD-map between s.h.D-spaces and suppose that g is homotopic to f; then g is an SHD-map and g is homotopic to f as an SHD-map.

(iii) Let $f: (X,\{\xi_n\}) \to (Y,\{\phi_n\})$ be an SHD-map between s.h.D-spaces and a homotopy equivalence with homotopy inverse g; then g is an SHD-map and both $f \circ g$ is homotopic to 1_Y and $g \circ f$ is homotopic to 1_X as SHD-maps.

<u>Proof</u>:

(i) Apply 6.1 (i) to the homotopy equivalence $\eta \circ f$: $X \to UY$. Thus X is an s.h.D-space and $\eta \circ f$ is an s.h.D-map. By 5.1, there exists a D-map $\tilde{f}$: $UX \to UY$ such that $\eta \circ f = \tilde{f} \circ \eta$. Then $f = r \circ \eta \circ f = r \circ \tilde{f} \circ \eta$ and $\tilde{f}$ represents f as an SHD-map by Definition 3.1.

(ii) Let F: $UX \to UY$ be a D-map representing f as an SHD-map. Thus $f = r \circ F \circ \eta$. But then $\eta \circ g \circ r \simeq \eta \circ f \circ r = \eta \circ r \circ F \circ \eta \circ r \simeq F$. By 6.1(ii), $\eta \circ g \circ r$: $UX \to UY$ is an s.h.D-map and $\eta \circ g \circ r \simeq F$ as s.h.D-maps. By 5.3(a) we have $\tilde{F} = F \circ r$: $UUX \to UY$. By 5.1, we have $\widetilde{\eta \circ g \circ r} \simeq_D \tilde{F}$: $UUX \to UY$. By 5.3(c) $Ur \circ U\eta \simeq_D 1_{ux}$. We also have that $r_{ux} = \tilde{1}_{ux} \simeq_D \widetilde{\eta \circ r} = Ur$ by 5.3(a), 5.1 and 6.1(ii); i.e., $Ur \simeq_D r_{ux}$: $UUX \to UX$. Thus

$$g = r \circ (\eta \circ g \circ r) \circ \eta = r \circ \widetilde{\eta \circ g \circ r} \circ \eta \circ \eta$$

$$= r \circ (\widetilde{\eta \circ g \circ r} \circ U\eta) \circ \eta,$$

hence $\widetilde{\eta \circ g \circ r} \circ U\eta$ represents g as an SHD-map. The equality $\eta \circ \eta = U\eta \circ \eta$ comes from the commutative diagram

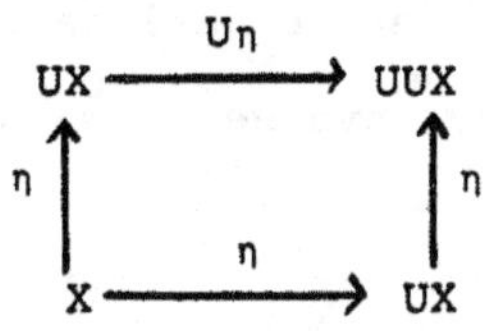

guaranteed by 5.1. In addition we have

$\widetilde{\eta\circ g\circ r}\circ U\eta \simeq_D \tilde{F}\circ U\eta = F\circ r\circ U\eta \simeq_D F\circ Ur\circ U\eta \simeq_D F.$

(iii) Again, let $F: UX \to UY$ represent f as an SHD-map, $f = r\circ F\circ\eta$. Then $\eta\circ g\circ r$ is a homotopy inverse to F since

$$\eta\circ g\circ r\circ F \simeq \eta\circ g\circ r\circ F\circ\eta\circ r = \eta\circ g\circ f\circ r \simeq \eta\circ r \simeq 1$$

and

$$F\circ\eta\circ g\circ r \simeq \eta\circ r\circ F\circ\eta\circ g\circ r = \eta\circ f\circ g\circ r \simeq \eta\circ r \simeq 1.$$

By 6.1(iii), $\eta\circ g\circ r: UY \to UX$ is an s.h.D-map and $F\circ\eta\circ g\circ r \simeq 1$ as s.h.D-maps. Exactly as in the previous argument, $\widetilde{\eta\circ g\circ r}\circ U\eta: UY \to UX$ represents g as an SHD-map. To see that the composition of these SHD-maps are homotopic to the respective identity maps, we have

$$F\circ(\widetilde{\eta\circ g\circ r})\circ U\eta = \widetilde{F\circ\eta\circ g\circ r}\circ U\eta \qquad \text{by 5.3(b)}$$

$$\simeq_D \tilde{1}\circ U\eta \text{ by 5.1 } = r\circ U\eta \qquad \text{by 5.3(a)}$$

$$\simeq_D Ur\circ U\eta \quad \text{as above} \quad \simeq_D 1 \qquad \text{by 5.3(c)}.$$

In the other direction, to show that $g\circ f \simeq 1$ as SHD-maps, since we now know that $g: Y \to X$ is an SHD-map between s.h.D-spaces as well as a homotopy equivalence with homotopy inverse f, we need only reverse the roles of f and g in the argument above and apply standard uniqueness of inverse arguments to conclude that the constructed representative $\widetilde{\eta\circ f\circ r}\circ U\eta$ of f as an SHD-map is homotopic as a D-map to the given representative F. ∎

7. Proof of Theorem 6.1(i)

In this section we prove Theorem 6.1(i): Let (Y,ϕ) be a D-space and $f\colon X \to Y$ a homotopy equivalence; then X is an s.h.D-space and f is an s.h.D-map.

Proof: Since f is a homotopy equivalence, there exists a homotopy inverse $g\colon Y \to X$ and a homotopy $k\colon I \times Y \to Y$ such that k_0 = identity on Y and $k_1 = f\circ g$. Also, as explained in [G, pp. 159-160], by slightly altering the underlying operad $\mathcal{D}$, we can replace f by its mapping cylinder and thus assume that $g\circ f$ is the identity map on X.

Define a map $\xi\colon DX \to X$ by the composition $\xi = g\circ\phi\circ D(f)$. By the naturality of η, we have $\xi\circ\eta = g\circ\phi\circ D(f)\circ\eta = g\circ\phi\circ\eta\circ f = g\circ f = 1$ on X.

We now need to construct a homotopy $\xi_1\colon I \times D^2X \to X$ which will render the diagram

$$\begin{array}{ccc} D^2X & \xrightarrow{\ \mu\ } & DX \\ {\scriptstyle D\xi}\downarrow & & \downarrow{\scriptstyle \xi} \\ DX & \xrightarrow{\ \xi\ } & X \end{array}$$

homotopy commutative. It is the failure of this diagram to be strictly commutative in general that causes X to fail to be a strict D-space. Define

$$\xi_1 = g\circ\phi\circ D(k)\circ D(\phi)\circ D^2(f)$$

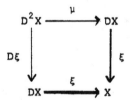

where $D(k): I \times DY \to DY$ is defined in the manner described in Section 1. To see that ξ_1 is the correct homotopy, note that

$$\begin{aligned}
\xi_1|_0 &= g\circ\phi\circ D(k)|_0\circ D(\phi)\circ D^2(f) \\
&= g\circ\phi\circ D(\phi)\circ D^2(f) && \text{since } D(k)|_0 = \text{id} \\
&= g\circ\phi\circ\mu\circ D^2(f) && \text{since } Y \text{ is a D-space} \\
&= g\circ\phi\circ D(f)\circ\mu && \text{since } \mu \text{ is natural} \\
&= \xi\circ\mu,
\end{aligned}$$

and that

$$\begin{aligned}
\xi_1|_1 &= g\circ\phi\circ D(k)|_1\circ D(\phi)\circ D^2(f) \\
&= g\circ\phi\circ D(f)\circ D(g)\circ D(\phi)\circ D^2(f) \\
&= \xi\circ D\xi.
\end{aligned}$$

This motivation for the s.h.D-structure on X leads us to define the requisite higher homotopies $\xi_q: I^q \times D^{q+1}X \to X$ by

$$\xi_q = g\circ\phi\circ D(k)\circ D(\phi)\circ\ldots\circ D^j(k)\circ D^j(\phi)\circ\ldots\circ D^q(k)\circ D^q(\phi)\circ D^{q+1}(f).$$

We have omitted the symbol "$1 \times$" which should preface each $D^j(k)$. Note also that we have a composition of q one-dimensional homotopies; we use each in succession on each of the q coordinates of I^q.

For the ξ_q to determine a valid s.h.D-structure on X, we have to verify the usual compatibility conditions. If, for example, $t_j = 0$, we have

$$\xi_q = g\circ\phi\circ D(k)\circ D(\phi)\circ\ldots\circ D^{j-1}(\phi)\circ D^j(k)|_0\circ D^j(\phi)\circ\ldots\circ D^q(k)\circ D^q(\phi)\circ D^{q+1}(f)$$

$$= g\circ\phi\circ\ldots\circ D^{j-1}(\phi)\circ D^j(\phi)\circ D^{j+1}(k)\circ\ldots\circ D^q(k)\circ D^q(\phi)\circ D^{q+1}(f)$$

$$= g\circ\phi\circ\ldots\circ D^{j-1}(\phi\circ D\phi)\circ D^{j+1}(k)\circ\ldots\circ D^q(k)\circ D^q(\phi)\circ D^{q+1}(f)$$

$$= g\circ\phi\circ\ldots\circ D^{j-1}(\phi\circ\mu)\circ D^{j+1}(k)\circ\ldots\circ D^q(k)\circ D^q(\phi)\circ D^{q+1}(f)$$

$$= g\circ\phi\circ\ldots\circ D^{j-1}(\phi)\circ D^j(k)\circ\ldots\circ D^{q-1}(k)\circ D^{q-1}(\phi)\circ D^q(f)\circ D^{j-1}\mu_{q-j}$$

$$= \xi_{q-1}\circ D^{j-1}\mu_{q-j}.$$

The next to last equality follows from repeated application of the naturality of μ.

If, on the other hand, $t_j = 1$, then

$$\xi_q = g\circ\phi\circ D(k)\circ D(\phi)\circ\ldots\circ D^{j-1}(\phi)\circ D^j(k)|_1\circ D^j(\phi)\circ\ldots\circ D^q(k)\circ D^q(\phi)\circ D^{q+1}(f)$$

$$= (g\circ\phi\circ D(k)\circ D(\phi)\circ\ldots\circ D^{j-1}(\phi)\circ D^j(f))\circ$$

$$(D^j(g)\circ D^j(\phi)\circ\ldots\circ D^q(k)\circ D^q(\phi)\circ D^{q+1}(f))$$

$$= \xi_j\circ D^j\xi_{q-j}.$$

Thus the $\{\xi_q\}$ as defined do indeed define an s.h.D-structure for X.

To see that $f\colon X \to Y$ is an s.h.D-map we define

$$f_n\colon I^n \times D^n X \to Y$$

by

$$f_n = k \circ \phi \circ D(k) \circ D(\phi) \circ \ldots \circ D^{n-1}(k) \circ D^{n-1}(\phi) \circ D^n(f).$$

To verify the compatibility conditions of Definition 3.2, note that if $t_1 = 0$,

$$f_n = k_o \circ \phi \circ D(k) \circ D(\phi) \circ \ldots \circ D^{n-1}(k) \circ D^{n-1}(\phi) \circ D^n(f)$$

$$= \phi \circ D(k) \circ D(\phi) \circ \ldots \circ D^{n-1}(k) \circ D^{n-1}(\phi) \circ D^n(f)$$

$$= \phi \circ Df_{n-1}.$$

If $t_1 = 1$,

$$f_n = k_1 \circ \phi \circ D(k) \circ D(\phi) \circ \ldots \circ D^{n-1}(k) \circ D^{n-1}(\phi) \circ D^n(f)$$

$$= f \circ g \circ \phi \circ D(k) \circ D(\phi) \circ \ldots \circ D^{n-1}(k) \circ D^{n-1}(\phi) \circ D^n(f)$$

$$= f \circ \xi_{n-1}.$$

When $t_j = 0$,

$$f_n = k\circ\phi\circ D(k)\circ D(\phi)\circ\ldots\circ D^{j-1}(k)|_0\circ D^{j-1}(\phi)\circ D^j(k)\circ D^j(\phi)\circ\ldots\circ D^n(f)$$

$$= k\circ\phi\circ D(k)\circ D(\phi)\circ\ldots\circ D^{j-2}(k)\circ D^{j-2}(\phi)\circ D^{j-1}(\phi)\circ D^j(k)\circ\ldots\circ D^n(f)$$

$$= k\circ\phi\circ D(k)\circ D(\phi)\circ\ldots\circ D^{j-2}(k)\circ D^{j-2}(\phi)\circ D^{j-2}(\mu)\circ D^j(k)\circ\ldots\circ D^n(f)$$

$$= f_{n-1}\circ D^{j-2}\mu_{n-j}.$$

Finally, if $t_j = 1$,

$$f_n = k\circ\phi\circ D(k)\circ D(\phi)\circ\ldots\circ D^{j-2}(k)\circ D^{j-2}(\phi)\circ D^{j-1}(k)|_1\circ D^{j-1}(\phi)\circ\ldots\circ D^n(f)$$

$$= (k\circ\phi\circ D(k)\circ D(\phi)\circ\ldots\circ D^{j-2}(k)\circ D^{j-2}(\phi)\circ D^{j-1}(f))\circ$$

$$(D^{j-1}(g)\circ D^{j-1}(\phi)\circ\ldots\circ D^n(f))$$

$$= f_{j-1}\circ D^{j-1}\xi_{n-j}. \quad \blacksquare$$

8. Proof of Theorem 6.1(ii)

In this section we prove Theorem 6.1(ii): Let $f: (X,\xi) \to (Y,\phi)$ be a D-map and suppose that g is homotopic to f; then g is an s.h.D-map and g is homotopic to f as an s.h.D-map.

Proof: We will define an s.h.D-structure for g by glueing together various homotopies gotten from the maps ϕ_n, ξ_n, and the functor D^i applied to the given homotopy between f and g. The procedure will be clear to the reader if he sketches the appropriate pictures for low values of n. To begin let $h: I \times X \to Y$ be a homotopy such that $h_o = f$ and $h_1 = g$.

We define an s.h.D-structure for g,

$$g_n: I^n \times D^nX \to Y,$$

by

$$g_n(t_1,\dots,t_n) = \begin{cases} h\circ\xi_{n-1}(2t_1 - 1,t_2,\dots,t_n) & \text{if } t_1 \geq \frac{1}{2} \\ \{(-\phi_{i-1}\circ D^ih\circ D^i\xi_{n-i-1})'\} & \text{if } t_1 \leq \frac{1}{2} \end{cases}$$

with $t_1,\dots,t_i \leq \frac{1}{2}$, $t_{i+1} \geq \frac{1}{2}$, $i = 1,\dots,n$ where

$$(-\phi_{i-1}\circ D^i h\circ D^i \xi_{n-i-1})' = \begin{cases} \begin{cases} (-\phi_{i-1}\circ D^i h\circ D^i \xi_{n-i-1})^r \text{ if } t_{i+1} \geq \frac{1}{2} + \sum_{j<i+1} t_j \\ (-\phi_{i-1}\circ D^i f\circ D^i \xi_{n-i-1}) \text{ if } t_{i+1} \leq \frac{1}{2} + \sum_{j<i+1} t_j \end{cases} & \text{if } i \leq n-1 \\ \begin{cases} (-\phi_{n-1}\circ D^n h)^r \text{ if } t_n \leq \frac{1}{2} - \sum_{i<n} t_i \\ (\phi_{n-1}\circ D^n f) \text{ if } t_n \geq \frac{1}{2} - \sum_{i<n} t_i \end{cases} & \text{if } i = n \end{cases}$$

A minus sign preceding a homotopy indicates that it should be evaluated in reverse direction on its first coordinate.

The following construction is used in the definition of g_n: let $h: I \times X \to Y$ be any homotopy, Δn an n-simplex, and v_0 a particular vertex of Δn. We may then construct a new homotopy $h^s: \Delta n \times X \to Y$ with $h^s(v_0,x) = h_0(x)$ and $h^s(v,x) = h_1(x)$ for all v on the face opposite to v_0 by defining h^s to be h on each line segment connecting v_0 to the face opposite v_0. It is clear that we may also extend h^s to a map $h^r: \Delta n \times I^m \times X \to Y$ by defining

$$h^r(s_0,\dots,s_n,t_1,\dots,t_m,x) = h^s(s_0,\dots,s_n,x).$$

In our definition of g_n, note that the homotopy $(-\phi_{i-1}\circ D^i h\circ D^i \xi_{n-i-1})^r$ is defined on $\Delta i{+}1 \times I^{n-i-1} \subset I^n$ where $\Delta i{+}1$ is determined by the intersection of the plane $t_{i+1} = \frac{1}{2} + \sum_{j<i+1} t_j$ with the cube $t_1,\dots,t_i \leq \frac{1}{2}$, $t_{i+1} \geq \frac{1}{2}$ and the inequality $t_{i+1} \geq \frac{1}{2} + \sum_{j<i+1} t_j$. We need to know the vertex

of the simplex opposite the face $t_{i+1} = \frac{1}{2} + \sum_{j<i+1} t_j$ in order to evaluate the map $(-\phi_{i-1}\circ D^i h\circ D^i \xi_{n-i-1})^r$; it is easy to check that the point $(0,\ldots,0,1,0,\ldots,0)$ with 1 in the $i+1$ st coordinate satisfies the requirements.

We now show that each $(-\phi_{i-1}\circ D^i h\circ D^i \xi_{n-i-1})'$ is well-defined. Suppose that $t_n = \frac{1}{2} - \sum_{i<n} t_i$. Then $(-\phi_{i-1}\circ D^i h\circ D^i \xi_{n-i-1})'$ $= (-\phi_{n-1}\circ D^n h)^s = \phi_{n-1}\circ D^n f$ since the vertex opposite the face $t_n = \frac{1}{2} - \sum_{i<n} t_i$ is $(0,\ldots,0)$ and $-h|_o = g$; we thus have agreement in this case. Now suppose that $t_{i+1} = \frac{1}{2} + \sum_{j<i+1} t_j$. Then $(-\phi_{i-1}\circ D^i h\circ D^i \xi_{n-i-1})' = (-\phi_{i-1}\circ D^i h\circ D^i \xi_{n-i-1})^r$ is defined on $\Delta i+1 \times I^{n-i-1}$ and thus equals $(-\phi_{i-1}\circ D^i h\circ D^i \xi_{n-i-1})^s(t_1,\ldots,t_{i+1})$ $= \phi_{i-1}\circ D^i f\circ D^i \xi_{n-i-1}$ since the vertex $(0,\ldots,0,1,0,\ldots,0)$ with $i+1$ coordinate equal to 1 is opposite the face in question, and $-\phi_{i-1}\circ D^i h\circ D^i \xi_{n-i-1}$ evaluated on this vertex is $\phi_{i-1}\circ D^i g\circ D^i \xi_{n-i-1}$. Thus this map is well-defined on each cube $t_1,\ldots,t_i \leq \frac{1}{2}$, $t_{i+1} \geq \frac{1}{2}$.

Next we show that it makes sense to glue together all the cubes that occur when $t_1 \leq \frac{1}{2}$. Suppose that $t_{i+1} = \frac{1}{2}$, $i \geq 1$, and $t_{i+2} > \frac{1}{2}$. Then

1) $\quad g_n = (-\phi_{i-1}\circ D^i h\circ D^i \xi_{n-i-1})'$ where $t_1,\ldots,t_i \leq \frac{1}{2}$, $t_{i+1} \geq \frac{1}{2}$

$$= \phi_{i-1}\circ D^i f\circ D^i \xi_{n-i-1} \text{ since } t_{i+1} = \frac{1}{2} \leq \frac{1}{2} + \sum_{j<i+1} t_j$$

and

2) $g_n = (-\phi_i \circ D^{i+1}h \circ D^{i+1}\xi_{n-i-2})'$ where $t_1,\ldots,t_{i+1} \leq \frac{1}{2}$, $t_{i+2} \geq \frac{1}{2}$

$$= \phi_i \circ D^{i+1}f \circ D^{i+1}\xi_{n-i-2} \text{ since } t_{i+2} = \frac{1}{2} \leq \frac{1}{2} + \sum_{j<i+2} t_j.$$

But

$$\phi_i \circ D^{i+1}f \circ D^{i+1}\xi_{n-i-2}$$

$$= \phi_{i-1} \circ \mu_{i-1} \circ D^{i+1}f \circ D^{i+1}\xi_{n-i-2} \quad \text{since } Y \text{ is a D-space}$$

$$= \phi_{i-1} \circ D^{i}f \circ D^{i}\xi_{n-i-2} \circ \mu_n \quad \text{by the naturality of } \mu$$

$$= \phi_{i-1} \circ D^{i}f \circ D^{i}\xi_{n-i-1} \quad \text{since } X \text{ is a D-space,}$$

and we have agreement between 1) and 2).

The final glueing process occurs when $t_1 = \frac{1}{2}$; we think of the parameter t_1 as running from left to right. The right hand face of the cube $t_1 \leq \frac{1}{2}$ is given by

$$\{(-\phi_{i-1} \circ D^{i}h \circ D^{i}\xi_{n-i-1})'\} \text{ where } t_1,\ldots,t_i \leq \frac{1}{2},\ t_{i+1} \geq \frac{1}{2},\ i = 2,\ldots,n$$

$$= \{\phi_{i-1} \circ D^{i}f \circ D^{i}\xi_{n-i-1}\} \quad \text{by previous calculations}$$

$$= \{\phi \circ D\phi \circ \ldots \circ D^{i-1}\phi \circ D^{i}f \circ D^{i}\xi_{n-i-1}\}$$

$$= \{\phi \circ D\phi \circ \ldots \circ D^{i-2}\phi \circ D^{i-1}f \circ D^{i-1}\xi \circ D^{i}\xi_{n-i-1}\} \quad \text{since } f \text{ is a D-map}$$

$$= \{\phi \circ D\phi \circ \ldots \circ D^{i-2}\phi \circ D^{i-1}f \circ D^{i-1}\xi \circ D^{i}\xi \circ \ldots \circ D^{n-1}\xi\}$$

$$= \{f \circ \xi_{n-1}\} \quad \text{by repeated application of the above procedure.}$$

On the other hand, the left hand edge of the cube $t_1 \geq \frac{1}{2}$ is given by $h \circ \xi_{n-1}(0,t_2,\ldots,t_n) = f \circ \xi_{n-1}$ and we have agreement. The verification of the fact that g_n is well-defined is now complete.

That $\{g_n\}$ is an s.h.D-map will now be demonstrated. Suppose $t_1 = 0$; then $g_n(0,t_2,\ldots,t_n) = \{(-\phi_{i-1} \circ D^i h \circ D^i \xi_{n-i-1})'\}$ where $t_2,\ldots,t_i \leq \frac{1}{2}$, $t_{i+1} \geq \frac{1}{2}$, $i = 1,\ldots,n$. If $t_2 \geq \frac{1}{2}$,

$$*) \qquad g_n(0,t_2,\ldots,t_n) = (-\phi \circ Dh \circ D\xi_{n-2})'(0,t_2,\ldots,t_n)$$

$$= \phi \circ Dh \circ D\xi_{n-2} = \phi \circ D(h \circ \xi_{n-2})$$

since the vertex $(0,1,0,\ldots,0)$ is opposite the face on which $(-\phi \circ Dh \circ D\xi_{n-2})'$ has value $\phi \circ Df \circ D\xi_{n-2}$, and thus when $t_2 = 1$, $(-\phi \circ Dh \circ D\xi_{n-2})(0,1,t_3,\ldots,t_n)$ has value $\phi \circ Dg \circ D\xi_{n-2}$. This has the effect of negating the negative sign above. Now if $t_2 \leq \frac{1}{2}$,

$$g_n(0,t_2,\ldots,t_n) = \{(-\phi_{i-1} \circ D^i h \circ D^i \xi_{n-i-1})'\}$$
$$\text{with } t_2,\ldots,t_i \leq \tfrac{1}{2},\ t_{i+1} \geq \tfrac{1}{2},\ i = 2,\ldots,n;$$

letting $k = i - 1$, we have

$$g_n(0,t_2,\ldots,t_n) = \{-\phi_k \circ D^{k+1} h \circ D^{k+1} \xi_{n-k-2}\}, \quad k = 1,\ldots,n-1$$

$$**) \qquad = \{(-\phi \circ D(\phi_{k-1} \circ D^k h \circ D^k \xi_{(n-1)-k-1})'\}.$$

Thus *) and **) yield

$$g_n(0,t_2,\ldots,t_n) = \phi \circ Dg_{n-1}.$$

If $t_1 = 1$,

$$g_n(1,t_2,\ldots,t_n) = h\circ\xi_{n-1}(1,t_2,\ldots,t_n) = g\circ\xi_{n-1}.$$

Now suppose that $t_j = 0$; if $t_1 \geq \frac{1}{2}$,

$$g_n(t_1,\ldots,0,\ldots,t_n) = h\circ\xi_{n-1}(2t_1 - 1,t_2,\ldots,0,\ldots,t_n)$$

$$= h\circ\xi_{n-1} = h\circ\xi_{n-2}\circ D^{j-2}\mu_{n-j}$$

since X is a D-space; if, however, $t_1 \leq \frac{1}{2}$

$$g_n(t_1,\ldots,0,\ldots,t_n) = \{(-\phi_{i-1}\circ D^i h\circ D^i\xi_{n-i-1})'\}$$

where $t_1,\ldots,t_i \leq \frac{1}{2}$, and we have two cases to consider: the first is $j < i + 1$; then

$$g_n = \{(-\phi_{i-1}\circ D^i h\circ D^i\xi_{n-i-1})'\}$$

$$= \{(-\phi\circ D\phi\circ\ldots\circ D^{i-1}\phi\circ D^i h\circ D^i\xi_{n-i-1})'\}$$

$$= \{(-\phi\circ D\phi\circ\ldots\circ\mu_{i-2}\circ D^i h\circ D^i\xi_{n-i-1})'\}$$

$$= \{(-\phi_{i-2}\circ D^{i-1}h\circ\mu_{i-1}\circ D^i\xi_{n-i-1})'\} \quad \text{by the naturality of } \mu$$

$$= \{(-\phi_{i-2}\circ D^{i-1}h\circ D^{i-1}\xi_{n-i-1}\circ\mu_{n-2})'\}$$

$$= \{(-\phi_{i-2}\circ D^{i-1}h\circ D^{i-1}\xi_{n-i-1}\circ D^{j-2}\mu_{n-j})'\} \quad \text{since X is a D-space}$$

$$= g_{n-1}\circ D^{j-2}\mu_{n-j}.$$

On the other hand, if $j > i + 1$, let $j = i + k$, $k = 1,\dots,n - i$, and then

$$g_n = \{(-\phi_{i-1} \circ D^i h \circ D^i \xi_{n-i-1})'\}$$

$$= \{(-\phi_{i-1} \circ D^i h \circ D^i (\xi_{n-i-2} \circ D^{k-2} \mu_{n-i-k})'\} \text{ when } t_k = 0,\ k > 1$$

by the definition of an s.h.D-structure

$$= \{(-\phi_{i-1} \circ D^i h \circ D^i \xi_{n-i-2} \circ D^{j-2} \mu_{n-j})'\} \quad i = 1,\dots,n - 1$$

$$= g_{n-1} \circ D^{j-2} \mu_{n-j}.$$

It remains to check the situation when $t_j = 1$. If $t_1 \geq \frac{1}{2}$,

$$g_n = h \circ \xi_{n-1}(2t_1 - 1,\dots,1,\dots,t_n)$$

$$= h \circ (\xi \circ D\xi \circ \dots \circ D^{j-2}\xi) \circ D^{j-1}\xi \circ \dots \circ D^{n-1}\xi$$

$$= h \circ \xi_{j-2} \circ D^{j-1}(\xi \circ D\xi \circ \dots \circ D^{n-1-j+1}\xi)$$

$$= h \circ \xi_{j-2} \circ D^{j-1} \xi_{n-j}$$

$$= g_{j-1} \circ D^{j-1} \xi_{n-j},$$

finally, if $t_1 \leq \frac{1}{2}$ and $t_j = 1$, we have $j \geq i + 1$ by the definition of g_n. Thus if we let $j = i + k$, $k = 1,\dots,n - i - 1$, we have

$$g_n = \{(-\phi_{i-1}\circ D^i h\circ D^i \xi_{n-i-1})'\}$$

$$= \{(-\phi_{i-1}\circ D^i h\circ D^i(\xi\circ D\xi\circ\ldots\circ D^{k-2}\xi\circ D^{k-1}\xi\circ\ldots\circ D^{n-i-1}\xi))'\}$$

$$= \{(-\phi_{i-1}\circ D^i h\circ D^i\xi_{k-2}\circ D^i D^{k-1}(\xi\circ\ldots\circ D^{n-i-k}\xi))'\}$$

$$= \{(-\phi_{i-1}\circ D^i h\circ D^i\xi_{k-2}\circ D^{i+k-1}\xi_{n-i-k})'\}$$

$$= \{(-\phi_{i-1}\circ D^i h\circ D^i\xi_{j-i-2}\circ D^{j-1}\xi_{n-j})'\}$$

$$= \{(-\phi_{i-1}\circ D^i h\circ D^i\xi_{(j-1)-i-1}\circ D^{j-1}\xi_{n-j})'\}$$

$$= g_{j-1}\circ D^{j-1}\xi_{n-j}.$$

The final element of the proof is the verification of the fact that f is homotopic to g as an s.h.D-map. For this, we will define an s.h.D-homotopy

$$h_n: I^{n+1} \times D^n X \to Y.$$

The case n = 1 will serve as a good illustration of our construction. We define h_1 on $I^2 \times DX$ as follows:

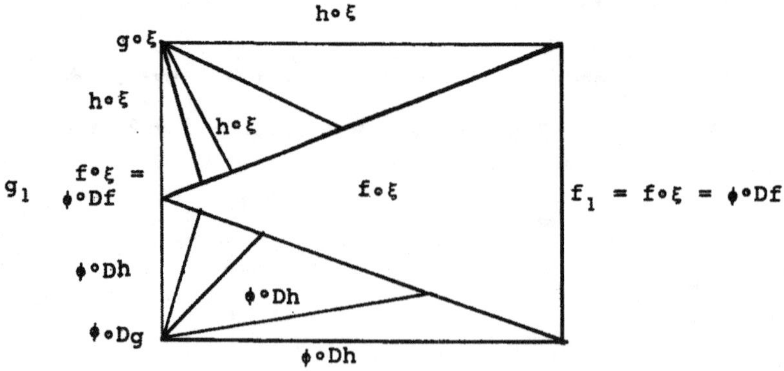

Note that, for example, $(\phi \circ Dh)'$ is defined on a simplex determined by the original simplex in the definition of g, and the additional vertex $(1,0)$.

In general, we define

$$h_n = \begin{cases} (h \circ \xi_{n-1})^r & \text{if } t_2 \geq \frac{1}{2}(1 + t_1) \\ -(\phi_{i-2} \circ D^{i-1} h \circ D^{i-1} \xi_{n-1})^r & \text{if } t_{i+1} \geq \frac{1}{2}(1 + t_1) + \sum_{j<i+1} t_j \\ (-\phi_{n-1} \circ D^n h)^r & \text{if } t_{n+1} \leq \frac{1}{2}(1 - t_1) - \sum_{j<i+1} t_j \\ f \circ \xi_{n-1} & \text{otherwise} \end{cases}$$

Here, the domain of $(-\phi_{i-2} \circ D^{i-1} h \circ D^{i-1} \xi_{n-i})^r$ is the simplex obtained from the simplex defined by $t_1 = 0$ (in the definition of g_n) together with the vertex $(1,0,\ldots,1,0,\ldots,0)$ where the second 1 is in the $i + 1$ coordinate. The proof that h_n is well-defined and is in fact an s.h.D-homotopy is similar to the proof that g_n is well-defined and is an s.h.D-map and is omitted. ∎

9. Proof of Theorem 6.1(iii)

In order to complete the proof of the geometrical homotopy invariance theorem, we shall utilize the following lemma due to Fuchs [7, p. 337]:

Lemma 9.1: Let $f\colon X \to Y$ be a homotopy equivalence, g a homotopy inverse of f, $k\colon I \times X \to X$ the homotopy such that $k_o = id_X$, $k_1 = g\circ f$; then a homotopy $h\colon I \times Y \to Y$ such that $h_o = id_Y$, $h_1 = f\circ g$ may be chosen so that $f\circ k \simeq h\circ f$ as homotopies between f and $f\circ g\circ f$.

Proof: Consider the diagram of path spaces

$$\begin{array}{ccccc} \Omega(Y^X, f, f\circ g\circ f) & \xrightarrow{g^*} & \Omega(Y^Y, f\circ g, f\circ g\circ f\circ g) & \xrightarrow{\phi} & \Omega(Y^Y, id_Y, f\circ g) \\ & & & & \Big\downarrow f^* \\ & & & & \Omega(Y^X, f, f\circ g\circ f) \end{array}$$

where f^* and g^* are induced by the maps f, g and ϕ is induced by any path from id_Y to $f\circ g\circ f\circ g$ (such a path exists since there is certainly a path α from id_Y to $f\circ g$ under the assumption that f is a homotopy equivalence and thus the claimed path may be taken to be $g^*[f\circ k] * \alpha$). Since f and g are homotopy equivalences, so are f^* and g^*; moreover, ϕ is a homotopy equivalence; thus each map in the diagram induces a one to one correspondence between path components. We may now choose h to be any path in the class $\phi\circ g^*[f\circ k]$. We then have that $f^*[h] = [h\circ f] = [f\circ k]$. Thus $f\circ k$ and $h\circ f$ lie in the same path component of $\Omega(Y^X, f, f\circ g\circ f)$ and are thus homotopic as paths. ∎

We proceed now with the proof of Theorem 6.1(iii): Let $f: (X,\xi) \to (Y,\phi)$ be a D-map and a homotopy equivalence with homotopy inverse g; then g is an s.h.D-map and $f\circ g$ is homotopic to 1_Y as an s.h.D-map.

<u>Proof</u>: Let $k: I \times X \to X$ be a homotopy such that $k_o = id_X$ and $k_1 = g\circ f$, and let $h: I \times Y \to Y$ be chosen as in Lemma 9.1 so that $h_o = id_Y$ and $h_1 = f\circ g$. We define an s.h.D-structure for g, $g_n: I^n \times D^nY \to X$, by

$$g_n = k\circ\xi\circ Dg_{n-1} \# -[(g\circ\phi\circ Dh\circ\ldots\circ D^{n-1}\phi\circ D^n h)'].$$

We evaluate such a composition of homotopies by evaluating the ith homotopy on the ith coordinate.

<u>Notation (i)</u>: Let p_n and $q_n: I^n \times Z \to X$ be homotopies. Define $p_n \# q_n: I^n \times Z \to Y$ to be the homotopy,

$$p_n \# q_n(t_1,\ldots,t_n,z) = \begin{cases} p_n(2t_1,t_2,\ldots,t_n) & \text{if } t_1 \leq \frac{1}{2} \\ q_n(2t_1 - 1,t_2,\ldots,t_n) & \text{if } t_1 \geq \frac{1}{2} \end{cases}$$

Of course, for this to make sense, $p_n|_{t_1=1}$ should equal $q_n|_{t_1=0}$ so that we may glue these faces together. Actually, one only needs $p_n|_{t_1=1}$ homotopic to $q_n|_{t_1=0}$ relative to $\{0,1\} \times Z$; if this possibility occurs as it will in the following considerations, we assume that the resulting homotopy is properly reparametrized.

(ii) Let $p_n: I^n \times Z \to X$ be any homotopy. We wish to reparametrize p_n inductively to obtain a new homotopy

p_n': $I^n \times Z \to X$ in the following manner: if $n = 2$, define p_2': $I^2 \times Z \to X$ by

$$p_2'|_{t_1=0} = p_2|_{(0,0)}, \quad p_2'|_{t_2=0} = p_2(t_1,0)$$

$$p_2'|_{t_2=1} = p_2|_{t_1=0}, \quad p_2'|_{t_1=1} = p_2|_{t_1=1} \# p_2|_{t_2=1}$$

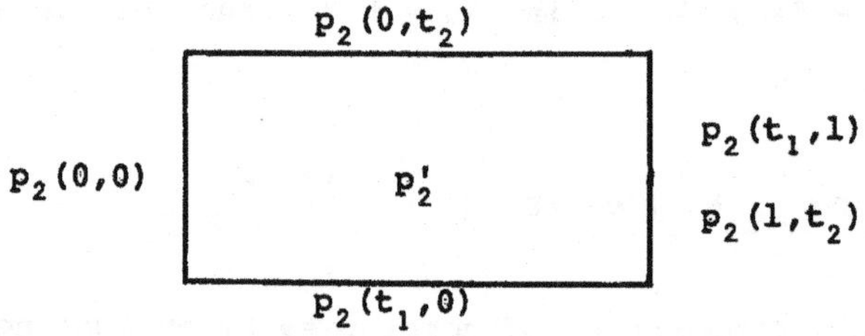

Now assuming that we have p_{n-1} reparametrized in this fashion, we define

$$p_n'|_{t_1=0} = p_n|_{(0,\ldots,0)}$$

$$p_n'|_{t_i=0} = (p_n|_{(t_1,\ldots,t_{i-1},0,\ldots,0)})'$$

$$p_n'|_{t_{i+1}=1} = (p_n|_{t_i=0})'$$

$$p_n'|_{t_1=1} = \# \, p_n|_{t_i=1} \qquad i = 1,\ldots,n$$

where $\# \, p_n|_{t_i=1}$ means fit together the n homotopies in a manner consistent with the edges of this face already defined; e.g., $n = 3$:

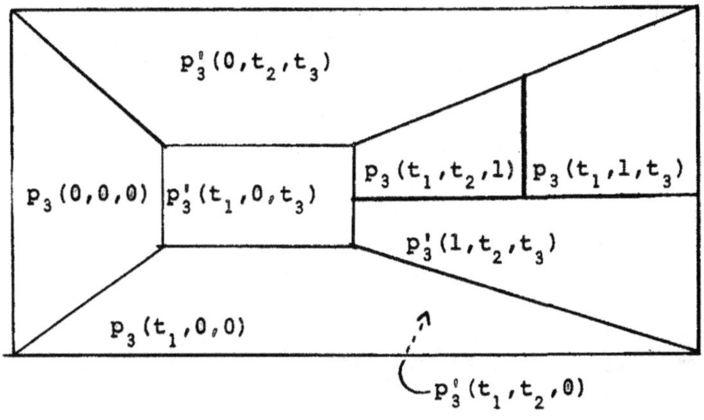

(iii) contrary to our use of the minus sign in the previous section a - preceding a homotopy indicates that the homotopy should be evaluated in reverse direction on each coordinate.

We proceed with the proof by first showing that each g_n is well-defined. First let n = 2; then

$$g_2 = k\circ\xi\circ Dg_1 \;\#\; -[(g\circ\phi\circ Dh\circ D\phi\circ D^2h)'].$$

To see that this is well-defined, consider first

$$k\circ\xi\circ Dg_1|_{t_1=1} = k\circ\xi\circ D[k\circ\xi\circ Dg \;\#\; -g\circ\phi\circ Dh]|_{t_1=1}$$

$$= g\circ f\circ\xi\circ Dk\circ D\xi\circ D^2g \;\#\; - g\circ f\circ\xi\circ Dg\circ D\phi\circ D^2h$$

$$= g\circ\phi\circ Df\circ Dk\circ D\xi\circ D^2g \;\#\; - g\circ\phi\circ Df\circ Dg\circ D\phi\circ D^2h$$

since f is a D-map

$$\simeq g\circ\phi\circ Dh\circ Df\circ D\xi\circ D^2g \;\#\; - g\circ\phi\circ Df\circ Dg\circ D\phi\circ D^2h \quad \text{by Lemma 9.1}$$

$$= g\circ\phi\circ Dh\circ D\phi\circ D^2f\circ D^2g \;\#\; - g\circ\phi\circ Df\circ Dg\circ D\phi\circ D^2h$$

$$= -[(g\circ\phi\circ Dh\circ D\phi\circ D^2h)']|_{t_1=0}.$$

Now let us assume that g_{n-1} is well-defined; to define g_n we have to fit together $k\circ\xi\circ Dg_{n-1}|_{t_1=1}$ and $-[(g\circ\phi\circ Dh\circ D\phi\circ\ldots\circ D^{n-1}\phi\circ D^n h)']|_{t_1=0}$. But

$$k\circ\xi\circ Dg_{n-1}|_{t_1=1}$$

$$= g\circ f\circ\xi\circ Dg_{n-1}$$

$$= g\circ f\circ\xi\circ D\{k\circ\xi\circ Dg_{n-2} \# -[(g\circ\phi\circ D\phi\circ\ldots\circ D^{n-2}\phi\circ D^{n-1}h)']\}$$

$$= g\circ f\circ\xi\circ Dk\circ D\xi\circ D^2 g_{n-2} \# -[(g\circ f\circ\xi\circ Dg\circ D\phi\circ\ldots\circ D^{n-1}\phi\circ D^n h)']$$

$$= g\circ\phi\circ Df\circ Dk\circ D\xi\circ D^2 g_{n-2} \# -[(g\circ\phi\circ Df\circ Dg\circ D\phi\circ D^2 h\circ\ldots\circ D^{n-1}\phi\circ D^n h)']$$

$$\simeq g\circ\phi\circ Dh\circ Df\circ D\xi\circ D^2 g_{n-2} \# -[(g\circ\phi\circ Df\circ Dg\circ D\phi\circ D^2 h\circ\ldots\circ D^{n-1}\phi\circ D^n h)']$$

$$= g\circ\phi\circ Dh\circ D(f\circ\xi\circ Dg_{n-2}) \# -[(g\circ\phi\circ Df\circ Dg\circ D\phi\circ D^2 h\circ\ldots\circ D^{n-1}\phi\circ D^n h)']$$

and the homotopy on the left fits together with $g\circ\phi\circ Dh\circ D(-[(g\circ\phi\circ Dh\circ\ldots\circ D^{n-1}h)'])$ by induction and the homotopy on the right is the remaining piece.

We now verify that the g_n satisfy the equalities needed for an s.h.D-map: If $t_1 = 0$,

$$g_n(0,t_2,\ldots,t_n) = k\circ\xi\circ Dg_{n-1}|_{t_1=0} = k_0\circ\xi\circ Dg_{n-1}$$

$$= \xi\circ Dg_{n-1}$$

If $t_1 = 1$, $g_n(1,t_2,\ldots,t_n) = -[(g\circ\phi\circ Dh\circ D\phi\circ\ldots\circ D^{n-1}\phi\circ D^n h)']|_{t_1=1}$

$$= (g\circ\phi\circ Dh\circ D\phi\circ\ldots\circ D^{n-1}\phi\circ D^n h)'|_{t_1=0}$$

$$= (g\circ\phi\circ Dh\circ D\phi\circ\ldots\circ D^{n-1}\phi\circ D^n h)|_{(0,\ldots,0)}$$

$$= g\circ\phi\circ D\phi\circ\ldots\circ D^{n-1}\phi$$

$$= g\circ\phi_{n-1}.$$

If $t_j = 0$, $g_n(t_1,\ldots,0,\ldots,t_n)$

$$= k\circ\xi\circ Dg_{n-1} \# -[(g\circ\phi\circ Dh\circ\ldots\circ D^{n-1}\phi\circ D^n h)']|_{t_j=0}$$

$$= k\circ\xi\circ Dg_{n-1}|_{t_j=0} \# -[(g\circ\phi\circ Dh\circ D\phi\circ\ldots\circ D^n h)']|_{t_j=0}$$

$$= k\circ\xi\circ Dg_{n-1}|_{t_j=0} \# -[(g\circ\phi\circ Dh\circ D\phi\circ\ldots\circ D^{n-1}\phi\circ D^n h)'|_{t_j=1}]$$

$$= k\circ\xi\circ Dg_{n-1}|_{t_j=0} \# -[(g\circ\phi\circ Dh\circ D\phi\circ\ldots\circ D^{n-1}\phi\circ D^n h|_{t_{j-1}=0})']$$

$$= k\circ\xi\circ Dg_{n-1}|_{t_j=0} \# -[(g\circ\phi\circ Dh\circ D\phi\circ\ldots\circ D^{j-2}\phi\circ D^{j-1}h_0\circ D^{j-1}\phi\circ\ldots)']$$

$$= k\circ\xi\circ Dg_{n-1}|_{t_j=0} \# -[(g\circ\phi\circ Dh\circ D\phi\circ\ldots\circ D^{j-2}\phi\circ D^{j-1}\phi\circ\ldots\circ D^n h)']$$

$$= k\circ\xi\circ Dg_{n-1}|_{t_j=0} \# -[(g\circ\phi\circ Dh\circ D\phi\circ\ldots\circ D^{n-1}h\circ D^{j-2}\mu_{n-j})']$$

$$= k\circ\xi\circ Dg_{n-2}\circ D^{j-2}\mu_{n-j} \# -[(g\circ\phi\circ Dh\circ D\phi\circ\ldots\circ D^{n-1}h\circ D^{j-2}\mu_{n-j})']$$

The next to last equality follows from the fact that (Y,ϕ) is a D-space and that μ is a natural transformation. Finally, if $t_j = 1$

$g_n(t_1,\ldots,1,\ldots,t_n)$

$= k\circ\xi\circ Dg_{n-1}|_{t_j=1} \# -[(g\circ\phi\circ Dh\circ D\phi\circ\ldots\circ D^{n-1}\phi\circ D^n h)']|_{t_j=1}$

$= k\circ\xi\circ Dg_{n-1}|_{t_j=1} \# -[(g\circ\phi\circ Dh\circ D\phi\circ\ldots\circ D^{n-1}\phi\circ D^n h)'|_{t_j=0}]$

$= k\circ\xi\circ Dg_{n-1}|_{t_j=1} \# -[(g\circ\phi\circ Dh\circ D\phi\circ\ldots\circ D^{n-1}\phi\circ D^n h|_{(t_1,\ldots,t_{j-1},0,\ldots,0)})']$

$= k\circ\xi\circ Dg_{n-1}|_{t_j=1} \# -[(g\circ\phi\circ Dh\circ D\phi\circ\ldots\circ D^{j-1}h\circ D^{j-1}\phi\circ D^j\phi\circ\ldots\circ D^{n-1}\phi)']$

$= k\circ\xi\circ Dg_{j-2}\circ D^{j-1}\phi_{n-j} \# -[(g\circ\phi\circ Dh\circ D\phi\circ\ldots\circ D^{j-1}h)']\circ D^{j-1}\phi_{n-j}.$

We conclude this proof by showing that $f\circ g$ is homotopic to 1_Y as an s.h.D-map. Define an s.h.D-homotopy $p_n: I^{n+1}\times D^nY \to Y$ by

$$p_n = -[(h\circ\phi\circ Dh\circ\ldots\circ D^{n-1}\phi\circ D^n h)'].$$

If $t_1 = 1$, we have

$$\begin{aligned} p_n &= -[(h\circ\phi\circ\ldots\circ D^n h)']|_{t_1=1} \\ &= (h\circ\phi\circ\ldots\circ D^n h)'|_{t_1=0} \\ &= (h\circ\phi\circ\ldots\circ D^n h)|_{(0,\ldots,0)} \\ &= \phi\circ\ldots\circ D^{n-1}\phi \end{aligned}$$

which is the canonical s.h.D-structure for the identity map. If $t_1 = 0$, we have

$$n = -[(h\circ\phi\circ\ldots\circ D^n h)']|_{t_1=0}$$

$$-[(h\circ\phi\circ\ldots\circ D^n h)'|_{t_1=1}]$$

$$-[\#(h\circ\phi\circ\ldots\circ D^n h|_{t_i=1})']$$

$$\# -[(h\circ\phi\circ\ldots\circ D^{i-2}\phi\circ D^{i-1}f\circ D^{i-1}g\circ\ldots\circ D^n h)'] \# -[(f\circ g\circ\phi\circ\ldots\circ D^n h)']$$

$$\# -[(h\circ\phi\circ\ldots\circ D^{i-2}h\circ D^{i-2}f\circ D^{i-1}\xi\circ D^{i-1}g\circ\ldots\circ D^n h)']\#-[(f\circ g\circ\phi\circ\ldots\circ D^n h)']$$

since f is a D-map

$$\# -[(h\circ\phi\circ\ldots\circ D^{i-3}\phi\circ D^{i-2}f\circ D^{i-2}k\circ D^{i-1}\xi\circ\ldots\circ D^n h)']\#-[(f\circ g\circ\phi\circ\ldots\circ D^n h)']$$

by Lemma 9.1

$$\# -[f\circ k\circ\xi\circ Dk\circ D\xi\circ\ldots\circ D^{i-2}k\circ D^{i-1}\xi\circ D^{i-1}g\circ\ldots\circ D^n h)']\#-[(f\circ g\circ\phi\circ\ldots\circ D^n h)']$$

by iteration of the two previous steps

$$f\circ\{\# -[k\circ\xi\circ Dk\circ D\xi\circ\ldots\circ D^{i-2}k\circ D^{i-1}\xi\circ D^{i-1}g\circ\ldots\circ D^n h)']\#-[(g\circ\phi\circ\ldots\circ D^n h)']\}$$

$$f\circ\{k\circ\xi\circ Dg_{n-1} \# -[(g\circ\phi\circ Dh\circ\ldots\circ D^n h)']\}$$

$$f\circ g_n$$

hich is our canonical s.h.D-structure for a composition of a)-map with an s.h.D-map.

The other equalities may be checked by a similar argument and thus our proof is completed. $\blacksquare$

Appendix

We present here a geometrical alternative to Definition 3.1 and discuss some of its consequences.

Definition A1: Let $(X,\{\xi_n\})$ and $(Y,\{\phi_n\})$ be s.h.D-spaces and let $f: X \to Y$ be a map. Then f is an s.h.D-map if there exists a collection of homotopies

$$f_n: I^n \times D^nX \to Y$$

such that

$$f_n(t_1,\ldots,t_n,z) = \begin{cases} f\circ\xi_{n-1}(t_2,\ldots,t_n,z) & \text{if } t_1 = 1 \\ f_{j-1}(t_1,\ldots,t_{j-1},D^{j-1}\xi_{n-j}(t_{j+1},\ldots,t_n,z)) & \text{if } t_j = 1 \\ f_{n-1}(t_1,\ldots,\hat{t}_j,\ldots,t_n,D^{j-2}\mu_{n-j}(z)) & \text{if } t_j = 0 \\ \{\phi_{i-1}\circ D^i f_{n-i}(2t_2,\ldots,2t_i,2t_{i+1}-1,t_{i+2},\ldots,t_n,z)\} & \\ \quad \text{where } t_1,\ldots,t_i \le \frac{1}{2},\ t_{i+1} \ge \frac{1}{2},\ i = 1,\ldots,n & \\ \quad \text{if } t_1 = 0. & \end{cases}$$

The concept of a homotopy between two such maps is a direct generalization of Definition 3.3.

We may also generalize Corollary 5.2 to conclude that there is a one to one correspondence between homotopy classes of s.h.D-maps from X to Y and homotopy classes of D-maps from UX to UY where X and Y are s.h.D-spaces. It can further be shown that there is a well-defined category whose objects are s.h.D-spaces and whose morphisms are homotopy classes of s.h.D-maps; in addition, U is a fully faithful functor from this category to hD[*T*]. This machinery may be used to provide an alternate proof of the homotopy invariance Theorem 6.2.

Throughout this work, we utilized geometrical complexes whose vertices indexed homotopic maps in various s.h.D-structures. We present here a typical counting argument used in formulating an s.h.D-structure. We analyze Definition A1, that of an s.h.D-map between s.h.D-spaces. Recall that the parameter space in the definition is a cube with one face subdivided. The reason for this is that a one-to-one correspondence between the vertices of the complex and the distinct maps $D^nX \to Y$ is needed. The argument that follows should be considered as joint work with Bob Ramsay.

Let us first count the distinct maps $D^nX \to Y$ inductively. Let #(n) denote this number and assume that we know #(n - 1). Now compose each of the #(n - 1) maps with first $D^{n-1}\xi$; then compose each of the original #(n - 1) maps with $D^{n-2}\mu$. This gives us 2#(n - 1) distinct maps $D^nX \to Y$. This, however, does not account for all possible such maps. To obtain the remaining ones, consider all compositions of the form

$$\phi \circ D\phi \circ \_ \circ \_ \circ \ldots \circ \_ \circ D^n f$$

where in the ith blank we may use either $D^{i+1}\phi$ or $D^{i-1}\mu_{n-i-3}$, $i = 1,\ldots,n - 2$. We thus have 2^{n-2} additional maps which are all distinct. We claim that $\#(n) = 2\#(n - 1) + 2^{n-2}$. The verification that all of the above maps are distinct and form the complete set of such is tedious but straightforward; one utilizes the properties of μ in the definition of a monad and the fact that there are n distinct maps $D^nX \to D^{n-1}X$ given by $D^{n-1}\xi$ and $D^i\mu_{n-2-i}$ with $i = 0,\ldots,n - 2$.

We now turn our attention to the subdivided cubes. Recall that we subdivided the $t_1 = 0$ face of I^n into n cubes of dimension $n - 1$ according to the relation

$$\{t_1,\ldots,t_i \leq \frac{1}{2}, t_{i+1} \geq \frac{1}{2}, t_{i+k} \text{ is arbitrary}\}$$

where $i = 1,\ldots,n$ and $k = 2,\ldots,n - i$. We want to count the number of vertices of this new complex. First, there are the 2^n vertices from I^n where each of the n coordinates is either 0 or 1. To count the others, fix i as above and let $t_{i+1} = \frac{1}{2}$. Since $t_1 = 0$, the coordinates of a vertex are given by $(0,\ldots,\frac{1}{2},\ldots)$ where there are $i - 1$ blank coordinates to the left of $\frac{1}{2}$ and $n - i - 1$ to the right of $\frac{1}{2}$. To the left, we may fill in either 0 or $\frac{1}{2}$; to the right we may fill in either 0 or 1. At any rate, we get 2^{n-2} vertices this way. But i may vary over $n - 1$ positions. Thus we have $(n - 1)2^{n-2}$ vertices of this type and thus have a total of $2^n + (n - 1)2^{n-2}$ vertices.

To see that the number of vertices agrees with the number of distinct maps $D^nX \to Y$, it is easy to check that the number $2^n + (n - 1)2^{n-2}$ satisfies the equality $\#(n) = 2\#(n - 1) + 2^{n-2}$.

Bibliography

1. Beck, J. On H-spaces and infinite loop spaces, Category Theory, Homology Theory and Their Applications, III, Springer-Verlag, 1969.

2. Boardman, J. M. and Vogt, R. M. Homotopy-everything H-spaces, Bull. AMS., 74(1968), 1117-1122.

3. ______________________________. Homotopy Invariant Algebraic Structures on Topological Spaces, Springer-Verlag, 1973.

4. Dold, A. and Lashof, R. Principal quasifibrations and fibre homotopy equivalence of bundles, Illinois J. of Math., 3(1959), 285-305.

5. Dold, A. and Thom, R. Quasifaserungen und unendliche symmetriche produkte, Ann. of Math., (2) 67(1958), 239-281.

6. Fuchs, Martin. A modified Dold-Lashof construction that does classify H-principal fibrations, Math. Ann., 192(1971), 328-340.

7. James, I. M. On H-spaces and their homotopy groups, Quart. J. Math. Oxford (2) 11(1960), 161-179.

8. Lada, T. and Malraison, P. Categories of triple algebras, (preprint).

G. May, J. P. The Geometry of Iterated Loop Spaces, Springer-Verlag, 1972.

G'. _________. E_∞ spaces, group completions, and permutative categories, London Mathematical Society Lecture Note Series 11, p. 61-94, 1974, Cambridge University Press.

9. Milnor, J. Construction of universal bundles, I, Ann. of Math., (2) 63(1956), 272-284.

10. Segal, G. Categories and cohomology theories, Topology, 13(1974), 293-312.

11. Stasheff, J. D. Homotopy associativity of H-spaces I, II, Trans. Amer. Math. Soc., 108(1963), 275-292, ibid 108(1963), 292-312.

12. _____________. H-spaces From a Homotopy Point of View, Springer-Verlag, 1970.

13. Steenrod, N. A convenient category of topological spaces, Mich. Math. J., 14(1967), 133-152.

INDEX

Errata and Addenda to [A], [G], and [G']

The cited papers contain a number of misprints, mistakes, and results since generalized. We indicate where changes should be made in the following list. Minor errors are indicated by line, with the material to be changed underlined (but mathematically irrelevant typographical errors have generally been ignored). The list also includes references to work by other authors which adds to the results of [G] and [G'].

1. [A, p. 158, line 11]: $\cdots \{\underline{x_{\gamma(j_1)} \otimes \cdots \otimes x_{\gamma(j_p)}} \mid \gamma \in G,\ j_1 \leq \cdots \leq j_p,\ j_1 < j_p\}. \cdots$

2. In view of the geometric construction of the homology operations analyzed by Cohen in III, there is no longer any real reason to use the categories $\mathcal{C}(p, n)$ for $n < \infty$ in [A, §2 and §3]. Restriction to the case $n = \infty$ would allow some simplification of notations.

3. [A, p. 161, line -4]: $\ldots D_i(x) = \theta_*(e_i \otimes \underline{x^p}), \ldots$

4. In the cohomology of spaces, [A, 6.8 (p. 188)] was first proven in

   T. Yamanoshita. On certain cohomological operations. J. Math. Soc. Japan 8(1956), 300-344.

5. [A, p. 193, line 14]: $\ldots (D \times \underline{1)D} = (1 \times D)D \ldots$

6. [A, 10.2 (p. 214)] is clearly false in the case $p = 2$ and $t > 1$, where $H^*(Z_{2^t}, 1, Z_2) = E(i_1) \otimes P(\beta_t i_1)$ just as in the pase $p > 2$. Therefore, when $n = 1$ and $p = 2$, [A, 10.3] only holds for $t = 1$.

7. The letter S was used for suspension in [G]. This standard notation is very awkward, and Σ has been used in [G'], [R], and the present volume.

8. The weak Hausdorff rather than the Hausdorff property should be required of spaces in $\mathcal{J}$ and $\mathcal{U}$ [G, p. 1] in order to validate some of the limit arguments used in [G].

9. [G, p. 4, line -2] : ... highly ..

10. An elaboration of the proof of [G, 1.9 (p. 8)] should show that if $\mathcal{C}$ is an E_∞ operad, then the product on a $\mathcal{C}$-space is an s.h. C-map (in the sense defined by Lada in V).

11. [G, 3.4 (p. 22)] is improved in [G', A. 2].

12. The proof of [G, 4.8 (p. 35)] is not quite correct since the specified homotopy $h: 1 \simeq fg$ is not a homotopy through points of $\mathcal{C}_n(j)$: the disjoint image requirement can be violated. The remedy is to first linearly shrink points $c \in \mathcal{C}_n(j)$ to their maximal inscribed equidiameter points and then linearly expand the resulting points to the maximal equidiameter points $fg(c)$.

13. The category $\mathcal{L}_\infty$ specified on [G, p. 40] is obviously appropriate to infinite loop space theory. It is now known that this category (or better, its coordinate-free equivalent) is also the appropriate starting point for the construction of a good stable homotopy category. See [R, ch. II] for a summary. Full details will appear in

J. P. May. The stable homotopy category and its applications. (Part of a forthcoming Monograph of the London Math. Soc.)

14. The proof of the approximation theorem in [G, §6-7] is not particularly attractive. An alternative proof, together with the appropriate generalization to the case of non-connected spaces, has been given in

G. Segal. Configuration spaces and iterated loop spaces. Inventiones Math. 21(1973), 213-221.

The cited generalization asserts that $\alpha_n : C_n X \to \Omega^n \Sigma^n X$ is a group completion in the sense of [G', 1.3], and this statement is also an immediate consequence of Cohen's calculations in III.

15. An interesting generalization of the approximation theorem (from a statement about the configuration spaces of R^n to a statement about the configuration spaces of smooth manifolds) has been given in

D. McDuff. Configuration spaces of positive and negative particles. Topology 14(1975), p. 91-107.

16. A basic application of the approximation theorem has been given in

V. P. Snaith. A stable decomposation of $\Omega^n S^n X$. J. London Math. Soc. 7(1974), 577-583.

Snaith shows that, for a connected space X, the suspension spectrum of $C_n X \simeq \Omega^n \Sigma^n X$ splits as the wedge of the suspension spectra of $F_j C_n X / F_{j-1} C_n X = e[\mathcal{C}_n(j), \Sigma_j, X]$. (See [G, p. 14] for the notation.) In his thesis (Northwestern Univ. 1975), P. O. Kirley proves that, for $n \geq 2$, there is no finite r such that $\Sigma^r \Omega^n \Sigma^n X$ splits as $\bigvee_j \Sigma^r e[\mathcal{C}_n(j), \Sigma_j, X]$.

17. As explained in [G', A. 5], the notion of strict propriety introduced in [G, 11.2 (p. 102)] is unnecessary.

18. [G, p. 103, line -3]: ...$X, Y \in \underline{\mathcal{S}}\mathcal{U}$,...

19. [G, p. 104, line 3]: ... $v = (\underline{t'_0}, \dots, t'_q) \in \Delta_q$.

20. [G, 11.13 (p. 109)] is improved in [G', A. 4].

21. As will be discussed in 30, when $n \geq 2$, [G, 13.1(ii) (p. 129)] generalizes in the non-connected case to the assertion that $B(\alpha_n \pi, 1, 1)$ is a group completion. It follows that [G, 13.1(iii)] remains true when Y is (n-1)-connected, that [G, 13.2 (p. 132, misnumbered as 13.3)] remains true when g is a group completion rather than a weak equivalence, and that [G, 13.4] remains true when X is grouplike.

22. In [G, 13.5(ii) (p. 134)], the connectivity hypothesis on X is unnecessary. To see this, merely use [G, 3.7] and [G', A. 2(ii) and A. 4] in place of [G, 3.4 and 11.13] in the proof.

23. [G, p. 136, line -5]: The reference should be to [G'], not [21].

24. As was proven in [G', 2.3], [G, 14.4(ii) (p. 144)] generalizes in the non-connected case to the assertion that $B(\alpha_\infty \pi_\infty, 1, 1)$ is a group completion. It follows that [G, 14.4(vi)] remains true when each Y_i is (i-1)-connected, that [G, 14.5] remains true when g is a group completion, and that [G, 14.6] remains true when $\Omega^j Y$ is connective and (in its second part) when X is arbitrary . Again, by [G', 3.1], X need not be connected in [G, 14.8], and [G, 14.7 and 14.9] remain true when X is (i-1)-connected.

25. The discussion of connectivity hypotheses and homotopy invariance in [G, p. 156-160] are of course obsolete.

26. [G, p. 166, line -8]:

$$(Su)[x,s] = \begin{cases} 4su(x) & \text{if } 0 \leq s \leq 1/4 \\ u(x) & \text{if } 1/4 \leq s \leq 3/4 \\ (4-4s)u(x) & \text{if } 3/4 \leq s \leq 1 \end{cases}$$

27. In [G', §1], it is asserted that $\zeta: G \to \Omega BG$ is a group completion if the monoid G and the H-space ΩBG are both admissible in the sense of [G', 1.3]. This restriction allows the simple proof given in

J. P. May. Classifying spaces and fibrations. Memoirs Amer. Math. Soc. 155 (1975).

A convincing, and not very much more difficult, proof assuming only that $\pi_0 G$ is central in $H_* G$ has since been given in

D. McDuff and G. Segal. Homology fibrations and the "group completion" Theorem. Preprint.

(Both proofs were suggested by unpublished arguments of Quillen.)

28. The proof of [G', 2.1] is incomplete, since the assertion that X or $B(M, C \times M, X)$ is strongly homotopy commutative is not obvious. The verification is unnecessary if one is willing to use the strengthened version of the group completion theorem cited in 27. Alternatively, a simple rigorous proof of [G', 2.1] is given in [R, VI 2.7(iv)].

30. By 14 and a comparison of the proofs of [G, 13.1] and [G', 2.3], the generalization of the former result cited in 21 requires only the appropriate analog of [G', 2.1]. Here I know of no construction to which the weaker form of the group completion theorem applies. For a local equivalence $\mathcal{D} \to \mathcal{C}_n$ of Σ-free operads with $n \geq 2$, we require a functor G and natural group completion $g: X \to GX$ for $\mathcal{D}$-spaces X. We assume that $\mathcal{D} = \mathcal{C} \times \mathcal{C}_n$ where $\mathcal{C}$ is locally contractible (since all known examples are of this form). We define

$$GX = \Omega BB(M, C \times C_1, X)$$

and let $g: X \to GX$ be the following composite:

$$X \xrightarrow{\tau(\eta)} B(C \times C_1, C \times C_1, X) \xrightarrow{B(\varepsilon, 1, 1)} B(M, C \times C_1, X) \xrightarrow{\zeta} GX .$$

Then $\tau(\eta)$ and $B(\varepsilon, 1, 1)$ are equivalences of H-spaces by [G, 9.8] and [G, 3.7 and G' A.2(ii) and A.4(ii)], while ζ is a group completion by 27. Here X is regarded as a $\mathcal{C} \times \mathcal{C}_1$-space by pullback along the inclusion $\mathcal{C} \times \mathcal{C}_1 \subset \mathcal{C} \times \mathcal{C}_n$.

31. [G', 3.7 (p. 76-the second result labeled 3.7)] is incorrect. The error occurs on line -7, from which a factor $\sigma(j_1, \ldots, j_k)$ has been omitted on the right side of the equation. With this factor, ν on line -5 depends on σ and the argument collapses. (See also [R, VI. 2.7(v) and (vi)].)

32. A more structured version of [G', 4.2] is given in [R, VI. 3.2].

33. [G', p. 82, line 1]: $\tilde{\Sigma}_k \times \tilde{\Sigma}_{j_1} \times \dots \times \tilde{\Sigma}_{j_k} \times \underline{\mathcal{A}}^j \xrightarrow{\gamma \times 1} \tilde{\Sigma}_j \times \mathcal{A}^j$.

34. [G', p. 82, line 5]: $\tilde{\gamma}(\sigma; \tau_1, \dots, \tau_k) = \tau_{\sigma^{-1}(1)} \oplus \dots \oplus \tau_{\sigma^{-1}(k)} \cdot \underline{\sigma(j_1, \dots, j_k)}$

35. The consistency with Bott periodicity asserted in the next to last paragraph of [G', p. 85] is proven rigorously in [R, VIII §1].

36. A quick proof of [G', A. 1] will appear in

J. P. May. On duality and completions in homotopy theory. (Part of a forthcoming Monograph of the London Math. Soc.)

37. [G', p. 90 line 3 (of top diagram)]: $X_{q+1} \times \underline{\Delta}_{q+1} \to F_{q+1}|X|$.

38. [G', p. 90 line 4]: where $g(\underline{s}_i x, u) = |x, \sigma_i u|$ and $g(x, \underline{\delta}_i v) = |\partial_i x, v|$ ...

ol. 399: Functional Analysis and its Applications. Proceedings 973. Edited by H. G. Garnir, K. R. Unni and J. H. Williamson. , 584 pages. 1974.

'ol. 400: A Crash Course on Kleinian Groups. Proceedings 1974. :dited by L. Bers and I. Kra. VII, 130 pages. 1974.

'ol. 401: M. F. Atiyah, Elliptic Operators and Compact Groups. , 93 pages. 1974.

'ol. 402: M. Waldschmidt, Nombres Transcendants. VIII, 277 ages. 1974.

ol. 403: Combinatorial Mathematics. Proceedings 1972. Edited y D. A. Holton. VIII, 148 pages. 1974.

ol. 404: Théorie du Potentiel et Analyse Harmonique. Edité par Faraut. V, 245 pages. 1974.

ol. 405: K. J. Devlin and H. Johnsbråten, The Souslin Problem. 'III, 132 pages. 1974.

ol. 406: Graphs and Combinatorics. Proceedings 1973. Edited y R. A. Bari and F. Harary. VIII, 355 pages. 1974.

'ol. 407: P. Berthelot, Cohomologie Cristalline des Schémas de :aracteristique p > o. II, 604 pages. 1974.

ol. 408: J. Wermer, Potential Theory. VIII, 146 pages. 1974.

ol. 409: Fonctions de Plusieurs Variables Complexes, Séminaire rançois Norguet 1970–1973. XIII, 612 pages. 1974.

ol. 410: Séminaire Pierre Lelong (Analyse) Année 1972–1973. I, 181 pages. 1974.

ol. 411: Hypergraph Seminar. Ohio State University, 1972. dited by C. Berge and D. Ray-Chaudhuri. IX, 287 pages. 1974.

ol. 412: Classification of Algebraic Varieties and Compact :omplex Manifolds. Proceedings 1974. Edited by H. Popp. V, 33 pages. 1974.

'ol. 413: M. Bruneau, Variation Totale d'une Fonction. XIV, 332 ages. 1974.

ol. 414: T. Kambayashi, M. Miyanishi and M. Takeuchi, Uniotent Algebraic Groups. VI, 165 pages. 1974.

'ol. 415: Ordinary and Partial Differential Equations. Proceedings 974. XVII, 447 pages. 1974.

ol. 416: M. E. Taylor, Pseudo Differential Operators. IV, 155 ages. 1974.

ol. 417: H. H. Keller, Differential Calculus in Locally Convex ;paces. XVI, 131 pages. 1974.

'ol. 418: Localization in Group Theory and Homotopy Theory nd Related Topics. Battelle Seattle 1974 Seminar. Edited by P. J. lilton. VI, 172 pages 1974.

'ol. 419: Topics in Analysis. Proceedings 1970. Edited by O. E. .ehto, I. S. Louhivaara, and R. H. Nevanlinna. XIII, 392 pages. 1974.

'ol. 420: Category Seminar. Proceedings 1972/73. Edited by G. M. :elly. VI, 375 pages. 1974.

'ol. 421: V. Poénaru, Groupes Discrets. VI, 216 pages. 1974.

'ol. 422: J.-M. Lemaire, Algèbres Connexes et Homologie des :spaces de Lacets. XIV, 133 pages. 1974.

'ol. 423: S. S. Abhyankar and A. M. Sathaye, Geometric Theory f Algebraic Space Curves. XIV, 302 pages. 1974.

'ol. 424: L. Weiss and J. Wolfowitz, Maximum Probability :stimators and Related Topics. V, 106 pages. 1974.

'ol. 425: P. R. Chernoff and J. E. Marsden, Properties of Infinite)imensional Hamiltonian Systems. IV, 160 pages. 1974.

ol. 426: M. L. Silverstein, Symmetric Markov Processes. X, 287 ages. 1974.

ol. 427: H. Omori, Infinite Dimensional Lie Transformation ;roups. XII, 149 pages. 1974.

ol. 428: Algebraic and Geometrical Methods in Topology, Proeedings 1973. Edited by L. F. McAuley. XI, 280 pages. 1974.

Vol. 429: L. Cohn, Analytic Theory of the Harish-Chandra C-Function. III, 154 pages. 1974.

Vol. 430: Constructive and Computational Methods for Differential and Integral Equations. Proceedings 1974. Edited by D. L. Colton and R. P. Gilbert. VII, 476 pages. 1974.

Vol. 431: Séminaire Bourbaki – vol. 1973/74. Exposés 436–452. IV, 347 pages. 1975.

Vol. 432: R. P. Pflug, Holomorphiegebiete, pseudokonvexe Gebiete und das Levi-Problem. VI, 210 Seiten. 1975.

Vol. 433: W. G. Faris, Self-Adjoint Operators. VII, 115 pages. 1975.

Vol. 434: P. Brenner, V. Thomée, and L. B. Wahlbin, Besov Spaces and Applications to Difference Methods for Initial Value Problems. II, 154 pages. 1975.

Vol. 435: C. F. Dunkl and D. E. Ramirez, Representations of Commutative Semitopological Semigroups. VI, 181 pages. 1975.

Vol. 436: L. Auslander and R. Tolimieri, Abelian Harmonic Analysis, Theta Functions and Function Algebras on a Nilmanifold. V, 99 pages. 1975.

Vol. 437: D. W. Masser, Elliptic Functions and Transcendence. XIV, 143 pages. 1975.

Vol. 438: Geometric Topology. Proceedings 1974. Edited by L. C. Glaser and T. B. Rushing. X, 459 pages. 1975.

Vol. 439: K. Ueno, Classification Theory of Algebraic Varieties and Compact Complex Spaces. XIX, 278 pages. 1975.

Vol. 440: R. K. Getoor, Markov Processes: Ray Processes and Right Processes. V, 118 pages. 1975.

Vol. 441: N. Jacobson, PI-Algebras. An Introduction. V, 115 pages. 1975.

Vol. 442: C. H. Wilcox, Scattering Theory for the d'Alembert Equation in Exterior Domains. III, 184 pages. 1975.

Vol. 443: M. Lazard, Commutative Formal Groups. II, 236 pages. 1975.

Vol. 444: F. van Oystaeyen, Prime Spectra in Non-Commutative Algebra. V, 128 pages. 1975.

Vol. 445: Model Theory and Topoi. Edited by F. W. Lawvere, C. Maurer, and G. C. Wraith. III, 354 pages. 1975.

Vol. 446: Partial Differential Equations and Related Topics. Proceedings 1974. Edited by J. A. Goldstein. IV, 389 pages. 1975.

Vol. 447: S. Toledo, Tableau Systems for First Order Number Theory and Certain Higher Order Theories. III, 339 pages. 1975.

Vol. 448: Spectral Theory and Differential Equations. Proceedings 1974. Edited by W. N. Everitt. XII, 321 pages. 1975.

Vol. 449: Hyperfunctions and Theoretical Physics. Proceedings 1973. Edited by F. Pham. IV, 218 pages. 1975.

Vol. 450: Algebra and Logic. Proceedings 1974. Edited by J. N. Crossley. VIII, 307 pages. 1975.

Vol. 451: Probabilistic Methods in Differential Equations. Proceedings 1974. Edited by M. A. Pinsky. VII, 162 pages. 1975.

Vol. 452: Combinatorial Mathematics III. Proceedings 1974. Edited by Anne Penfold Street and W. D. Wallis. IX, 233 pages. 1975.

Vol. 453: Logic Colloquium. Symposium on Logic Held at Boston, 1972–73. Edited by R. Parikh. IV, 251 pages. 1975.

Vol. 454: J. Hirschfeld and W. H. Wheeler, Forcing, Arithmetic, Division Rings. VII, 266 pages. 1975.

Vol. 455: H. Kraft, Kommutative algebraische Gruppen und Ringe. III, 163 Seiten. 1975.

Vol. 456: R. M. Fossum, P. A. Griffith, and I. Reiten, Trivial Extensions of Abelian Categories. Homological Algebra of Trivial Extensions of Abelian Categories with Applications to Ring Theory. XI, 122 pages. 1975.

Vol. 457: Fractional Calculus and Its Applications. Proceedings 1974. Edited by B. Ross. VI, 381 pages. 1975.

Vol. 458: P. Walters, Ergodic Theory – Introductory Lectures. VI, 198 pages. 1975.

Vol. 459: Fourier Integral Operators and Partial Differential Equations. Proceedings 1974. Edited by J. Chazarain. VI, 372 pages. 1975.

Vol. 460: O. Loos, Jordan Pairs. XVI, 218 pages. 1975.

Vol. 461: Computational Mechanics. Proceedings 1974. Edited by J. T. Oden. VII, 328 pages. 1975.

Vol. 462: P. Gérardin, Construction de Séries Discrètes p-adiques. »Sur les séries discrètes non ramifiées des groupes réductifs déployés p-adiques«. III, 180 pages. 1975.

Vol. 463: H.-H. Kuo, Gaussian Measures in Banach Spaces. VI, 224 pages. 1975.

Vol. 464: C. Rockland, Hypoellipticity and Eigenvalue Asymptotics. III, 171 pages. 1975.

Vol. 465: Séminaire de Probabilités IX. Proceedings 1973/74. Edité par P. A. Meyer. IV, 589 pages. 1975.

Vol. 466: Non-Commutative Harmonic Analysis. Proceedings 1974. Edited by J. Carmona, J. Dixmier and M. Vergne. VI, 231 pages. 1975.

Vol. 467: M. R. Essén, The Cos $\pi\lambda$ Theorem. With a paper by Christer Borell. VII, 112 pages. 1975.

Vol. 468: Dynamical Systems – Warwick 1974. Proceedings 1973/74. Edited by A. Manning. X, 405 pages. 1975.

Vol. 469: E. Binz, Continuous Convergence on C(X). IX, 140 pages. 1975.

Vol. 470: R. Bowen, Equilibrium States and the Ergodic Theory of Anosov Diffeomorphisms. III, 108 pages. 1975.

Vol. 471: R. S. Hamilton, Harmonic Maps of Manifolds with Boundary. III, 168 pages. 1975.

Vol. 472: Probability-Winter School. Proceedings 1975. Edited by Z. Ciesielski, K. Urbanik, and W. A. Woyczyński. VI, 283 pages. 1975.

Vol. 473: D. Burghelea, R. Lashof, and M. Rothenberg, Groups of Automorphisms of Manifolds. (with an appendix by E. Pedersen) VII, 156 pages. 1975.

Vol. 474: Séminaire Pierre Lelong (Analyse) Année 1973/74. Edité par P. Lelong. VI, 182 pages. 1975.

Vol. 475: Répartition Modulo 1. Actes du Colloque de Marseille-Luminy, 4 au 7 Juin 1974. Edité par G. Rauzy. V, 258 pages. 1975. 1975.

Vol. 476: Modular Functions of One Variable IV. Proceedings 1972. Edited by B. J. Birch and W. Kuyk. V, 151 pages. 1975.

Vol. 477: Optimization and Optimal Control. Proceedings 1974. Edited by R. Bulirsch, W. Oettli, and J. Stoer. VII, 294 pages. 1975.

Vol. 478: G. Schober, Univalent Functions – Selected Topics. V, 200 pages. 1975.

Vol. 479: S. D. Fisher and J. W. Jerome, Minimum Norm Extremals in Function Spaces. With Applications to Classical and Modern Analysis. VIII, 209 pages. 1975.

Vol. 480: X. M. Fernique, J. P. Conze et J. Gani, Ecole d'Eté de Probabilités de Saint-Flour IV-1974. Edité par P.-L. Hennequin. XI, 293 pages. 1975.

Vol. 481: M. de Guzmán, Differentiation of Integrals in R^n. XII, 226 pages. 1975.

Vol. 482: Fonctions de Plusieurs Variables Complexes II. Séminaire François Norguet 1974–1975. IX, 367 pages. 1975.

Vol. 483: R. D. M. Accola, Riemann Surfaces, Theta Functions, and Abelian Automorphisms Groups. III, 105 pages. 1975.

Vol. 484: Differential Topology and Geometry. Proceedings 1974. Edited by G. P. Joubert, R. P. Moussu, and R. H. Roussarie. IX, 287 pages. 1975.

Vol. 485: J. Diestel, Geometry of Banach Spaces – Selected Topics. XI, 282 pages. 1975.

Vol. 486: S. Stratila and D. Voiculescu, Representations of AF-Algebras and of the Group U (∞). IX, 169 pages. 1975.

Vol. 487: H. M. Reimann und T. Rychener, Funktionen beschränkter mittlerer Oszillation. VI, 141 Seiten. 1975.

Vol. 488: Representations of Algebras, Ottawa 1974. Proceedings 1974. Edited by V. Dlab and P. Gabriel. XII, 378 pages. 1975.

Vol. 489: J. Bair and R. Fourneau, Etude Géométrique des Espaces Vectoriels. Une Introduction. VII, 185 pages. 1975.

Vol. 490: The Geometry of Metric and Linear Spaces. Proceedings 1974. Edited by L. M. Kelly. X, 244 pages. 1975.

Vol. 491: K. A. Broughan, Invariants for Real-Generated Uniform Topological and Algebraic Categories. X, 197 pages. 1975.

Vol. 492: Infinitary Logic: In Memoriam Carol Karp. Edited by D. W. Kueker. VI, 206 pages. 1975.

Vol. 493: F. W. Kamber and P. Tondeur, Foliated Bundles and Characteristic Classes. XIII, 208 pages. 1975.

Vol. 494: A Cornea and G. Licea. Order and Potential Resolvent Families of Kernels. IV, 154 pages. 1975.

Vol. 495: A. Kerber, Representations of Permutation Groups II. V, 175 pages. 1975.

Vol. 496: L. H. Hodgkin and V. P. Snaith, Topics in K-Theory. Two Independent Contributions. III, 294 pages. 1975.

Vol. 497: Analyse Harmonique sur les Groupes de Lie. Proceedings 1973–75. Edité par P. Eymard et al. VI, 710 pages. 1975.

Vol. 498: Model Theory and Algebra. A Memorial Tribute to Abraham Robinson. Edited by D. H. Saracino and V. B. Weispfenning. X, 463 pages. 1975.

Vol. 499: Logic Conference, Kiel 1974. Proceedings. Edited by G. H. Müller, A. Oberschelp, and K. Potthoff. V, 651 pages 1975.

Vol. 500: Proof Theory Symposion, Kiel 1974. Proceedings. Edited by J. Diller and G. H. Müller. VIII, 383 pages. 1975.

Vol. 501: Spline Functions, Karlsruhe 1975. Proceedings. Edited by K. Böhmer, G. Meinardus, and W. Schempp. VI, 421 pages. 1976.

Vol. 502: János Galambos, Representations of Real Numbers by Infinite Series. VI, 146 pages. 1976.

Vol. 503: Applications of Methods of Functional Analysis to Problems in Mechanics. Proceedings 1975. Edited by P. Germain and B. Nayroles. XIX, 531 pages. 1976.

Vol. 504: S. Lang and H. F. Trotter, Frobenius Distributions in GL_2-Extensions. III, 274 pages. 1976.

Vol. 505: Advances in Complex Function Theory. Proceedings 1973/74. Edited by W. E. Kirwan and L. Zalcman. VIII, 203 pages. 1976.

Vol. 506: Numerical Analysis, Dundee 1975. Proceedings. Edited by G. A. Watson. X, 201 pages. 1976.

Vol. 507: M. C. Reed, Abstract Non-Linear Wave Equations. VI, 128 pages. 1976.

Vol. 508: E. Seneta, Regularly Varying Functions. V, 112 pages. 1976.

Vol. 509: D. E. Blair, Contact Manifolds in Riemannian Geometry. VI, 146 pages. 1976.

Vol. 510: V. Poènaru, Singularités C^∞ en Présence de Symétrie. V, 174 pages. 1976.

Vol. 511: Séminaire de Probabilités X. Proceedings 1974/75. Edité par P. A. Meyer. VI, 593 pages. 1976.

Vol. 512: Spaces of Analytic Functions, Kristiansand, Norway 197… Proceedings. Edited by O. B. Bekken, B. K. Øksendal, and A. Stra… VIII, 204 pages. 1976.

Vol. 513: R. B. Warfield, Jr. Nilpotent Groups. VIII, 115 pages. 197…

Vol. 514: Séminaire Bourbaki vol. 1974/75. Exposés 453 – 470. … 276 pages. 1976.

Vol. 515: Bäcklund Transformations. Nashville, Tennessee 197… Proceedings. Edited by R. M. Miura. VIII, 295 pages. 1976.